3rd May, 2011.
W. g. Y.

Applied Numerical Linear Algebra

William W. Hager
Pennsylvania State University

Prentice Hall, Englewood Cliffs, New Jersey 07632

Library of Congress Cataloging-in-Publication Data

HAGER, WILLIAM W., (date)
Applied numerical linear algebra.
Bibliography: p.
Includes index.
1. Algebras, Linear. 2. Numerical calculations.
I. Title.
QA184.H33 1988 512'.5 87-7303
ISBN 0-13-041294-5

Editorial/production supervision
and interior design: Jean Hunter
Cover design: Lundgren Graphics, Ltd.
Manufacturing buyer: Paula Massenaro

Printed in the United States of America

10 9 8 7 6 5 4 3 2 1

ISBN 0-13-041294-5 01

PRENTICE-HALL INTERNATIONAL (UK) LIMITED, *London*
PRENTICE-HALL OF AUSTRALIA PTY. LIMITED, *Sydney*
PRENTICE-HALL CANADA INC., *Toronto*
PRENTICE-HALL HISPANOAMERICANA, S.A., *Mexico*
PRENTICE-HALL OF INDIA PRIVATE LIMITED, *New Delhi*
PRENTICE-HALL OF JAPAN, INC., *Tokyo*
SIMON & SCHUSTER ASIA PTE. LTD., *Singapore*
EDITORA PRENTICE-HALL DO BRASIL, LTDA., *Rio de Janeiro*

Contents

Preface

This book emanates from a course taught first at Carnegie-Mellon University (1977–1980) and then at Pennsylvania State University (1980–present). At Pennsylvania State University, the undergraduate numerical analysis sequence consists of two one semester courses (taken in either order). One course deals with interpolation, numerical differentiation, and numerical integration and the other course deals with linear equations, nonlinear equations, and eigenvalue problems. This book, which is tailored to the latter course, is accessible to juniors, seniors, and first year graduate students. Since there are many references to advanced material as well as some new insights presented here for the first time, it is hoped that the book will also be a useful research tool. A software package called NAPACK can be used in conjunction with the book. Instructions for obtaining NAPACK software using either Netlib or Psulib appear in Chapter 8. A supplement is also available which gives an overview of *Applied Numerical Linear Algebra* as well as the solutions to the review exercises.

The author is grateful to the many people who have sent reprints and preprints of their work and to those who have helped with the development of the book. Some of those who have contributed references for special topics include

Eugene Allgower (continuation method references),
Ivo Babuska (finite element software references),
Iain Duff (information concerning the Harwell Subroutine Library),
Molly Mahaffy (information concerning supercomputers), and
Stephen McCormick (multigrid references).

Comments by the reviewers were also much appreciated. The reviewers include

Robert D. Adams (University of Kansas),
Joseph E. D'Atri (Rutgers University),
George Phillip Barker (University of Missouri, Kansas City),
Murray Eisenberg (University of Massachusetts, Amherst),
David Lesley (San Diego State University),
Anne L. Ludington (Loyola College),
Ancel C. Mewborn (University of North Carolina, Chapel Hill),
Gilbert Strang (Massachusetts Institute of Technology), and
Robert C. Thompson (University of California, Santa Barbara).

Nilotpal Ghosh, who taught with me on several occasions, has contributed some exercises as indicated in the text.

I wish to thank the many people involved with the production of the book. These people include

Jean Hunter (production editor),
Audrey Marshall (supplement editor),
Michael Mays, West Virginia University (proof reader),
David Ostrow (mathematics editor),
Grafacon, Inc. (composition), and
Reproduction Drawings, Ltd. (art).

Finally, I wish to thank my family, my wife Georgine and my children Billy and Ann as well as my Mother and Father, for their support and tolerance. And I thank the Lord for the health, strength, and perseverance needed to complete this project.

William W. Hager
Pennsylvania State University

1

Introduction

1-1. THE ISSUES

We begin with the following problem: Find the value of x satisfying the equation

$$x = e^{-x}. \tag{1}$$

Although this equation cannot be solved algebraically, it is easy to find the solution numerically. Let us generate a sequence $x_1, x_2, x_3, \cdots$ of approximations to the true solution x with each x_{k+1} constructed from the previous x_k. To construct x_{k+1}, we use the rule

$$x_{k+1} = e^{-x_k}, \tag{2}$$

where k is an integer that takes on the values $1, 2, 3, \cdots$. When k is 1, we have $x_2 = e^{-x_1}$. Making the initial choice $x_1 = 1$, we can compute x_2:

$$x_2 = e^{-x_1} = e^{-1.} \approx .368.$$

Then setting $k = 2$ in (2) yields $x_3 = e^{-x_2}$ and substituting the value $x_2 = .368$ just computed gives us

$$x_3 = e^{-x_2} = e^{-.368} \approx .692.$$

For the next iteration, k is 3 and $x_4 = e^{-x_3} = e^{-.692} \approx .500$.

The first 20 iterations appear in Table 1-1. These iterations are implemented by the Fortran program in Table 1-2. Notice that the program statement

```
X = EXP(-X)
```

is the actual iteration (2). This statement overwrites the old iterate x_k, which is stored in X, with the new iterate x_{k+1}. Often we will express a relation like (2) relating the old iterate to the new iterate with the notation

$$x^{\text{new}} = e^{-x^{\text{old}}}.$$

The x_k in Table 1-1 settle around .567. Of course, the root of equation (1) is

TABLE 1-1 ITERATION (2) STARTING FROM $x_1 = 1$.

k	x_k	k	x_k
1	1.000	11	0.568
2	0.368	12	0.566
3	0.692	13	0.568
4	0.500	14	0.567
5	0.606	15	0.567
6	0.545	16	0.567
7	0.580	17	0.567
8	0.560	18	0.567
9	0.571	19	0.567
10	0.565	20	0.567

not exactly .567. Since the code in Table 1-2 prints just three significant digits, only three-place accuracy is possible. Computing and printing with higher precision, we obtain

$$x = .567143290409783 \cdots .$$

At any step, the error $x_k - x$ is estimated by the **residual,** the difference between the right side and the left side of the equation. At a point p, the residual $r(p)$ is defined by

$$r(p) = e^{-p} - p.$$

The residual measures how well the equation is satisfied. If the residual is zero, then $e^{-p} - p = 0$ and $p = x$, the solution of (1). If $r(p)$ is not zero, then p is not the solution, and the size of the residual measures the **error** $p - x$. Of course, the connection between the residual and the error may be complicated. Nonetheless, except for pathological examples, the error approaches zero as the residual approaches zero. The residual at x_k is

$$r(x_k) = e^{-x_k} - x_k.$$

Since $x_{k+1} = e^{-x_k}$, the residual at x_k can be written

$$r(x_k) = x_{k+1} - x_k,$$

TABLE 1-2 FORTRAN PROGRAM FOR THE ITERATION (2).

```
         K = 1
         X = 1.
10       WRITE(6,20) K,X
20       FORMAT(F10.3)
         IF ( K .GE. 20 ) STOP
         K = K + 1
         X = EXP(-X)
         GOTO 10
         END
```

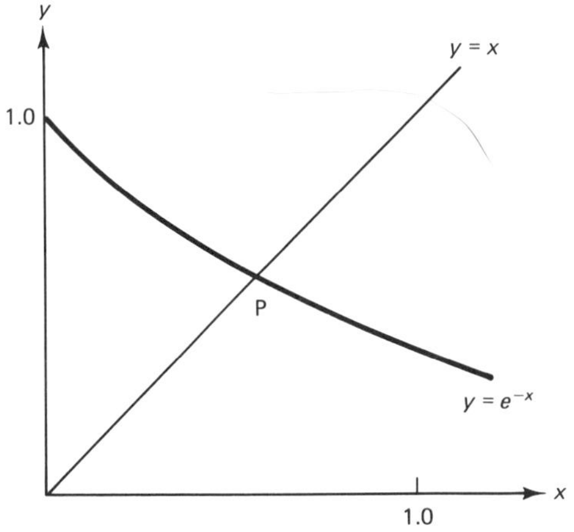

Figure 1-1 The line $y = x$ and the curve $y = e^{-x}$.

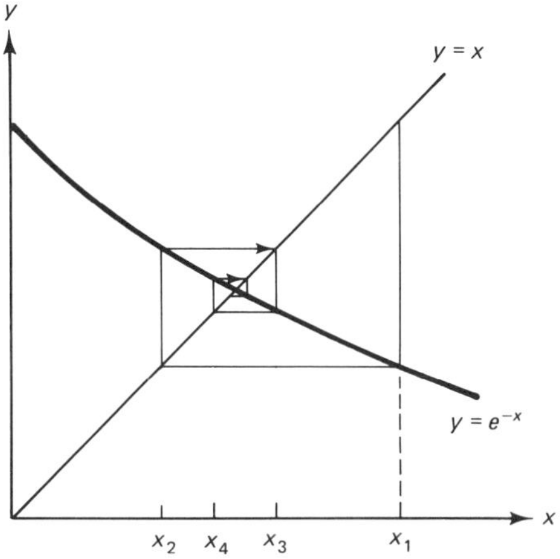

Figure 1-2 An iteration cobweb.

the difference in successive iterates. When the iterate difference approaches zero, the iterations converge to the root.

The iteration (2) has a geometric interpretation in terms of the line $y = x$ and the curve $y = e^{-x}$. At the intersection point P in Figure 1-1, both of the relations $y = x$ and $y = e^{-x}$ hold. Therefore, at P we have $x = y = e^{-x}$ and equation (1) is satisfied. The x_k are generated graphically as follows: Beginning at the point with x-coordinate x_1 on the line $y = x$, move vertically to the curve $y = e^{-x}$, then move horizontally to the point with x-coordinate x_2 on the line $y = x$; move vertically to the curve $y = e^{-x}$, then move horizontally to the point with x-coordinate x_3 on the line $y = x$. This construction generates the cobweb in Figure 1-2 which meets the line $y = x$ at the x_k. Using the equation $x = e^{-x}$ and the iteration $x_{k+1} = e^{-x_k}$ for illustration, we now discuss three issues that must be considered when any problem is solved numerically.

1-1.1 Conditioning

The concept of problem conditioning is somewhat vague. Loosely speaking, a problem is **well conditioned** if small changes in problem parameters produce small changes in the outcome. Thus to decide whether a problem is well conditioned, we make a list of the problem's parameters, we change each parameter slightly, and we observe the outcome. If the new outcome deviates from the old one "just a little," the problem is well conditioned. This definition of problem conditioning is vague since phrases like "small change" are vague—a small change in one situation is a big change in another situation. A precise definition of problem conditioning can only be given in specific applications.

To illustrate conditioning, let us consider a block of steel suspended by a wire from the ceiling (see Figure 1-3). The weight of the steel is often called the *load*. As the load increases, the wire stretches and the end of the wire is displaced downward. Let us regard the load as the problem parameter and let

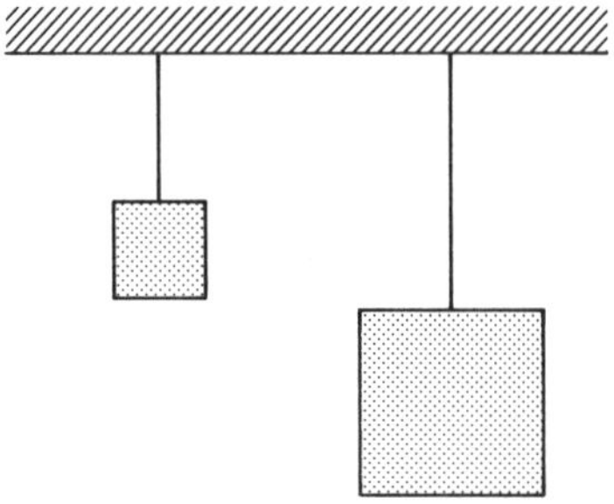

Figure 1-3 Blocks of steel suspended by wires.

us regard the displacement of the wire as the outcome. For small loads, the wire's displacement is proportional to the load and we say that the wire is *elastic*. But when the load reaches a critical *yield* value, the wire becomes *plastic* and pulls apart. The problem of stretching a wire is well conditioned for small loads, but near the transition between the elastic and the plastic region, the problem is ill conditioned, and a small change in the load can cause a large change in displacement. Obviously, the method for computing the displacement in the elastic region is different from the method for computing the displacement near the elastic-plastic transition. In general, techniques for solving ill-conditioned problems are different from techniques for solving well-conditioned problems.

For a second example of conditioning, consider the equation $x = e^{-x}$. Although this equation has no physical parameter like the load, there is a numerical parameter which is encountered when we try to solve the equation on the computer. Even though the computer finds a x such that $x = e^{-x}$, the equation is not exactly solved since the computer's value for e^{-x} contains a small error. Theoretically, the value of the exponential function is obtained by summing an infinite series. But on the computer, this infinite sum must be truncated. Thus the computer solves the equation $e^{-x} - x = \varepsilon$, where ε is the error in the value of the exponential function. Often an equation is considered well conditioned if the solution to the computer's equation $e^{-x} - x = \varepsilon$ is near the solution to the original equation $x = e^{-x}$ for small ε. Since $e^{-x} - x$ is the residual, we can also say that *an equation is well conditioned if a relatively small residual implies a relatively small error*. The absolute residual corresponding to the computed solution $x = .567$ is $|e^{-.567} - .567| \approx .00022$, while the absolute error is $|.56714 \cdots - .567| \approx .00014$. This matching of small error with small residual suggests that the equation $x = e^{-x}$ is well conditioned.

On the other hand, consider the equation

$$x = \tfrac{1}{2}(x^2 + 1) \tag{3}$$

with the root $x = 1$. The approximation $x = 1.0001$ is in error by 10^{-4} while the residual is $.5(1.0001^2 + 1) - 1.0001 = 5 \times 10^{-9}$. Since the error is much larger than the residual, the equation is ill conditioned. Equation (3) is ill conditioned since $x = 1$ is a multiple root. Rearranging (3), we obtain the quadratic equation $x^2 - 2x + 1 = 0$, which factors into $(x - 1)^2 = 0$. Since the factor $x - 1$ is squared, we say that $x = 1$ is a root with multiplicity 2. Methods for solving an equation like (3) with a multiple root are different from methods for solving an equation like (1) with a simple root.

1-1.2 Stability

The algorithm must give the right answer, even if small errors are made. Loosely speaking, an algorithm is **stable** if small changes in algorithm parameters have a small effect on the algorithm's output. For the iteration (2), a natural parameter is the starting guess x_1. Figure 1-2 shows that the iteration (2) converges to the intersection point, where x equals e^{-x}, and the scheme is stable—make small (or big) changes in x_1 and the algorithm still converges to the root x. On the other hand, consider the equation

$$x = -\log x,$$

which is equivalent to the original equation $x = e^{-x}$. That is, taking the logarithm of each side of the original equation yields $\log x = \log e^{-x}$ or $x = -\log x$. However, except for the special starting guess $x_1 = x$, the successive substitutions

$$x_{k+1} = -\log x_k \tag{4}$$

diverge from the root (see Figure 1-4). Algorithm (4) is unstable since each iteration amplifies the error.

Another illustration of an unstable algorithm is developed by Dahlquist and Björck [D1]. They consider the problem of computing the integral I_n defined by

$$I_n = \int_0^1 \frac{x^n}{x+5}\,dx.$$

The integral I_0 is easy to evaluate:

$$I_0 = \int_0^1 \frac{1}{x+5}\,dx = \log(x+5)\Big|_0^1 = \log 6 - \log 5 \approx .182.$$

To evaluate $I_1, I_2, \cdots$, we utilize the following observation:

$$\begin{aligned} I_n + 5I_{n-1} &= \int_0^1 \frac{x^n + 5x^{n-1}}{x+5}\,dx = \int_0^1 \frac{x^{n-1}(x+5)}{x+5}\,dx \\ &= \int_0^1 x^{n-1}\,dx = \frac{1}{n}. \end{aligned}$$

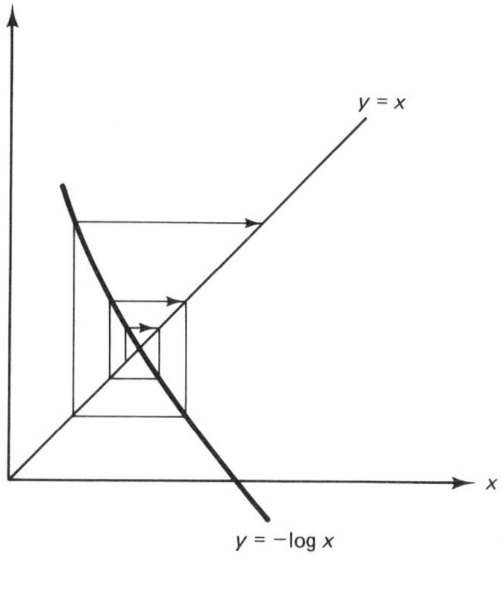

Figure 1-4 The cobweb generated by the iteration $x_{k+1} = -\log x_k$.

Hence, I_n and I_{n-1} satisfy the recurrence

$$I_n = \frac{1}{n} - 5I_{n-1}, \tag{5}$$

and I_n is determined from I_{n-1}. Since I_0 has been evaluated, we put $n = 1$ in (5) to compute I_1:

$$I_1 = 1 - 5I_0 \approx 1 - 5 \times .182 = .090.$$

In a similar fashion, substituting $n = 2$, $n = 3$, and $n = 4$ in (5) yields

$$I_2 = \frac{1}{2} - 5I_1 \approx \quad .050,$$

$$I_3 = \frac{1}{3} - 5I_2 \approx \quad .083,$$

$$I_4 = \frac{1}{4} - 5I_3 \approx -.165.$$

On the other hand, the correct I_4 is about .034. This algorithm is **unstable** since small arithmetic errors are amplified each step. The approximation $I_0 \approx .182$ is wrong in the fourth digit right of the decimal point, and when $I_1 = 1 - 5I_0$ is computed, the error is multiplied by 5. Similarly, when I_2 is evaluated, the error in I_1 is multiplied by 5. Thus the total error in I_2 is $5 \times 5 = 25$ times the error in I_0. After computing I_4, the total error is $5^4 = 625$ times the initial error in I_0. Even though the recurrence (5) is theoretically correct, numerical errors are amplified quickly.

1-1.3 Cost

Before implementing an algorithm, we study its feasibility by estimating the cost of program development and computer facilities. There is a trade-off between human cost and machine cost. Investing more hours in a program can increase its efficiency. But the total cost must be minimized. A program that is used frequently or that runs for a long time, warrants careful development, while human cost is crucial for a short program that is run just once.

Machine cost has two components, **storage** and **time.** The cost of storage depends on the storage device. Typical storage devices include cache memory, main memory, disks, drums, and tapes. As depicted in Figure 1-5, storing data in cache memory is much more expensive than storing data on a tape. On the other hand, data in cache memory are processed much faster than data on a tape since the access time associated with cache memory is much smaller than the access time associated with a tape. When developing a computer program, we must utilize storage devices efficiently. Data accessed frequently should be stored in main memory, while data accessed seldom should be stored on a disk

Fast access	Cache memory	Expensive storage
	Main memory	
	Drum	
	Disk	
Slow access	Tape	Cheap storage

Figure 1-5 Trade-off between access time and storage cost.

TABLE 1-3 EXECUTION TIMES IN MICROSECONDS FOR AN IBM 370 MODEL 3090 COMPUTER USING THE FORTRAN IV-G COMPILER.

	Precision		
Statement	REAL*4	REAL*8	COMPLEX*8
X = Y + Z	.04	.06	.21
X = Y − Z	.04	.06	.21
X = Y * Z	.07	.08	.93
X = Y / Z	.26	.43	2.02
X = EXP(Y)	2.86	2.99	7.00
X = ALOG(Y)	2.81	4.30	6.80
X = SIN(Y)	1.97	2.80	8.91
X = SQRT(Y)	2.36	3.81	4.37

or on a tape. Storage is insignificant for the code in Table 1-2 while running time is proportional to the number of iterations. Examining Table 1-1, accuracy improves about one digit every five iterations, so six-place accuracy requires 30 iterations. Using an IBM 370 model 3090 computer, we estimate the time for common Fortran statements in Table 1-3. These times are given in microseconds (a microsecond is 10^{-6} second) and correspond to the values Y = 3 and Z = 1. Clearly, the time for 30 iterations of the code in Table 1-2 is insignificant.

In summary, when solving a problem numerically, we must consider problem conditioning, algorithm stability, human cost, and machine cost. One goal in this book is to study the relative merits of algorithms so that one is able to select the best numerical approach to a problem.

Exercises

1-1.1. Consider the equation $x = 1 + \tan x$, where $-1.5 \leqq x \leqq .5$. Using the successive substitution formula $x_{k+1} = 1 + \tan x_k$ and the starting guess $x_1 = 0$, compute x_2 and x_3. Sketch the iterations as in Figure 1-2.

1-1.2. Rewrite the equation of Exercise 1-1.1 in the form $x = \arctan(x - 1)$ and consider the substitution formula $x_{k+1} = \arctan(x_k - 1)$. Starting from $x_1 = 0$, compute several iterations and sketch the analog of Figure 1-2.

1-2. STOPPING CRITERIA

Ideally, we would like to stop a sequence of iterations when the error is smaller than some specified tolerance. In practice, it is often difficult to estimate the error. To illustrate some of the difficulties involved with error estimation, let us consider two equivalent equations: (a) $x = e^{-x}$ and (b) $10^{-6}x = 10^{-6}e^{-x}$. Suppose that each of these equations is solved by an iterative technique and the iterations are terminated when the magnitude of the residual is at most 10^{-6}. Since the residual $r(p) = 10^{-6}(e^{-p} - p)$ corresponding to the second equation is 10^{-6} times the residual corresponding to the first equation, the iteration associated with the second equation terminates much sooner than the iteration associated with the first equation. More precisely, for the second equation, the residual is smaller than the desired tolerance when $0 \leqq x_k \leqq 1.2$, while for the first equation, the residual is smaller than the desired tolerance when $.5671427 \leqq x_k \leqq .5671439$. Even though the equations are equivalent, the iteration associated with the second equation can terminate with a much larger error than the iteration associated with the first equation. Thus termination criteria based on the magnitude of the residual must be formulated carefully.

When it is difficult to perform a detailed error analysis, we often stop the iterations when they converge in some qualitative sense. If a program is executed interactively, it can be interrupted when the iterations "settle down." But if a program is executed off-line in batch mode, the computer must be programmed to stop. Let us consider the iterations in Table 1-1. The program in Table 1-2 contains a statement that stops execution after 20 iterations. Inspecting the output in Table 1-1, we see that the three printed digits repeat starting at iteration 15, so the program could have been stopped at iteration 14. The decision to stop after 20 iterations is okay, although it is better to stop when the three printed digits repeat.

Designing a "black box" that will inspect the output of an algorithm and decide the best stopping point seems hopeless. For example, the best stopping point for an algorithm which generates the output shown in Table 1-4 is debatable—the iterations oscillate about 2, but they never converge to a specific number. Although it is impossible to design a black box that works in every case, three stopping criteria apply to most situations. We stop in each of the following cases:

1. Successive iterations nearly agree.
2. The algorithm is too time consuming.
3. Iterations oscillate about an average value without converging.

Some people like to add a fourth criterion:

4. Stop when the iterations diverge.

TABLE 1-4 ITERATIONS THAT DO NOT CONVERGE.

k	x_k	k	x_k	k	x_k
1	3.0000	11	2.0074	21	1.9956
2	2.6667	12	2.0033	22	1.8406
3	2.4000	13	1.9996	23	1.9959
4	2.2500	14	1.9892	24	1.9962
5	2.1538	15	2.0022	25	2.0026
6	2.0951	16	2.0062	26	2.0332
7	2.0587	17	1.9981	27	2.0016
8	2.0361	18	1.9907	28	2.0007
9	2.0220	19	2.0010	29	1.9867
10	2.0132	20	2.0047	30	2.0030

But if the iterations diverge, either our program has an error or the algorithm does not work, and these problems are normally resolved when a code is tested. And if an error slips by us and the iterations diverge quickly, the computer interrupts the program and prints a message such as "OVERFLOW." Conversely, if the iterations diverge slowly, the program stops by criterion 2 since the algorithm is too time consuming. For these reasons, we feel that there is no point in checking for divergence.

The second criterion is the easiest to implement. Since running time is related to the number of iterations, criterion 2 can be expressed in terms of the number of iterations. For example, if we find that an algorithm usually converges within 100 iterations, our program should halt automatically after 100 iterations. During the initial tests of a program, we often stop the computations after a few iterations until we feel confident in the program.

Now, let us state the first criterion more precisely. One measure of agreement between x and y is the **relative difference** R defined by

$$R(x, y) = \frac{|x - y|}{\text{maximum}\{|x|, |y|\}}.$$

For example, the relative difference between 683 and 687 is

$$\frac{687 - 683}{687} = \frac{4}{687} = .00582 \cdots = 10^{-2.23 \cdots}.$$

The exponent -2.23, the common logarithm of the relative difference, indicates that the numbers 687 and 683 differ in the third significant digit. If d is an integer, then at most the $(d + 1)$st significant digit of x and y differ if

$$R(x, y) \leqq 10^{-d}.$$

Loosely speaking, criterion 1 can be stated: To obtain d digits of agreement in successive iterations, stop when the relative difference in successive iterations is less than 10^{-d}.

Finally, let us consider the third criterion. If the iterations oscillate without settling down, then the **absolute difference** $|x_{k+1} - x_k|$ typically does not approach zero. Although one is tempted to stop whenever the absolute difference increases, some algorithms exhibit the damped oscillation depicted in Figure 1-6. When

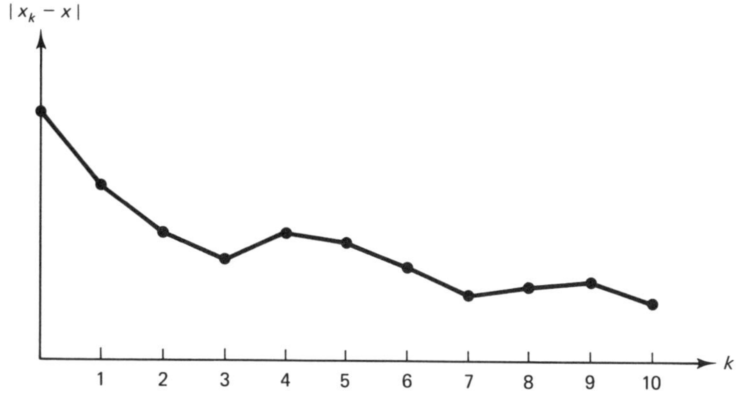

Figure 1-6 Oscillating convergence.

TABLE 1-5 AVERAGE DIFFERENCES ASSOCIATED WITH TABLE 1-4.

Iterations	Average absolute difference
1–4	.4375
5–8	.0535
9–12	.0082
13–16	.0078
17–20	.0074
21–24	.0799

absolute differences decay on the average, the algorithm should continue. For example, using the iterations from Table 1-4, we average over groups of four iterations, obtaining Table 1-5. Notice that the average increases after 21 iterations, roughly the same place that our intuition says "stop."

Fortunately, the oscillation observed in Table 1-4 is uncommon, but when it occurs, a special stopping routine is required. On the other hand, criteria 1 and 2 are easily implemented in a subroutine. We call this procedure STOPIT and its arguments are

DIF: difference in successive iterations

SIZE: x_{k+1}

NDIGIT: desired number of correct digits

LIMIT: maximum number of iterations

If either stopping criterion 1 or stopping criterion 2 is satisfied, STOPIT sets DIF = −ABS(DIF). Conversely, if neither criterion 1 nor criterion 2 is satisfied, STOPIT sets DIF = ABS(DIF). Thus the iterations should continue until the value of DIF returned by STOPIT is less than or equal to zero. To illustrate, the program in Table 1-6 incorporates STOPIT in its implementation of iteration (2).

The program in Table 1-6 stops the iterations whenever three "correct digits" or 40 iterations are achieved. Another subroutine called WHATIS helps us monitor the convergence speed by printing the iteration number, absolute

TABLE 1-6 ITERATION (2).

```
      X0 = 1.
10    X1 = EXP(-X0)
      DIF = X1 - X0
      SIZE = X1
      X0 = X1
      CALL STOPIT(DIF,SIZE,3,40)
      IF ( DIF .GT. 0. ) GOTO 10
      WRITE(6,*) X1
      END
```

difference, and stopping criterion. Inserting the statement

```
CALL WHATIS(DIF,SIZE)
```

just after the "CALL STOPIT" statement in Table 1-6 produces the following output:

```
ITERATION:    1   DIF:   0.63E+00
ITERATION:    2   DIF:   0.32E+00
ITERATION:    3   DIF:   0.19E+00
ITERATION:    4   DIF:   0.11E+00
ITERATION:    5   DIF:   0.61E-01
ITERATION:    6   DIF:   0.34E-01
ITERATION:    7   DIF:   0.19E-01
ITERATION:    8   DIF:   0.11E-01
ITERATION:    9   DIF:   0.63E-02
ITERATION:   10   DIF:   0.35E-02
ITERATION:   11   DIF:   0.20E-02
ITERATION:   12   DIF:   0.11E-02
ITERATION:   13   DIF:   0.65E-03
ITERATION:   14   DIF:   0.37E-03
ERROR BELOW TOLERANCE
0.567
```

Inserting the "CALL WHATIS" statement just before the "END" statement, we obtain only the final iteration number, the final absolute difference, and the stopping criterion. (Note that subroutine STOPIT returns the iteration number in the variable SIZE; hence subroutine WHATIS can extract the iteration number from the input variable SIZE.)

Exercise

1-2.1. Explain why a stopping criterion such as "stop when $|x^{\text{new}} - x^{\text{old}}| \leqq 10^{-6}$" is not good in general. (*Hint:* Suppose that x is either the diameter of a hydrogen atom in meters or the diameter of the universe in meters.)

1-3. MACHINE PRECISION

The program in Table 1-6 requests three correct digits. Although we might request 100 correct digits, the code generated by most compilers never achieves this accuracy. What is the most accuracy that can be achieved? The answer to this question is connected with the computer's representation of numbers. Today's computers have fixed-point (integer) and floating-point (exponential) numbers. Scientific computations typically use floating-point numbers. The three-digit decimal floating-point representation of $\pi = 3.1415926\ldots$ is

$$+.314 \times 10^1.$$

The fraction .314 is the **mantissa,** 10 is the **base,** and 1 is the **exponent.** Generally, n-digit decimal floating-point numbers have the form

$$\pm .d_1 d_2 \cdots d_n \times 10^p,$$

where the digits d_1 through d_n are integers between 0 and 9, and d_1 is never zero except in the special case $d_1 = d_2 = \cdots = d_n = 0$. Although base 10 is fine for human beings and pocket calculators, it is more efficient to store and process large amounts of data using either base 2 (binary) or a base that is a power of 2, such as $2^3 = 8$ (octal) or $2^4 = 16$ (hexadecimal). Base b floating-point numbers have the form

$$\pm .d_1 d_2 \cdots d_n \times b^p, \tag{6}$$

where each digit is between 0 and $b - 1$. For the decimal system, b is 10 and the digits are between 0 and 9, while b is 2 in the binary system and the digits are 0 or 1.

There are two common methods for converting a number

$$\pm .d_1 d_2 \cdots d_n d_{n+1} \cdots$$

into the floating-point form (6). The first method is to chop after digit n, obtaining $\pm .d_1 d_2 \cdots d_n$. The second method is to chop after digit n, but round d_n up or down depending on whether $d_{n+1} \geqq b/2$ or $d_{n+1} < b/2$. For example, with three-digit decimal floating-point numbers, we have

$$.2499 \longrightarrow \begin{cases} .249 \text{ (chopping)} \\ .250 \text{ (rounding)} \end{cases}$$

and

$$-.2499 \longrightarrow \begin{cases} -.249 \text{ (chopping)} \\ -.250 \text{ (rounding)}. \end{cases}$$

The IBM 370 computer chops while the DEC 11/780 VAX computer rounds.

Notice that the machine's representation of numbers limits the attainable accuracy. For n-digit floating-point numbers, no more than n digits of accuracy can be achieved, so a sequence of iterations should stop soon after the first $n - 1$ digits converge. If an algorithm is programmed to run on a specific computer, we can look up the mantissa length n and the base b, and we can select an appropriate value for STOPIT's parameter NDIGIT. But if a program is run on several different computers, looking up b and n for each machine can be a nuisance. For this reason, it is better to have the program compute the machine precision. The **machine epsilon** E is a common measure of computing precision. By definition E is the smallest floating-point number with the property that $1 + E > 1$. As we will soon see, the machine epsilon is related to the length of the mantissa, the machine base, and the machine arithmetic. If the computer chops, then $E = b^{1-n}$ and if the computer rounds, then $E = .5b^{1-n}$.

The code in Table 1-7 estimates the machine epsilon. This code assigns A the initial value 1 and then successively divides A by 2 until the value of A is so small that due to arithmetic errors, the computed sum 1. + A is actually equal to 1. This code just approximates the machine epsilon since it does not necessarily find the smallest number with the property that 1. + A > 1; the code in Table 1-7 finds the smallest multiple of 2 with the property that 1. + A > 1. To understand how this code works, let us assume that the computer has a two-digit decimal mantissa and that numbers are chopped when converted to floating-point form. The code first initializes A to 1 and then divides by 2 to

TABLE 1-7 PROGRAM TO ESTIMATE THE MACHINE EPSILON.

```
      A = 1.
10    A = A/2.
      T = 1. + A
      IF ( T .GT. 1. ) GOTO 10
      A = A + A
```

obtain A = 1./2. = .5; T is assigned 1. + A = 1. + .5 = 1.5. Since 1.5 is greater than 1, the program returns to statement 10, where A is assigned the value A/2. = .5/2. = .25 and T is assigned 1. + A = 1. + .25.

To evaluate the floating-point sum 1. + .25, we must discuss machine arithmetic. Different computers add floating-point numbers in different ways, but as a general rule, the sum of two floating-point numbers x and y is around $\mathrm{fl}(x + y)$, the floating-point representation of the exact sum "x plus y." That is, if $\oplus$ denotes the computer's sum, then

$$x \oplus y \approx \mathrm{fl}(x + y).$$

Assuming that $x \oplus y$ is exactly $\mathrm{fl}(x + y)$, the two-digit decimal sum of 1. and .25 is

$$1. \oplus .25 = \mathrm{fl}(1.25) = \mathrm{fl}(.125 \times 10^1) = .12 \times 10^1.$$

Thus the assigned value of T is 1.2, which is greater than 1, and the code returns to statement 10, where A is assigned the value A/2. = .25/2. Like addition, the quotient of floating-point numbers x and y is around $\mathrm{fl}(x/y)$. That is, if $\oplus\!\!\!\!\div$ denotes the computer's quotient, then

$$x \,⊕\!\!\!\!\div\, y \approx \mathrm{fl}(x/y).$$

Again, assuming that $x \,⊕\!\!\!\!\div\, y$ is exactly $\mathrm{fl}(x/y)$, the assigned value of A is

$$.25 \,⊕\!\!\!\!\div\, 2. = \mathrm{fl}(.125) = \mathrm{fl}(.125 \times 10^0) = .12 \times 10^0.$$

Then 1 is added to A giving T:

$$\mathrm{T} = 1. \oplus \mathrm{A} = 1. \oplus .12 = \mathrm{fl}(1.12) = \mathrm{fl}(.112 \times 10^1) = .11 \times 10^1.$$

Since 1.1 is greater than 1, the program returns to statement 10. The next values of A and T are

$$\mathrm{A} = \mathrm{A}/2. = .12 \,⊕\!\!\!\!\div\, 2. = \mathrm{fl}(.06) = \mathrm{fl}(.60 \times 10^{-1}) = .60 \times 10^{-1}$$

and

$$\mathrm{T} = 1. + \mathrm{A} = 1. \oplus .06 = \mathrm{fl}(1.06) = \mathrm{fl}(.106 \times 10^1) = .10 \times 10^1.$$

Since $\mathrm{fl}(1.06)$ is 1, the assigned value of T is 1 and the looping stops. The final statement in Table 1-7 sets

$$\mathrm{A} = \mathrm{A} + \mathrm{A} = .06 \oplus .06 = \mathrm{fl}(.12) = \mathrm{fl}(.12 \times 10^0) = .12 \times 10^0.$$

In summary, a computer that employs a two-digit decimal mantissa and chopping arithmetic finds that A is .12, which roughly equals 10^{-1}. As stated earlier, for a computer that employs a n-digit base b mantissa, $E = b^{1-n}$ with

TABLE 1-8 FLOATING-POINT PARAMETERS.†

	Single precision			Double precision		
Computer	b	n	b^{1-n}	b	n	b^{1-n}
IBM 360 and 370	16	6	$.95 \times 10^{-6}$	16	14	$.22 \times 10^{-15}$
CDC 6500, Cyber 205, ETA 10	2	47	$.14 \times 10^{-13}$	2	95	$.50 \times 10^{-28}$
Cray	2	48	$.71 \times 10^{-14}$	2	96	$.25 \times 10^{-28}$
UNIVAC 1108	2	27	$.15 \times 10^{-7}$	2	60	$.17 \times 10^{-17}$
DEC 11/780 VAX	2	24	$.12 \times 10^{-6}$	2	56	$.28 \times 10^{-16}$
DEC 2060	2	27	$.15 \times 10^{-7}$	2	62	$.43 \times 10^{-18}$

chopping arithmetic while $E = .5b^{1-n}$ with rounding arithmetic. In general, the program in Table 1-7 computes an A that differs from the true machine epsilon by at most a factor of 2. Moreover, if the machine base is 2 or a power of 2, the A computed by this program is exactly equal to the machine epsilon. Table 1-8 lists the base, the mantissa length, and the parameter b^{1-n} for several different computers.

Exercises

1-3.1. Using single precision, evaluate the following expression on your computer (or pocket calculator):

```
X = (1. - (1./41.)*41.)*10.**18
```

Theoretically, X is zero. Explain why the computed X is not zero. Now replace the factor 10^{18} by b^n, where b is the base and n is the mantissa length of your computer and print the new X. Explain your result. Finally, change 41 to 64 and print the new X. Explain your result.

1-3.2. If a computer employs a two-digit decimal mantissa and chopping arithmetic, what is the final value of A for the following code? If the computer works in base 2 or a base that is a power of 2, what is the final value of A?

```
       A = .5
       B = A
10     A = A/2.
       B = B + A
       IF (A+B .GT. B) GOTO 10
       A = 2.*A
```

1-3.3. Write a program that estimates the smallest positive floating-point number on a computer.

† Some of the difference between floating point parameters for say an IBM 370 computer and a Cyber 205 computer relates to semantics. On the IBM computer, a floating point number stored in a 32 bit word is considered single precision while on the Cyber 205 computer, a floating point number stored in a 32 bit word is considered half precision. Although hardware (or firmware) for almost the same computing precisions are available for both computers, these precisions are labeled "single, double, and quadruple" in the IBM environment while in the Cyber environment, the corresponding precisions are labeled "half, full, and double."

1-4. ERROR PROPAGATION

Let us evaluate the expression

$$y = \frac{(x + 1)^2 - 1}{x} \tag{7}$$

at $x = .1, .01, .001, \cdots$ using the program in Table 1-9. The output from an IBM 370 computer appears in Table 1-10. Expanding $(x + 1)^2$, we see that y simplifies to $x + 2$:

$$y = \frac{x^2 + 2x + 1 - 1}{x} = x + 2.$$

TABLE 1-9 PROGRAM TO EVALUATE (7).

```
      DO 10 I = 1,7
         X = 10.**(-I)
         Y = ((X+1.)**2 - 1.)/X
         WRITE(6,*)X,Y
10    CONTINUE
```

Thus for $x = .1, .01, .001, \cdots, y$ equals $2.1, 2.01, 2.001, \cdots$. In Table 1-10 the difference between the computed y and the correct y is small when x is .1, but the difference increases when x decreases. In particular, when x is 10^{-7}, the computed y has no correct digits.

TABLE 1-10 VALUE OF y COMPUTED BY THE PROGRAM IN TABLE 1-9.

x	Computed y	Correct y
10^{-1}	2.1000	2.1000
10^{-2}	2.0098	2.0100
10^{-3}	1.9999	2.0010
10^{-4}	1.9836	2.0001
10^{-5}	1.9073	2.0000
10^{-6}	1.9073	2.0000
10^{-7}	0.0000	2.0000

Before we can understand how errors propagate in a complicated expression like (7), the errors in the basic operations $\oplus$, $\ominus$, $\otimes$, $\oplus$, and $\mathrm{fl}(\cdot)$ must be understood. First let us consider the error in converting a number into floating-point form. For three-digit decimal floating-point numbers, recall that

$$.24999\cdots \longrightarrow \begin{cases} .249 \text{ (chopping)} \\ .250 \text{ (rounding)} \end{cases}$$

and

$$.24950\cdots \longrightarrow \begin{cases} .249 \text{ (chopping)} \\ .250 \text{ (rounding).} \end{cases}$$

In general, the difference between a number x and its floating-point representation is a small multiple of x. For example, chopping to convert the number $.24999\cdots$ into three-digit decimal floating-point form, we have

$$.24999\cdots - \mathrm{fl}(.24999\cdots) = .24999\cdots - .249 = .00099\cdots$$
$$= .004(.24999\cdots).$$

In this example, $x - \mathrm{fl}(x) = .004x$. In general, $x - \mathrm{fl}(x) = \delta x$, or equivalently, $\mathrm{fl}(x) = x(1 - \delta)$, where δ can be bounded in terms of the machine epsilon:

$$0 \leqq \delta \leqq E \tag{8}$$

for chopping computers and

$$-E \leqq \delta \leqq E \tag{9}$$

for rounding computers.

Now let us consider the basic arithmetic operations beginning with subtraction. In Section 1-3 we note that an operation between floating-point numbers often equals the floating-point representation of the exact result. In other words, if x and y are floating-point numbers, then

$$x \ominus y = \mathrm{fl}(x - y).$$

When x and y are not floating-point numbers, they must be converted to floating-point form before subtracting y from x. For arbitrary x and y, the computed "x minus y" is

$$x \ominus y = \mathrm{fl}(\mathrm{fl}(x) - \mathrm{fl}(y)).$$

Since each floating-point operation is equivalent to multiplying by $1 - \delta$, $x \ominus y$ can be written

$$x \ominus y = (1 - \delta_1)((1 - \delta_2)x - (1 - \delta_3)y),$$

where δ_1, δ_2, and δ_3 satisfy (8) if the computer chops or (9) if the computer rounds. Subtracting the exact "x minus y" from the computed "x minus y," neglecting the small terms containing the factors $\delta_1\delta_2$ and $\delta_1\delta_3$, and taking the worst possible values for the δ_i which satisfy the rounding inequality (9), we obtain

$$|(x \ominus y) - (x - y)| \approx \begin{cases} E(|x - y| + |x + y|) & \text{if } xy > 0, \\ 2E|x - y| & \text{if } xy < 0. \end{cases} \tag{10}$$

That is, the top estimate holds when x and y have the same sign, while the bottom estimate holds when x and y have opposite sign. Of course, errors can cancel when $x \ominus y$ is evaluated, so we emphasize that (10) reflects a rounding computer in the worst case.

Recall that the digits of accuracy in a computation are related to the relative difference between the correct result and the computed result. The absolute difference between the correct result and the computed result divided by the magnitude of the correct result is called the **relative error.** If the relative error in the computed difference $x \ominus y$ is 10^{-d}, the computed $x \ominus y$ just differs from the correct $x - y$ in digit $d + 1$. Dividing (10) by $|x - y|$, we see that in the worst case, the relative error in $x \ominus y$ has the approximation

$$\frac{|(x \ominus y) - (x - y)|}{|x - y|} \approx \begin{cases} E\left(1 + \dfrac{|x + y|}{|x - y|}\right) & \text{if } xy > 0, \\ 2E & \text{if } xy < 0. \end{cases} \tag{11}$$

Observe that the relative error blows up as x approaches y; equivalently, the number of correct digits in the computed difference $x \ominus y$ approaches zero as x approaches y. *In summary, subtracting nearly equal numbers can be dangerous and the computed difference $x \ominus y$ may have no correct digits when x is sufficiently close to y.*

To illustrate the danger involved with subtracting nearly equal numbers, let us compute the difference .2500 − .2499 using three-digit decimal chopping arithmetic:

$$.2500 \ominus .2499 = \mathrm{fl}(\mathrm{fl}(.2500) - \mathrm{fl}(.2499)) = \mathrm{fl}(.250 - .249) = .001,$$

while the correct difference is .0001. The computed $x \ominus y$ is 10 times the correct $x - y$. This observation explains the inaccurate y in Table 1-10. The program in Table 1-9 subtracts 1 from $(x + 1)^2$ when x is near zero. But when x is near zero, $(x + 1)^2$ is near 1 and subtracting 1 from $(x + 1)^2$ is dangerous. In fact, if x is 10^{-6}, then (11) tells us that the relative error in the computed difference "$(x + 1)^2$ minus 1" is about 10^6E in the worst case. For the IBM 370 computer, E is around 10^{-6} in single precision and $10^6E \approx 1$. Thus the computed difference has (potentially) no correct digits when x is 10^{-6}.

The computer's $\oplus$ behaves like its $\ominus$. The only difference is that the qualifiers "$xy > 0$" and "$xy < 0$" in (11) are interchanged. In contrast to addition and subtraction, multiplication and division are relatively safe. It can be shown that for rounding arithmetic in the worst case,

$$\frac{|x \otimes y - x \times y|}{|x \times y|} \approx 3E \qquad \text{and} \qquad \frac{|x \oplus\!\!\!\!\div\, y - x/y|}{|x/y|} \approx 3E.$$

Unlike addition and subtraction, there are no exceptional x and y where the relative error in multiplication or division blows up. In particular, the division by x in (7) is "safe" even when x is near zero since the relative error in this division is on the order of the machine epsilon.

The propagation of errors in the product or in the sum of several numbers can be analyzed in a similar manner. It turns out that the relative error in the computed product of m numbers is around $(2m - 1)E$. Therefore, if mE is small, the relative error in the computed product is small. The error in a sum depends on the partial sums generated by the computation. Let us assume for simplicity that we are adding m positive numbers $x_1, x_2, \cdots, x_m$. If x_1 is added to x_2, x_3 is added to the partial sum $x_1 + x_2$, and so on, then the error in the computed sum satisfies the relation

$$|(x_1 \oplus \cdots \oplus x_m) - (x_1 + \cdots + x_m)| \leqq 2E(S_2 + S_3 + \cdots + S_m), \qquad (12)$$

where S_n, called the nth **partial sum,** is the sum of the first n terms in the series:

$$S_n = x_1 + x_2 + \cdots + x_n.$$

The error in the computed sum is minimized when the terms are arranged to minimize $S_2 + S_3 + \cdots + S_m$. Observe that each partial sum $S_2, \cdots, S_m$ contains both x_1 and x_2, each partial sum $S_3, \cdots, S_m$ contains x_3, each partial sum $S_4, \cdots, S_m$ contains x_4, and so on. Hence, x_1 and x_2 appear $m - 1$ times, x_3 appears $m - 2$ times, x_4 appears $m - 3$ times, and so on. *The right side of (12) is minimized when the terms in the sum $x_1 + x_2 + \cdots + x_m$ are arranged in order from smallest to largest.*

TABLE 1-11 COMPUTED VALUE OF $y = x^2 - 2x + 1$ NEAR $x = 1$.

x	Computed y times 10^6	Correct y times 10^6
.9990	1.79	1.00
.9994	.95	.36
.9998	.48	.04
1.0002	.00	.04
1.0006	.00	.36
1.0010	.95	1.00

To illustrate the importance of ordering terms in a sum, consider the series

$$1 + .001 + .001 + \cdots + .001, \tag{13}$$

where .001 appears 1000 times. Using three-digit decimal floating-point arithmetic, the sum (13) is 1 when computed from left to right since

$$1 \oplus .001 = \mathrm{fl}(1.001) = 1.00.$$

On the other hand, if the order of terms in (13) is reversed, we get the correct sum 2.00: $.001 \oplus .001 = .002$, $.002 \oplus .001 = .003$, $\cdots$, $.999 \oplus .001 = 1.00$, and $1.00 \oplus 1.00 = 2.00$. In summary, when adding positive numbers, add the smallest numbers first.

This section concludes with two more applications where error propagation is important. Let us evaluate the expression

$$y = x^2 - 2x + 1$$

near $x = 1$. Computing y in single precision with an IBM 370 computer, we get Table 1-11. Observe that the computed y agrees poorly with the correct y—in fact, the computed y is zero at $x = 1.0006$ while $y = x^2 - 2x + 1 = (x - 1)^2$ just vanishes at $x = 1$. The inaccurate values for y in Table 1-11 resulted from subtracting nearly equal numbers when evaluating the expression $x^2 - 2x + 1$. For x near 1, $x^2 - 2x$ roughly equals -1, so the sum "$(x^2 - 2x)$ plus 1" is really the difference between 1 and a number near 1.

Since $y = x^2 - 2x + 1$ contains the factor $(x - 1)$ squared, we say that $x = 1$ is a zero of y with multiplicity 2. The evaluation of any polynomial near a multiple zero involves the subtraction of nearly equal numbers and the computed value of the polynomial is relatively inaccurate. If $p(x)$ is the polynomial of degree m with a zero z of multiplicity m, then in the worst case, the relative error in the computed value of $p(x)$ near $x = z$ roughly satisfies

$$\frac{|p(x) - \text{computed } p(x)|}{|p(x)|} \approx \frac{E(2z)^m}{|x - z|^m}. \tag{14}$$

As x approaches the zero z, the relative error in the computed value of $p(x)$ approaches infinity. For the polynomial $p(x) = x^2 - 2x + 1$ with the zero $x = 1$ of multiplicity 2 and for a computer with machine epsilon $E = 10^{-6}$, (14) tells us that the relative error in the computed value of $p(x)$ at $x = .999$ is roughly 4, so the computed value of $p(.999)$ may have no correct digits.

Finally, we illustrate the propagation of errors using Euler's finite difference approximation to the derivative. Recall from calculus that the derivative of f evaluated at x is defined by

$$f'(x) = \lim_{\Delta x \to 0} \frac{f(x + \Delta x) - f(x)}{\Delta x}.$$

Thus for any fixed Δx, we have the approximation:

$$\boxed{f'(x) \approx \frac{f(x + \Delta x) - f(x)}{\Delta x}.} \tag{15}$$

Pictorially, the left side of (15) is the slope of the tangent at x while the right side is the slope of the chord depicted in Figure 1-7. As Δx approaches zero, the slope of the chord approaches the slope of the tangent.

For example, let us estimate the derivative of $f(x) = x^2$ at $x = 1$ using (15):

$$f'(1) \approx \frac{(1 + \Delta x)^2 - 1}{\Delta x}. \tag{16}$$

As Δx tends to zero, the right side of (16) approaches $f'(1)$. Observe that (16) is essentially the expression y in (7) except that the variable x in (7) is replaced by Δx. Thus the program in Table 1-9 evaluates the right side of (16) for a sequence of Δx tending to zero. In Table 1-10, notice that y approaches 2, then falls off to zero as Δx (the first column) decreases. On the other hand, the derivative of $f(x) = x^2$ is $f'(x) = 2x$, which is 2 at $x = 1$. Theoretically, the right side of (16) approaches 2 as Δx decreases. In fact, expanding $(1 + \Delta x)^2$, we see that

$$\frac{(1 + \Delta x)^2 - 1}{\Delta x} = \frac{1 + 2\Delta x + \Delta x^2 - 1}{\Delta x} = 2 + \Delta x. \tag{17}$$

So the error in the approximation (16) to the derivative is just Δx. What caused the discrepancy between the actual derivative $f'(1) = 2$ and the approximation in Table 1-10?

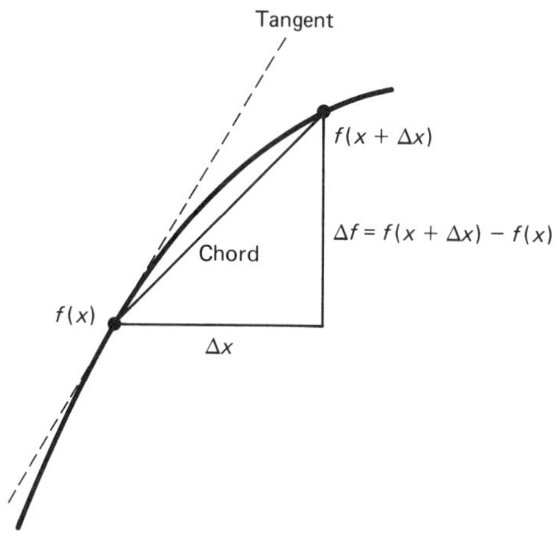

Figure 1-7 The tangent and the chord.

When the derivative of a function is estimated by a finite difference, we must contend with two different errors. First, there is the error in the approximation (15) associated with the choice of Δx. For the function $f(x) = x^2$, we see in (17) that this error is Δx. In general, the error in Euler's finite difference approximation is related to the second derivative of f:

$$f'(x) - \frac{f(x + \Delta x) - f(x)}{\Delta x} \approx f''(x)\frac{\Delta x}{2}. \tag{18}$$

If $f(x) = x^2$, then $f''(x)$ is 2 and the error is $f''(x)\Delta x/2 = \Delta x$. As Δx tends to zero, the error in the finite difference approximation tends to zero. Second, there is the error in evaluating the expression

$$\frac{f(x + \Delta x) - f(x)}{\Delta x}. \tag{19}$$

When Δx is near zero, $f(x + \Delta x)$ is near $f(x)$ and (19)'s numerator is the difference of nearly equal numbers. By (10) the error in this subtraction is roughly $2E|f(x)|$ assuming that $|f(x + \Delta x) - f(x)|$ is much smaller than $|f(x)|$. When (19) is computed, this error is divided by Δx so that the computational error is roughly

$$\frac{2E|f(x)|}{|\Delta x|}. \tag{20}$$

In summary, as Δx tends to zero, the computational error (20) increases while the approximation error (18) decreases. The best Δx balances these two errors. Equating (18) and (20) and solving for Δx yields

$$\Delta x_{\text{best}} \approx 2\sqrt{\frac{E|f(x)|}{|f''(x)|}} \tag{21}$$

for rounding arithmetic. The analogous result for chopping arithmetic is

$$\Delta x_{\text{best}} \approx \sqrt{\frac{2E|f(x)|}{|f''(x)|}}.$$

For the function $f(x) = x^2$, the point $x = 1$, and the machine epsilon $E = 10^{-6}$, we have

$$\Delta x_{\text{best}} \approx \sqrt{\frac{2 \times 10^{-6}}{2}} = 10^{-3},$$

which agrees with the best Δx observed in Table 1-10. (In Table 1-10, the best Δx corresponds to the entry in column 1 for which the adjacent entry in column 2 is closest to 2.000.)

In some sense, formula (21) for the best Δx is not very practical since it contains $f''(x)$—if we can differentiate twice, why use a finite difference to approximate the first derivative? Despite the presence of the second derivative in (21), this formula is helpful since it provides qualitative information about Δx_{best}. Notice that the best Δx is proportional to the square root of E and hence by (18), the error corresponding to Δx_{best} is also proportional to the square root of E. Thus if we want to use (15) to estimate $f'(x)$ with single precision accuracy, the difference $f(x + \Delta x) - f(x)$ should be computed in double precision.

Exercise

1-4.1. (by N. Ghosh) Using single precision, evaluate the expression

$$a = 1000\left(\frac{c}{\sqrt{b^2 + c} - b} - 2b\right)$$

when b is 1 and c is .004004. Compare the computed value of a with the exact value $a = 2$. Show that a can be written

$$a = \frac{1000c}{\sqrt{b^2 + c} + b}.$$

Now evaluate a when b is 1 and c is .004004. Explain why this second expression is more accurate.

1-5. THE SEMICONDUCTOR EQUATION

The solution to the equation $x = e^{-x}$ is simply a number, $x = .567 \cdots$. We now consider a problem whose solution $x(t)$ is a function depending on a parameter t which takes values between 0 and 1. This function must satisfy the following conditions: $x(t)$ vanishes at $t = 0$ and at $t = 1$, and for every t between 0 and 1, we have

$$\frac{d^2x(t)}{dt^2} = -e^{-x(t)}. \tag{22}$$

Equation (22) is related to semiconductors and the solution is sketched in Figure 1-8. This problem is a **boundary-value problem** and the requirement that $x(t)$ vanish at $t = 0$ and at $t = 1$ is the **boundary condition.** The equations modeling electric and magnetic fields, stress and strain in structures, and many other phenomena can be formulated as boundary-value problems.

The solution to the semiconductor equation will be approximated numerically. Our strategy is to replace the second derivative by a finite-difference approximation and to generate a system of equations whose solution approximates the actual solution to the semiconductor equation. Equation (15) gives a finite-difference approximation to the first derivative. Similarly, the second derivative has the approximation:

$$\boxed{\frac{d^2x(t)}{dt^2} \approx \frac{x(t - \Delta t) - 2x(t) + x(t + \Delta t)}{(\Delta t)^2}} \tag{23}$$

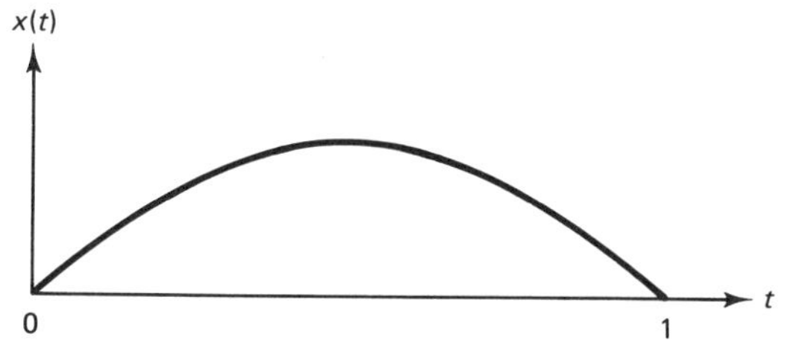

Figure 1-8 Solution to the semiconductor equation.

This approximation to the second derivative is a **centered difference** approximation since it involves the symmetric points $t + \Delta t$ and $t - \Delta t$. Geometrically, (23) is derived by taking the second derivative of the parabola that passes through the curve $x(t)$ at three points. The error in (23) is related to the deviation between $x(t)$ and this parabolic approximation. Recall that the error in Euler's approximation to the first derivative can be expressed in terms of the second derivative. In a similar manner, the error in the approximation (23) can be expressed in terms of the fourth derivative:

$$\frac{d^2x(t)}{dt^2} - \frac{x(t - \Delta t) - 2x(t) + x(t + \Delta t)}{(\Delta t)^2} \approx \frac{d^4x(t)}{dt^4}\frac{(\Delta t)^2}{12}. \tag{24}$$

Let us assume that $x(t)$ satisfies the semiconductor equation so that the second derivative of $x(t)$ is $-e^{-x(t)}$, and let us neglect the error term, treating (23) as an equality—we will take Δt small, so that $(\Delta t)^2$ is tiny and the error term (24) is much smaller than the second derivative in (23). Under these assumptions, equations (22) and (23) yield the following **finite-difference** relation:

$$-x(t - \Delta t) + 2x(t) - x(t + \Delta t) = (\Delta t)^2 e^{-x(t)}. \tag{25}$$

Remember that the solution to the semiconductor equation does not satisfy (25) exactly since we have neglected the error in the finite-difference approximation to the second derivative. But when Δt is small, the error (24) is tiny so we hope that the solution to (25) is near the solution to (22). To avoid confusing the solution to the boundary-value problem (22) with the solution to the finite-difference relation, the solution to (25) is labeled y. That is, y satisfies the equation

$$-y(t - \Delta t) + 2y(t) - y(t + \Delta t) = (\Delta t)^2 e^{-y(t)}. \tag{26}$$

Since y is an approximation to x and x vanishes at $t = 0$ and at $t = 1$, the natural boundary condition for y is $y(0) = 0$ and $y(1) = 0$.

Taking a fixed Δt, we now formulate a system of equations for the value of y at a specific set of points. We begin by partitioning the interval [0, 1] into subintervals of equal length. Four subintervals are pictured in Figure 1-9. The points t_0, t_1, t_2, t_3, and t_4 are called the **grid** or the **mesh,** and mesh spacing is $\Delta t = \frac{1}{4} = .25$. For N subdivisions, the mesh spacing is $\Delta t = 1/N$ and the mesh points are $t_0 = 0$, $t_1 = \Delta t$, $t_2 = 2\Delta t, \cdots, t_N = N\,\Delta t = 1$ (see Figure 1-10). In general, the ith grid point is $t_i = i\,\Delta t$.

Notice that t_0 is 0 and t_N is 1. Since $y(0) = y(1) = 0$ by the boundary condition, both $y(t_0)$ and $y(t_N)$ are zero while $y(t_1)$ through $y(t_{N-1})$ are unknown. Taking $N = 4$, we use the finite difference equation (26) to generate equations for the three unknowns $y(t_1)$, $y(t_2)$, and $y(t_3)$. When N is 4, Δt is $\frac{1}{4}$. Setting $t = t_1$ in (26) gives us the relation

$$-y(t_1 - \Delta t) + 2y(t_1) - y(t_1 + \Delta t) = \tfrac{1}{16}e^{-y(t_1)}.$$

Since $t_1 - \Delta t = t_0$ and $t_1 + \Delta t = t_2$, this reduces to

$$-y_0 + 2y_1 - y_2 = \tfrac{1}{16}e^{-y_1},$$

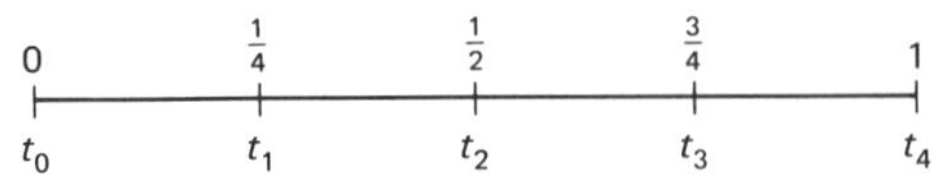

Figure 1-9 Four mesh intervals.

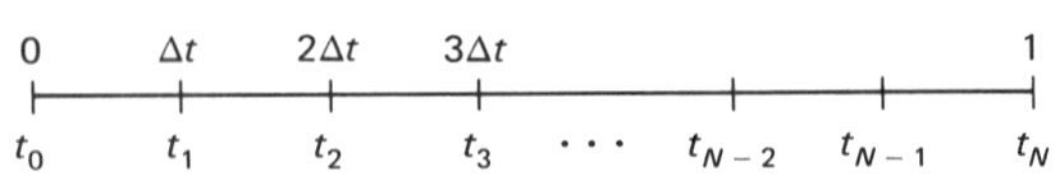

Figure 1-10 N mesh intervals.

where y_0, y_1, and y_2 denote $y(t_0)$, $y(t_1)$, and $y(t_2)$, respectively. Similarly, setting $t = t_2$ and $t = t_3$ in (26), gives two additional relations:

$$-y_1 + 2y_2 - y_3 = \tfrac{1}{16}e^{-y_2},$$
$$-y_2 + 2y_3 - y_4 = \tfrac{1}{16}e^{-y_3}.$$

Since the boundary condition implies that $y_0 = y_4 = 0$, these three relations reduce to

$$\begin{aligned} 2y_1 - y_2 \qquad &= \tfrac{1}{16}e^{-y_1}, \\ -y_1 + 2y_2 - y_3 &= \tfrac{1}{16}e^{-y_2}, \\ - y_2 + 2y_3 &= \tfrac{1}{16}e^{-y_3}. \end{aligned} \tag{27}$$

There are three equations corresponding to four mesh intervals. In general, for N mesh intervals, there are $N - 1$ equations which are compactly written

$$-y_{i-1} + 2y_i - y_{i+1} = (\Delta t)^2 e^{-y_i}, \tag{28}$$

where i assumes the values $1, 2, \cdots, N - 1$. The boundary condition is expressed $y_0 = y_N = 0$.

Exercises

1-5.1. In the special case $N = 2$, what is the single equation generated by (28)? In the special case $N = 3$, what are the two equations generated by (28)?

1-5.2. Give the analog of (28) for the boundary-value problem

$$\frac{d^2x}{dt^2}(t) = e^{-t}, \qquad 0 \leqq t \leqq 1,$$

and the boundary condition $x(0) = x(1) = 0$.

1-5.3. Give the analog of (28) for the boundary-value problem

$$\frac{d^2x}{dt^2}(t) + \frac{dx}{dt}(t) - 2x(t) = e^{-t}, \qquad 0 \leqq t \leqq 1,$$

and the boundary condition $x(0) = x(1) = 0$. [*Hint:* Approximate the second derivative using the centered-difference scheme (23) and approximate the first derivative using Euler's forward difference (15).]

1-5.4. Using single precision arithmetic, estimate the second derivative of e^t evaluated at $t = 0$ by inserting $\Delta t = .5, .25, .125, \cdots, 2^{-30}$ into (23). Theoretically, the second derivative of e^t is 1 at $t = 0$. Inspecting your computer output, which Δt yields the best approximation to the second derivative? Now repeat the computations using double precision. Which Δt is the best?

1-5.5. Referring to the analysis at the end of Section 1-4, estimate (in general) the best Δt in the approximation (23) to the second derivative.

1-6. AN ALGORITHM FOR THE SEMICONDUCTOR EQUATION

We now propose an iterative method to solve the system of equations (28). When N is 2, there is a single unknown: $y_0 = 0$ and $y_N = y_2 = 0$ are known, Δt is $\frac{1}{2}$, and the only equation is

$$2y_1 = \tfrac{1}{4}e^{-y_1}.$$

Dividing by 2, this has the familiar form

$$y_1 = \tfrac{1}{8}e^{-y_1}.$$

[Compare this to equation (1).] Motivated by our previous success, we try the iteration

$$y_1^{\text{new}} = \tfrac{1}{8}e^{-y_1^{\text{old}}}.$$

Starting from the initial guess $y_1 = 0$, the iterations converge quickly to the root $y_1 = .1117 \cdots$. Similarly, the system of three equations (27) can be written

$$y_1 = \tfrac{1}{2}(y_2 + \tfrac{1}{16}e^{-y_1}),$$
$$y_2 = \tfrac{1}{2}(y_1 + y_3 + \tfrac{1}{16}e^{-y_2}),$$
$$y_3 = \tfrac{1}{2}(y_2 + \tfrac{1}{16}e^{-y_3}),$$

which suggests the iteration

$$y_1^{\text{new}} = \tfrac{1}{2}(y_2^{\text{old}} + \tfrac{1}{16}e^{-y_1^{\text{old}}}),$$
$$y_2^{\text{new}} = \tfrac{1}{2}(y_1^{\text{old}} + y_3^{\text{old}} + \tfrac{1}{16}e^{-y_2^{\text{old}}}),$$
$$y_3^{\text{new}} = \tfrac{1}{2}(y_2^{\text{old}} + \tfrac{1}{16}e^{-y_3^{\text{old}}}).$$

Starting from the initial guess $y_1 = y_2 = y_3 = 0$, these iterations again converge quickly to the solution

$$y_1 = .0852 \cdots,$$
$$y_2 = .1132 \cdots,$$
$$y_3 = .0852 \cdots.$$

The general system of $N - 1$ equations is solved by the iteration

$$y_i^{\text{new}} = \tfrac{1}{2}(y_{i-1}^{\text{old}} + y_{i+1}^{\text{old}} + (\Delta t)^2 e^{-y_i^{\text{old}}}), \tag{29}$$

where $i = 1, 2, \cdots, N - 1$ and $y_0 = y_N = 0$.

We now develop a computer program to implement the iteration (29). The first step in developing a program is to organize array and variable names. In implementing (29), we need arrays to store y_i^{new} and y_i^{old}, and we need variables to store the parameters N, $N - 1$, $\Delta t = 1/N$, and $(\Delta t)^2$. Our naming convention is summarized in Table 1-12.

After organizing storage, it helps to write a short outline of the program. For many programs, this outline takes the form:

1. Initialization
2. Iteration
3. Output

TABLE 1-12 VARIABLE NAMES.

Problem parameter	Fortran variable
y_i^{new}	NEW(I)
y_i^{old}	OLD(I)
N	N
$N - 1$	NM1
Δt	D
$(\Delta t)^2$	S

For the iteration (29), these three program segments can be refined further and a more detailed outline follows:

1. Initialization
   a. Read number of intervals N.
   b. Initialize OLD to zero and set NM1 = N − 1, D = 1./N, and S = D*D.
2. Iteration
   a. Code (29) paying special attention to $i = 1$ and $i = N - 1$.
   b. Set OLD = NEW and test for convergence.
3. Output
   a. Print N followed by t_i and y_i for i between 1 and $N - 1$.
   b. Branch to the start and read a new value for N.

The actual computer program (stored under the name SEMCON in NAPACK) is listed below.

```
        ________________________________________________
       |                                                |
       |   APPROXIMATE THE SEMICONDUCTOR EQUATION       |
       |          AT EVENLY SPACED POINTS               |
       |________________________________________________|

      REAL NEW(100),OLD(100),D,DIF,S,SIZE,T
      INTEGER I,N,NM1
C     BUILTIN FUNCTIONS: EXP,FLOAT

C                 *** INPUT NUMBER OF INTERVALS ***

10    READ(5,*) N
      IF ( N .LT. 2 ) STOP
      IF ( N .LE. 100 ) GOTO 20
      WRITE(6,*) 'ARRAY DIMENSION EXCEEDED'
      STOP

C         *** PREPROCESS AND INITIALIZE VARIABLES ***

20    NM1 = N - 1
      D = 1./FLOAT(N)
      S = D*D
      DO 30 I = 1,N
         OLD(I) = 0.
30    CONTINUE

C                  *** COMPUTE NEXT ITERATION ***

40    NEW(1) = .5*(OLD(2) + S*EXP(-OLD(1)))
      IF ( N .LE. 2 ) GOTO 60
      DO 50 I = 2,NM1
         NEW(I) = .5*(OLD(I-1) + OLD(I+1) + S*EXP(-OLD(I)))
50    CONTINUE
```

```
C                    *** UPDATE AND TEST ERROR ***

60    CALL UPDATE(DIF,SIZE,NEW,OLD,NM1)
      CALL STOPIT(DIF,SIZE,4,5000)
      IF ( DIF .GT. 0 ) GOTO 40
      CALL WHATIS(DIF,SIZE)

C                    *** PRINT THE SOLUTION ***

      WRITE(6,70) N
70    FORMAT(I10,' INTERVALS',/'    T        X(T)'/)
      DO 90 I = 1,NM1
         T = I*D
         WRITE(6,80) T,NEW(I)
80       FORMAT(F10.4,F15.4)
90    CONTINUE
      WRITE(6,100)
100   FORMAT(//)
      GOTO 10
      END
```

Observe that the variables D, DIF, S, SIZE, and T are declared real at the start of the program even though these variables are real by default. Similarly, the variables I, N, and NM1 are declared integer at the start of the program even though these variables are integer by default. There are two reasons for declaring each variable at the start of the program. First, it is easy to convert the program to double precision when greater accuracy is needed: Change REAL to DOUBLE PRECISION and change the built-in functions such as FLOAT and EXP to their double precision analogs DFLOAT and DEXP. Second, when a program is modified, it helps to have a list of the variables that appear in the original program so that duplications and conflicts in the names of variables can be avoided.

Also observe that the parameter S = D*D is evaluated near the start of the program. Although the statement

```
NEW(I) = .5*(OLD(I-1) + OLD(I+1) + D*D*EXP(-OLD(I)))
```

is equivalent to

```
NEW(I) = .5*(OLD(I-1) + OLD(I+1) + S*EXP(-OLD(I))),
```

the second statement is better. The first statement makes the computer evaluate the product D*D for each I while the second statement utilizes the quantity S = D*D, which is evaluated once and for all in the **preprocessing** segment of the program. Even though optimizing compilers do some preprocessing automatically, many programs are more readable when the programmer preprocesses some quantities.

The iteration segment of the code computes y_i^{new} using (29), sets OLD = NEW, and tests for convergence. For the iteration $x^{\text{new}} = e^{-x^{\text{old}}}$, DIF and SIZE denote $x^{\text{new}} - x^{\text{old}}$ and x^{new}, respectively. For the semiconductor program, the iteration involves arrays and any measure of the iteration difference must incorporate every array element. Moreover, the iteration difference should be zero only when y_i^{new} equals y_i^{old} for every i. Subroutine UPDATE, which is invoked in the program, defines the iteration difference to be the sum of the absolute difference between respective elements of NEW and OLD. Similarly, the iteration size is the absolute sum of NEW's elements. In other words, UPDATE computes the value of DIF and the value of SIZE by the rule

$$\begin{aligned} \text{DIF} &= \Sigma\ \text{ABS(OLD(I)} - \text{NEW(I))}, \\ \text{SIZE} &= \Sigma\ \text{ABS(NEW(I))}. \end{aligned}$$

After computing DIF and SIZE, UPDATE sets OLD = NEW.

The final step in developing a program is to find programming errors. Although systematic programming techniques minimize errors, a code nearly always has bugs. Since the *total program works only after each component works,* the first step in isolating program bugs is to partition the code into small, natural components (like the subroutines or the three program segments), and test them separately. Using the simplest parameters such as N = 1 and LIMIT = 3, check the value of variables at the beginning and at the end of loops. Then gradually increase the parameters taking say N = 2 or N = 3 and taking LIMIT = 10. If each component in the program seems to work correctly, then assemble the complete program and monitor the execution using print statements at the beginning and at the end of components. Also output can be passed to another program for more testing. For example, in the last segment of the semiconductor program, NEW can be fed to a subroutine that computes the expression

```
NEW(I+1) - 2.*NEW(I) + NEW(I-1) + S*EXP(NEW(I)),
```

which should be near zero.

Exercises

1-6.1. Propose other iterations similar to (29) for solving the system of equations (28). (*Hint:* Consider changing either y_{i-1}^{old} to y_{i-1}^{new} or changing y_{i+1}^{old} to y_{i+1}^{new}.)

1-6.2. Suppose you are given a copy of the program developed above and you are asked to solve the system of equations derived in Exercise 1-5.2 (or Exercise 1-5.3). What changes must be made in the program to handle the new system of equations?

1-7. ALGORITHM EVALUATION

In this section we evaluate the algorithm (29) to see whether this is a practical method for solving the semiconductor equation. When evaluating an algorithm, the major concerns are computer time and computer storage. Let us determine the time and storage needed to obtain an approximation y_i to $x(t_i)$ with six-place accuracy. In the process of determining the time and storage, the following questions are addressed:

1. How does the error $y_i - x(t_i)$ depend on the number of mesh intervals N?
2. How big an N is needed for six-place accuracy?

3. How does the program running time depend on N?
4. What is the running time for six-place accuracy?

In analyzing algorithms, one often finds that running time, number of iterations, and accuracy are proportional to a parameter N, which measures the problem size, raised to a power. The iteration (29) is no exception. Recall that the finite-difference approximation (26) was derived by neglecting an error term (24) proportional to $(\Delta t)^2$. Thus the solution y_i to the finite-difference system (28) is not exactly $x(t_i)$, but it is near $x(t_i)$. Mathematicians have proved that the error $E_N = y_i - x(t_i)$ has the form CN^p, where p is a fixed constant independent of N and C slowly varies with N. For example, C might equal $2 + N^{-1}$, which is nearly 2 for large N. Although $C = 2 + N^{-1}$ depends on N, C is essentially 2 when N is large.

In analyzing the algorithm (29), we focus on the specific grid point $t_i = \frac{1}{2}$, although the analysis and the conclusions are quite similar for the other grid points. Since $t_i = i\,\Delta t = i/N$, it follows that t_i is $\frac{1}{2}$ if and only if i is $N/2$. Running the program from Section 1-6 using several different values of N, we obtain y_i at $i = N/2$, which is listed in Table 1-13. These computations were made with double-precision arithmetic and with the stopping parameter NDIGIT = 14.

As expected, the approximations to $x(.5)$ in Table 1-13 are converging to a limit .11370 · · · . Using an extrapolation technique developed in the next section and using the data of Table 1-13, the constants C and p in the error relation $y_i - x(t_i) \approx CN^p$ can be estimated at $i = N/2$. From Section 1-8 we have $C \approx -.008$, $p \approx -2$, and $y_i - x(t_i) \approx -.008N^{-2}$ at $i = N/2$. In summary, if E_N denotes the error $y_i - x(t_i)$ at $i = N/2$, then $E_N \approx -.008N^{-2}$.

We have answered the first question: How does the error depend on N? Using this formula for the error, let us estimate the computer storage needed for six-place accuracy. Most of the program storage is consumed by the arrays NEW and OLD. Since the dimension of NEW or OLD is at least N, we must determine the value of N where the computed y_i agrees with $x(t_i)$ to six places. The relation $E_N = CN^p$ implies that

$$N = \left(\frac{|E_N|}{|C|}\right)^{1/p}.$$

Recall that six-place accuracy means that the relative error $|E_N|/x(.5)$ is 10^{-6}, or equivalently, $|E_N| = x(.5) \times 10^{-6}$. Substituting $C = -.008$, $p = -2$, and

TABLE 1-13 SOLUTION y_i TO FINITE DIFFERENCE SYSTEM (28) AT $i = N/2$.

N	$y_{N/2}$
4	.11320077125781
8	.11357641845693
16	.11367174958781
32	.11369567361486
64	.11370166036576

TABLE 1-14 N VERSUS THE NUMBER OF ITERATIONS.

N	I_N
4	98
8	373
16	1351
32	5119
64	19442

$|E_N| = x(.5) \times 10^{-6}$, where $x(.5) \approx .114$ (see Table 1-13), we obtain

$$N \approx \left(\frac{.114 \times 10^{-6}}{.008}\right)^{-1/2} \approx 260.$$

Thus the array storage required for either NEW or OLD is 260 elements, which is a small storage requirement for most computers.

We now consider the third question: How does running time depend on N? First note that running time T_N is the product between the number of iterations I_N and the time per iteration T_I:

$$T_N = I_N T_I.$$

For the semiconductor program, the number of iterations depends on the number of mesh intervals. Table 1-14 gives the number of iterations corresponding to various N. Like the error E_N, the number of iterations I_N has an approximation of the form $I_N \approx CN^p$, where C and p are different from their previous values. To evaluate C and p, let us consider the ratio I_{2N}/I_N:

$$\frac{I_{2N}}{I_N} = \frac{C(2N)^p}{CN^p} = 2^p.$$

Solving for p, we have

$$p = \log_2\left(\frac{I_{2N}}{I_N}\right). \tag{30}$$

Knowing p, we can solve for C from the relation $I_N = CN^p$:

$$C = N^{-p} I_N. \tag{31}$$

Inserting the data of Table 1-14 into formulas (30) and (31), we obtain the estimates for p and C given in Table 1-15. Since $p \approx 2$ and $C \approx 6.5$, the number

TABLE 1-15 ESTIMATED p AND C.

N	p	C
4	1.928	6.765
8	1.857	7.850
16	1.921	6.554
32	1.925	6.478

of iterations I_N is approximately $6.5N^2$. Observe in Table 1-15 that there is a different p and C corresponding to each N. These differences are due to the fact that I_N is *approximately* equal to CN^p. If I_N is identically equal to CN^p, then each p and C in Table 1-15 would be the same. Typically, as N tends to infinity, the estimated p and C approach a limit. If we wish to analyze the performance of an algorithm for large N, then the p and C corresponding to the largest N should be utilized in our analysis.

Now let us consider the dependence of the running time on N. Since $I_N \approx 6.5N^2$, the number of iterations is proportional to N^2. Since each iteration of the program in Section 1-6 involves $N - 1$ executions of the statement

```
NEW(I) = .5*(OLD(I-1) + OLD(I+1) + S*EXP(-OLD(I))),
```

the time T_I per iteration is proportional to N. Therefore, $T_N = I_N T_I$ is proportional to $N^2 \times N = N^3$ and there exists a constant C such that $T_N = CN^3$. To estimate T_{260}, the computing time for six-place accuracy, we form the ratio T_{260}/T_{64}; the C cancels giving

$$\frac{T_{260}}{T_{64}} = \left(\frac{260}{64}\right)^3 \approx 67.$$

This shows that the program corresponding to 260 mesh intervals is 67 times slower than the program corresponding to 64 mesh intervals. Since execution time on an IBM 370 model 3081 computer is about 18 seconds when N is 64, it follows that

$$T_{260} \approx 67 \times 18 \text{ seconds} = 1206 \text{ seconds},$$

or more than 20 minutes. This algorithm is *terrible*. In contrast, for another method developed in Chapter 2, T_{260} is less than 1 second.

1-8. EXTRAPOLATION TECHNIQUES

Suppose that a sequence of numbers $z_1, z_2, z_3, \cdots$ is converging to a limit z_∞. If the structure of the error $E_n = z_n - z_\infty$ is known, extrapolation techniques may be used to estimate both the limit of the sequence and various constants that appear in the expression for the error. For illustration, let us estimate the limit of the sequence of numbers appearing in Table 1-13. (Recall that Table 1-13 gives the finite-difference approximation at $t = \frac{1}{2}$ to the solution of the semiconductor equation for various choices of the mesh parameter N.)

In many finite-difference applications, one finds that the error has the form Cn^p, where n is the number of mesh intervals. In other words, $E_n = z_n - z_\infty \approx Cn^p$. Assuming that this approximation to the error is actually an equality, we can use three elements in the sequence, say z_2, z_4, and z_8, to solve for the three unknowns p, C, and z_∞. Inserting $n = 2$, $n = 4$, and $n = 8$ into the error relation $z_n - z_\infty = E_n = Cn^p$, we have

$$z_2 - z_\infty = C2^p, \tag{32a}$$

$$z_4 - z_\infty = C4^p, \tag{32b}$$

$$z_8 - z_\infty = C8^p. \tag{32c}$$

Since z_2, z_4, and z_8 are known, this system provides three equations for the three unknowns. To solve for p, subtract (32b) from (32a) and subtract (32c) from (32b), giving us two relations:

$$z_2 - z_4 = C2^p - C4^p = C2^p(1 - 2^p),$$

$$z_4 - z_8 = C4^p - C8^p = C4^p(1 - 2^p).$$

Then divide each side of the second relation by the corresponding side of the first relation to obtain

$$2^p = \frac{z_4 - z_8}{z_2 - z_4}, \tag{33}$$

which implies that

$$p = \log_2\left(\frac{z_4 - z_8}{z_2 - z_4}\right).$$

To find z_∞, (32b) is multiplied by 2^p and subtracted from (32c), giving

$$z_8 - z_\infty - 2^p(z_4 - z_\infty) = 0.$$

Solving for z_∞, we find that

$$z_\infty = \frac{2^p z_4 - z_8}{2^p - 1}.$$

Substituting for 2^p using (33) and simplifying the resulting expression yields

$$z_\infty = z_8 - \frac{(z_8 - z_4)^2}{z_2 - 2z_4 + z_8}.$$

Finally, knowing p and z_∞, we can solve for C from (32a):

$$C = (z_2 - z_\infty)2^{-p}.$$

Although z_2, z_4, and z_8 were used to compute the unknowns z_∞, p, and C, any three elements z_i, z_j, and z_k with the property that $j^2 = ik$ can be employed in the same way. For the sequence z_2, z_4, and z_8, i is 2, j is 4, k is 8, and $j^2 = 4^2 = 16 = 2 \times 8 = i \times k$. If r denotes the ratio j/i, the general formulas are

$$z_\infty = z_k - \frac{(z_k - z_j)^2}{z_i - 2z_j + z_k}, \qquad p = \log_r\left(\frac{z_j - z_k}{z_i - z_j}\right), \qquad C = (z_i - z_\infty)i^{-p}. \tag{34}$$

This estimate for z_∞ is called **Aitken's extrapolation.** In utilizing (34), it is important to know z_i, z_j, and z_k accurately. Since the sequence $z_1, z_2, \cdots$ approaches the limit z_∞, both the numerator and the denominator in Aitken's extrapolation approach zero: The numerator contains $z_k - z_j \approx z_\infty - z_\infty = 0$ and the denominator contains $z_i - 2z_j + z_k \approx z_\infty - 2z_\infty + z_\infty = 0$. Since both the numerator and the denominator involve the difference of nearly equal numbers, Aitken's estimate for z_∞ is polluted by rounding error unless z_i, z_j, and z_k are accurate.

Returning to Table 1-13 and identifying the second column of Table 1-13 with $z_4, z_8, z_{16}, \cdots$, we apply (34) to groups of three consecutive numbers from the second column to obtain the estimates for z_∞, p, and C given in Table 1-16. Notice that each group of three numbers from the second column yields a different estimate for z_∞, p, and C. The reason for these differences is that the error E_n is approximately equal to Cn^p. Typically, the error in this approximation becomes smaller as n increases and the estimates generated by (34) approach a

TABLE 1-16 ESTIMATES GENERATED BY (34).

n	z_n	z_∞	p	C
4	.11320077125781			
8	.11357641845693	.11370412428380	−1.97863981553721	−.00781847034369
16	.11367174958781	.11370372817032	−1.99190253230140	−.00801216921919
32	.11369567361486	.11370361522334	−2.00719550140420	−.00832970228905
64	.11370166036576			

limit as n tends to infinity. From the third and fourth columns of Table 1-16, we see that $p \approx -2$ and $C \approx -.008$.

As we proceed down the second column of Table 1-16, the z_n approach the limit z_∞. Similarly, proceeding down the third column, the successive extrapolations approach z_∞. This observation suggests a strategy for obtaining an even better approximation to the limit z_∞: Apply Aitken's extrapolation to the extrapolated values. These computations are summarized by the following picture:

$$\begin{array}{lllll}
.11320077125781 & & & & \\
.11357641845693 & \longrightarrow & .11370412428380 & & \\
.11367174958781 & \longrightarrow & .11370372817032 & \longrightarrow & .11370356082916 \\
.11369567361486 & \longrightarrow & .11370361522334 & & \\
.11370166036576 & & & &
\end{array}$$

The first column comes from Table 1-13. Each number in the second and third columns is Aitken's extrapolation of the three closest numbers in the preceding column, and the final extrapolation is the best estimate for z_∞. Of course, these three columns can be extended to four or five columns. If we compute in high precision and if the approximation $E_n \approx Cn^p$ for the error is valid, this estimate for z_∞ is very accurate. (To test the validity of the relation $E_n \approx Cn^p$, we can compute a sequence of p's and C's as we did in Table 1-16 to see whether their values are relatively constant.)

If additional information is available concerning the error $E_n = z_n - z_\infty$, then the extrapolation procedure may be simplified. For example, if $E_n = Cn^p$, where p is a known constant, there are just two unknowns: z_∞ and C. Similar to the derivation of (34), it is possible to express z_∞ and C in terms of p and the value of z_n for two different n's. In particular, if r denotes the ratio j/i, then you will show in Exercise 1-8.1 that

$$z_\infty = \frac{z_j - r^p z_i}{1 - r^p}, \qquad C = i^{-p}(z_i - z_\infty). \tag{35}$$

This formula for z_∞ is called **Richardson's extrapolation.**

Another common situation is where the sequence $z_1, z_2, z_3, \cdots$ approaches a limit z_∞ and the error $E_n = z_n - z_\infty$ has the form Ca^n. Here a, like C, is a fixed unknown constant. Again, the three unknowns z_∞, a, and C can be expressed in terms of three elements in the sequence. In Exercise 1-8.2 you use z_n, z_{n-1}, and z_{n-2} to obtain the formulas

$$z_\infty = z_n - \frac{(z_n - z_{n-1})^2}{z_n - 2z_{n-1} + z_{n-2}}, \qquad a = \frac{z_n - z_{n-1}}{z_{n-1} - z_{n-2}}, \qquad C = (z_n - z_\infty)a^{-n}. \tag{36}$$

Exercises

1-8.1. Verify formula (35).

1-8.2. Verify formula (36).

1-8.3. Use formula (36) to determine the limit of the sequence $1, \frac{3}{2}, \frac{7}{4}, \frac{15}{8}, \frac{31}{16}, \cdots$.

REVIEW PROBLEMS

1-1. Consider the equation

$$x^3 - x - 2 = 0.$$

From the graph in Figure 1-11, we conclude that there is a root between $x = 1$ and $x = 2$. Let us examine two different methods for finding the root. First, rearrange the equation to get $x = (x + 2)^{1/3}$ and use the iteration

$$x^{\text{new}} = (x^{\text{old}} + 2)^{1/3}$$

with the initial guess $x = 1$. Second, rearrange the equation to get $x = (x + 2)/x^2$ and use the iteration

$$x^{\text{new}} = (x^{\text{old}} + 2)/(x^{\text{old}})^2.$$

Sketch the iteration cobweb for each method. Which method works? How many iterations are needed to achieve five-place accuracy?

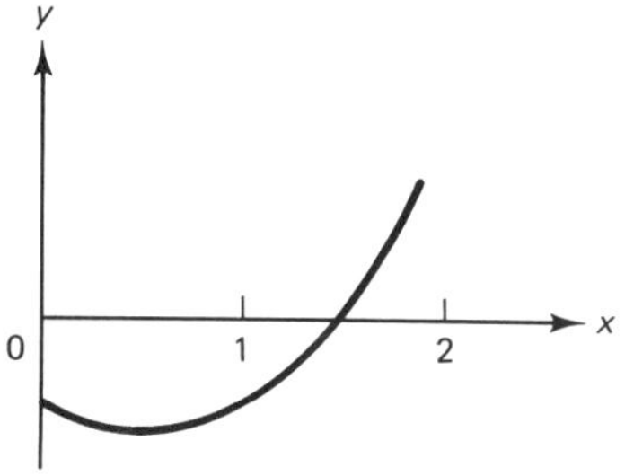

Figure 1-11 The graph of $y = x^3 - x - 2$.

1-2. In Section 1-1 we compute the integral

$$I_n = \int_0^1 \frac{x^n}{x + 5}\,dx$$

using the recurrence $I_n = n^{-1} - 5I_{n-1}$, where $I_0 = \log 6 - \log 5 = .182 \cdots$. The computed I_n are very inaccurate since the recurrence multiplies the error by 5 each step. Now rewrite the recurrence in the form

$$I_{n-1} = \frac{1}{5n} - \frac{1}{5} I_n, \tag{37}$$

so that each step multiplies the error by $\frac{1}{5}$ instead of by 5. Since the error is not amplified, the iteration (37) is called **stable.** If say I_{20} is known, then (37) lets us compute $I_{19}, I_{18}, \cdots, I_0$. For x between 0 and 1, $x^n/(x + 5)$ is between 0 and $\frac{1}{5}$. Hence I_n is between 0 and $\frac{1}{5}$ for every n. Using the starting condition $I_{20} = 0$, compute $I_{19}, I_{18}, \cdots, I_0$. Then try the starting condition $I_{20} = .2$. Determine the error in the computed I_0 by comparing it to the exact value $I_0 = \log 6 - \log 5 = .182 \cdots$. The recurrence (37) is so stable that the error due to the starting condition quickly disappears.

1-3. The following program computes a computer's base and mantissa length, and decides whether the machine rounds or chops. Explain how the program works assuming the computer employs a two-digit decimal mantissa and rounding arithmetic. Append a statement to the program which prints the base, the mantissa length, and a message like "THIS MACHINE ROUNDS" or "THIS MACHINE CHOPS." Run the program on your computer.

```
         X = 1
10       X = X + X
         IF ((1.+X)-X .EQ. 1.) GOTO 10
         B = 0.
20       B = B + 1.
         IF ( X+B .EQ. X ) GOTO 20
         R = 0.
         IF ( X+2.*B .GT. X+B ) GOTO 30
         R = 1.
         B = B + B
30       N = 0
40       N = N + 1
         X = X/B
         IF ( X .GE. B ) GOTO 40
```

1-4. If X(I) is a function that computes the Ith term in a series, then the usual code for computing the sum X(1) + X(2) + · · · + X(N) is the following:

```
      REAL S,X
      S = 0.                                   (38)
      DO 10 I = 1,N
10       S = S + X(I)
```

Consider the series

$$S = \sum_{i=1}^{10,000} i^{-2} = 1 + \frac{1}{4} + \frac{1}{9} + \frac{1}{16} + \cdots + \frac{1}{100,000,000}. \tag{39}$$

The function X in (38) which generates the series (39) is

```
FUNCTION X(I)
X = 1./FLOAT(I)**2
RETURN
END
```

Compute S using the code (38). In Section 1-4 we saw that it is better to add the small terms before the big terms. Modify the code(s) above so that the computation of S starts with the smallest terms in the sum. Estimate the error in both the forward sum and the backward sum by repeating the computation in double precision. (That is, change REAL to DOUBLE PRECISION, FLOAT to DFLOAT, FUNCTION to DOUBLE PRECISION FUNCTION, and rerun the program.) The error in the forward sum is how many times the error in the backward sum?

1-5. Although accuracy can be improved by ordering the terms in a sum, Kahan has devised an algorithm (see Table 1-17) that gives much greater accuracy than the usual code (38). Kahan observes that when X(I) is added to S, there is an error equal to the difference between the exact "S + X(I)" and the computed "S $\oplus$ X(I)." The variable C in Table 1-17 estimates this error, so the final sum is the computed sum S plus the correction C. Use Kahan's function SUM to compute the forward sum (39). Compare the error in this sum to the error in the single-precision sums computed in Problem 1-4.

TABLE 1-17 KAHAN'S ALGORITHM FOR COMPUTING A SUM.

```
      FUNCTION SUM(X,N)
      REAL C,F,S,T,X,Y
      S = 0.
      C = 0.
      DO 10 I = 1,N
         Y = C + X(I)
         T = S + Y
         F = 0.
         IF ( ABS(T) .GT. ABS(S) ) F = (.46*T-T) + T
         C = ((S-F) - (T-F)) + Y
10    S = T
      SUM = S + C
      RETURN
      END
```

1-6. Consider the two roots

$$x_{\pm} = \frac{-b \pm \sqrt{b^2 - 4c}}{2} \tag{40}$$

of the quadratic equation

$$x^2 + bx + c = 0.$$

Based on your observation in Exercise 1-4.1, derive a new formula for x_+ which is more accurate when b is positive and $4|c|$ is small relative to b^2. How is this new formula related to x_-? Derive an analogous formula for x_- which is accurate when b is negative and $4|c|$ is small relative to b^2. Using both (40) and the new formulas for $x_{\pm}$, write a program to compute the roots of the quadratic equation.

1-7. Let us consider the computer's graph for the function (see [C4])

$$y = 10^6(x^6 - 6x^5 + 15x^4 - 20x^3 + 15x^2 - 6x + 1)$$

near $x = 1$. If y is evaluated near $x = 1$ using an IBM 370 computer and if the computed points on the graph are connected with straight line segments using a Versatec plotter, we obtain Figure 1-12 on page 36. Theoretically, y equals $10^6(x - 1)^6$, which is always positive for $x \neq 1$. But in Figure 1-12 we observe a fuzzy region near $x = 1$ where y is both positive and negative. In the fuzzy region where the computed value of y is negative, what can you say about the relative error in the computed y? Use relation (14) to estimate theoretically the diameter of the region where the computed value of y is negative. [*Hint:* The machine epsilon for the IBM 370 appears in Table 1-8. Of course, the theoretical diameter is larger than the actual diameter since (14) assumes the worst possible propagation of errors.]

1-8. Consider the boundary-value problem

$$\frac{d^2x}{dt^2}(t) + \frac{dx}{dt}(t) - 2x(t) = 3e^t, \qquad 0 \leqq t \leqq 1, \tag{41}$$

and the boundary condition

$$x(1) = 0, \qquad x'(0) = 0.$$

Using the finite-difference technique developed in Section 1-5, derive a system of equations whose solution approximates the solution to (41). [*Hint:* The first derivative term in (41) and in the boundary condition at $t = 0$ can be replaced by the Euler

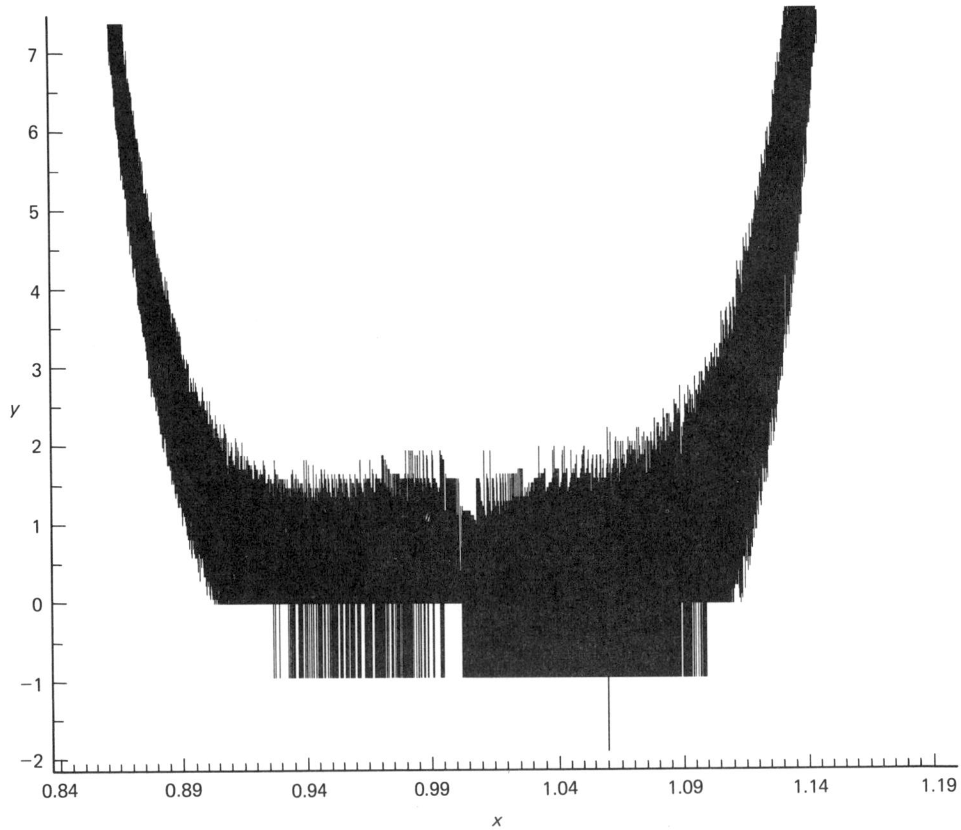

Figure 1-12 The computed graph of y versus x for Problem 1-7.

approximation (15). In particular, the approximation to the boundary condition $x'(0) = 0$ is

$$x'(0) \approx \frac{x(\Delta t) - x(0)}{\Delta t} = 0. \tag{42}$$

As a consequence of (42), the variables y_0 and y_1 in your finite-difference system are equal and y_0 can be replaced by y_1.] Formulate an iterative method for solving your system of equations by moving the y_i term to the left side of equation i and by moving the other terms to the right side. Modify the program in Section 1-6 to handle this new system and run the program using 10 mesh intervals.

1-9. Repeat Problem 1-8 for the boundary-value problem

$$-\frac{d^2x}{dt^2}(t) + 9x(t) = 10 \sin t, \qquad 0 \leqq t \leqq 2\pi, \tag{43}$$

with the periodic boundary condition

$$x(0) = x(2\pi), \qquad x'(0) = x'(2\pi).$$

The derivative $x'(0)$ in the boundary condition can be approximated by the forward Euler scheme (42), while the derivative $x'(2\pi)$ is approximated by the backward Euler scheme

$$x'(2\pi) \approx \frac{x(2\pi) - x(2\pi - \Delta t)}{\Delta t}.$$

From the boundary condition, you will find that $y_0 = y_N = (y_1 + y_{N-1})/2$, which can be inserted in the first and the last equations of the finite-difference system.

1-10. Write a subroutine AITKEN(Z,N) to compute an N-column Aitken extrapolation table. The first 2N − 1 elements of the single subscripted, double precision array Z contain the values to be extrapolated. The subroutine computes the next N − 1 columns of the extrapolation table and stores them in Z starting at location 2N. Thus the final extrapolation is stored in location N^2.

1-11. The constant π is given by the series

$$\pi = \sum_{i=1}^{\infty} \frac{8}{(4i-1)(4i-3)} = \frac{8}{3} + \frac{8}{35} + \frac{8}{99} + \cdots. \tag{44}$$

If the series is chopped after N terms, then the error due to the neglected terms is about $1/(2N)$. Hence, to obtain an estimate for π with error at most $.5 \times 10^{-14}$, we must add the first 10^{14} terms in the series. On the basis of the execution times in Table 1-3, computing this partial sum will take several years on an IBM 370 computer. On the other hand, using Aitken's extrapolation, the series limit can be estimated with far fewer terms. Let S_N denote the sum of the first N terms in the series (44). Using double precision, compute S_8, S_{16}, S_{32}, S_{64}, S_{128}, S_{256}, and S_{512}. Then form a four column Aitken extrapolation table using the program developed in Problem 1-10. What is the difference between the final extrapolation and π? [In Fortran, you can always determine π by evaluating the arc cosine of −1. That is, the statement "P = ARCOS(−1.)" puts the single precision value of π in P. Similarly, if P is a double precision variable, then the statement "P = DARCOS(−1.D0)" puts the double precision value of π in P. Note that the function ARCOS is replaced by ACOS in Fortran 77.]

1-12. The trapezoidal rule estimates an integral by approximating the area of the integration region with the sum of the areas of small trapezoids. Applying the trapezoidal rule to the integral

$$\int_0^1 x^{.25} e^{-x}\, dx, \tag{45}$$

we obtain the estimates given in Table 1-18. These estimates depend on the number N of trapezoids employed in the approximation. How many trapezoids are needed to obtain an estimate for the integral (45) with relative error at most 10^{-10}?

TABLE 1-18
ESTIMATES FOR THE INTEGRAL (45).

N	Estimate
8	.4532397135
16	.4669511005
32	.4727464066
64	.4751875058
128	.4762143356

REFERENCES

The text [D1] of Dahlquist and Björck mentioned in Section 1-1 is an excellent numerical methods reference which contains a wealth of examples, exercises, and algorithms. The iterations that appear in Table 1-4 are generated by the secant method of Chapter 4 applied to the equation $x^2 - 4x + 4.00003 = 0$. Some of the earliest papers dealing with the estimation of machine parameters include [M2] by Malcolm and [G4] by Gentleman and Marovich. A list of machine parameters is given in MINPACK [M10]. Detailed studies of rounding errors are found in Wilkinson's book [W5]. Fast and accurate routines to evaluate the elementary functions are developed by Cody and Waite in [C3]. The boundary-value problem referred to as the semiconductor equation in Section 1-5 is a simplified version of the true semiconductor equation, which is the focus of the September 1983 issue of the *SIAM Journal on Scientific and Statistical Computing*.

2

Elimination Schemes

2-1. GAUSSIAN ELIMINATION

This chapter studies one of the most important topics in computing: *linear systems of equations*. Many problems in physics and engineering are formulated in terms of a linear system of equations. For example, the steady-state voltage or current in an electrical circuit is the solution to a linear system. But the importance of linear systems transcends these linear applications. Often numerical methods for solving nonlinear problems are iterative procedures for which each iteration involves solving a linear system. This principle is illustrated in Section 2-7, where we develop a fast iterative method for solving the nonlinear semiconductor system (28) of Chapter 1. There are two strategies for solving linear systems of equations: direct methods like Gaussian elimination and iterative methods that can be traced back to Gauss, Jacobi, and Seidel. This chapter examines elimination schemes and their implementation on the computer.

To illustrate Gaussian elimination, let us solve the following system of three equations in three unknowns:

$$\begin{aligned} 2x_1 + 4x_2 + 2x_3 &= 4 \\ 4x_1 + 7x_2 + 7x_3 &= 13 \\ -2x_1 - 7x_2 + 5x_3 &= 7. \end{aligned} \tag{1}$$

The unknown x_1 is eliminated from the second and third equations by subtracting 2 times the first equation from the second equation, and by subtracting -1 times the first equation from the third equation to obtain

$$\begin{aligned} 2x_1 + 4x_2 + 2x_3 &= 4 \\ - x_2 + 3x_3 &= 5 \\ - 3x_2 + 7x_3 &= 11. \end{aligned}$$

Then x_2 is eliminated from the third equation by subtracting 3 times the second equation from the third equation, generating a triangular structure:

$$\begin{aligned} 2x_1 + 4x_2 + 2x_3 &= 4 \qquad &(2a)\\ - x_2 + 3x_3 &= 5 \qquad &(2b)\\ - 2x_3 &= -4. \qquad &(2c) \end{aligned}$$

The next phase, called **back substitution**, successively computes x_3, x_2, and x_1. Dividing equation (2c) by -2 gives

$$x_3 = \frac{-4}{-2} = 2,$$

and substituting $x_3 = 2$ into equation (2b) yields

$$x_2 = \frac{5 - 3x_3}{-1} = \frac{5 - 3 \times 2}{-1} = 1.$$

Finally, putting $x_2 = 1$ and $x_3 = 2$ in equation (2a), we get

$$x_1 = \frac{4 - 4x_2 - 2x_3}{2} = \frac{4 - 4 \times 1 - 2 \times 2}{2} = -2.$$

We now show how to express linear equations and Gaussian elimination using matrix and vector notation. The right side of system (1) forms a three-component column vector which we label **b**:

$$\mathbf{b} = \begin{bmatrix} 4 \\ 13 \\ 7 \end{bmatrix}.$$

Other column vectors associated with the system (1) are the unknowns, which we label **x**:

$$\mathbf{x} = \begin{bmatrix} x_1 \\ x_2 \\ x_3 \end{bmatrix},$$

and the solution

$$\begin{bmatrix} -2 \\ 1 \\ 2 \end{bmatrix}. \qquad (3)$$

Two vectors are **equal** if their respective components are equal. Thus the vector equality

$$\begin{bmatrix} x_1 \\ x_2 \\ x_3 \end{bmatrix} = \begin{bmatrix} -2 \\ 1 \\ 2 \end{bmatrix}$$

relating the unknown vector to the solution vector is equivalent to the three separate equalities $x_1 = -2$, $x_2 = 1$, and $x_3 = 2$ obtained by equating the respective vector components.

The coefficients of each equation in (1) form a three component row vector. The coefficients of the first equation in (1) form the row vector

$$[2 \quad 4 \quad 2]. \qquad (4)$$

The coefficients of the entire system (1) form a 3 × 3 square matrix which we label **A**:

$$\mathbf{A} = \begin{bmatrix} 2 & 4 & 2 \\ 4 & 7 & 7 \\ -2 & -7 & 5 \end{bmatrix}.$$

There are two 3 × 3 matrices associated with the elimination of (1):

$$\mathbf{U} = \begin{bmatrix} 2 & 4 & 2 \\ 0 & -1 & 3 \\ 0 & 0 & -2 \end{bmatrix}$$

and

$$\mathbf{L} = \begin{bmatrix} 1 & 0 & 0 \\ 2 & 1 & 0 \\ -1 & 3 & 1 \end{bmatrix}.$$

The first matrix contains the coefficients of the eliminated system (2). The matrix elements 2, −1, and −2 form the **diagonal** of **U**. This matrix is **upper triangular** since every element beneath the diagonal is zero. The second matrix **L** represents the elimination steps. During elimination, we subtracted

2 times the first equation from the second equation,
−1 times the first equation from the third equation,
3 times the second equation from the third equation.

The elimination factors 2, −1, and 3 are called the **multipliers**. The multipliers are stored in **L** beneath the diagonal. For example, the −1 in row 3 and column 1 is the multiplier used to eliminate x_1 from the third equation. A matrix is **lower triangular** if each element above the diagonal is zero. **L** is **unit lower triangular** since it is lower triangular and every diagonal element is 1.

Now let us consider matrix and vector operations. The product between a row vector and a column vector, called the **dot product**, is the sum of the products between respective components. The dot product between the row vector (4) and the solution vector (3) is

$$[2 \quad 4 \quad 2]\begin{bmatrix} -2 \\ 1 \\ 2 \end{bmatrix} = 2\times -2 + 4\times 1 + 2\times 2 = 4.$$

The dot product between the row vector (4) and the unknown vector **x** is

$$[2 \quad 4 \quad 2]\begin{bmatrix} x_1 \\ x_2 \\ x_3 \end{bmatrix} = 2x_1 + 4x_2 + 2x_3.$$

The product between a row vector and a column vector is a scalar. The product between a matrix and a column vector is a column vector. Two different (but equivalent) ways to multiply a matrix and a vector will be illustrated using the product between the coefficient matrix and the solution vector:

$$\begin{bmatrix} 2 & 4 & 2 \\ 4 & 7 & 7 \\ -2 & -7 & 5 \end{bmatrix}\begin{bmatrix} -2 \\ 1 \\ 2 \end{bmatrix}. \qquad (5)$$

The first way to compute the product vector is row oriented. Each component of the product vector is the dot product between a row from the matrix and the vector. For (5) there are three dot products:

$$\begin{bmatrix} 2 & 4 & 2 \end{bmatrix} \begin{bmatrix} -2 \\ 1 \\ 2 \end{bmatrix} = 4,$$

$$\begin{bmatrix} 4 & 7 & 7 \end{bmatrix} \begin{bmatrix} -2 \\ 1 \\ 2 \end{bmatrix} = 13,$$

$$\begin{bmatrix} -2 & -7 & 5 \end{bmatrix} \begin{bmatrix} -2 \\ 1 \\ 2 \end{bmatrix} = 7.$$

These three dot products yield the three components of the product vector. The matrix-vector product (5) can be expressed

$$\begin{bmatrix} 2 & 4 & 2 \\ 4 & 7 & 7 \\ -2 & -7 & 5 \end{bmatrix} \begin{bmatrix} -2 \\ 1 \\ 2 \end{bmatrix} = \begin{bmatrix} 4 \\ 13 \\ 7 \end{bmatrix}. \tag{6}$$

In other words, the coefficient matrix **A** times the solution vector equals the right side **b**.

The second way to compute a matrix-vector product is column oriented. We multiply each column of the matrix by the corresponding component of the vector and add:

$$\begin{bmatrix} 2 & 4 & 2 \\ 4 & 7 & 7 \\ -2 & -7 & 5 \end{bmatrix} \begin{bmatrix} -2 \\ 1 \\ 2 \end{bmatrix} = \begin{bmatrix} 2 \\ 4 \\ -2 \end{bmatrix}(-2) + \begin{bmatrix} 4 \\ 7 \\ -7 \end{bmatrix}(1) + \begin{bmatrix} 2 \\ 7 \\ 5 \end{bmatrix}(2). \tag{7}$$

To evaluate the vector-scalar products on the right side of (7), multiply each component of the vector by the scalar. For example, the first vector-scalar product is

$$\begin{bmatrix} 2 \\ 4 \\ -2 \end{bmatrix}(-2) = \begin{bmatrix} -4 \\ -8 \\ 4 \end{bmatrix}.$$

Hence the right side of (7) can be simplified as follows:

$$\begin{bmatrix} 2 \\ 4 \\ -2 \end{bmatrix}(-2) + \begin{bmatrix} 4 \\ 7 \\ -7 \end{bmatrix}(1) + \begin{bmatrix} 2 \\ 7 \\ 5 \end{bmatrix}(2)$$

$$= \begin{bmatrix} -4 \\ -8 \\ 4 \end{bmatrix} + \begin{bmatrix} 4 \\ 7 \\ -7 \end{bmatrix} + \begin{bmatrix} 4 \\ 14 \\ 10 \end{bmatrix} = \begin{bmatrix} -4 + 4 + 4 \\ -8 + 7 + 14 \\ 4 - 7 + 10 \end{bmatrix} = \begin{bmatrix} 4 \\ 13 \\ 7 \end{bmatrix}.$$

Both the row oriented and the column oriented method to compute a matrix-vector product give the same final result (6).

The product between the coefficient matrix **A** and the unknown vector **x** is

$$\mathbf{Ax} = \begin{bmatrix} 2 & 4 & 2 \\ 4 & 7 & 7 \\ -2 & -7 & 5 \end{bmatrix} \begin{bmatrix} x_1 \\ x_2 \\ x_3 \end{bmatrix} = \begin{bmatrix} 2x_1 + 4x_2 + 2x_3 \\ 4x_1 + 7x_2 + 7x_3 \\ -2x_1 - 7x_2 + 5x_3 \end{bmatrix}. \tag{8}$$

Hence equation (1) can be written $\mathbf{Ax} = \mathbf{b}$:

$$\underset{\mathbf{A}}{\begin{bmatrix} 2 & 4 & 2 \\ 4 & 7 & 7 \\ -2 & -7 & 5 \end{bmatrix}} \underset{\mathbf{x}}{\begin{bmatrix} x_1 \\ x_2 \\ x_3 \end{bmatrix}} = \underset{\mathbf{b}}{\begin{bmatrix} 4 \\ 13 \\ 7 \end{bmatrix}}.$$

The left side $\mathbf{Ax}$ of this equation is the vector given by (8). Equating the respective components of the vector (8) to the right side $\mathbf{b}$ gives us equation (1).

The rules for multiplying two matrices are similar to the rules for multiplying a matrix by a vector. In the following example, we multiply a 4×3 matrix $\mathbf{B}$ by a 3×2 matrix $\mathbf{C}$ to get the 4×2 product $\mathbf{P}$:

$$\underset{\mathbf{B}}{\begin{bmatrix} 2 & -1 & 1 \\ 4 & 0 & -1 \\ -3 & 1 & 2 \\ 2 & 4 & 0 \end{bmatrix}} \underset{\mathbf{C}}{\begin{bmatrix} -1 & -2 \\ 1 & 1 \\ 2 & 0 \end{bmatrix}} = \underset{\mathbf{P}}{\begin{bmatrix} -1 & -5 \\ -6 & -8 \\ 8 & 7 \\ 2 & 0 \end{bmatrix}}.$$

Each element of $\mathbf{P}$ is the dot product between a row from $\mathbf{B}$ and a column from $\mathbf{C}$. The 7 in row 3 and column 2 of $\mathbf{P}$ is the dot product between row 3 from $\mathbf{B}$ and column 2 from $\mathbf{C}$. In general, the element in row i and column j of $\mathbf{P}$ is the dot product between row i from $\mathbf{B}$ and column j from $\mathbf{C}$. Thus the number of rows in the product matches the number of rows in $\mathbf{B}$ and the number of columns in the product matches the number of columns in $\mathbf{C}$. Moreover, the product $\mathbf{P} = \mathbf{BC}$ is only defined when the number of columns in $\mathbf{B}$ matches the number of rows in $\mathbf{C}$.

Using dot products, we can compute the individual elements of $\mathbf{P}$. Using matrix-vector products, we can compute an entire row or an entire column of $\mathbf{P}$. Each column of $\mathbf{P}$ is the product between $\mathbf{B}$ and the corresponding column of $\mathbf{C}$. In particular, the first column of $\mathbf{P}$ is given by the product

$$\begin{bmatrix} 2 & -1 & 1 \\ 4 & 0 & -1 \\ -3 & 1 & 2 \\ 2 & 4 & 0 \end{bmatrix} \begin{bmatrix} -1 \\ 1 \\ 2 \end{bmatrix} = \begin{bmatrix} 2 \\ 4 \\ -3 \\ 2 \end{bmatrix}(-1) + \begin{bmatrix} -1 \\ 0 \\ 1 \\ 4 \end{bmatrix}(1) + \begin{bmatrix} 1 \\ -1 \\ 2 \\ 0 \end{bmatrix}(2) = \begin{bmatrix} -1 \\ -6 \\ 8 \\ 2 \end{bmatrix}.$$

Similarly, each row of $\mathbf{P}$ is the product between the corresponding row of $\mathbf{B}$ and matrix $\mathbf{C}$. For example, the first row of $\mathbf{P}$ is given by

$$\begin{aligned} [2 \quad -1 \quad 1] \begin{bmatrix} -1 & -2 \\ 1 & 1 \\ 2 & 0 \end{bmatrix} &= (2)[-1 \quad -2] + (-1)[1 \quad 1] + (1)[2 \quad 0] \\ &= [-1 \quad -5]. \end{aligned}$$

The coefficient matrix $\mathbf{A}$ is related to the triangular matrices $\mathbf{L}$ and $\mathbf{U}$ through the identity $\mathbf{A} = \mathbf{LU}$:

$$\underset{\mathbf{A}}{\begin{bmatrix} 2 & 4 & 2 \\ 4 & 7 & 7 \\ -2 & -7 & 5 \end{bmatrix}} = \underset{\mathbf{L}}{\begin{bmatrix} 1 & 0 & 0 \\ 2 & 1 & 0 \\ -1 & 3 & 1 \end{bmatrix}} \underset{\mathbf{U}}{\begin{bmatrix} 2 & 4 & 2 \\ 0 & -1 & 3 \\ 0 & 0 & -2 \end{bmatrix}}.$$

To verify this identity, let $\mathbf{a}_1$, $\mathbf{a}_2$, and $\mathbf{a}_3$ be the three rows of **A**, and let $\mathbf{u}_1$, $\mathbf{u}_2$, and $\mathbf{u}_3$ be the three rows of **U**. Since the first row of **A** never changes during elimination, $\mathbf{u}_1 = \mathbf{a}_1$. During the first elimination step, the first row of **A** is multiplied by 2 and subtracted from the second row of **A** to obtain the second row of **U**:

$$\mathbf{u}_2 = \mathbf{a}_2 - 2\mathbf{a}_1 = \mathbf{a}_2 - 2\mathbf{u}_1.$$

Finally, the third row $\mathbf{u}_3$ of the eliminated matrix was generated in two steps. We subtracted -1 times the first row of **A** from the third row of **A**, and we subtracted from the third row of the resulting matrix, 3 times the second row of the partly eliminated matrix. Since the first row of **A** is $\mathbf{u}_1$ and since the second row of the partly eliminated matrix is $\mathbf{u}_2$, we have

$$\mathbf{u}_3 = \mathbf{a}_3 - (-1\mathbf{u}_1) - 3\mathbf{u}_2.$$

To summarize, the rows of **A** and **U** satisfy the relations

$$\begin{aligned} \mathbf{u}_1 &= \mathbf{a}_1, \\ \mathbf{u}_2 &= \mathbf{a}_2 - \quad 2\mathbf{u}_1, \\ \mathbf{u}_3 &= \mathbf{a}_3 - (-1\mathbf{u}_1) - 3\mathbf{u}_2. \end{aligned}$$

Notice that the coefficients 2 and -1 of $\mathbf{u}_1$ and 3 of $\mathbf{u}_2$ are the multipliers. The first equation says that the first row of **A** and the first row of **U** are identical. The second equation implies that the second row of **U** equals the second row of **A** minus two times the first row of **U**. The third equation says that in generating $\mathbf{u}_3$, we subtract from $\mathbf{a}_3$, -1 times $\mathbf{u}_1$ and 3 times $\mathbf{u}_2$. Solving these three equations for $\mathbf{a}_1$, $\mathbf{a}_2$, and $\mathbf{a}_3$, we have

$$\begin{aligned} \mathbf{a}_1 &= \quad \mathbf{u}_1, \\ \mathbf{a}_2 &= \quad 2\mathbf{u}_1 + \mathbf{u}_2, \\ \mathbf{a}_3 &= -\mathbf{u}_1 + 3\mathbf{u}_2 + \mathbf{u}_3. \end{aligned}$$

Since two matrices are equal if and only if their respective rows are equal, these three equations can be written as one matrix equality:

$$\begin{bmatrix} \mathbf{a}_1 \\ \mathbf{a}_2 \\ \mathbf{a}_3 \end{bmatrix} = \begin{bmatrix} \mathbf{u}_1 \\ 2\mathbf{u}_1 + \mathbf{u}_2 \\ -\mathbf{u}_1 + 3\mathbf{u}_2 + \mathbf{u}_3 \end{bmatrix} \tag{9}$$

The left side of this matrix equality is **A**. We now show that the right side is **LU**.

By the definition of **L**, the product **LU** can be written

$$\mathbf{LU} = \begin{bmatrix} 1 & 0 & 0 \\ 2 & 1 & 0 \\ -1 & 3 & 1 \end{bmatrix} \begin{bmatrix} \mathbf{u}_1 \\ \mathbf{u}_2 \\ \mathbf{u}_3 \end{bmatrix}.$$

Each row of **LU** is the product between a row from **L** and matrix **U**. For example, the third row of **LU** equals the third row of **L** times **U**:

$$[-1 \quad 3 \quad 1]\begin{bmatrix} \mathbf{u}_1 \\ \mathbf{u}_2 \\ \mathbf{u}_3 \end{bmatrix}.$$

To compute this vector-matrix product, we multiply each row of the matrix by the corresponding component of the vector and add. Thus the third row of **LU** equals $-\mathbf{u}_1 + 3\mathbf{u}_2 + \mathbf{u}_3$, which is identical to the third row of (9). To summarize, since the right side of (9) equals **LU**, it follows that $\mathbf{A} = \mathbf{LU}$. As this example indicates,

Gaussian elimination is equivalent to factoring a matrix into the product between a unit lower triangular matrix and an upper triangular matrix.

Although we do not show it, the decomposition $\mathbf{A} = \mathbf{LU}$ is unique in the following sense: If $\mathbf{A} = \mathbf{LU}$, where **L** is unit lower triangular and **U** is upper triangular with nonzero diagonal elements, then **L** is the only unit lower triangular matrix and **U** is the only upper triangular matrix for which $\mathbf{A} = \mathbf{LU}$.

Gaussian elimination is **row elimination** since multiples of one row are subtracted from another row. We now study **column elimination** where multiples of one column are subtracted from another column. Returning to the coefficient matrix **A** of equation (1), let us subtract 2 times the first column from the second column, and let us subtract 1 times the first column from the third column to get

$$\begin{bmatrix} 2 & 0 & 0 \\ 4 & -1 & 3 \\ -2 & -3 & 7 \end{bmatrix}.$$

For the second elimination step, we subtract -3 times the second column from the third column, generating the lower triangular matrix

$$\begin{bmatrix} 2 & 0 & 0 \\ 4 & -1 & 0 \\ -2 & -3 & -2 \end{bmatrix}. \tag{10}$$

Using the multipliers 2, 1, and -3, we form the unit upper triangular matrix

$$\begin{bmatrix} 1 & 2 & 1 \\ 0 & 1 & -3 \\ 0 & 0 & 1 \end{bmatrix}. \tag{11}$$

For example, the -3 in row 2 and column 3 is the multiplier used to annihilate the coefficient in row 2 and column 3. Again, **A** is the product between the lower triangular matrix (10) and the upper triangular matrix (11):

$$\mathbf{A} = \begin{bmatrix} 2 & 4 & 2 \\ 4 & 7 & 7 \\ -2 & -7 & 5 \end{bmatrix} = \begin{bmatrix} 2 & 0 & 0 \\ 4 & -1 & 0 \\ -2 & -3 & -2 \end{bmatrix}\begin{bmatrix} 1 & 2 & 1 \\ 0 & 1 & -3 \\ 0 & 0 & 1 \end{bmatrix}.$$

Thus column elimination factors **A** into the product between a lower triangular matrix and a unit upper triangular matrix while row elimination factors **A** into the product between a unit lower triangular matrix and an upper triangular matrix. To avoid ambiguity, we sometimes attach the subscript 1 to the unit triangular factor. With this notation, row elimination produces the $\mathbf{L}_1\mathbf{U}$ factorization of **A** where $\mathbf{L}_1$ is a lower triangular matrix with ones on its diagonal while column elimination produces the $\mathbf{LU}_1$ factorization of $\mathbf{A}$ where $\mathbf{U}_1$ is an upper triangular matrix with ones on its diagonal.

The triangular factors computed by row elimination or by column elimination are closely connected. Observe that the diagonal of $\mathbf{L}$ generated by column elimination and the diagonal of $\mathbf{U}$ generated by row elimination are the same. Using the diagonal elements 2, −1, and −2, we form the matrix

$$\mathbf{D} = \begin{bmatrix} 2 & 0 & 0 \\ 0 & -1 & 0 \\ 0 & 0 & -2 \end{bmatrix}.$$

A matrix like $\mathbf{D}$ which is entirely zero off the diagonal is called a **diagonal matrix**. Multiplying each column of $\mathbf{L}_1$ by the corresponding diagonal element of $\mathbf{D}$ yields $\mathbf{L}$. Since multiplying each column of $\mathbf{L}_1$ by the corresponding diagonal element of $\mathbf{D}$ is equivalent to postmultiplying $\mathbf{L}_1$ by $\mathbf{D}$, the connection between $\mathbf{L}_1$, $\mathbf{D}$, and $\mathbf{L}$ can be stated "$\mathbf{L}$ equals $\mathbf{L}_1$ times $\mathbf{D}$":

$$\underset{\mathbf{L}}{\begin{bmatrix} 2 & 0 & 0 \\ 4 & -1 & 0 \\ -2 & -3 & -2 \end{bmatrix}} = \underset{\mathbf{L}_1}{\begin{bmatrix} 1 & 0 & 0 \\ 2 & 1 & 0 \\ -1 & 3 & 1 \end{bmatrix}} \underset{\mathbf{D}}{\begin{bmatrix} 2 & 0 & 0 \\ 0 & -1 & 0 \\ 0 & 0 & -2 \end{bmatrix}}.$$

Similarly, multiplying each row of $\mathbf{U}_1$ by the corresponding diagonal element of $\mathbf{D}$ yields $\mathbf{U}$. Since multiplying each row of $\mathbf{U}_1$ by the corresponding diagonal element of $\mathbf{D}$ is equivalent to premultiplying $\mathbf{U}_1$ by $\mathbf{D}$, it follows that $\mathbf{U}$ equals $\mathbf{D}$ times $\mathbf{U}_1$:

$$\underset{\mathbf{U}}{\begin{bmatrix} 2 & 4 & 2 \\ 0 & -1 & 3 \\ 0 & 0 & -2 \end{bmatrix}} = \underset{\mathbf{D}}{\begin{bmatrix} 2 & 0 & 0 \\ 0 & -1 & 0 \\ 0 & 0 & -2 \end{bmatrix}} \underset{\mathbf{U}_1}{\begin{bmatrix} 1 & 2 & 1 \\ 0 & 1 & -3 \\ 0 & 0 & 1 \end{bmatrix}}.$$

To summarize, $\mathbf{A} = \mathbf{L}_1\mathbf{U} = \mathbf{LU}_1$, where $\mathbf{U} = \mathbf{DU}_1$ and $\mathbf{L} = \mathbf{L}_1\mathbf{D}$. Combining these relations, we see that

$$\mathbf{A} = \mathbf{L}_1\mathbf{U} = \mathbf{L}_1(\mathbf{DU}_1)$$

and

$$\mathbf{A} = \mathbf{LU}_1 = (\mathbf{L}_1\mathbf{D})\mathbf{U}_1.$$

The factorization $\mathbf{A} = \mathbf{L}_1(\mathbf{DU}_1) = (\mathbf{L}_1\mathbf{D})\mathbf{U}_1$ into the product between a unit lower triangular matrix, a diagonal matrix, and a unit upper triangular matrix is called the **LDU** factorization. For the coefficient matrix $\mathbf{A}$ of equation (1), we have

$$\underset{\mathbf{A}}{\begin{bmatrix} 2 & 4 & 2 \\ 4 & 7 & 7 \\ -2 & -7 & 5 \end{bmatrix}} = \underset{\mathbf{L}_1}{\begin{bmatrix} 1 & 0 & 0 \\ 2 & 1 & 0 \\ -1 & 3 & 1 \end{bmatrix}} \underset{\mathbf{D}}{\begin{bmatrix} 2 & 0 & 0 \\ 0 & -1 & 0 \\ 0 & 0 & -2 \end{bmatrix}} \underset{\mathbf{U}_1}{\begin{bmatrix} 1 & 2 & 1 \\ 0 & 1 & -3 \\ 0 & 0 & 1 \end{bmatrix}}.$$

In writing this **LDU** factorization, we dropped the parentheses grouping terms of the product. The **associative law** states that *the terms in a matrix product can be grouped in any manner if order is preserved*. Since $\mathbf{L}_1(\mathbf{DU}_1) = (\mathbf{L}_1\mathbf{D})\mathbf{U}_1$, the parentheses can be placed either way and they are often omitted.

Exercises

2-1.1. Compute the following products:

$$\text{(a)} \begin{bmatrix} 1 & -1 \\ 0 & 1 \\ 1 & 0 \end{bmatrix} \begin{bmatrix} 1 & 1 & 0 \\ 1 & -1 & 1 \end{bmatrix} \qquad \text{(b)} \begin{bmatrix} 1 & 2 & 3 \end{bmatrix} \begin{bmatrix} 1 \\ 2 \\ 3 \end{bmatrix} \qquad \text{(c)} \begin{bmatrix} 1 \\ 2 \\ 3 \end{bmatrix} \begin{bmatrix} 1 & 2 & 3 \end{bmatrix}$$

$$\text{(d)} \begin{bmatrix} 2 & -1 & 0 & 0 \\ -1 & 2 & -1 & 0 \\ 0 & -1 & 2 & -1 \\ 0 & 0 & -1 & 2 \end{bmatrix} \begin{bmatrix} 1 \\ 1 \\ 1 \\ 1 \end{bmatrix}$$

2-1.2. Factor the following matrices into the product between a lower triangular matrix and an upper triangular matrix. Also give the **LDU** factorization in each example.

$$\text{(a)} \begin{bmatrix} 2 & 4 \\ 6 & 9 \end{bmatrix} \qquad \text{(b)} \begin{bmatrix} 4 & 12 & 16 \\ 16 & 45 & 58 \\ 8 & 15 & 12 \end{bmatrix} \qquad \text{(c)} \begin{bmatrix} 2 & 4 & 6 & 8 \\ 6 & 10 & 22 & 20 \\ 4 & 10 & 7 & 21 \\ 2 & 0 & 10 & 7 \end{bmatrix}$$

$$\text{(d)} \begin{bmatrix} 2 & 2 & 2 & 4 \\ 4 & 3 & 3 & 1 \\ 0 & 2 & 3 & 5 \\ 0 & 0 & 3 & 2 \end{bmatrix} \qquad \text{(e)} \begin{bmatrix} 2 & 0 & 0 & 4 \\ 0 & 3 & 3 & 0 \\ 0 & 9 & 11 & 0 \\ 4 & 0 & 0 & 2 \end{bmatrix} \qquad \text{(f)} \begin{bmatrix} 2 & 0 & 0 & 0 & 4 \\ 0 & 3 & 0 & 9 & 0 \\ 0 & 0 & 2 & 0 & 0 \\ 0 & 9 & 0 & 29 & 0 \\ 4 & 0 & 0 & 0 & 12 \end{bmatrix}$$

2-1.3. If **A** is a $m \times n$ matrix and **x** is a vector with n components, how many scalar-scalar multiplications and additions are needed to compute the matrix-vector product **Ax**?

2-1.4. If **A** is $l \times m$ and **B** is $m \times n$, how many scalar-scalar multiplications and additions are involved in the computation of the matrix-matrix product **AB**?

2-1.5. By the associative law for matrix multiplication, (**AB**)**C** = **A**(**BC**) = **ABC**. Although the placement of parentheses does not affect the product, it does affect the computing time. If **A** is $1 \times n$, **B** is $n \times 1$, and **C** is $1 \times n$, which of the products (**AB**)**C** or **A**(**BC**) can be computed faster? (*Hint:* Look at some examples.)

2-1.6. Suppose that **A** is $k \times l$, **B** is $l \times m$, and **C** is $m \times n$. How many multiplications and additions are involved in the computation of (**AB**)**C**? How many multiplications and additions are involved in the computation of **A**(**BC**)? In the special case $k = 3$, $l = 5$, $m = 3$, and $n = 2$, which of the products (**AB**)**C** or **A**(**BC**) can be computed faster?

2-2. TRIANGULAR SYSTEMS

Given the **L** and **U** factors of the coefficient matrix, a linear system can be decomposed into two triangular equations. Replacing the coefficient matrix **A** in the system **Ax** = **b** by its **LU** factorization and utilizing the associative law for matrix multiplication, we have

$$\mathbf{Ax} = (\mathbf{LU})\mathbf{x} = \mathbf{L}(\mathbf{Ux}) = \mathbf{Ly},$$

where **y** = **Ux**. Hence **Ax** = **b** is equivalent to the two equations:

$$\mathbf{Ly} = \mathbf{b} \qquad \text{and} \qquad \mathbf{Ux} = \mathbf{y}.$$

Since **Ax** equals **Ly**, the first equation is the original one, **Ax** = **b**, expressed in terms of **y**. The second equation is the definition of **y**. The solution to the first equation is **y**. Knowing **y**, we can solve the second equation for **x**.

To illustrate, consider the system

$$\begin{bmatrix} 2 & 4 & 2 \\ 4 & 7 & 7 \\ -2 & -7 & 5 \end{bmatrix} \begin{bmatrix} x_1 \\ x_2 \\ x_3 \end{bmatrix} = \begin{bmatrix} 4 \\ 13 \\ 7 \end{bmatrix}$$

from Section 2-1 and the factorization

$$\begin{bmatrix} 2 & 4 & 2 \\ 4 & 7 & 7 \\ -2 & -7 & 5 \end{bmatrix} = \begin{bmatrix} 1 & 0 & 0 \\ 2 & 1 & 0 \\ -1 & 3 & 1 \end{bmatrix} \begin{bmatrix} 2 & 4 & 2 \\ 0 & -1 & 3 \\ 0 & 0 & -2 \end{bmatrix}.$$

In this case, the relation $\mathbf{Ly} = \mathbf{b}$ is

$$\begin{bmatrix} 1 & 0 & 0 \\ 2 & 1 & 0 \\ -1 & 3 & 1 \end{bmatrix} \begin{bmatrix} y_1 \\ y_2 \\ y_3 \end{bmatrix} = \begin{bmatrix} 4 \\ 13 \\ 7 \end{bmatrix},$$

which is the matrix representation of the following system of three equations:

$$\begin{aligned} y_1 \qquad\qquad\quad &= 4, \\ 2y_1 + \ y_2 \qquad\ &= 13, \\ -y_1 + 3y_2 + y_3 &= \ 7. \end{aligned} \tag{12}$$

This lower triangular system is solved by **forward substitution**. The first equation implies that y_1 is 4. Replacing y_1 by 4 in the second equation yields

$$y_2 = 13 - 2y_1 = 13 - 2 \times 4 = 5.$$

Then replacing y_1 by 4 and y_2 by 5, the third equation yields

$$y_3 = 7 + y_1 - 3y_2 = 7 + 4 - 3 \times 5 = -4.$$

Having determined

$$\mathbf{y} = \begin{bmatrix} 4 \\ 5 \\ -4 \end{bmatrix},$$

we must now solve the upper triangular system $\mathbf{Ux} = \mathbf{y}$:

$$\begin{bmatrix} 2 & 4 & 2 \\ 0 & -1 & 3 \\ 0 & 0 & -2 \end{bmatrix} \begin{bmatrix} x_1 \\ x_2 \\ x_3 \end{bmatrix} = \begin{bmatrix} 4 \\ 5 \\ -4 \end{bmatrix}.$$

Or equivalently, we must solve the three equations

$$\begin{aligned} 2x_1 + 4x_2 + 2x_3 &= \ \ 4, \\ - \ x_2 + 3x_3 &= \ \ 5, \\ - 2x_3 &= -4. \end{aligned}$$

But this system is exactly equation (2), which we solved by back substitution in Section 2-1.

In summary, $\mathbf{Ax} = \mathbf{b}$ is processed in three steps:

1. **LU** factor **A**.
2. Solve $\mathbf{Ly} = \mathbf{b}$ by forward substitution.
3. Solve $\mathbf{Ux} = \mathbf{y}$ by back substitution.

Is this algorithm less efficient than the method discussed in Section 2-1, where the right side and the coefficients are processed simultaneously and step 2 is omitted? It can be shown that the number of arithmetic operations involved in the three-step process is the same as the number of arithmetic operations associated with the method of Section 2-1. On the other hand, when we consider applications, there is a clear advantage to the three-step process. For example, in a problem that requires the solution of several linear systems that have identical coefficients but different right sides, step 1 is executed once at the beginning while step 2 and step 3 are applied to each right side. Hence, the arithmetic corresponding to step 1 of the three-step process is not duplicated when we change the right side.

Let us now develop computer codes for forward substitution and for back substitution. For a system of three equations and three unknowns, the unit lower triangular system $\mathbf{Ly} = \mathbf{b}$ has the form

$$\begin{bmatrix} 1 & 0 & 0 \\ l_{21} & 1 & 0 \\ l_{31} & l_{32} & 1 \end{bmatrix} \begin{bmatrix} y_1 \\ y_2 \\ y_3 \end{bmatrix} = \begin{bmatrix} b_1 \\ b_2 \\ b_3 \end{bmatrix},$$

where l_{ij} denotes the coefficient of $\mathbf{L}$ in row i and column j. The corresponding system of three equations is

$$\begin{aligned} y_1 \qquad\qquad\qquad &= b_1, \\ l_{21}y_1 + \quad y_2 \qquad\quad &= b_2, \\ l_{31}y_1 + l_{32}y_2 + y_3 &= b_3. \end{aligned}$$

Solving successively for y_1, y_2, and y_3, we have

$$\begin{aligned} y_1 &= b_1, \\ y_2 &= b_2 - l_{21}y_1, \\ y_3 &= b_3 - l_{31}y_1 - l_{32}y_2. \end{aligned}$$

Observe that y_i is expressed in terms of b_i and the multipliers from $\mathbf{L}$'s row i. In particular, $y_1 = b_1$ and for $i > 1$, y_i is given by

$$y_i = b_i - \sum_{j=1}^{i-1} l_{ij}\, y_j. \tag{13}$$

That is, y_i is obtained by subtracting from b_i the quantities $l_{ij}y_j$ for j between 1 and $i - 1$. Formula (13) is also valid for a general unit lower triangular system of n equations. For three equations in three unknowns, we use (13) to compute y_2 and y_3. For a general system of n equations, (13) is used to compute y_2 through y_n.

If L is a real array which contains the multipliers, and B and Y are real arrays that contain the components of $\mathbf{b}$ and $\mathbf{y}$, respectively, the following Fortran code does forward substitution:

```
Y(1) = B(1)
DO 20 I = 2,N
   IM1 = I - 1
   T = B(I)
   DO 10 J = 1,IM1
      T = T - L(I,J)*Y(J)
```

```
10          CONTINUE
            Y(I) = T
20      CONTINUE
```

This code subtracts from T = B(I) the quantity L(I,J)*Y(J) for J between 1 and IM1 = I − 1, and then sets Y(I) = T.

In the streamlined code that appears in Table 2-1, the Y array is eliminated and the components of **y** overwrite the elements of B. For example, if this code is stopped just after I is assigned the value 4, then the first three components of **y** are stored in B(1), B(2), and B(3), while B(4) through B(N) store the original components of **b**. When execution continues, L(4,J)*B(J) is subtracted from B(4) for J = 1, 2, and 3 inside the J loop. Since B(1) = y_1, B(2) = y_2, and B(3) = y_3, it follows that L(4,J)*B(J) = L(4,J)*Y(J) for J = 1, 2, and 3. At the end of the I = 4 cycle of the I loop, y_4 is stored in B(4). After executing the entire code, array B contains the solution **y** to the system **Ly** = **b**.

The code in Table 2-1 is summarized by the following symbolic program:

$$
\boxed{\begin{array}{l} i = 2 \text{ to } n \\ \quad j = 1 \text{ to } i-1 \\ \qquad b_i \leftarrow b_i - l_{ij} b_j \\ \quad \text{next } j \\ \text{next } i \end{array}} \tag{14}
$$

In other words, we subtract $l_{ij}b_j$ from b_i for j between 1 and $i - 1$, where i increments from 2 to n. A symbolic program such as (14) describes the essential parts of an algorithm without concerning us with compiler quirks. For example, the loop limit I − 1 is not allowed in some versions of Fortran and a variable IM1 = I − 1 must be introduced before executing the j-loop in (14). Besides computing IM1, the program in Table 2-1 also preprocesses B(I) outside the J loop.

Now let us consider back substitution for a system of three equations. The relation **Ux** = **y** takes the form

$$
\begin{bmatrix} u_{11} & u_{12} & u_{13} \\ 0 & u_{22} & u_{23} \\ 0 & 0 & u_{33} \end{bmatrix} \begin{bmatrix} x_1 \\ x_2 \\ x_3 \end{bmatrix} = \begin{bmatrix} y_1 \\ y_2 \\ y_3 \end{bmatrix},
$$

TABLE 2-1 FORWARD SUBSTITUTION

```
        DO 20 I = 2,N
           IM1 = I - 1
           T = B(I)
           DO 10 J = 1,IM1
              T = T - L(I,J)*B(J)
10         CONTINUE
           B(I) = T
20      CONTINUE
```

which is equivalent to the three equations

$$\begin{aligned} u_{11}x_1 + u_{12}x_2 + u_{13}x_3 &= y_1, \\ u_{22}x_2 + u_{23}x_3 &= y_2, \\ u_{33}x_3 &= y_3. \end{aligned}$$

Starting with the last equation and solving for x_3, x_2, and x_1 in this order, we have

$$x_3 = \frac{y_3}{u_{33}},$$

$$x_2 = \frac{y_2 - u_{23}x_3}{u_{22}},$$

$$x_1 = \frac{y_1 - u_{12}x_2 - u_{13}x_3}{u_{11}}.$$

In computing x_i, the product between elements in row i of **U** and the corresponding components of **x** is subtracted from y_i and divided by u_{ii}. For a general system of n equations, back substitution starts with $x_n = y_n/u_n$, and then for $i = n - 1$ down to 1, x_i is given by the rule

$$x_i = \frac{y_i - \sum_{j=i+1}^{n} u_{ij}x_j}{u_{ii}}.$$

If the elements of **U** and **y** are stored in arrays U and Y, respectively, the code in Table 2-2 implements back substitution, overwriting Y with the solution **x** to the system **Ux** = **y**. The symbolic program corresponding to Table 2-2 is

$i = n$ down to 1
   $j = i+1$ to n
      $y_i \leftarrow y_i - u_{ij}\, y_j$
   next j
   $y_i \leftarrow y_i/u_{ii}$
next i

(15)

TABLE 2-2 BACK SUBSTITUTION

```
      Y(N) = Y(N)/U(N,N)
      NM1 = N - 1
      DO 20 IB = 1,NM1
         I = N - IB
         IP1 = I + 1
         T = Y(I)
         DO 10 J = IP1,N
            T = T - U(I,J)*Y(J)
10       CONTINUE
         Y(I) = T/U(I,I)
20    CONTINUE
```

When i is n, the range for the j index in our symbolic program is "$n + 1$ up to n." Since this range is empty, the statement $y_i \leftarrow y_i - u_{ij} y_j$ is skipped when i is n and we just execute the statement $y_n \leftarrow y_n/u_{nn}$. Since many versions of Fortran execute the statements in a DO loop at least once, regardless of the loop limits, the statement $y_n \leftarrow y_n/u_{nn}$ is treated separately in Table 2-2.

The substitution algorithms (14) and (15) employ row operations. For example, the deepest loop in (15) involves the product between elements in row i of $\mathbf{U}$ and the corresponding components of $\mathbf{y}$. Now let us modify (14) and (15) to use column operations. We start with the special system $\mathbf{Ly} = \mathbf{b}$ given by equation (12). Moving all terms to the right side except the diagonal terms, we have

$$\begin{aligned} y_1 &= 4, \\ y_2 &= 13 - 2y_1, \\ y_3 &= 7 + y_1 - 3y_2. \end{aligned} \tag{16}$$

The first equation tells us that y_1 is 4. Inserting $y_1 = 4$ in the second and third equations yields

$$\begin{aligned} y_1 & & &= 4, \\ y_2 &= 13 - 2 \times 4 &&= 5, \\ y_3 &= 7 + 4 - 3y_2 &&= 11 - 3y_2. \end{aligned} \tag{17}$$

The second equation implies that y_2 is 5. Inserting $y_2 = 5$ in the third equation gives

$$\begin{aligned} y_1 & &&= 4, \\ y_2 & &&= 5, \\ y_3 &= 11 - 3 \times 5 &&= -4. \end{aligned}$$

To see that this algorithm is processing the columns of $\mathbf{L}$, let us express (16) using vector notation:

$$\begin{bmatrix} y_1 \\ y_2 \\ y_3 \end{bmatrix} = \begin{bmatrix} 4 \\ 13 \\ 7 \end{bmatrix} - y_1 \begin{bmatrix} 0 \\ 2 \\ -1 \end{bmatrix} - y_2 \begin{bmatrix} 0 \\ 0 \\ 3 \end{bmatrix}. \tag{18}$$

The vectors multiplying y_1 and y_2 are formed from columns 1 and 2 of $\mathbf{L}$, respectively. The first component of these column vectors is zero so (18) implies that $y_1 = 4$, the first component of the right-side vector. Equation (17) is derived by replacing y_1 by 4 in equation (18), to obtain

$$\begin{bmatrix} y_1 \\ y_2 \\ y_3 \end{bmatrix} = \begin{bmatrix} 4 \\ 13 \\ 7 \end{bmatrix} - 4 \begin{bmatrix} 0 \\ 2 \\ -1 \end{bmatrix} - y_2 \begin{bmatrix} 0 \\ 0 \\ 3 \end{bmatrix} = \begin{bmatrix} 4 \\ 5 \\ 11 \end{bmatrix} - y_2 \begin{bmatrix} 0 \\ 0 \\ 3 \end{bmatrix}. \tag{19}$$

The first two components of the vector multiplying y_2 are zero. Hence (19) implies that $y_1 = 4$ and $y_2 = 5$. Finally, replacing y_2 by 5 yields

$$\begin{bmatrix} y_1 \\ y_2 \\ y_3 \end{bmatrix} = \begin{bmatrix} 4 \\ 5 \\ 11 \end{bmatrix} - 5 \begin{bmatrix} 0 \\ 0 \\ 3 \end{bmatrix} = \begin{bmatrix} 4 \\ 5 \\ -4 \end{bmatrix}. \tag{20}$$

Column-oriented forward substitution is exactly the elimination process of Section 2-1 applied to the right side of equation (1). The successive right sides generated between equation (1) and the upper triangular system (2) also appear on the right sides of (18), (19), and (20). For this reason, forward substitution by columns is usually called **forward elimination**.

For a general system of three equations, the analog of equation (18) is

$$\begin{bmatrix} y_1 \\ y_2 \\ y_3 \end{bmatrix} = \begin{bmatrix} b_1 \\ b_2 \\ b_3 \end{bmatrix} - y_1 \begin{bmatrix} 0 \\ l_{21} \\ l_{31} \end{bmatrix} - y_2 \begin{bmatrix} 0 \\ 0 \\ l_{32} \end{bmatrix}.$$

We subtract from **b** both the product between y_1 and the multipliers in **L**'s column 1, and the product between y_2 and the multipliers in **L**'s column 2. The following symbolic program subtracts from **b** the product between y_j and the multipliers in **L**'s column j for j between 1 and $n - 1$, overwriting **b** with the solution **y** to the lower triangular system **Ly** = **b**.

```
j = 1 to n-1
    i = j+1 to n
        b_i ← b_i − l_ij b_j
    next i
next j
```
(21)

As i increments in the deepest loop, the product between b_j $(= y_j)$ and the multiplier l_{ij} from **L**'s column j is subtracted from b_i. The program in Table 2-3 implements (21) by embedding the statement "B(I) = B(I) − L(I,J)*B(J)" inside two loops where J sweeps from 1 to N − 1 and I increases from J + 1 to N.

Now let us develop a column strategy for back substitution using the special upper triangular system (2) for illustration. Moving all terms to the right side except the diagonal terms, we obtain

$$\begin{aligned} 2x_1 &= 4 - 4x_2 - 2x_3, \\ -x_2 &= 5 \qquad\quad - 3x_3, \\ -2x_3 &= -4. \end{aligned} \tag{22}$$

The last equation tells us that x_3 is $-4/-2 = 2$. Substituting $x_3 = 2$ into the first and second equation gives

$$\begin{aligned} 2x_1 &= 4 - 4x_2 - 2 \times 2 = -4x_2, \\ -x_2 &= 5 \qquad\quad - 3 \times 2 = -1, \\ x_3 &\qquad\qquad\qquad\quad = 2. \end{aligned}$$

TABLE 2-3 FORWARD ELIMINATION

```
      NM1 = N - 1
      DO 20 J = 1,NM1
         JP1 = J + 1
         T = B(J)
         DO 10 I = JP1,N
            B(I) = B(I) - L(I,J)*T
10       CONTINUE
20    CONTINUE
```

The second equation implies that x_2 is $-1/-1 = 1$. Inserting $x_2 = 1$ in the first equation, we have

$$\begin{aligned} 2x_1 &= -4, \\ x_2 &= 1, \\ x_3 &= 2. \end{aligned}$$

Dividing the first equation by 2 yields $x_1 = -2$, so the solution is $x_1 = -2$, $x_2 = 1$, and $x_3 = 2$.

To clarify the column strategy for back substitution, it helps to write equation (22) using vector notation:

$$\begin{bmatrix} 2x_1 \\ -x_2 \\ -2x_3 \end{bmatrix} = \begin{bmatrix} 4 \\ 5 \\ -4 \end{bmatrix} - x_2 \begin{bmatrix} 4 \\ 0 \\ 0 \end{bmatrix} - x_3 \begin{bmatrix} 2 \\ 3 \\ 0 \end{bmatrix}. \tag{23}$$

The vectors multiplying x_2 and x_3 correspond to the second and third columns of **U**. In the column strategy, we start with the last component of (23) and divide by -2 to get $x_3 = -4/-2 = 2$. Then inserting $x_3 = 2$ in (23) gives

$$\begin{bmatrix} 2x_1 \\ -x_2 \\ x_3 \end{bmatrix} = \begin{bmatrix} 4 \\ 5 \\ 2 \end{bmatrix} - x_2 \begin{bmatrix} 4 \\ 0 \\ 0 \end{bmatrix} - 2 \begin{bmatrix} 2 \\ 3 \\ 0 \end{bmatrix} = \begin{bmatrix} 0 \\ -1 \\ 2 \end{bmatrix} - x_2 \begin{bmatrix} 4 \\ 0 \\ 0 \end{bmatrix}. \tag{24}$$

Dividing the second component of (24) by -1, we see that $x_2 = -1/-1 = 1$, and inserting $x_2 = 1$ in (24) yields

$$\begin{bmatrix} 2x_1 \\ x_2 \\ x_3 \end{bmatrix} = \begin{bmatrix} 0 \\ 1 \\ 2 \end{bmatrix} - 1 \begin{bmatrix} 4 \\ 0 \\ 0 \end{bmatrix} = \begin{bmatrix} -4 \\ 1 \\ 2 \end{bmatrix}. \tag{25}$$

Dividing the first component in (25) by 2 yields $x_1 = -4/2 = -2$, so the final solution is

$$\begin{bmatrix} x_1 \\ x_2 \\ x_3 \end{bmatrix} = \begin{bmatrix} -2 \\ 1 \\ 2 \end{bmatrix}.$$

For a general system of three equations, the analog of (23) is

$$\begin{bmatrix} u_{11}x_1 \\ u_{22}x_2 \\ u_{33}x_3 \end{bmatrix} = \begin{bmatrix} y_1 \\ y_2 \\ y_3 \end{bmatrix} - x_2 \begin{bmatrix} u_{12} \\ 0 \\ 0 \end{bmatrix} - x_3 \begin{bmatrix} u_{13} \\ u_{23} \\ 0 \end{bmatrix}.$$

To compute x_3, the third component of this identity is divided by u_{33}. Then the product between x_3 and elements in the third column of **U** is subtracted from **y**. After computing x_2, we subtract from the new **y** the product between x_2 and elements from the second column of **U**. Finally, to compute x_1, we divide the first component of the new **y** by u_{11}. For a general system of n upper triangular equations, the following symbolic program overwrites the right side **y** of the system $\mathbf{Ux} = \mathbf{y}$ with the solution **x**:

$$\boxed{\begin{array}{l} j = n \text{ down to } 1 \\ \quad y_j \leftarrow y_j/u_{jj} \\ \quad i = 1 \text{ to } j-1 \\ \qquad y_i \leftarrow y_i - u_{ij}\,y_j \\ \quad \text{next } i \\ \text{next } j \end{array}} \tag{26}$$

Back substitution by columns will be called **back elimination**. If the elements of **U** and **y** are stored in the arrays U and Y, respectively, then the computer program in Table 2-4 implements (26), overwriting Y with the solution **x** to the system **Ux** = **y**.

Is it better to process rows or columns? Moler notes in reference [M7] that the answer to this question may depend on the storage structure for a double-subscripted array. Conceptually, computer memory is a long linear list. Even though a program employs a two dimensional array, the array elements are stored in a one-dimensional structure. Fortran stores a double-subscripted array by columns while PL/I stores it by rows. For example, Fortran stores the 3×3 array

$$\begin{bmatrix} 2 & 0 & 6 \\ 1 & 3 & 2 \\ 5 & 4 & 7 \end{bmatrix}$$

as the linear list

$$2 \quad 1 \quad 5 \quad 0 \quad 3 \quad 4 \quad 6 \quad 2 \quad 7.$$

For a computer where all memory locations have the same access time, rows and columns are processed equally fast. On the other hand, in many computing environments, there are a variety of memory devices. One possible memory configuration is depicted in Figure 2-1 on page 56. As mentioned in Chapter 1, main memory is limited and expensive compared to disk storage or tape storage. If a program is processing a large amount of data, many operating systems automatically partition the data into *blocks* or *pages* which are divvied among storage devices. A few pages are stored in main memory, while most of the

TABLE 2-4 BACK ELIMINATION

```
      NM1 = N - 1
      DO 20 JB = 1,NM1
         JM1 = N - JB
         J = JM1 + 1
         T = Y(J)/U(J,J)
         Y(J) = T
         DO 10 I = 1,JM1
            Y(I) = Y(I) - U(I,J)*T
10       CONTINUE
20    CONTINUE
      Y(1) = Y(1)/U(1,1)
```

Cache memory	Fast
Fast memory	
Slow memory	
Disk	Slow

Figure 2-1 Memory speed.

data may reside on a disk. If a program is processing a matrix and the computer cannot find an array element in fast memory, the entire page that contains the missing number is retrieved from the slower storage device. If the matrix is stored and processed by columns, this page contains column elements which are used soon in the computations. In contrast, suppose that the matrix is stored by columns, but the program processes the matrix by rows. As the program works along a row, elements are retrieved from the slow storage device. But each element brings with it the entire page that contains that element. Since this page contains column elements which are not used when we process the row, the computer wastes time by moving the entire page into fast memory. Fortran stores matrices by columns, so it is better to process columns in Fortran. Row-oriented algorithms execute quicker in PL/I.

Exercises

2-2.1. Use a factorization computed in Exercise 2-1.2 to solve the factored system $(\mathbf{LU})\mathbf{x} = \mathbf{b}$ for the following choices of $\mathbf{b}$ (corresponding to each matrix in Exercise 2-1.2):

$$\text{(a)} \begin{bmatrix} 2 \\ 9 \end{bmatrix} \quad \text{(b)} \begin{bmatrix} 4 \\ 16 \\ 6 \end{bmatrix} \quad \text{(c)} \begin{bmatrix} 4 \\ 26 \\ -9 \\ 17 \end{bmatrix} \quad \text{(d)} \begin{bmatrix} 2 \\ -3 \\ 4 \\ -1 \end{bmatrix} \quad \text{(e)} \begin{bmatrix} 2 \\ 0 \\ -2 \\ -2 \end{bmatrix} \quad \text{(f)} \begin{bmatrix} 6 \\ -12 \\ 2 \\ -38 \\ 16 \end{bmatrix}$$

2-2.2. Write a symbolic program to solve $\mathbf{Ly} = \mathbf{b}$ and $\mathbf{Ux} = \mathbf{y}$, where $\mathbf{L}$ is lower triangular but each diagonal element is not necessarily 1, and $\mathbf{U}$ is upper triangular with every diagonal element equal to 1. Write both a row-oriented algorithm and a column-oriented algorithm.

2-2.3. Given the **LDU** factorization of $\mathbf{A}$, express the solution to $\mathbf{Ax} = \mathbf{b}$ in terms of the solution to three separate equations involving $\mathbf{L}$, $\mathbf{D}$, and $\mathbf{U}$, respectively. Write a symbolic program to solve $\mathbf{Ax} = \mathbf{b}$ assuming that the elements of $\mathbf{A}$ have been replaced by the corresponding nonzero elements of $\mathbf{L}$, $\mathbf{D}$, and $\mathbf{U}$.

2-2.4. Let $\mathbf{L}$ be a lower triangular matrix and let $\mathbf{e}_j$ denote the vector with every component equal to zero except for component j, which is 1.

(a) Which components of the vector $\mathbf{Le}_j$ must be zero?

(b) Let $\mathbf{x}$ be a vector whose first k components are zero. Which components of $\mathbf{Lx}$ must be zero?

(c) Using your answer to part (b), explain why the product between two lower triangular matrices is lower triangular.

2-2.5. Let $\mathbf{L}$ be a lower triangular matrix with nonzero diagonal elements and let $\mathbf{e}_j$ denote the vector with every component equal to zero except for component j, which is 1.

(a) If $\mathbf{x}$ is the solution to $\mathbf{Lx} = \mathbf{e}_j$, then which components of $\mathbf{x}$ must be zero?

(b) Using your answer to part (a), explain why the inverse of a lower triangular matrix is lower triangular.

2-3. FACTORIZATION

For the example in Section 2-1, Gaussian elimination generates the following sequence of coefficient matrices:

$$\underset{\mathbf{A}}{\begin{bmatrix} 2 & 4 & 2 \\ 4 & 7 & 7 \\ -2 & -7 & 5 \end{bmatrix}} \longrightarrow \begin{bmatrix} 2 & 4 & 2 \\ 0 & -1 & 3 \\ 0 & -3 & 7 \end{bmatrix} \longrightarrow \underset{\mathbf{U}}{\begin{bmatrix} 2 & 4 & 2 \\ 0 & -1 & 3 \\ 0 & 0 & -2 \end{bmatrix}}.$$

For a general matrix, the first elimination step annihilates coefficients in column 1 beneath the diagonal, the second elimination step annihilates coefficients in column 2 beneath the diagonal, and so on. A partly eliminated matrix has the structure sketched in Figure 2-2. In more detail, before annihilating elements in column k, the partly processed coefficient matrix resembles Figure 2-3. (In displaying a matrix, our convention is that empty portions of the matrix are always zero. Hence the lower left corner of the matrix in Figure 2-3 is zero by assumption.) During the next elimination step, multiples of row k are subtracted from rows beneath it. The element a_{ik} in row i and column k is annihilated when each element in row k is multiplied by $l_{ik} = a_{ik}/a_{kk}$ and subtracted from the corresponding element in row i. In particular, a_{kj} is multiplied by l_{ik} and subtracted from a_{ij}. The new entry in row i and column j is $a_{ij} - l_{ik}a_{kj}$. The column index j sweeps from k to n since both column k and the columns to its right are affected. Since each element in column k beneath the diagonal is annihilated, i steps from $k + 1$ to n. Since the subdiagonal elements in columns 1 through $n - 1$ are eliminated, k steps from 1 to $n - 1$. In summary, the following symbolic program implements row elimination:

```
k = 1 to n-1
    i = k+1 to n
        l_ik ← a_ik/a_kk
        j = k to n
            a_ij ← a_ij - l_ik a_kj
        next j
    next i
next k
```
(27)

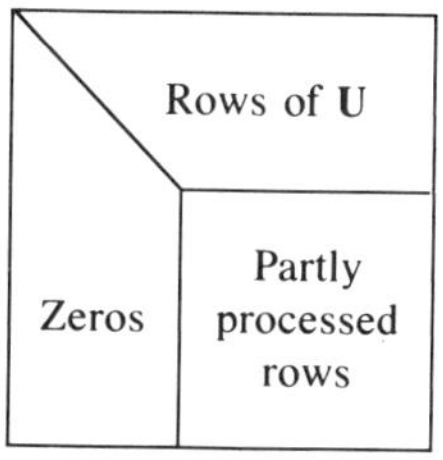

Figure 2-2 Partly eliminated matrix.

$$\begin{bmatrix} a_{11} & a_{12} & \cdots & a_{1k} & \cdots & a_{1j} & \cdots & a_{1n} \\ & a_{22} & \cdots & a_{2k} & \cdots & a_{2j} & \cdots & a_{2n} \\ & & \ddots & \vdots & & \vdots & & \vdots \\ & & & a_{kk} & \cdots & a_{kj} & \cdots & a_{kn} \\ & & & \vdots & & \vdots & & \vdots \\ & & & a_{ik} & \cdots & a_{ij} & \cdots & a_{in} \\ & & & \vdots & & \vdots & & \vdots \\ & & & a_{nk} & \cdots & a_{nj} & \cdots & a_{nn} \end{bmatrix}$$

Figure 2-3 $n \times n$ matrix before the kth elimination step.

The multiplier $l_{ik} = a_{ik}/a_{kk}$ is computed before the j loop. Inside the j loop, l_{ik} times row k is subtracted from row i, where i ranges from $k+1$ to n.

This algorithm employs row operations since a multiple of row k is subtracted from row i in the deepest loop. We now reindex (27) to perform the same elimination step while accessing the columns of the matrix rather than the rows. In the deepest loop, i is the row and j is the column. Holding i fixed and incrementing j takes us along a row. Conversely, holding j fixed and incrementing i takes us down a column. Therefore, a column-oriented version of the row-elimination scheme (27) is obtained by incrementing i before j, as in the following algorithm.

```
k = 1 to n-1
    i = k+1 to n
        l_ik ← a_ik/a_kk
    next i
    j = k to n
        i = k+1 to n
            a_ij ← a_ij - l_ik a_kj
        next i
    next j
next k
```
(28)

Observe that the index range in (28) is the same as the index range in (27): i still increases from $k+1$ to n, and j still sweeps from k to n—our only change is to increment the i loop inside the j loop and to compute the multipliers before entering the j loop. We emphasize that even though algorithm (28) is column oriented, it still performs row elimination. Each step of algorithm (28) multiplies row k by l_{ik} and subtracts it from row i, although the array elements are processed by columns.

In summary, there are two different types of elimination strategies: row elimination, where multiples of one row are subtracted from another row, and column elimination, where multiples of one column are subtracted from another column. Each of these strategies can be implemented using either a row-oriented scheme or a column-oriented scheme. The orientation is related to how the algorithm accesses the coefficients inside the deepest loop. If the columns of a matrix are accessed inside the deepest loop, the algorithm is column oriented. If the rows of a matrix are accessed inside the deepest loop, the algorithm is row oriented.

When implementing (27) or (28), the multipliers can be stored in the space occupied by annihilated coefficients. Whenever a coefficient is annihilated, the multiplier is stored in place of zero. After elimination, the original **A** is replaced by the storage structure depicted in Figure 2-4 where the multipliers are stored

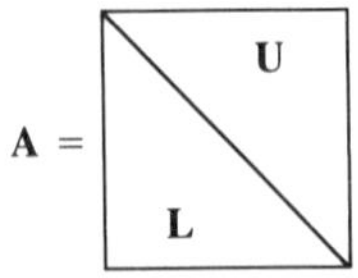

Figure 2-4 Storage structure for elimination.

beneath the diagonal and the elements of **U** are stored on the diagonal and above the diagonal. The program in Table 2-5 implements (28) using this storage structure. Observe that the index range for the J loop in Table 2-5 is K + 1 to N, while the index range for the corresponding loop in (28) is k to n. This difference is related to the multiplier storage. During the kth elimination step, multipliers are stored in column k, where zeros would be generated by the elimination process. Rather than generate these zeros and wipe out the multipliers, elimination starts at column $k + 1$.

Recall from Chapter 1 that an algorithm must not only work, but be insensitive to small rounding errors. To illustrate a potential pitfall of Gaussian elimination, let us solve the system

$$\begin{aligned} .001x_1 + x_2 &= 1, \\ x_1 + x_2 &= 2 \end{aligned} \tag{29}$$

on a computer that employs two-digit decimal floating-point arithmetic with rounding. The first equation is multiplied by 1000 and subtracted from the second equation, giving us

$$\begin{aligned} .001x_1 + \qquad x_2 &= 1, \\ (1 - 1000)x_2 &= 2 - 1000. \end{aligned}$$

In evaluating the difference $1 \ominus 1000$, the computer gets

$$1 \ominus 1000 = \text{fl}(1 - 1000) = \text{fl}(-999) = \text{fl}(-.999 \times 10^3) = -1.00 \times 10^3.$$

Since $1 \ominus 1000 = -1000$ and $2 \ominus 1000 = -1000$ using two-digit floating-point arithmetic with rounding, the computer obtains the following upper triangular system:

$$\begin{aligned} .001x_1 + \qquad x_2 &= \quad 1, \\ -1000\,x_2 &= -1000. \end{aligned}$$

TABLE 2-5 ROW ELIMINATION BY COLUMNS

```
      NM1 = N - 1
      DO 20 K = 1,NM1
         KP1 = K + 1
         T = A(K,K)
C
C     *** COMPUTE MULTIPLIERS ***
C
         DO 10 I = KP1,N
10          A(I,K) = A(I,K)/T
C
C     *** ELIMINATE BY COLUMNS ***
C
         DO 20 J = KP1,N
            T = A(K,J)
            DO 20 I = KP1,N
20             A(I,J) = A(I,J) - A(I,K)*T
```

Back substitution gives

$$x_2 = \frac{-1000}{-1000} = 1,$$

$$x_1 = \frac{1 - x_2}{.001} = 0.$$

On the other hand, with exact arithmetic, we obtain the upper triangular system

$$\begin{aligned} .001x_1 + \quad x_2 &= \quad 1, \\ -\ 999x_2 &= -998, \end{aligned}$$

whose solution is

$$\begin{aligned} x_2 &= \frac{998}{999} && \approx .999, \\ x_1 &= \frac{1 - 998/999}{.001} = \frac{1000}{999} && \approx 1.001. \end{aligned} \tag{30}$$

The error in x_1 is 100% since the computed x_1 is zero while the correct x_1 is near 1.

Observe that the computed upper triangular system agrees well with the correct system. The large error in the computed solution arose in back substitution when we evaluated $(1 - x_2)/.001$. Since x_2 is near 1, $1 - x_2$ is the difference between 1 and a number near 1. We saw in Chapter 1 that the relative error in the difference of nearly equal numbers can be large. Thus the relative error in $1 \ominus x_2$ can be large. One way to improve the accuracy in the computed solution is to interchange the equations in (29) and work with the system:

$$\begin{aligned} x_1 + x_2 &= 2, \\ .001x_1 + x_2 &= 1. \end{aligned}$$

Using two-digit decimal floating-point arithmetic with rounding, we obtain the upper triangular system

$$\begin{aligned} x_1 + x_2 &= 2, \\ x_2 &= 1, \end{aligned}$$

and back substitution yields $x_2 = x_1 = 1$, which agrees well with the exact solution.

The inaccurate solution obtained during our first attempt at solving (29) is ultimately connected with the small coefficient of x_1 in the first equation. Since the true x_1 is near 1, $.001x_1$ is near .001, and from the first equation in (29), $1 - x_2 = .001x_1 \approx .001$. Thus $1 - x_2$ is forced to be the difference of nearly equal numbers. Small diagonal elements can be avoided using **row pivoting:**

> *Before the kth elimination step when x_k is eliminated from equations $k + 1$ through n, interchange equation k with the equation beneath it which has the largest absolute coefficient for x_k.*

This largest coefficient is called the **pivot**. If the pivot is zero, then Gaussian elimination must stop since a_{kk} is zero and the multiplier $l_{ik} = a_{ik}/a_{kk}$ is undefined. A system with a zero pivot is called **singular**, and except for very special right sides, a singular system has no solution (see Exercise 2-3.5).

Row pivoting is illustrated using the system

$$
\begin{aligned}
2x_1 + 2x_2 - 2x_3 &= 8,\\
-4x_1 - 2x_2 + 2x_3 &= -14,\\
-2x_1 + 3x_2 + 9x_3 &= 9.
\end{aligned} \tag{31}
$$

Since -4 is the largest absolute coefficient of x_1, the first equation and the second equation are interchanged, giving us

$$
\begin{aligned}
-4x_1 - 2x_2 + 2x_3 &= -14,\\
2x_1 + 2x_2 - 2x_3 &= 8,\\
-2x_1 + 3x_2 + 9x_3 &= 9.
\end{aligned}
$$

Eliminating x_1 from the second equation and from the third equation, we obtain

$$
\begin{aligned}
-4x_1 - 2x_2 + 2x_3 &= -14,\\
x_2 - x_3 &= 1,\\
4x_2 + 8x_3 &= 16.
\end{aligned}
$$

For the second elimination step, 4 is the absolute largest coefficient of x_2 beneath the diagonal so the second and third equations are interchanged, giving us

$$
\begin{aligned}
-4x_1 - 2x_2 + 2x_3 &= -14,\\
4x_2 + 8x_3 &= 16,\\
x_2 - x_3 &= 1.
\end{aligned}
$$

Eliminating x_2 from the third equation yields the upper triangular system:

$$
\begin{aligned}
-4x_1 - 2x_2 + 2x_3 &= -14,\\
+ 4x_2 + 8x_3 &= 16,\\
- 3x_3 &= -3,
\end{aligned}
$$

which is solved by back substitution to get the solution $x_3 = 1$, $x_2 = 2$, and $x_1 = 3$.

Often row pivoting is used in conjunction with row elimination. Similarly, for column elimination, the analogous pivoting scheme is **column pivoting**:

> *Before the kth elimination step when coefficients in row k to the right of the diagonal are annihilated, interchange column k with the column to its right which has the largest absolute coefficient in row k.*

Row pivoting and column pivoting are sometimes called **partial pivoting** since part of the matrix is inspected before selecting a pivot. Partial pivoting is better than no pivoting. However, in some cases, even partial pivoting can generate an inaccurate solution. For example, let us multiply the first equation in (29) by 2000 to obtain the equivalent system:

$$
\begin{aligned}
2x_1 + 2000x_2 &= 2000\\
x_1 + x_2 &= 2.
\end{aligned} \tag{32}
$$

If (32) is solved with row elimination and row pivoting, no interchange is needed since the coefficient 2 of x_1 is the largest coefficient in column 1. But we observed earlier that unless the equations in (32) are interchanged, the computed solution is inaccurate. One strategy for handling a system similar to (32) is to divide each equation by a scale factor, like the equation's biggest coefficient, before starting the elimination. A better, but more complicated strategy called **complete pivoting** is to inspect all the coefficients in the partly processed portion of the matrix (see Figure 2-3) before each elimination step. Before the kth elimination step, interchange two rows and two columns to move the largest absolute coefficient in the partly processed portion of the matrix to the kth diagonal position.

Let us illustrate complete pivoting using the system (32). Since the absolute largest coefficient is 2000, we interchange the first and second columns to obtain the equivalent system

$$\begin{aligned} 2000x_2 + 2x_1 &= 2000, \\ x_2 + x_1 &= 2. \end{aligned}$$

Eliminating x_2 from the second equation using two-digit decimal floating-point arithmetic with rounding, we obtain

$$\begin{aligned} 2000x_2 + 2x_1 &= 2000, \\ x_1 &= 1. \end{aligned}$$

The second equation tells us that $x_1 = 1$ and substituting for x_1 in the first equation yields $x_2 = 1$. Hence the computed solution agrees well with the exact solution (30).

When solving a system of equations using complete pivoting, each row interchange is equivalent to interchanging two equations while each column interchange is equivalent to interchanging two unknowns. Often, the program overhead associated with complete pivoting is more significant than the improvement in accuracy. For this reason, partial pivoting is used more frequently than complete pivoting.

A convenient storage structure for partial pivoting emanates from work by Forsythe and Moler [F3], Moler [M7], and Forsythe, Malcolm, and Moler [F2]. This structure, which is illustrated using the system (31), involves an array to store the coefficients and an array to record the pivots. Below we trace the elimination steps and the storage structure. The multipliers generated during an elimination step are listed to the left of the coefficient matrix.

Original matrix

Matrix:
$$\begin{bmatrix} 2 & 2 & -2 \\ -4 & -2 & 2 \\ -2 & 3 & 9 \end{bmatrix}$$

Array:

2	2	−2
−4	−2	2
−2	3	9

Pivot:

Interchange first equation and second equation

Matrix:
$$\begin{bmatrix} -4 & -2 & 2 \\ 2 & 2 & -2 \\ -2 & 3 & 9 \end{bmatrix}$$

Array:

−4	−2	2
2	2	−2
−2	3	9

Pivot:

2

Eliminate x_1

Matrix	Array	Pivot
$\begin{bmatrix} & -4 & -2 & 2 \\ -.5 & & 1 & -1 \\ .5 & & 4 & 8 \end{bmatrix}$	$\begin{array}{ccc} -4 & -2 & 2 \\ -.5 & 1 & -1 \\ .5 & 4 & 8 \end{array}$	2

Interchange second equation and third equation

Matrix	Array	Pivot
$\begin{bmatrix} -4 & -2 & 2 \\ & 4 & 8 \\ & 1 & -1 \end{bmatrix}$	$\begin{array}{ccc} -4 & -2 & 2 \\ -.5 & 4 & 8 \\ .5 & 1 & -1 \end{array}$	2 3

Eliminate x_2

Matrix	Array	Pivot
$\begin{bmatrix} & -4 & -2 & 2 \\ & & 4 & 8 \\ .25 & & 1 & -1 \end{bmatrix}$	$\begin{array}{ccc} -4 & -2 & 2 \\ -.5 & 4 & 8 \\ .5 & .25 & -3 \end{array}$	2 3

In general, the kth element of the pivot array is the number of the equation which is interchanged with equation k during the kth elimination step. If there is no interchange, the kth element of the pivot array is k since equation k is essentially interchanged with itself. In this storage structure, the coefficients are interchanged as the elimination progresses, but the previously computed multipliers are untouched. In particular, for the example above, the multipliers stored in column 1 are not touched when row 2 and row 3 of the coefficient matrix are interchanged.

To implement column-oriented row elimination with partial pivoting and with the Forsythe-Malcolm-Moler storage structure, we start with algorithm (28) and we insert statements to locate the pivot and interchange rows of the coefficient matrix:

$$
\begin{aligned}
&k = 1 \text{ to } n-1 \\
&\quad m = \arg\max \{|a_{ik}| : i = k, \cdots, n\} \\
&\quad p_k \leftarrow m \\
&\quad a_{mk} \leftrightarrow a_{kk} \\
&\quad a_{ik} \leftarrow a_{ik}/a_{kk} \text{ for } i = k+1 \text{ to } n \\
&\quad j = k+1 \text{ to } n \\
&\quad\quad a_{mj} \leftrightarrow a_{kj} \\
&\quad\quad a_{ij} \leftarrow a_{ij} - a_{ik}a_{kj} \text{ for } i = k+1 \text{ to } n \\
&\quad \text{next } j \\
&\text{next } k
\end{aligned}
$$

Here the notation $m = \arg\max \{|a_{ij}| : i = k, \cdots, n\}$ means that m is any row index i with the property that

$$|a_{mk}| = \max\{|a_{ik}| : i = k, \cdots, n\}.$$

The statement "$p_k \leftarrow m$" records the pivot row number in component k of $\mathbf{p}$. The double arrow in (33) means interchange the contents of variables. For example, $x \leftrightarrow y$ means interchange the contents of x and y. The following three statements will interchange the contents of the variables X and Y:

```
T = X
X = Y
Y = T.
```

Observe that the interchange of row m (the pivot row) with row k is accomplished in two different parts of (33). At the top of the program, the statement "$a_{mk} \leftrightarrow a_{kk}$" moves the pivot to the kth diagonal position while just inside the j-loop, the statement "$a_{mj} \leftrightarrow a_{kj}$" interchanges element j of row m with element j of row k for $j > k$. Although row k and row m could be interchanged near the start of algorithm (33), it is more efficient to postpone this interchange until the elimination step.

After factoring a matrix with (33), how do we solve the system $\mathbf{Ax} = \mathbf{b}$? We return to system (31) for insight. When performing row elimination with row pivoting, each elimination step is preceded by an interchange operation. Except for this interchange operation, the way we process the right side of (31) is exactly the same as the way we processed the right side before introducing pivots. The following implementation of forward and back elimination utilizes the factorization computed in (33). The statement "$b_j \leftrightarrow b_{p_j}$" interchanges components j and p_j on the right side. The remaining statements in this algorithm are identical to algorithm (21) (forward elimination) followed by (26) (back substitution).

```
j = 1 to n−1
    b_j ↔ b_{p_j}
    i = j+1 to n
        b_i ← b_i − a_ij b_j
    next i
next j
j = n down to 1
    b_j ← b_j/a_jj
    i = 1 to j−1
        b_i ← b_i − a_ij b_j
    next i
next j
```
(34)

We now provide a different explanation for (34) using elementary matrices. First note that either a pivot or an elimination step is equivalent to multiplication by a matrix. For example, multiplying the left side of a matrix by

$$\begin{bmatrix} 0 & 1 \\ 1 & 0 \end{bmatrix}$$

is equivalent to interchanging two rows as we see below:

$$\begin{bmatrix} 0 & 1 \\ 1 & 0 \end{bmatrix}\begin{bmatrix} 1 & 3 \\ 2 & 4 \end{bmatrix} = \begin{bmatrix} 2 & 4 \\ 1 & 3 \end{bmatrix}.$$

Similarly, multiplying the left side of a matrix by

$$\begin{bmatrix} 1 & 0 \\ 2 & 1 \end{bmatrix}$$

is equivalent to adding 2 times the first row to the second row:

$$\begin{bmatrix} 1 & 0 \\ 2 & 1 \end{bmatrix}\begin{bmatrix} 1 & 3 \\ 2 & 4 \end{bmatrix} = \begin{bmatrix} 1 & 3 \\ 4 & 10 \end{bmatrix}.$$

When factoring a matrix using row pivoting and row elimination, we perform a pivot step, an elimination step, a pivot step, an elimination step, and so on. Each of these steps is equivalent to multiplication by an appropriate matrix. Let $\mathbf{P}_i$ be the matrix associated with the ith pivot step and let $\mathbf{L}_i$ be the matrix associated with the ith elimination step. Then the reduction of $\mathbf{A}$ to the upper triangular matrix $\mathbf{U}$ is obtained in the following way:

$$\mathbf{L}_{n-1}\mathbf{P}_{n-1} \cdots \mathbf{L}_2\mathbf{P}_2\mathbf{L}_1\mathbf{P}_1\mathbf{A} = \mathbf{U}.$$

There are $n - 1$ pivot steps and $n - 1$ elimination steps since we annihilate coefficients in column 1 through column $n - 1$ beneath the diagonal. Solving for $\mathbf{A}$, we have

$$\mathbf{A} = \mathbf{P}_1^{-1}\mathbf{L}_1^{-1} \cdots \mathbf{P}_{n-1}^{-1}\mathbf{L}_{n-1}^{-1}\mathbf{U}.$$

Now consider the linear system $\mathbf{Ax} = \mathbf{b}$. As in Section 2-2, we use the associate law for matrix multiplication to decompose the factored system $(\mathbf{P}_1^{-1}\mathbf{L}_1^{-1} \cdots \mathbf{P}_{n-1}^{-1}\mathbf{L}_{n-1}^{-1}\mathbf{U})\mathbf{x} = \mathbf{b}$ into two separate equations:

$$\mathbf{P}_1^{-1}\mathbf{L}_1^{-1} \cdots \mathbf{P}_{n-1}^{-1}\mathbf{L}_{n-1}^{-1}\mathbf{y} = \mathbf{b} \qquad \text{and} \qquad \mathbf{Ux} = \mathbf{y}.$$

The second equation is solved by back substitution in the usual way. To solve the first equation, we multiply by $\mathbf{L}_{n-1}\mathbf{P}_{n-1} \cdots \mathbf{L}_1\mathbf{P}_1$ to get

$$\mathbf{y} = \mathbf{L}_{n-1}\mathbf{P}_{n-1} \cdots \mathbf{L}_1\mathbf{P}_1\mathbf{b}.$$

Therefore, to obtain $\mathbf{y}$ we multiply $\mathbf{b}$ by $\mathbf{P}_1$ followed by $\mathbf{L}_1$ followed by $\mathbf{P}_2$ and so on. In algorithm (34), the loop where i ranges from $j + 1$ to n corresponds to an elimination step. Since each elimination step is preceded by the pivot "$b_j \leftrightarrow b_{pj}$," the first part of (34) essentially computes the product

$$\mathbf{L}_{n-1}\mathbf{P}_{n-1} \cdots \mathbf{L}_1\mathbf{P}_1\mathbf{b}.$$

The second part of algorithm (34) essentially solves $\mathbf{Ux} = \mathbf{y}$ by back elimination.

The factorization scheme (33) is tailored to a compiler that stores a matrix by columns. If the matrix is stored by rows, then it is better to factor $\mathbf{A}$ using row-oriented column elimination and column pivoting. An easy way to obtain this row-oriented algorithm is to reverse the indices in (33). That is, ik is replaced by ki, mk is replaced by km, and so on. By reversing the indices, the operations that were applied to rows previously are now applied to columns giving us column elimination. In stating the resulting algorithm (35), we also interchange the "dummy" indices i and j throughout the program. After this interchange, i refers to a row and j refers to a column.

```
k = 1 to n-1
    m = arg max {|a_kj| : j = k, ..., n}
    p_k ← m
    a_km ↔ a_kk
    j = k+1 to n
        a_kj ← a_kj / a_kk
    next j
    i = k+1 to n
        a_im ↔ a_ik
        j = k+1 to n
            a_ij ← a_ij - a_kj a_ik
        next j
    next i
next k
```

(35)

We also present for reference the corresponding row-oriented algorithm that solves $\mathbf{Ax} = \mathbf{b}$ using the factorization computed by (35). An explanation for this algorithm in terms of elementary matrices is developed in Exercise 2-3.4.

```
i = 1 to n
    j = 1 to i-1
        b_i ← b_i - a_ij b_j
    next j
    b_i ← b_i / a_ii
next i
i = n-1 down to 1
    j = i+1 to n
        b_i ← b_i - a_ij b_j
    next j
    b_i ↔ b_(p_i)
next i
```

(36)

Subroutine FACT in NAPACK implements the partial pivoting scheme (33). The subroutine has three arguments:

A: array containing coefficient matrix
LA: leading (row) dimension of array A
N: dimension of matrix stored in A

The parameters LA and N are often confused. The value of LA depends on how the array A is dimensioned. If the dimension statement reads "REAL

A(5,10)," then LA is 5. The value of N is the number of rows and columns associated with the matrix stored in array A. Although N is never larger than LA, N may be strictly smaller than LA. Another subroutine called SOLVE implements the solution scheme (34). The arguments of SOLVE are

X: solution (can be identified with B)
A: output of subroutine FACT
B: right side

If the B array is also used for the X argument, the original right side is destroyed. That is, the subroutine overwrites the right side with the solution.

Table 2-6 gives a program to solve equation (29) using FACT and SOLVE. Note that the original matrix is 2×2 while the array C is dimensioned 3×4. When a matrix is factored, we also save the pivot rows and we perform other bookkeeping that will help us later to estimate the accuracy in the computed solution. The input array must be large enough to store the pivot rows and these extra data. For a $n \times n$ matrix, subroutine FACT needs at least $3 + n(n + 1)$ elements altogether. Hence, the product between the row dimension and the column dimension of the input array must be at least $3 + n(n + 1)$. In the example above, $n = 2$ and $3 + n(n + 1) = 9$ while C has $3 \times 4 = 12$ elements. The smallest possible dimension for C is 3×3, which yields exactly 9 array elements. Appendix 1 summarizes the storage requirement for subroutines contained in NAPACK. Before using a subroutine, be sure that the coefficient array is large enough to carry out the factorization.

To conclude this section, we now expand our discussion of the parameter LA in the argument list for subroutine FACT. Consider the following program:

```
REAL A(4,3)
A(1,1) = 1
A(2,1) = 2
A(1,2) = 3
A(2,2) = 4
```
(37)

TABLE 2-6 PROGRAM TO SOLVE EQUATION (29)

```
REAL C(3,4),B(2)
C(1,1) = .001
C(2,1) = 1.
C(1,2) = 1.
C(2,2) = 1.
CALL FACT(C,3,2)
B(1) = 1.
B(2) = 2.
CALL SOLVE(B,C,B)
WRITE(6,*) B(1),B(2)
END
```

This code creates a 4 × 3 array A and stores inside it the 2 × 2 matrix

$$\begin{bmatrix} 1 & 3 \\ 2 & 4 \end{bmatrix}.$$

Since computer memory is linear and Fortran stores an array by columns, A is the following linear list in computer memory:

1	2	*	*	3	4	*	*	*	*	*	*
Column 1				Column 2				Column 3			

Here the *'s denote undefined numbers. When a subroutine is invoked in Fortran, the subroutine is given the location of the start of storage for each subroutine argument. Thus when the statement "CALL FACT(A,LA,N)" is executed, FACT is given the location of A(1,1) but not the actual dimension of array A. Argument LA is needed by FACT in order to determine where the columns associated with array A begin and end.

Although in many applications the coefficient matrix is stored in a two-dimensional array, there are some cases where it is more convenient to store the coefficients in a one-dimensional array or in a three-dimensional array. For a two-dimensional array, the LA argument of FACT is the leading (row) dimension of the array. When the array dimension is not two, LA is the number of elements in memory between the start of one column of the matrix and the start of the next column of the matrix. With this insight, we solve equation (31) using a single-subscripted array:

```
REAL A(15),B(3)
DATA A/2.,-4.,-2.,2.,-2.,3.,-2.,2.,9./
DATA B/8.,-14.,9./
CALL FACT(A,3,3)
CALL SOLVE(B,A,B)
WRITE(6,*) B(1),B(2),B(3)
END
```

This program creates a single-subscripted array A with the following 15 elements:

2. −4. −2. 2. −2. 3. −2. 2. 9. * * * * * *

Since there are three elements in memory between the start of one column and the start of the next column of the matrix, LA is 3.

Exercises

2-3.1. What is the output of the following program?

```
REAL A(3,3)
DATA A/1.,2.,3.,4.,5.,6.,7.,8.,9./
CALL OUT(A)
END
```

```
SUBROUTINE OUT(A)
REAL A(2,1)
WRITE(6,*) A(1,3)
RETURN
END
```

(Note that when a subroutine argument is a two-dimensional array inside the subroutine, the column dimension is ignored. The dimension statement in the main program allocates storage while the dimension statement in the subroutine tells the subroutine how to interpret storage. In the previous program, the dimension statement "REAL A(2,1)" specifies that there are two elements in each column of the A array. The column dimension "1" is ignored since A is a subroutine argument. Consequently, A can have more than one column inside the subroutine.)

2-3.2. Write a subroutine PACK(A,LA,N) that pushes a square N × N matrix stored in a double-subscripted array A with leading dimension LA to the beginning of the array so that the columns of the matrix are stored sequentially in computer memory. For example, the output of the following program is "1, 2, · · · , 9".

```
      REAL A(8,10)
      DO 10 J=1,3
      DO 10 I=1,3
10       A(I,J) = I + 3*(J-1)
      CALL PACK(A,8,3)
      CALL LIST(A,9)
      END
      SUBROUTINE LIST(A,N)
      REAL A(1)
      WRITE(6,*) (A(I), I=1,N)
      RETURN
      END
```

[Hint: Even though A is a double-subscripted array in the main program, you can use the dimension "REAL A(1)" in PACK. This dimension implies that A will be a single-subscripted array throughout the subroutine. Now push the matrix columns to the start of the single-subscripted array.]

2-3.3. Use subroutines FACT and SOLVE to compute the solution to the systems that you solved by hand in Exercise 2-2.1. Write one program which contains a loop that reads a set of coefficients, reads the corresponding right side, calls FACT, calls SOLVE, and branches back to read the next set of coefficients.

2-3.4. Use pivot and elimination matrices to describe column elimination with column pivoting [algorithm (35)] and the corresponding solution routine [algorithm (36)]. [Note that the analogous discussion for row elimination and row pivoting follows algorithm (34).]

2-3.5. Find a solution to the system

$$\begin{aligned} 2x_1 + x_2 - 2x_3 &= 4, \\ 4x_1 + x_2 - x_3 &= 7, \\ 6x_1 - x_2 + 6x_3 &= 8. \end{aligned}$$

Now change 8 on the right side to 7 and try to solve the system. Why is there no solution?

2-4. SPECIAL MATRICES

When the coefficient matrix for a linear system has special structure, Gaussian elimination often simplifies. We will examine elimination schemes for classes of matrices that arise frequently in applications.

2-4.1 Tridiagonal Matrix

A matrix is *tridiagonal* if its nonzero elements are either on the diagonal or adjacent to the diagonal. Some tridiagonal matrices appear below:

$$\begin{bmatrix} 0 & 0 \\ 0 & 0 \end{bmatrix} \quad \begin{bmatrix} 1 & 0 & 0 \\ 0 & 2 & 0 \\ 0 & 0 & 3 \end{bmatrix} \quad \begin{bmatrix} 1 & 6 & 0 \\ 4 & 2 & 7 \\ 0 & 5 & 3 \end{bmatrix} \quad \begin{bmatrix} 2 & 1 & 0 & 0 \\ 2 & 2 & 2 & 0 \\ 0 & 1 & 0 & 2 \\ 0 & 0 & 2 & 1 \end{bmatrix}.$$

In general, the nonzero elements of a tridiagonal matrix lie in three bands: the superdiagonal, diagonal, and subdiagonal (see Figure 2-5). Applications that lead to tridiagonal coefficient matrices include an electrical circuit where each node is connected to its predecessor and its successor and second-order boundary-value problems such as the semiconductor equation of Chapter 1.

Let us factor the following tridiagonal matrix $\mathbf{A}$:

$$\mathbf{A} = \begin{bmatrix} 2 & 1 & 0 & 0 \\ 4 & 5 & 2 & 0 \\ 0 & -6 & 0 & 3 \\ 0 & 0 & -4 & 2 \end{bmatrix}.$$

We subtract 2 times the first row from the second row to obtain

$$\begin{bmatrix} 2 & 1 & 0 & 0 \\ 0 & 3 & 2 & 0 \\ 0 & -6 & 0 & 3 \\ 0 & 0 & -4 & 2 \end{bmatrix}.$$

Then we subtract -2 times the second row from the third row to get

$$\begin{bmatrix} 2 & 1 & 0 & 0 \\ 0 & 3 & 2 & 0 \\ 0 & 0 & 4 & 3 \\ 0 & 0 & -4 & 2 \end{bmatrix}.$$

Finally, subtracting -1 times the third row from the fourth row gives us the upper triangular matrix

$$\begin{bmatrix} 2 & 1 & 0 & 0 \\ 0 & 3 & 2 & 0 \\ 0 & 0 & 4 & 3 \\ 0 & 0 & 0 & 5 \end{bmatrix}.$$

$$\begin{bmatrix} * & * & & & \\ * & * & * & & \\ & * & * & * & \\ & & * & * & * \\ & & & * & * \end{bmatrix}$$

Figure 2-5 Structure of a tridiagonal matrix.

Thus the factorization of **A** is

$$\begin{bmatrix} 2 & 1 & 0 & 0 \\ 4 & 5 & 2 & 0 \\ 0 & -6 & 0 & 3 \\ 0 & 0 & -4 & 2 \end{bmatrix} = \begin{bmatrix} 1 & 0 & 0 & 0 \\ 2 & 1 & 0 & 0 \\ 0 & -2 & 1 & 0 \\ 0 & 0 & -1 & 1 \end{bmatrix} \begin{bmatrix} 2 & 1 & 0 & 0 \\ 0 & 3 & 2 & 0 \\ 0 & 0 & 4 & 3 \\ 0 & 0 & 0 & 5 \end{bmatrix}.$$

This example reveals four properties of tridiagonal matrices and Gaussian elimination without pivots:

1. There is at most one nonzero multiplier each elimination step. This multiplier corresponds to the annihilated subdiagonal element.
2. The superdiagonal is invariant during elimination and it becomes the superdiagonal of the upper triangular factor.
3. Each factor has at most two nonzero bands.
4. Each elimination step processes a 2 × 2 submatrix.

To expound the fourth property, the first elimination step processes the submatrix formed by the intersection of the first two rows and the first two columns:

$$\begin{bmatrix} 2 & 1 \\ 4 & 5 \end{bmatrix} \longleftarrow \begin{bmatrix} 2 & 1 & 0 & 0 \\ 4 & 5 & 2 & 0 \\ 0 & -6 & 0 & 3 \\ 0 & 0 & -4 & 2 \end{bmatrix}.$$

In other words, to carry out the first elimination step, we only need to know the coefficients in this 2 × 2 submatrix. Subtracting 2 times the first row from the second row yields

$$\begin{bmatrix} 2 & 1 \\ 0 & 3 \end{bmatrix} \longleftarrow \begin{bmatrix} 2 & 1 & 0 & 0 \\ 0 & 3 & 2 & 0 \\ 0 & -6 & 0 & 3 \\ 0 & 0 & -4 & 2 \end{bmatrix}$$

The next elimination step processes the submatrix formed from the "3" in the last elimination step, and the three elements −6, 0, and 2 from the original **A**:

$$\begin{bmatrix} 3 & 2 \\ -6 & 0 \end{bmatrix} \longleftarrow \begin{bmatrix} 2 & 1 & 0 & 0 \\ 0 & 3 & 2 & 0 \\ 0 & -6 & 0 & 3 \\ 0 & 0 & -4 & 2 \end{bmatrix}$$

The final elimination step processes the submatrix formed by the intersection of the last two rows and the last two columns:

$$\begin{bmatrix} 4 & 3 \\ -4 & 2 \end{bmatrix} \quad \begin{bmatrix} 2 & 1 & 0 & 0 \\ 0 & 3 & 2 & 0 \\ 0 & 0 & 4 & 3 \\ 0 & 0 & -4 & 2 \end{bmatrix}.$$

Based on these observations, a tridiagonal matrix is factored using three single-subscripted arrays which store the coefficients on the diagonal, on the subdiagonal, and on the superdiagonal. Consider the general 3×3 tridiagonal matrix:

$$\begin{bmatrix} d_1 & u_1 & 0 \\ l_1 & d_2 & u_2 \\ 0 & l_2 & d_3 \end{bmatrix}.$$

The first elimination step processes the 2×2 submatrix

$$\begin{bmatrix} d_1 & u_1 \\ l_1 & d_2 \end{bmatrix}.$$

To annihilate l_1, the first row is multiplied by l_1/d_1 and subtracted from the second row. Hence d_2 is replaced by $d_2 - u_1 l_1/d_1$. The second elimination step processes the 2×2 submatrix

$$\begin{bmatrix} d_2 & u_2 \\ l_2 & d_3 \end{bmatrix},$$

which contains the new d_2 and three elements l_2, d_3, and u_2 from the original matrix. To annihilate l_2, the first row is multiplied by l_2/d_2 and subtracted from the second row. Thus d_3 is replaced by $d_3 - u_2 l_2/d_2$.

For a $n \times n$ tridiagonal matrix, the elimination proceeds the same way. During the kth step, we process the 2×2 submatrix

$$\begin{bmatrix} d_k & u_k \\ l_k & d_{k+1} \end{bmatrix}.$$

The first row of the submatrix is multiplied by l_k/d_k and subtracted from the second row. The diagonal element d_{k+1} is replaced by $d_{k+1} - u_k l_k/d_k$. As k steps from 1 to $n - 1$, we obtain the following algorithm:

$k = 1$ to $n-1$
  $l_k \leftarrow l_k/d_k$
  $d_{k+1} \leftarrow d_{k+1} - l_k u_k$
next k

(38)

Note that this algorithm overwrites the subdiagonal elements stored in **l** with the multipliers. Also, the diagonal of the upper triangular factor overwrites the original diagonal coefficients stored in **d**.

Given the factorization **LU** of a tridiagonal matrix, let us now solve the triangular systems $\mathbf{Ly} = \mathbf{b}$ and $\mathbf{Ux} = \mathbf{y}$. The factorization of a 3×3 tridiagonal matrix has the form

$$\underset{\mathbf{L}}{\begin{bmatrix} 1 & 0 & 0 \\ l_1 & 1 & 0 \\ 0 & l_2 & 1 \end{bmatrix}} \underset{\mathbf{U}}{\begin{bmatrix} d_1 & u_1 & 0 \\ 0 & d_2 & u_2 \\ 0 & 0 & d_3 \end{bmatrix}}.$$

The corresponding lower triangular system $\mathbf{Ly} = \mathbf{b}$ is

$$\begin{bmatrix} 1 & 0 & 0 \\ l_1 & 1 & 0 \\ 0 & l_2 & 1 \end{bmatrix} \begin{bmatrix} y_1 \\ y_2 \\ y_3 \end{bmatrix} = \begin{bmatrix} b_1 \\ b_2 \\ b_3 \end{bmatrix},$$

which is equivalent to the three equations

$$\begin{aligned} y_1 \qquad\qquad &= b_1, \\ l_1 y_1 + \quad y_2 \qquad &= b_2, \\ l_2 y_2 + y_3 &= b_3. \end{aligned}$$

Starting with the first equation, we see that $y_1 = b_1$, $y_2 = b_2 - l_1 y_1$, and $y_3 = b_3 - l_2 y_2$. The general rule is

$$y_k = b_k - l_{k-1} y_{k-1},$$

where k increments from 2 to n. To overwrite the starting $\mathbf{b}$ with the solution $\mathbf{y}$, we replace y_k and y_{k-1} with b_k and b_{k-1}, respectively, to obtain the following implementation of forward substitution for a tridiagonal matrix:

```
k = 2 to n
    b_k ← b_k − l_{k−1} b_{k−1}
next k
```
(39)

Now let us consider back substitution. For three tridiagonal equations, the system $\mathbf{Ux} = \mathbf{y}$ has the form

$$\begin{aligned} d_1 x_1 + u_1 x_2 \qquad\qquad &= y_1, \\ d_2 x_2 + u_2 x_3 &= y_2, \\ d_3 x_3 &= y_3. \end{aligned}$$

Starting with the last equation, we have

$$x_3 = \frac{y_3}{d_3},$$

$$x_2 = \frac{y_2 - u_2 x_3}{d_2},$$

$$x_1 = \frac{y_1 - u_1 x_2}{d_1}.$$

The relationship between x_k and x_{k+1} can be stated

$$x_k = \frac{y_k - u_k x_{k+1}}{d_k},$$

where k increments from $n - 1$ down to 1. Finally, to overwrite the starting $\mathbf{y}$ with the solution $\mathbf{x}$, we replace x_k and x_{k+1} with y_k and y_{k+1} to obtain

$$\boxed{\begin{array}{l} y_n \leftarrow y_n/u_n \\ k = n-1 \text{ down to } 1 \\ \qquad y_k \leftarrow (y_k - u_k y_{k+1})/d_k \\ \text{next } k \end{array}} \tag{40}$$

Subroutine TFACT, which factors a tridiagonal matrix without pivots, has the following arguments:

L: subdiagonal (can be identified with U)
D: diagonal (length at least 3 + N)
U: superdiagonal
N: matrix dimension

After factoring the coefficient matrix using TFACT, a system of equations is solved using TSOLVE. The argument list is

X: solution (can be identified with B)
L,D,U: output of TFACT
B: right side

When partial pivoting is employed, an extra band can be generated during elimination. Consider the tridiagonal matrix

$$\begin{bmatrix} 1 & 3 & 0 & 0 \\ 2 & 4 & 4 & 0 \\ 0 & 2 & -2 & -2 \\ 0 & 0 & 1 & 2 \end{bmatrix}.$$

The elimination process is summarized below.

$$\begin{bmatrix} 1 & 3 & 0 & 0 \\ 2 & 4 & 4 & 0 \\ 0 & 2 & -2 & -2 \\ 0 & 0 & 1 & 2 \end{bmatrix} \xrightarrow[\longrightarrow]{\text{interchange rows 1 \& 2}} \begin{bmatrix} 2 & 4 & 4 & 0 \\ 1 & 3 & 0 & 0 \\ 0 & 2 & -2 & -2 \\ 0 & 0 & 1 & 2 \end{bmatrix} \xrightarrow[\longrightarrow]{\text{eliminate column 1}}$$

$$\begin{bmatrix} 2 & 4 & 4 & 0 \\ 0 & 1 & -2 & 0 \\ 0 & 2 & -2 & -2 \\ 0 & 0 & 1 & 2 \end{bmatrix} \xrightarrow[\longrightarrow]{\text{interchange rows 2 \& 3}} \begin{bmatrix} 2 & 4 & 4 & 0 \\ 0 & 2 & -2 & -2 \\ 0 & 1 & -2 & 0 \\ 0 & 0 & 1 & 2 \end{bmatrix} \xrightarrow[\longrightarrow]{\text{eliminate column 2}}$$

$$\begin{bmatrix} 2 & 4 & 4 & 0 \\ 0 & 2 & -2 & -2 \\ 0 & 0 & -1 & 1 \\ 0 & 0 & 1 & 2 \end{bmatrix} \xrightarrow[\longrightarrow]{\text{eliminate column 3}} \begin{bmatrix} 2 & 4 & 4 & 0 \\ 0 & 2 & -2 & -2 \\ 0 & 0 & -1 & 1 \\ 0 & 0 & 0 & 3 \end{bmatrix}$$

For a tridiagonal matrix, row pivoting will only interchange two adjacent rows, and this interchange can generate a nonzero entry above the superdiagonal. The band just above the superdiagonal is called the second superdiagonal. In the example above, rows 1 and 2 are interchanged before the first elimination step and the superdiagonal element "4" in row 2 is moved to the second superdiagonal in row 1. Due to pivoting, the upper triangular factor has three bands: the diagonal, the superdiagonal, and the second superdiagonal. Although pivoting creates an extra band in the upper triangular factor, the multiplier structure is unchanged; each elimination step generates at most one nonzero multiplier corresponding to the annihilated subdiagonal element.

When employing partial pivoting, observe that each elimination step processes a 2×3 submatrix until we reach the lower right corner where a 2×2 submatrix is processed. In the example above, the first elimination step processes the submatrix formed by the intersection of the first two rows and the first three columns:

$$\begin{bmatrix} 1 & 3 & 0 \\ 2 & 4 & 4 \end{bmatrix} \qquad \begin{bmatrix} 1 & 3 & 0 & 0 \\ 2 & 4 & 4 & 0 \\ 0 & 2 & -2 & -2 \\ 0 & 0 & 1 & 2 \end{bmatrix}.$$

Interchanging the first two rows gives us

$$\begin{bmatrix} 2 & 4 & 4 \\ 1 & 3 & 0 \end{bmatrix},$$

and subtracting $\frac{1}{2}$ times the first row from the second row yields

$$\begin{bmatrix} 2 & 4 & 4 \\ 0 & 1 & -2 \end{bmatrix} \qquad \begin{bmatrix} 2 & 4 & 4 & 0 \\ 0 & 1 & -2 & 0 \\ 0 & 2 & -2 & -2 \\ 0 & 0 & 1 & 2 \end{bmatrix}.$$

The second elimination step processes the submatrix formed by the intersection of rows 2 and 3 with columns 2, 3, and 4:

$$\begin{bmatrix} 1 & -2 & 0 \\ 2 & -2 & -2 \end{bmatrix} \qquad \begin{bmatrix} 2 & 4 & 4 & 0 \\ 0 & 1 & -2 & 0 \\ 0 & 2 & -2 & -2 \\ 0 & 0 & 1 & 2 \end{bmatrix}.$$

The last elimination step works with the 2×2 submatrix

$$\begin{bmatrix} -1 & 1 \\ 1 & 2 \end{bmatrix} \qquad \begin{bmatrix} 2 & 4 & 4 & 0 \\ 0 & 2 & -2 & 2 \\ 0 & 0 & -1 & 1 \\ 0 & 0 & 1 & 2 \end{bmatrix}.$$

We now develop a partial pivoting program to factor a general tridiagonal matrix with subdiagonal, diagonal, superdiagonal, and second superdiagonal denoted

l, **d**, **u**, and **s**, respectively. The kth elimination step processes the 2×3 submatrix

$$\begin{bmatrix} d_k & u_k & s_k \\ l_k & d_{k+1} & u_{k+1} \end{bmatrix}$$

consisting of elements from rows k and $k+1$. If $|l_k| > |d_k|$, then rows k and $k+1$ are interchanged. After this interchange, the multiplier $m = l_k/d_k$ is computed and m times row k is subtracted from row $k+1$. In particular, m times u_k is subtracted from d_{k+1} and m times s_k is subtracted from u_{k+1}. The new d_{k+1} is $d_{k+1} - mu_k$ and the new u_{k+1} is $u_{k+1} - ms_k$. The following algorithm factors a tridiagonal matrix with partial pivoting, overwriting **l** with the multipliers and overwriting **d**, **u**, and **s** with the bands of the upper triangular factor.

$k = 1$ to $n-1$
    if $|l_k| > |d_k|$ then $d_k \leftrightarrow l_k$, $u_k \leftrightarrow d_{k+1}$, and $s_k \leftrightarrow u_{k+1}$
    $l_k \leftarrow l_k/d_k$, $d_{k+1} \leftarrow d_{k+1} - l_k u_k$, $u_{k+1} \leftarrow u_{k+1} - l_k s_k$
next k

(41)

Subroutine PFACT implements this algorithm, and like FACT, it has three arguments:

A: array containing bands of coefficient matrix (length at least 3 + 4N)
LA: leading (row) dimension of array A
N: matrix dimension

But unlike FACT, just the superdiagonal, diagonal, and subdiagonal (in this order) are packed into the first three rows of A. Another subroutine called PSOLVE solves a factored system. For illustration, the system

$$\begin{bmatrix} 2 & 1 & 0 & 0 \\ 3 & 2 & 1 & 0 \\ 0 & 3 & 2 & 1 \\ 0 & 0 & 3 & 2 \end{bmatrix} \begin{bmatrix} x_1 \\ x_2 \\ x_3 \\ x_4 \end{bmatrix} = \begin{bmatrix} 2 \\ 2 \\ 0 \\ 1 \end{bmatrix} \tag{42}$$

is solved by the following program:

```
      REAL A(5,6),B(4)
      DATA B/2.,2.,0.,1./
      DO 10 I = 1,4
         A(1,I) = 1.
         A(2,I) = 2.
         A(3,I) = 3.
10    CONTINUE
      CALL PFACT(A,5,4)
      CALL PSOLVE(B,A,B)
      WRITE(6,*) (B(I), I=1,4)
      END
```

As noted above, the storage needed to factor a $n \times n$ tridiagonal matrix is at least $3 + 4n$. In our case, n is 4 and the required storage is $3 + 4 \times 4 = 19$, which is less than $5 \times 6 = 30$, the product between the row and column dimension of A. Of course, A could have been dimensioned 3×7—this array has $3 \times 7 = 21$ elements, which is still larger than the minimum size 19.

Exercises

2-4.1. Solve system (42) using subroutines TFACT and TSOLVE.

2-4.2. By hand, **LU** factor the following tridiagonal matrix without pivots:

$$\begin{bmatrix} 2 & 1 & 0 & 0 \\ 4 & -2 & 1 & 0 \\ 0 & 4 & 0 & 1 \\ 0 & 0 & 4 & 7 \end{bmatrix}.$$

How do the nonzero multipliers relate to the diagonal of **U** and the subdiagonal of the original matrix? Consider a $n \times n$ tridiagonal matrix with constant superdiagonal and with constant subdiagonal. That is, there exist two constants u and l such that $a_{i,i+1} = u$ and $a_{i+1,i} = l$ for $i = 1, 2, \cdots, n - 1$. A tridiagonal matrix with this structure can be stored using an array D that contains the diagonal elements and two variables L and U that contain l and u, respectively. Write a subroutine to factor the tridiagonal matrix stored in L, D, and U, and write a subroutine to solve the factored system.

2-4.2 Band Matrix

A square matrix whose nonzero elements are concentrated near the diagonal is called a *band matrix*. A (l, u)-band matrix is zero below the lth subdiagonal and above the uth superdiagonal. A diagonal matrix is (0, 0)-band, a tridiagonal matrix is (1, 1)-band, and the following matrix is (2, 1)-band:

$$\begin{bmatrix} 2 & 1 & 0 & 0 & 0 \\ 4 & 4 & 1 & 0 & 0 \\ 6 & 5 & 3 & 1 & 0 \\ 0 & 6 & 5 & 3 & 1 \\ 0 & 0 & 6 & 5 & 3 \end{bmatrix}. \tag{43}$$

Finite-element or finite-difference approximations to a boundary-value problem lead to a system of equations whose coefficient matrix has a band structure. Also, an electrical network where each node is connected to nearby nodes is modeled by a system of equations whose coefficient matrix has a band structure.

Whenever the number of bands is small relative to the matrix dimension, the matrix is predominantly zero. The number of elements in the bands of (43) is $3 + 4 + 5 + 4 = 16$. The number of zero elements is $25 - 16 = 9$, so the proportion of zeros is $9/25 = .36$ and the matrix is 36% zero. On the other hand, a 100×100 matrix with a (2, 1)-band structure has at most 396 nonzero elements. There are 100 elements on the diagonal, 99 elements on the superdiagonal, 99 elements on the subdiagonal, and 98 elements on the second subdiagonal. Altogether there are $100 + 99 + 99 + 98 = 396$ elements forming the four bands. Since a 100×100 matrix has 10,000 elements altogether, there are $10{,}000 - 396 = 9604$ zero elements, and the matrix is 96.04% zero.

A (l, u)-band matrix is stored compactly in a $(l + u + 1) \times n$ rectangle. This rectangle is exhibited in Table 2-7 for the matrix (43). Although the right corners of the rectangle in Table 2-7 are undefined, this "wasted" region is

TABLE 2-7 BAND STORAGE FOR THE MATRIX (43)

1	1	1	1	*	superdiagonal
2	4	3	3	3	diagonal
4	5	5	5	*	subdiagonal
6	6	6	*	*	second subdiagonal

relatively small when $l + u$ is much less than the matrix dimension n. For a (2, 1)-band matrix like (43), there are four bands so the storage rectangle contains $4n$ elements and just 4 of these elements are wasted, independent of n. For example, if n is 100, 4 out of the 400 elements in the storage rectangle are not utilized. Of course, this waste can be eliminated if the bands are packed inside a single-subscripted array. But storage savings must be balanced against program complexity. Processing a matrix which is stored in this supercompressed form is awkward and the increased program complexity usually outweighs the storage savings.

Now let us factor a band matrix, and we begin by factoring the matrix (43) without pivots. The successive elimination steps are displayed below:

$$\begin{bmatrix} 2 & 1 & 0 & 0 & 0 \\ 4 & 4 & 1 & 0 & 0 \\ 6 & 5 & 3 & 1 & 0 \\ 0 & 6 & 5 & 3 & 1 \\ 0 & 0 & 6 & 5 & 3 \end{bmatrix} \longrightarrow \begin{bmatrix} 2 & 1 & 0 & 0 & 0 \\ 0 & 2 & 1 & 0 & 0 \\ 0 & 2 & 3 & 1 & 0 \\ 0 & 6 & 5 & 3 & 1 \\ 0 & 0 & 6 & 5 & 3 \end{bmatrix} \longrightarrow \begin{bmatrix} 2 & 1 & 0 & 0 & 0 \\ 0 & 2 & 1 & 0 & 0 \\ 0 & 0 & 2 & 1 & 0 \\ 0 & 0 & 2 & 3 & 1 \\ 0 & 0 & 6 & 5 & 3 \end{bmatrix} \longrightarrow$$

$$\begin{bmatrix} 2 & 1 & 0 & 0 & 0 \\ 0 & 2 & 1 & 0 & 0 \\ 0 & 0 & 2 & 1 & 0 \\ 0 & 0 & 0 & 2 & 1 \\ 0 & 0 & 0 & 2 & 3 \end{bmatrix} \longrightarrow \begin{bmatrix} 2 & 1 & 0 & 0 & 0 \\ 0 & 2 & 1 & 0 & 0 \\ 0 & 0 & 2 & 1 & 0 \\ 0 & 0 & 0 & 2 & 1 \\ 0 & 0 & 0 & 0 & 2 \end{bmatrix}.$$

Collecting the multipliers, we obtain the factorization:

$$\begin{bmatrix} 2 & 1 & 0 & 0 & 0 \\ 4 & 4 & 1 & 0 & 0 \\ 6 & 5 & 3 & 1 & 0 \\ 0 & 6 & 5 & 3 & 1 \\ 0 & 0 & 6 & 5 & 3 \end{bmatrix} = \begin{bmatrix} 1 & 0 & 0 & 0 & 0 \\ 2 & 1 & 0 & 0 & 0 \\ 3 & 1 & 1 & 0 & 0 \\ 0 & 3 & 1 & 1 & 0 \\ 0 & 0 & 3 & 1 & 1 \end{bmatrix} \begin{bmatrix} 2 & 1 & 0 & 0 & 0 \\ 0 & 2 & 1 & 0 & 0 \\ 0 & 0 & 2 & 1 & 0 \\ 0 & 0 & 0 & 2 & 1 \\ 0 & 0 & 0 & 0 & 2 \end{bmatrix}.$$

Observe that if the matrix (43) is stored in the rectangle of Table 2-7, the factorization can be performed inside the rectangle. Whenever a coefficient beneath the diagonal is annihilated, the corresponding multiplier overwrites the annihilated coefficient. Thus elimination transforms Table 2-7 into Table 2-8, the factorization.

TABLE 2-8 FACTORS OF THE BAND MATRIX (43)

1	1	1	1	*	superdiagonal of **U**
2	2	2	2	2	diagonal of **U**
2	1	1	1	*	subdiagonal of **L**
3	3	3	*	*	second subdiagonal of **L**

Now consider each elimination step. The first elimination step processes the 3 × 2 submatrix formed by the intersection of the first three rows and the first two columns:

$$\begin{bmatrix} 2 & 1 \\ 4 & 4 \\ 6 & 5 \end{bmatrix} \qquad \begin{bmatrix} 2 & 1 & 0 & 0 & 0 \\ 4 & 4 & 1 & 0 & 0 \\ 6 & 5 & 3 & 1 & 0 \\ 0 & 6 & 5 & 3 & 1 \\ 0 & 0 & 6 & 5 & 3 \end{bmatrix}.$$

We subtract 2 times the first row from the second row, and we subtract 3 times the first row from the third row giving us

$$\begin{bmatrix} 2 & 1 \\ 0 & 2 \\ 0 & 2 \end{bmatrix} \qquad \begin{bmatrix} 2 & 1 & 0 & 0 & 0 \\ 0 & 2 & 1 & 0 & 0 \\ 0 & 2 & 3 & 1 & 0 \\ 0 & 6 & 5 & 3 & 1 \\ 0 & 0 & 6 & 5 & 3 \end{bmatrix}.$$

The second elimination step processes the submatrix formed from two 2's in the previous submatrix, and the elements 6, 5, 3, and 1 from the original matrix (43):

$$\begin{bmatrix} 2 & 1 \\ 2 & 3 \\ 6 & 5 \end{bmatrix} \qquad \begin{bmatrix} 2 & 1 & 0 & 0 & 0 \\ 0 & 2 & 1 & 0 & 0 \\ 0 & 2 & 3 & 1 & 0 \\ 0 & 6 & 5 & 3 & 1 \\ 0 & 0 & 6 & 5 & 3 \end{bmatrix}.$$

The elimination continues the same way, each step processing a 3 × 2 submatrix until we reach the lower right corner, where a 2 × 2 submatrix is processed.

For the (2, 1)-band matrix studied above, each elimination step processes a 3 × 2 submatrix. In general, for a (l, u)-band matrix, each elimination step processes a $(l + 1) \times (u + 1)$ submatrix. To derive the general factorization algorithm, it helps to visualize the $(l + 1) \times (u + 1)$ block of elements involved in the kth elimination step (see Figure 2-6). In elimination step k, the coefficients in column k beneath the diagonal are annihilated. To annihilate the coefficient a_{ik} in row i and column k, row k is multiplied by $l_{ik} = a_{ik}/a_{kk}$ and subtracted from row i. The new element in row i and column j is $a_{ij} - l_{ik}a_{kj}$. This formula applies to rows $k + 1$ through $k + l$ and columns k through $k + u$. When the limit $k + l$ or $k + u$ is bigger than n, it is replaced by n since elimination stops

$$\begin{bmatrix} a_{kk} & \cdots & a_{kj} & \cdots & a_{k,k+u} \\ \vdots & & \vdots & & \vdots \\ a_{ik} & \cdots & a_{ij} & \cdots & a_{i,k+u} \\ \vdots & & \vdots & & \vdots \\ a_{k+l,k} & \cdots & a_{k+l,j} & \cdots & a_{k+l,k+u} \end{bmatrix}$$

Figure 2-6 Elements processed during the kth elimination step for a (l, u)-band matrix.

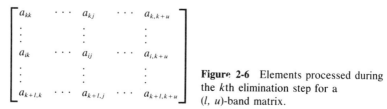

at the borders of the matrix. In summary, column-oriented factorization of a (l, u)-band matrix can be stated:

$$
\begin{array}{l}
k = 1 \text{ to } n-1 \\
\quad i = k+1 \text{ to } \min\{n, k+l\} \\
\quad\quad l_{ik} \leftarrow a_{ik}/a_{kk} \\
\quad \text{next } i \\
\quad j = k \text{ to } \min\{n, k+u\} \\
\quad\quad i = k+1 \text{ to } \min\{n, k+l\} \\
\quad\quad\quad a_{ij} \leftarrow a_{ij} - l_{ik}a_{kj} \\
\quad\quad \text{next } i \\
\quad \text{next } j \\
\text{next } k
\end{array}
\tag{44}
$$

Comparing the factorization scheme (44) for a band matrix to the scheme (28) for a general matrix, the only difference is the upper limit for i and j. The row index i stops at n for a general matrix and at $\min\{n, k + l\}$ for a band matrix. Similarly, the column index j stops at n for a general matrix and at $\min\{n, k + u\}$ for a band matrix.

Forward and back elimination schemes can be developed in a similar fashion. Again, the only difference between these band schemes and the general algorithms (21) and (26) is the index range. For a (l, u)-band matrix, column-oriented forward elimination for $\mathbf{Ly} = \mathbf{b}$ is the following:

$$
\begin{array}{l}
j = 1 \text{ to } n-1 \\
\quad i = j+1 \text{ to } \min\{n, j+l\} \\
\quad\quad b_i \leftarrow b_i - l_{ij}b_j \\
\quad \text{next } i \\
\text{next } j
\end{array}
\tag{45}
$$

Hence, for a band matrix, the row index i stops at $i = j + l$ (corresponding to the edge of the bands) or at $i = n$ (corresponding to the bottom of the matrix), whichever comes first. In contrast, for an upper triangular system, the lower range of the index i is modified by the band structure. Back elimination for $\mathbf{Ux} = \mathbf{y}$ is the following:

$$
\begin{array}{l}
j = n \text{ down to } 1 \\
\quad y_j \leftarrow y_j/u_{jj} \\
\quad i = \max\{1, j-u\} \text{ to } j-1 \\
\quad\quad y_i \leftarrow y_i - u_{ij}y_j \\
\quad \text{next } i \\
\text{next } j
\end{array}
\tag{46}
$$

For a general matrix, the index i starts at $i = 1$ while for a band matrix, the starting point is either $i = j - u$ (corresponding to the edge of the bands) or $i = 1$ (corresponding to the top of the matrix), whichever comes first.

When a band matrix is factored using partial pivoting, the pivoting process generates extra bands above the diagonal. Recall that Gaussian elimination with row pivoting reduces a (1, 1)-band tridiagonal matrix to a (0, 2)-band upper triangular matrix. Similarly, factoring a (l, u)-band matrix with row pivoting, we obtain a $(0, l + u)$-band upper triangular factor whose first u superdiagonal bands correspond to the superdiagonal bands of the original matrix and whose next l superdiagonal bands are generated by pivoting. To illustrate the effects of partial pivoting, we eliminate a (2, 1)-band matrix:

$$\begin{bmatrix} 1 & 3 & 0 & 0 & 0 \\ 1 & 3 & 2 & 0 & 0 \\ 2 & 4 & 6 & -2 & 0 \\ 0 & 2 & -4 & 4 & 4 \\ 0 & 0 & 1 & 2 & 1 \end{bmatrix} \xrightarrow[\longrightarrow]{\text{interchange rows 1 \& 3}} \begin{bmatrix} 2 & 4 & 6 & -2 & 0 \\ 1 & 3 & 2 & 0 & 0 \\ 1 & 3 & 0 & 0 & 0 \\ 0 & 2 & -4 & 4 & 4 \\ 0 & 0 & 1 & 2 & 1 \end{bmatrix} \xrightarrow[\longrightarrow]{\text{eliminate column 1}}$$

$$\begin{bmatrix} 2 & 4 & 6 & -2 & 0 \\ 0 & 1 & -1 & 1 & 0 \\ 0 & 1 & -3 & 1 & 0 \\ 0 & 2 & -4 & 4 & 4 \\ 0 & 0 & 1 & 2 & 1 \end{bmatrix} \xrightarrow[\longrightarrow]{\text{interchange rows 2 \& 4}} \begin{bmatrix} 2 & 4 & 6 & -2 & 0 \\ 0 & 2 & -4 & 4 & 4 \\ 0 & 1 & -3 & 1 & 0 \\ 0 & 1 & -1 & 1 & 0 \\ 0 & 0 & 1 & 2 & 1 \end{bmatrix} \xrightarrow[\longrightarrow]{\text{eliminate column 2}}$$

$$\begin{bmatrix} 2 & 4 & 6 & -2 & 0 \\ 0 & 2 & -4 & 4 & 4 \\ 0 & 0 & -1 & -1 & -2 \\ 0 & 0 & 1 & -1 & -2 \\ 0 & 0 & 1 & 2 & 1 \end{bmatrix} \xrightarrow[\longrightarrow]{\text{eliminate column 3}} \begin{bmatrix} 2 & 4 & 6 & -2 & 0 \\ 0 & 2 & -4 & 4 & 4 \\ 0 & 0 & -1 & -1 & -2 \\ 0 & 0 & 0 & -2 & -4 \\ 0 & 0 & 0 & 1 & -1 \end{bmatrix} \xrightarrow[\longrightarrow]{\text{eliminate column 4}}$$

$$\begin{bmatrix} 2 & 4 & 6 & -2 & 0 \\ 0 & 2 & -4 & 4 & 4 \\ 0 & 0 & -1 & -1 & -2 \\ 0 & 0 & 0 & -2 & -4 \\ 0 & 0 & 0 & 0 & -3 \end{bmatrix}.$$

Before the first elimination step, row 1 and row 3 are interchanged, and before the second elimination step, row 2 and row 4 are interchanged. These interchanges generate two extra superdiagonal bands so that this (2, 1)-band matrix has a (0, 3)-band upper triangular factor.

When factoring a (2, 1)-band matrix with partial pivoting, each elimination step processes a 3×4 submatrix until we reach the lower right corner, where the size of the submatrix decreases. In particular, the first elimination step processes the submatrix formed by the intersection of the first three rows and the first four columns:

$$\begin{bmatrix} 1 & 3 & 0 & 0 \\ 1 & 3 & 2 & 0 \\ 2 & 4 & 6 & -2 \end{bmatrix} \longleftarrow \begin{bmatrix} 1 & 3 & 0 & 0 & 0 \\ 1 & 3 & 2 & 0 & 0 \\ 2 & 4 & 6 & -2 & 0 \\ 0 & 2 & -4 & 4 & 4 \\ 0 & 0 & 1 & 2 & 1 \end{bmatrix}.$$

The first and third rows are interchanged and elements in column 1 are annihilated, giving us

$$\begin{bmatrix} 2 & 4 & 6 & -2 \\ 0 & 1 & -1 & 1 \\ 0 & 1 & -3 & 1 \end{bmatrix} \qquad \begin{bmatrix} 2 & 4 & 6 & -2 & 0 \\ 0 & 1 & -1 & 1 & 0 \\ 0 & 1 & -3 & 1 & 0 \\ 0 & 2 & -4 & 4 & 4 \\ 0 & 0 & 1 & 2 & 1 \end{bmatrix}.$$

The second elimination step processes the 3×4 submatrix

$$\begin{bmatrix} 1 & -1 & 1 & 0 \\ 1 & -3 & 1 & 0 \\ 2 & -4 & 4 & 4 \end{bmatrix} \qquad \begin{bmatrix} 2 & 4 & 6 & -2 & 0 \\ 0 & 1 & -1 & 1 & 0 \\ 0 & 1 & -3 & 1 & 0 \\ 0 & 2 & -4 & 4 & 4 \\ 0 & 0 & 1 & 2 & 1 \end{bmatrix}.$$

For a (2, 1)-band matrix, each elimination step processes a 3×4 submatrix until we reach the lower right corner. In general, for a (l, u)-band matrix, each elimination step processes a $(l + 1) \times (l + u + 1)$ submatrix until we reach the lower right corner.

Recall that the scheme (44) for processing a band matrix without pivots is identical to the scheme (28) for processing a general matrix except that the row and column indices stop at the borders of a submatrix. Similarly, a partial pivoting scheme for a band matrix is derived from the general algorithm (33) by restricting the row and column indices to the appropriate submatrix. Since each elimination step for a (l, u)-band matrix processes a $(l + 1) \times (l + u + 1)$ submatrix, the kth elimination step processes the submatrix depicted in Figure 2-7.

Thus the kth step processes a_{ij} for i between $k + 1$ and $k + l$ and for j between k and $k + l + u$. Imposing these restrictions on the row and column indices in (33) gives us the following algorithm to factor a band matrix with row pivoting:

```
k = 1 to n-1
    m = arg max{|a_ik| : i = k, ..., min{n, k+l}}
    p_k <- m
    a_mk <-> a_kk
    i = k+1 to min{n, k+l}
        a_ik <- a_ik / a_kk
    next i
    j = k+1 to min{n, k+l+u}
        a_mj <-> a_kj
        i = k+1 to min{n, k+l}
            a_ij <- a_ij - a_ik a_kj
        next i
    next j
next k
```
(47)

Similarly, forward and back elimination for a band matrix are obtained from the general scheme (34) by restricting the index range to the bands:

$$\begin{bmatrix} a_{kk} & \cdots & a_{kj} & \cdots & a_{k,k+l+u} \\ \vdots & & \vdots & & \vdots \\ a_{ik} & \cdots & a_{ij} & \cdots & a_{i,k+l+u} \\ \vdots & & \vdots & & \vdots \\ a_{k+l,k} & \cdots & a_{k+l,j} & \cdots & a_{k+l,k+l+u} \end{bmatrix}$$

Figure 2-7 Elements processed during the kth step for a (l, u)-band matrix with row pivoting.

```
j = 1 to n-1
    b_j ↔ b_{p_j}
    i = j+1 to min{n, j+l}
        b_i ← b_i - a_ij b_j
    next i
next j
j = n down to 1
    b_j ← b_j/a_jj
    i = max{1, j-l-u} to j-1
        b_i ← b_i - a_ij b_j
    next i
next j
```
(48)

Subroutine BFACT implements the factorization scheme (47) while BSOLVE solves a factored system using (48). BFACT's arguments are

A: array containing matrix bands (length at least 5 + (2L + U + 2)N)

LA: leading (row) dimension of array A

N: matrix dimension

L: number of lower diagonal bands

U: number of upper diagonal bands

The order of the bands in the input array A should coincide with the order of the bands in the coefficient matrix: the superdiagonal bands are first followed by the diagonal followed by the subdiagonal bands. For illustration, the (2,1)-band system

$$\begin{bmatrix} 4 & 1 & 0 & 0 & 0 \\ 2 & 4 & 1 & 0 & 0 \\ 1 & 2 & 4 & 1 & 0 \\ 0 & 1 & 2 & 4 & 1 \\ 0 & 0 & 1 & 2 & 4 \end{bmatrix} \begin{bmatrix} x_1 \\ x_2 \\ x_3 \\ x_4 \\ x_5 \end{bmatrix} = \begin{bmatrix} 4 \\ 1 \\ -3 \\ -1 \\ 3 \end{bmatrix}$$

is solved by the following program:

```
REAL A(4,10),B(5)
DATA B/4.,1.,-3.,-1.,3./
DO 10 I = 1,5
   A(1,I) = 1
   A(2,I) = 4
```

```
          A(3,I) = 2
          A(4,I) = 1
   10     CONTINUE
          CALL BFACT(A,4,5,2,1)
          CALL BSOLVE(B,A,B)
          WRITE(6,*) (B(I), I=1,5)
          END
```

Exercises

2-4.3. Solve equation (42) using subroutines BFACT and BSOLVE.

2-4.4. Suppose that a band matrix is stored in an array A using the structure described in this section. What is the row and column in array A which corresponds to row i and column j of the matrix? What is the row and column of the matrix which corresponds to row I and column J of array A?

2-4.5. If a (l, u)-band matrix is factored with column elimination and column pivoting, what is the band structure of the factors?

2-4.6. Suppose that **A** and **B** are $n \times n$ matrices. If **A** is a (l_1, u_1)-band matrix and **B** is a (l_2, u_2)-band matrix, what are the band widths of the product $\mathbf{P} = \mathbf{AB}$?

2-4.7. A (2, 2)-band matrix is called a **pentadiagonal matrix**. A pentadiagonal matrix can be stored in five single-subscripted arrays L1, L2, D, U1, and U2 which contain the subdiagonal, second subdiagonal, diagonal, superdiagonal, and second superdiagonal, respectively. Write a subroutine to factor a pentadiagonal matrix without pivots and write a subroutine to solve a factored system. Devise some test problems to verify that your subroutines work. (*Hint:* One method for obtaining a test problem is to start with a known matrix **A** and a known vector **y** and compute $\mathbf{b} = \mathbf{Ay}$. Using your subroutines, solve $\mathbf{Ax} = \mathbf{b}$. If your program works, then $\mathbf{x} = \mathbf{y}$ to within rounding errors.)

2-4.3 Symmetric Matrix

The following matrix **A** is symmetric:

$$\mathbf{A} = \begin{bmatrix} 2 & 4 & 6 \\ 4 & 7 & 8 \\ 6 & 8 & 5 \end{bmatrix}. \tag{49}$$

A square matrix is *symmetric* if each element equals its reflection across the diagonal. For a general 3×3 matrix

$$\begin{bmatrix} a_{11} & a_{12} & a_{13} \\ a_{21} & a_{22} & a_{23} \\ a_{31} & a_{32} & a_{33} \end{bmatrix},$$

symmetry means that

$$a_{12} = a_{21}, \qquad a_{31} = a_{13}, \qquad a_{32} = a_{23},$$

which is equivalent to the condition

$$a_{ij} = a_{ji}$$

for each i and j.

Symmetry can also be characterized using the transpose. The **transpose** of a matrix, denoted by a superscript T, is obtained by interchanging the rows and columns. For example,

$$\begin{bmatrix} 2 & 3 & 6 \\ 8 & 2 & 4 \end{bmatrix}^T = \begin{bmatrix} 2 & 8 \\ 3 & 2 \\ 6 & 4 \end{bmatrix}$$

and

$$\begin{bmatrix} 1 & 2 & 3 \\ 4 & 5 & 3 \\ 2 & 1 & 6 \end{bmatrix}^T = \begin{bmatrix} 1 & 4 & 2 \\ 2 & 5 & 1 \\ 3 & 3 & 6 \end{bmatrix}.$$

For a square matrix, interchanging rows and columns is the same as reflecting elements across the diagonal, so symmetry is equivalent to the condition

$$\mathbf{A}^T = \mathbf{A}.$$

Symmetric matrices arise in many applications. The coefficient matrix associated with an electrical circuit or with an approximation to a "self-adjoint" boundary-value problem such as the semiconductor equation of Chapter 1 is usually symmetric.

As seen below, the symmetric matrix (49) is stored compactly in a single subscripted array:

2	4	6	7	8	5
Row 1			Row 2		Row 3

Elements beneath the diagonal, known from symmetry, are omitted. Similarly, a $n \times n$ symmetric matrix is stored in an array of length $n(n + 1)/2$. The matrix elements are stored row after row, starting with the diagonal element in each row. Thus the order of elements in storage is

$$a_{11}a_{12} \cdots a_{1n}a_{22}a_{23} \cdots a_{2n}a_{33}a_{34} \cdots a_{3n} \cdots\cdots a_{nn}. \tag{50}$$

To factor the symmetric matrix (49), we subtract 2 times the first row from the second row, and we subtract 3 times the first row from the third row, giving us

$$\begin{bmatrix} 2 & 4 & 6 \\ 0 & -1 & -4 \\ 0 & -4 & -13 \end{bmatrix}. \tag{51}$$

Then 4 times the second row is subtracted from the third row to obtain

$$\begin{bmatrix} 2 & 4 & 6 \\ 0 & -1 & -4 \\ 0 & 0 & 3 \end{bmatrix}.$$

The multipliers are $l_{21} = 2$, $l_{31} = 3$, and $l_{32} = 4$, and the **LU** factorization of (49) is

$$\begin{bmatrix} 2 & 4 & 6 \\ 4 & 7 & 8 \\ 6 & 8 & 5 \end{bmatrix} = \begin{bmatrix} 1 & 0 & 0 \\ 2 & 1 & 0 \\ 3 & 4 & 1 \end{bmatrix} \begin{bmatrix} 2 & 4 & 6 \\ 0 & -1 & -4 \\ 0 & 0 & 3 \end{bmatrix}.$$

Notice the connection between **L** and **U**. For the first elimination step, the multipliers are $l_{21} = 4/2 = u_{12}/u_{11}$ and $l_{31} = 6/2 = u_{13}/u_{11}$. Thus the multipliers in **L**'s first column are obtained by dividing **U**'s first row by the diagonal element.

Similarly, during the second elimination step, the multiplier is $l_{32} = -4/-1 = u_{23}/u_{22}$. In general, when a symmetric matrix is factored without pivots, l_{ij} is related to u_{ji} through the identity

$$l_{ij} = \frac{u_{ji}}{u_{jj}}. \tag{52}$$

In other words, each column of **L** equals the corresponding row of **U** divided by the diagonal element. This relation between **L** and **U** implies a special structure for the **LDU** factorization of a symmetric matrix. The **LDU** factorization of (49) is

$$\underset{\mathbf{A}}{\begin{bmatrix} 2 & 4 & 6 \\ 4 & 7 & 8 \\ 6 & 8 & 5 \end{bmatrix}} = \underset{\mathbf{L}_1}{\begin{bmatrix} 1 & 0 & 0 \\ 2 & 1 & 0 \\ 3 & 4 & 1 \end{bmatrix}} \underset{\mathbf{D}}{\begin{bmatrix} 2 & 0 & 0 \\ 0 & -1 & 0 \\ 0 & 0 & 3 \end{bmatrix}} \underset{\mathbf{U}_1}{\begin{bmatrix} 1 & 2 & 3 \\ 0 & 1 & 4 \\ 0 & 0 & 1 \end{bmatrix}}.$$

Observe that the rows of $\mathbf{U}_1$ and the columns of $\mathbf{L}_1$ are the same. Recall that $\mathbf{U}_1$ is obtained from **U** by dividing each row of **U** by the diagonal element. But (52) tells us that each row of **U** divided by the diagonal element equals the corresponding column of **L**. Thus each row of $\mathbf{U}_1$ equals the corresponding column of **L**, or equivalently, $\mathbf{U}_1$ is the transpose of **L**. In summary,

the **LDU** *factorization of a symmetric matrix has the form* $\mathbf{LDL}^T$.

For a special class of symmetric matrices, the $\mathbf{LDL}^T$ factorization can be further simplified. Suppose that every element of **D** is nonnegative and let **S** denote the diagonal matrix for which each element is the square root of the corresponding element of **D**. Since $\mathbf{D} = \mathbf{S}^2$, the matrix **S** is called the square root of **D**. Replacing **D** in the factorization $\mathbf{A} = \mathbf{LDL}^T$ with $\mathbf{S}^2$, we have $\mathbf{A} = \mathbf{LSSL}^T$. Since a diagonal matrix is symmetric and since the transpose of the product **LS** of two matrices is the product $\mathbf{S}^T\mathbf{L}^T$ of the transposes in the reverse order, it follows that $(\mathbf{LS})^T = \mathbf{S}^T\mathbf{L}^T = \mathbf{SL}^T$. Hence **A** can be expressed

$$\mathbf{A} = \mathbf{LSSL}^T = \mathbf{LS}(\mathbf{LS})^T.$$

If **C** denotes the product **LS**, then **A** is equal to $\mathbf{CC}^T$. The product $\mathbf{CC}^T$ is called the **Cholesky factorization** of **A**. In summary, if **A** is symmetric and $\mathbf{A} = \mathbf{LDL}^T$ where the elements of **D** are nonnegative, then **A** has the Cholesky factorization $\mathbf{CC}^T$ where **C** is a lower triangular matrix.

Property (52) relating the multipliers to the coefficients of **U** is connected with the symmetry condition $a_{ij} = a_{ji}$. For the first elimination step, the multipliers are $l_{i1} = a_{i1}/a_{11}$. But by symmetry, $a_{i1} = a_{1i}$ which implies that $l_{i1} = a_{i1}/a_{11} = a_{1i}/a_{11}$. Since the first row of **A** and of **U** are the same, $a_{1i} = u_{1i}$ and $a_{11} = u_{11}$. Hence $l_{i1} = a_{i1}/a_{11} = a_{1i}/a_{11} = u_{1i}/u_{11}$, which establishes (52) when the column index j of l_{ij} equals 1. After the first elimination step, the lower right corner of the partly processed matrix (51) is symmetric. Consequently, the multipliers l_{i2} generated during the second elimination step also satisfy (52). To check that symmetry is always preserved in the lower right corner after each elimination step, let us compare the element in row i and column j to the element in row j and column i. In Section 2-3, we saw that the kth elimination step replaces a_{ij} by

$$a_{ij} - a_{kj}l_{ik} = a_{ij} - \frac{a_{kj}a_{ik}}{a_{kk}}, \tag{53}$$

where $l_{ik} = a_{ik}/a_{kk}$ is the multiplier. Interchanging j and i, it follows that the element in row j and column i is $a_{ji} - a_{ki}a_{jk}/a_{kk}$. But for a symmetric matrix, we have $a_{ij} = a_{ji}$, $a_{ki} = a_{ik}$, and $a_{jk} = a_{kj}$. With these substitutions,

$$a_{ji} - \frac{a_{jk}a_{ki}}{a_{kk}} = a_{ij} - \frac{a_{kj}a_{ik}}{a_{kk}}.$$

Since the element in row j and column i is equal to the element (53) in row i and column j, elimination preserves symmetry.

The factorization of a symmetric matrix can be performed within the storage structure (50). As the elimination progresses, we only need to save matrix elements that lie on the diagonal or above the diagonal—the multipliers are discarded since they can be reconstructed from **U** using identity (52), and the elements beneath the diagonal of the partly processed matrix are known from symmetry. In formulating an elimination scheme for a symmetric matrix, we will modify the general algorithm (27) in three ways:

1. Since the multiplier a_{ik}/a_{kk} is discarded, it is saved in the single-subscripted variable t.
2. Since a_{ik} in the elimination step (53) lies beneath the diagonal when i is greater than k and since elements beneath the diagonal are not stored, a_{ik} is replaced by the symmetric element a_{ki} above the diagonal.
3. The index range is truncated so that elements beneath the diagonal are not touched.

Making these changes in (27), the factorization of a symmetric matrix without pivots can be stated:

```
k = 1 to n-1
    i = k+1 to n
        t ← a_ki/a_kk
        j = i to n
            a_ij ← a_ij - t a_kj
        next j
    next i
next k
```
(54)

Comparing (27) to (54), we see that j increments from k to n during the kth elimination step of (27) while j increments from i to n in (54). For a symmetric matrix, we just process elements between the diagonal and column n. Since a_{ii} is the ith diagonal element, the column index j in (54) increments from i to n instead of from k to n.

Subroutine SFACT factors a symmetric matrix using (54). The subroutine arguments are

A: coefficient matrix stored using (50) [length at least 3 + .5N(N + 1)]

N: matrix dimension

W: work array with at least N elements

The work array is used to estimate the "size" of the coefficient matrix. This auxiliary computation, which does not alter the program's execution time significantly, is used later (see Chapter 3) to estimate the conditioning of the linear system. Subroutine SSOLVE solves a factored system by forward elimination and back substitution. For example, the symmetric system

$$2x_1 + 4x_2 + 6x_3 = 14,$$

$$4x_1 + 7x_2 + 8x_3 = 21,$$

$$6x_1 + 8x_2 + 5x_3 = 20,$$

is solved by the following program:

```
REAL A(9),B(3)
DATA A/2.,4.,6.,7.,8.,5./
CALL SFACT(A,3,B)
DATA B/14.,21.,20./
CALL SSOLVE(B,A,B)
WRITE(6,*) B(1),B(2),B(3)
END
```

The elimination scheme developed above for a symmetric system does not incorporate pivoting. Although not every matrix can be processed without pivots, there are two classes of matrices that never require pivots: column diagonally dominant matrices and symmetric positive definite matrices. A matrix is **column diagonally dominant** if in each column, the diagonal element is larger in absolute value than the absolute sum of off-diagonal elements in the column. The following matrix satisfies this requirement since the absolute diagonal elements 5, 4, and 3 are larger than the corresponding off-diagonal column sums 4, 3, and 2, respectively:

$$\begin{bmatrix} -5 & -2 & -1 \\ 3 & 4 & 1 \\ 1 & 1 & -3 \end{bmatrix}.$$

In general, a matrix is column diagonally dominant if

$$|a_{jj}| > \sum_{\substack{i=1 \\ i \neq j}}^{n} |a_{ij}|$$

for $j = 1, \cdots, n$. This relation implies that each diagonal element is larger than the other elements in a column. Thus no pivot is needed before the first elimination step. Furthermore, after each elimination step, the lower right corner of the partly processed matrix (see Figure 2-2) remains column diagonally dominant and row pivoting is never required (see Exercise 2-4.11).

A symmetric matrix is **positive definite** if it has an **LU** factorization where every diagonal element of **U** is positive. An example appears below:

$$\underset{\mathbf{A}}{\begin{bmatrix} 3 & 6 \\ 6 & 14 \end{bmatrix}} = \underset{\mathbf{L}}{\begin{bmatrix} 1 & 0 \\ 2 & 1 \end{bmatrix}} \underset{\mathbf{U}}{\begin{bmatrix} 3 & 6 \\ 0 & 2 \end{bmatrix}}.$$

Since the diagonal elements 3 and 2 of **U** are positive, **A** is positive definite. It can be shown that a symmetric matrix is positive definite if and only if $\mathbf{x}^T\mathbf{A}\mathbf{x}$ is positive for every nonzero **x** (see Exercise 2-4.12). Symmetric positive definite matrices arise frequently in applications. For example, if the coefficient a_{ij} is expressed in the form $\mathbf{f}_i^T\mathbf{f}_j$, where $\mathbf{f}_1, \mathbf{f}_2, \cdots, \mathbf{f}_n$ are linearly independent vectors, then **A** is positive definite. To understand why pivoting is never needed when **A** is symmetric and positive definite, we must discuss the propagation of errors in Gaussian elimination.

Loosely speaking, Gaussian elimination without pivots is "reasonably" accurate if elimination does not amplify the coefficients too much. For the system

$$\begin{bmatrix} .001 & 1 \\ 1 & 1 \end{bmatrix}\begin{bmatrix} x_1 \\ x_2 \end{bmatrix} = \begin{bmatrix} 1 \\ 2 \end{bmatrix}$$

studied in Section 2-3, the biggest coefficient is 1. But after elimination, the biggest coefficient of **U** is 1000. Elimination amplified the coefficients by the factor 1000, and this amplification factor is connected with the 100% error we observed in the computed solution. It can be shown that the relative error in the computed solution to a linear system of n equations is bound by the expression $(n^3 + 3n^2)cgE$, where E is the machine epsilon, c is the matrix condition number, and g is the coefficient growth during elimination. The condition number, a parameter defined in Chapter 3, measures the problem's conditioning. A large condition number means that the equation $\mathbf{Ax} = \mathbf{b}$ is ill conditioned. The growth g is defined by

$$g = \frac{\max\limits_{i,j,k} |a_{ij}^k|}{\max\limits_{i,j} |a_{ij}^1|}, \tag{55}$$

where a_{ij}^k is the coefficient in row i and column j before the kth elimination step. The growth measures the amplification in the coefficients generated by elimination relative to the coefficients of the original matrix. Note that g is at least 1 since we can take $k = 1$ in the numerator of (55). The main effect of pivoting is to keep g near 1. But if g is near 1 without pivots, pivoting will not affect g significantly. You show in Exercise 2-4.16 that without pivoting, g is exactly 1 for a symmetric positive definite matrix; therefore, pivoting is not required.

Although some matrices can be factored without pivoting, the symmetric matrix

$$\begin{bmatrix} 0 & 1 \\ 1 & 1 \end{bmatrix}$$

illustrates one case where elimination is impossible. Rows must be interchanged, giving

$$\begin{bmatrix} 1 & 1 \\ 0 & 1 \end{bmatrix},$$

which is unsymmetric. To preserve symmetry, pivoting must be performed in a symmetric manner; if two rows are interchanged, then the corresponding columns are also interchanged. Interchanging rows i and j and columns i and j also interchanges the diagonal elements a_{ii} and a_{jj}. Thus symmetric pivoting can be used to move the biggest diagonal element to the pivot position. But if every diagonal element is zero, we can never remove the zeros from the diagonal using symmetric pivoting. The fundamental idea behind pivoting strategies for a symmetric matrix is the following: Instead of eliminating coefficients beneath the diagonal as we do in the standard elimination algorithm, eliminate coefficients beneath the subdiagonal. For example, consider the following matrix:

$$\mathbf{A} = \begin{bmatrix} 0 & 0 & 1 & 2 \\ 0 & 2 & 4 & 2 \\ 1 & 4 & 6 & 4 \\ 2 & 2 & 4 & 4 \end{bmatrix}.$$

The largest coefficient beneath the diagonal in the first column is the 2 in row 4. Interchanging rows 2 and 4 and columns 2 and 4 yields

$$\mathbf{A} = \begin{bmatrix} 0 & 2 & 1 & 0 \\ 2 & 4 & 4 & 2 \\ 1 & 4 & 6 & 4 \\ 0 & 2 & 4 & 2 \end{bmatrix}.$$

To annihilate coefficients in the first column beneath the subdiagonal, we multiply the second row by $\frac{1}{2}$ and subtract from the third row to obtain

$$\mathbf{A} = \begin{bmatrix} 0 & 2 & 1 & 0 \\ 2 & 4 & 4 & 2 \\ 0 & 2 & 4 & 3 \\ 0 & 2 & 4 & 2 \end{bmatrix}.$$

To restore symmetry, the same elimination step is applied to the columns. The second column is multiplied by $\frac{1}{2}$ and subtracted from the third column to get

$$\mathbf{A} = \begin{bmatrix} 0 & 2 & 0 & 0 \\ 2 & 4 & 2 & 2 \\ 0 & 2 & 3 & 3 \\ 0 & 2 & 3 & 2 \end{bmatrix}.$$

Without going into the details, using symmetric pivoting and symmetric elimination, we can annihilate coefficients beneath the subdiagonal to obtain a tridiagonal matrix. The resulting factorization can be expressed in the form $\mathbf{A} = (\mathbf{PL})\mathbf{T}(\mathbf{L}^T\mathbf{P}^T)$, where $\mathbf{T}$ is a symmetric tridiagonal matrix, $\mathbf{L}$ is a unit lower triangular matrix, and $\mathbf{P}$ is a permutation matrix which corresponds to the pivot operations (multiplying the left side of a matrix by $\mathbf{P}$ essentially interchanges various rows corresponding to the pivot operations). To solve the equation $\mathbf{Ax} = \mathbf{b}$ starting from the factorization $\mathbf{A} = (\mathbf{PL})\mathbf{T}(\mathbf{L}^T\mathbf{P}^T)$, we use the associative law to obtain the equivalent system:

$$\begin{aligned} (\mathbf{PL})\mathbf{z} &= \mathbf{b}, \\ \mathbf{Ty} &= \mathbf{z}, \\ (\mathbf{L}^T\mathbf{P}^T)\mathbf{x} &= \mathbf{y}. \end{aligned}$$

Solving the first equation $(\mathbf{PL})\mathbf{z} = \mathbf{b}$ yields $\mathbf{z}$, solving the second equation $\mathbf{Ty} = \mathbf{z}$ yields $\mathbf{y}$, and solving the third equation $(\mathbf{L}^T\mathbf{P}^T)\mathbf{x} = \mathbf{y}$ yields the final solution $\mathbf{x}$. Since the inverse of a permutation matrix is its transpose, we can multiply the first equation by $\mathbf{P}^T$ to obtain $\mathbf{Lz} = \mathbf{P}^T\mathbf{b}$ which is solved by forward elimination since $\mathbf{L}$ is lower triangular. To solve $\mathbf{Ty} = \mathbf{z}$, we can employ Gaussian elimination with partial pivoting. To solve the third equation, $(\mathbf{L}^T\mathbf{P}^T)\mathbf{x} = \mathbf{y}$ we substitute $\mathbf{w} = \mathbf{P}^T\mathbf{x}$ to obtain $\mathbf{L}^T\mathbf{w} = \mathbf{y}$, which is solved by back substitution. Finally, $\mathbf{x}$ is equal to $\mathbf{Pw}$.

There are several different algorithms to compute the factorization $\mathbf{A} = (\mathbf{PL})\mathbf{T}(\mathbf{L}^T\mathbf{P}^T)$. In each case, their derivation and analysis is a little technical. We state Aasen's method [A1], while references to schemes by Bunch, Kaufman, Parlett, and Reid appear at the end of the chapter.

$$
\begin{array}{l}
d_1 \leftarrow a_{11} \\
k = 1 \text{ to } n-1 \\
\quad l \leftarrow k + 1 \\
\quad m = \arg\max\{|a_{ik}| : i = l \text{ to } n\} \\
\quad a_{mj} \leftrightarrow a_{lj} \text{ for } j = 1 \text{ to } n \\
\quad a_{im} \leftrightarrow a_{il} \text{ for } i = l \text{ to } n \\
\quad p_l \leftarrow m,\ u_k \leftarrow a_{lk} \\
\quad a_{ik} \leftarrow a_{ik}/a_{lk} \text{ for } i = l+1 \text{ to } n \\
\quad t_i \leftarrow u_i a_{l,i-1} + d_{i+1} a_{li} + u_{i+1} a_{l,i+1} \text{ for } i = 1 \text{ to } k-2 \\
\quad t_i \leftarrow u_i a_{l,i-1} + d_{i+1} a_{li} + u_k \text{ for } i = k-1 \\
\quad t_k \leftarrow a_{ll} - \sum_{j=1}^{k-1} a_{lj} t_j \\
\quad d_l \leftarrow t_k - u_k a_{l,k-1} \\
\quad a_{il} \leftarrow a_{il} - \sum_{j=1}^{k} a_{ij} t_j \text{ for } i = l+1 \text{ to } n \\
\text{next } k
\end{array}
\tag{56}
$$

Our convention in stating this algorithm is that $a_{ij} = 0$ for $j \leqq 0$ and a summation with a vacuous index range is zero. The matrices $\mathbf{T}$, $\mathbf{L}$, and $\mathbf{P}$ are reconstructed from (56) in the following way: The diagonal and the superdiagonal of the symmetric matrix $\mathbf{T}$ are stored in $\mathbf{d}$ and $\mathbf{u}$, respectively. The first column of $\mathbf{L}$ contains a 1 followed by zeros and the remaining elements of $\mathbf{L}$ beneath the diagonal are stored beneath the subdiagonal of $\mathbf{A}$. Finally, the permutation matrix $\mathbf{P}$ is obtained by interchanging rows p_i and i of the identity matrix for $i = n$ down to 2 (the identity matrix is the diagonal matrix with every diagonal element equal to one). Subroutine IFACT implements (56), while the corresponding solution subroutine is ISOLVE.

Exercises

2-4.8. Consider a system $\mathbf{Ax} = \mathbf{b}$ of three equations in three unknowns where $a_{ij} = 1/(i + j - 1)$ and every component of $\mathbf{b}$ is 1. Solve this system using SFACT and SSOLVE.

2-4.9. Suppose that a $n \times n$ symmetric matrix is stored in an array A with leading dimension LA. Write a subroutine SMODE(A,LA,N) that transforms the storage into the symmetric mode depicted in (50). For example, the output of the following program is 1, 2, 3, 4, 5, 6.

```
REAL A(3,3)
DATA A/1.,2.,3.,2.,4.,5.,3.,5.,6./
CALL SMODE(A,3,3)
```

```
CALL LIST(A,6)
END
```

Subroutine LIST appears in Exercise 2-3.2.

2-4.10. If a symmetric matrix is stored in an array A using the structure depicted in (50), what is the storage location in A corresponding to row i and column j of the matrix? What is the row and the column of the matrix corresponding to storage location I in the array A?

2-4.11. Show that if a column diagonally dominant matrix is factored using row elimination, the lower right corner of the partly processed matrix (see Figure 2-2) is always column diagonally dominant. (*Hint:* The kth elimination step replaces a_{ij} by

$$a'_{ij} = a_{ij} - \frac{a_{kj}a_{ik}}{a_{kk}}.$$

The lower right corner is column diagonally dominant if

$$|a'_{jj}| > \sum_{\substack{i=k+1 \\ i \neq j}}^{n} |a'_{ij}|$$

for $j = k+1, \cdots, n$. Show that this inequality is satisfied whenever the coefficients in the lower right corner generated by the $(k-1)$st elimination step are column diagonally dominate.

2-4.12. Show that if $\mathbf{A}$ is a symmetric positive definite matrix, then $\mathbf{x}^T\mathbf{A}\mathbf{x} > 0$ for every nonzero vector $\mathbf{x}$. (*Hint:* Substitute $\mathbf{A} = \mathbf{LDL}^T$ in the expression $\mathbf{x}^T\mathbf{A}\mathbf{x}$ and define $\mathbf{y} = \mathbf{L}^T\mathbf{x}$. If $\mathbf{A}$ is positive definite, what can you say about $\mathbf{D}$?) Conversely, if $\mathbf{x}^T\mathbf{A}\mathbf{x} > 0$ for every nonzero $\mathbf{x}$ and $\mathbf{A} = \mathbf{LDL}^T$, then show that $\mathbf{A}$ is positive definite.

2-4.13. Show that the inequality $\mathbf{x}^T\mathbf{A}\mathbf{x} > 0$ for every nonzero $\mathbf{x}$ implies that each diagonal element of $\mathbf{A}$ is positive.

2-4.14. Suppose that $\mathbf{A}$ is a symmetric matrix and $\mathbf{x}^T\mathbf{A}\mathbf{x} > 0$ for every nonzero vector $\mathbf{x}$. Show that $\mathbf{A}$ is positive definite. (*Hint:* By Exercise 2-4.13, $u_{11} > 0$. Let k denote the first elimination step for which $u_{kk} \leqq 0$ and let $\mathbf{A}_k$ denote the submatrix of $\mathbf{A}$ formed by the intersection of its first k columns and its first k rows. Show that $\mathbf{x}^T\mathbf{A}_k\mathbf{x} > 0$ for every nonzero vector $\mathbf{x}$. How does the $\mathbf{LU}$ factorization of $\mathbf{A}_k$ relate to the $\mathbf{LU}$ factorization of $\mathbf{A}$?)

2-4.15. Suppose that $\mathbf{A} = \mathbf{B}^T\mathbf{B}$ where $\mathbf{B}$ is a $m \times n$ matrix with linearly independent columns. Show that $\mathbf{A}$ is positive definite. (*Hint:* Utilize Exercise 2-4.14.)

2-4.16. Show that when a symmetric positive definite matrix is factored without pivots, the absolute largest coefficient in the lower right corner of the partly processed matrix (see Figure 2-2) never exceeds the largest coefficient in the original matrix. This result is verified in three steps.

(a) Suppose that $\mathbf{A}$ is symmetric and $\mathbf{x}^T\mathbf{A}\mathbf{x} > 0$ for every nonzero vector $\mathbf{x}$. Show that the largest element in absolute value lies on the diagonal.

(b) Explain why the lower right corner of $\mathbf{A}$ is positive definite after each elimination step. (*Hint:* Let $\mathbf{A}_k$ denote the lower right corner of the partly processed matrix, where k is the number of rows and columns in this submatrix. If $\mathbf{L}_k$ and $\mathbf{D}_k$ denote the submatrix formed by the intersection of the last k rows and the last k columns of $\mathbf{L}$ and $\mathbf{D}$, respectively, then how are $\mathbf{A}_k$, $\mathbf{L}_k$, and $\mathbf{D}_k$ related?)

(c) Show that after each elimination step, the diagonal elements of the partly processed matrix decrease.

Since the lower right corner of the partly processed matrix is positive definite, the diagonal elements decrease in size, and the largest element in absolute value lies on the diagonal, we conclude that elimination causes no growth in the coefficients. More precisely, an elimination step does not increase the size of the absolute largest coefficient.

2-4.17. How is the diagonal of **C** in Cholesky's decomposition $\mathbf{CC}^T$ related to the **D** in the $\mathbf{LDL}^T$ factorization of a symmetric positive definite matrix?

2-4.4 Symmetric Band Matrix

A symmetric band matrix can be stored in a rectangle formed from the diagonal and off-diagonal bands. For example, the matrix

$$\begin{bmatrix} 4 & 2 & 1 & 0 & 0 \\ 2 & 4 & 2 & 2 & 0 \\ 1 & 2 & 4 & 2 & 3 \\ 0 & 2 & 2 & 4 & 2 \\ 0 & 0 & 3 & 2 & 4 \end{bmatrix} \tag{57}$$

is stored in the 3 × 5 rectangle of Table 2-9. A matrix which is zero below the hth subdiagonal and above the hth superdiagonal is called a h-band matrix or a band matrix with **half bandwidth** h. A diagonal matrix is 0-band, a tridiagonal matrix is 1-band, a pentadiagonal matrix is 2-band, and so on. To factor a symmetric band matrix, we combine the scheme (54) for factoring a symmetric matrix with the scheme (44) for factoring a band matrix. Due to symmetry, the multipliers are discarded and due to the band structure, the elimination is restricted to the bands. Thus elimination without pivots for a symmetric band matrix can be stated:

```
k = 1 to n−1
    i = k+1 to min{n, k+h}
        t ← a_ki/a_kk
        j = i to min{n, k+h}
            a_ij ← a_ij − t a_kj
        next j
    next i
next k
```
(58)

Subroutine HFACT factors a symmetric band matrix without pivots using (58). The subroutine arguments are

A: array containing matrix bands starting with diagonal [length at least 4 + N(H + 1)]

TABLE 2-9 SYMMETRIC BAND STORAGE FOR THE MATRIX (57)

4	4	4	4	4	Diagonal
2	2	2	2	*	Band 1
1	2	3	*	*	Band 2

LA: leading (row) dimension of array A
N: matrix dimension
H: half bandwidth

As usual, a system that has been factored with subroutine HFACT can be solved with subroutine HSOLVE. For example, the solution to the system

$$\begin{bmatrix} 3 & 1 & 0 & 0 \\ 1 & 3 & 1 & 0 \\ 0 & 1 & 3 & 1 \\ 0 & 0 & 1 & 3 \end{bmatrix} \begin{bmatrix} x_1 \\ x_2 \\ x_3 \\ x_4 \end{bmatrix} = \begin{bmatrix} 5 \\ 10 \\ 15 \\ 15 \end{bmatrix}, \tag{59}$$

is computed by the program

```
      REAL A(2,6),B(4)
      DATA B/5.,10.,15.,15./
      DO 10 I = 1,4
         A(1,I) = 3.
         A(2,I) = 1.
10    CONTINUE
      CALL HFACT(A,2,4,1)
      CALL HSOLVE(B,A,B)
      WRITE(6,*) (B(I), I=1,4)
      END
```

This program is just for illustration—since the coefficient matrix in (59) is both symmetric and tridiagonal, it would be more efficient to compute the solution using TFACT and TSOLVE.

Exercises

2-4.18. Using HFACT and HSOLVE, compute the solution to $\mathbf{Ax} = \mathbf{b}$, where $\mathbf{A}$ is the matrix in (57) and

$$\mathbf{b} = \begin{bmatrix} 4 \\ -2 \\ 9 \\ 6 \\ 14 \end{bmatrix}.$$

2-4.19. Suppose that a symmetric band matrix is stored in an array A using the structure described in this section. What is the row and column in array A that corresponds to row i and column j of the matrix? What is the row and column of the matrix that corresponds to row I and column J of array A?

2-4.5 Sparse Matrix

A matrix which is predominantly zero is called *sparse*. A tridiagonal matrix or a band matrix is sparse, but the nonzero elements of a sparse matrix are not necessarily near the diagonal. Any linear system with the property that each equation just involves a few of the unknowns will have a sparse coefficient matrix. Let us consider a specific class of sparse matrices called the bordered matrices. A bordered matrix is completely zero except possibly its diagonal, its first row, and its first column. A 4×4 bordered system appears below:

$$\begin{bmatrix} 1 & 2 & 2 & 2 \\ 1 & 1 & 0 & 0 \\ 1 & 0 & -1 & 0 \\ 1 & 0 & 0 & 1 \end{bmatrix} \begin{bmatrix} x_1 \\ x_2 \\ x_3 \\ x_4 \end{bmatrix} = \begin{bmatrix} 3 \\ 0 \\ 0 \\ 1 \end{bmatrix}. \tag{60}$$

Solving this system in the usual fashion, we subtract the first equation from the equations beneath it, giving us

$$\begin{bmatrix} 1 & 2 & 2 & 2 \\ 0 & -1 & -2 & -2 \\ 0 & -2 & -3 & -2 \\ 0 & -2 & -2 & -1 \end{bmatrix} \begin{bmatrix} x_1 \\ x_2 \\ x_3 \\ x_4 \end{bmatrix} = \begin{bmatrix} 3 \\ -3 \\ -3 \\ -2 \end{bmatrix}. \tag{61}$$

Continuing the elimination, we obtain the upper triangular system

$$\begin{bmatrix} 1 & 2 & 3 & 2 \\ 0 & -1 & -2 & -2 \\ 0 & 0 & 1 & 2 \\ 0 & 0 & 0 & -1 \end{bmatrix} \begin{bmatrix} x_1 \\ x_2 \\ x_3 \\ x_4 \end{bmatrix} = \begin{bmatrix} 3 \\ -3 \\ 3 \\ -2 \end{bmatrix},$$

which is solved by back substitution.

Notice that the first elimination step (61) generates nonzero coefficients in place of the zero coefficients in (60). [Although the first elimination step also puts zeros in column 1, remember that in computer memory the multipliers are stored in the space occupied by annihilated coefficients. Thus the computer replaces the zeros in the first column of the coefficient matrix (61) with the multipliers.] Nonzero elements generated during elimination are called **fill-in**. A bordered matrix like (60) suffers from complete fill-in during the first elimination step. We now show how this fill-in can be avoided. First, write (60) in its equivalent form

$$\begin{aligned} x_1 + 2x_2 + 2x_3 + 2x_4 &= 3, \\ x_1 + x_2 \qquad\qquad\quad &= 0, \\ x_1 \qquad - x_3 \qquad\quad &= 0, \\ x_1 \qquad\qquad\quad + x_4 &= 1. \end{aligned}$$

Interchange the first equation and the fourth equation to get

$$\begin{aligned} x_1 \qquad\qquad\quad + x_4 &= 1, \\ x_1 + x_2 \qquad\qquad\quad &= 0, \\ x_1 \qquad - x_3 \qquad\quad &= 0, \\ x_1 + 2x_2 + 2x_3 + 2x_4 &= 3, \end{aligned}$$

then interchange the x_1 terms with the x_4 terms to obtain

$$\begin{aligned} x_4 \qquad\qquad\quad + x_1 &= 1, \\ x_2 \qquad\quad + x_1 &= 0, \\ - x_3 + x_1 &= 0, \\ 2x_4 + 2x_2 + 2x_3 + x_1 &= 3. \end{aligned}$$

Using matrix notation, this system is

$$\begin{bmatrix} 1 & 0 & 0 & 1 \\ 0 & 1 & 0 & 1 \\ 0 & 0 & -1 & 1 \\ 2 & 2 & 2 & 1 \end{bmatrix} \begin{bmatrix} x_4 \\ x_2 \\ x_3 \\ x_1 \end{bmatrix} = \begin{bmatrix} 1 \\ 0 \\ 0 \\ 3 \end{bmatrix}. \tag{62}$$

Equations (60) and (62) are closely connected since each coefficient matrix is obtained from the other by interchanging the first and fourth rows and the first and fourth columns. But when (62) is solved by Gaussian elimination, there is no fill-in. We multiply the first three equations by 2, 2, and -2, respectively, and we subtract from the fourth equation to get the upper triangular system

$$\begin{bmatrix} 1 & 0 & 0 & 1 \\ 0 & 1 & 0 & 1 \\ 0 & 0 & -1 & 1 \\ 0 & 0 & 0 & -1 \end{bmatrix} \begin{bmatrix} x_4 \\ x_2 \\ x_3 \\ x_1 \end{bmatrix} = \begin{bmatrix} 1 \\ 0 \\ 0 \\ 1 \end{bmatrix},$$

which is solved by back substitution. This time the zeros were preserved during elimination.

In general, a bordered matrix like (62) can be factored using three arrays. Consider the 4×4 matrix

$$\begin{bmatrix} d_1 & 0 & 0 & u_1 \\ 0 & d_2 & 0 & u_2 \\ 0 & 0 & d_3 & u_3 \\ l_1 & l_2 & l_3 & d_4 \end{bmatrix}. \tag{63}$$

To factor this matrix, we multiply the first three rows by l_1/d_1, l_2/d_2, and l_3/d_3, respectively, and subtract from the fourth row. The elimination process annihilates the first three elements in row 4 and replaces d_4 by

$$d_4 - \sum_{i=1}^{3} \frac{u_i l_i}{d_i}.$$

The following algorithm implements this factorization scheme for a general $n \times n$ bordered matrix like (63), overwriting $\mathbf{l}$ with the multipliers and overwriting d_n with the final diagonal element of $\mathbf{U}$.

```
i = 1 to n−1
    l_i ← l_i/d_i
    d_n ← d_n − l_i u_i
next i
```
(64)

As this example illustrates, an efficient factorization routine for a sparse matrix has three parts:

1. Reorder the equations and the unknowns to reduce the fill-in.
2. Develop an economic storage structure for the new system.
3. Factor the matrix within this storage structure.

For the bordered system, we interchange the first and last equation and the first and last unknown to eliminate the fill-in. The storage structure is the three arrays in (63), and algorithm (64) carries out the factorization. George and Liu [G5] give a comprehensive analysis of general techniques to reorder, store, and factor a sparse symmetric positive definite matrix. More recently, methods for nonsymmetric matrices are developed in the book [D20] by Duff, Erisman, and Reid. Although we do not discuss reordering techniques, we will discuss several schemes for storing a sparse matrix.

Gustavson [G17] stores a sparse matrix using three arrays. An array A contains the nonzero elements in each row, an integer array C gives the column index for each nonzero element, and an integer array R gives the location in A of the first nonzero element in each row. For example, the matrix

$$\mathbf{A} = \begin{bmatrix} 6 & 0 & 1 & 0 & 0 \\ 1 & 2 & 0 & 0 & 0 \\ 0 & 0 & 4 & 1 & 0 \\ 0 & 0 & 0 & 6 & 1 \\ 0 & 5 & 1 & 0 & 2 \end{bmatrix} \tag{65}$$

is stored as follows:

I	1	2	3	4	5	6	7	8	9	10	11
A(I)	6	1	1	2	4	1	6	1	5	1	2
C(I)	1	3	1	2	3	4	4	5	2	3	5
R(I)	1	3	5	7	9	12					

There is one element in array A and one element in array C corresponding to each nonzero coefficient in **A**. For convenience, R(6) is set to 12, the number of elements in A plus 1. To locate a specific coefficient like a_{53} inside array A, we observe that R(5) is 9, which implies that elements from row 5 of matrix **A** are stored in array A starting at location 9. Beginning at location 9, we advance through C until we find the column index 3 or until we reach the end of storage for row 5 (in which case the coefficient is zero). Since C(10) is 3, a_{53} is stored in A(10). The following program locates a_{ij} and puts it in T:

```
            T = 0.
            L = R(I)
            M = R(I+1) - 1
            DO 10 K = L,M
10            IF ( C(K) .EQ. J ) GOTO 20
            GOTO 30
20          T = A(K)
30          CONTINUE
```

If the nonzero elements of a matrix are near each other in every row, then **profile storage** may be appropriate. This structure also involves three arrays. An array A stores the elements in each row between the first and last nonzero elements. The integer array C gives the column of the first nonzero element in each row, and the integer array R gives the location in A of the first nonzero element in each row. The storage structure for the matrix (65) is the following:

I	1	2	3	4	5	6	7	8	9	10	11	12	13
A(I)	6	0	1	1	2	4	1	6	1	5	1	0	2
C(I)	1	1	3	4	2								
R(I)	1	4	6	8	10	14							

Since the first and last nonzero coefficients in row 1 of **A** are a_{11} and a_{13}, respectively, the first three elements of array A contain $a_{11} = 6$, $a_{12} = 0$, and $a_{13} = 1$. Since some zero coefficients are stored in the profile structure, the A array in profile storage is larger than the corresponding array in Gustavson's structure. On the other hand, the C array in profile storage is smaller since we just store the column of the first nonzero element in each row. Again, the final element in the R array [R(6) above] stores one more than the number of elements in A. To locate a specific coefficient like a_{53} inside array A, we first inspect the R array. Since R(5) is 10, the first nonzero element in row 5 is stored in A(10). Since C(5) is 2, this element lies in column 2 of matrix **A** and corresponds to a_{52}. Thus a_{53} is stored in A(10 + 1) = A(11). The following program locates a_{ij} and puts it in T:

```
      T = 0.
      IF ( J .LT. C(I) ) GOTO 10
      K = J - C(I) + R(I)
      IF ( K .GE. R(I+1) ) GOTO 10
      T = A(K)
10    CONTINUE
```

If the nonzero elements of a matrix are near each other in every row and the matrix is symmetric, then Jennings' profile scheme [J4] can be utilized. Jennings stores in array A those elements in each row of the matrix which lie between the diagonal element and the first nonzero element left of the diagonal. The integer array R gives the location in A of the first nonzero element in each row. The symmetric matrix

$$\mathbf{A} = \begin{bmatrix} 2 & 3 & 5 & 0 & 0 \\ 3 & 6 & 0 & 0 & 0 \\ 5 & 0 & 4 & 1 & 0 \\ 0 & 0 & 1 & 8 & 7 \\ 0 & 0 & 0 & 7 & 1 \end{bmatrix} \tag{66}$$

is stored as follows:

I	1	2	3	4	5	6	7	8	9	10
A(I)	2	3	6	5	0	4	1	8	7	1
R(I)	1	2	4	7	9	11				

In row 4 of **A**, the first nonzero coefficient to the left of the diagonal is a_{43}. Hence, only the coefficients $a_{43} = 1$ and $a_{44} = 8$ are stored in array A. As usual, the final element in the R array stores one more than the number of elements in A. Jennings' symmetric profile scheme differs from the unsymmetric profile scheme in two respects: (a) the C array is eliminated and (b) coefficients above the diagonal are not stored. To locate a specific coefficient like a_{32} inside array A, we observe that R(4) is 7, which implies that a_{33}, the diagonal element in row 3, is stored in A(7 − 1) = A(6). Since a_{32} is just left of a_{33} in row 3, a_{32} is stored in A(6 − 1) = A(5). A program to locate a_{ij} and store it in T appears below:

```
      T = 0.
      IF ( I .GT. J ) GOTO 10
      K = R(J)
      L = R(J+1) + I - J - 1
      GOTO 20
10    K = R(I)
      L = R(I+1) + J - I - 1
20    IF ( L .LT. K ) GOTO 30
      T = A(L)
30    CONTINUE
```

A symmetric matrix which requires no pivoting can be factored inside Jennings' profile structure since the fill-in corresponds to matrix elements found in array A. For example, when the matrix (66) is factored, the zero in row 3 and column 2 is filled in. Since space has been allocated for a_{32} in array A, this fill-in is accounted for in Jennings' structure. On the other hand, a matrix stored using the first two structures cannot be factored "in place" unless the fill-in is accounted for when storage is created. For example, when the matrix (65) is factored, nonzero numbers are assigned to a_{23} and a_{54}, but neither of these elements appears in Gustavson's structure while just a_{54} appears in the unsymmetric profile structure. Therefore, when a sparse matrix is factored using a compact storage scheme, space must be allocated for both the nonzero matrix elements, and the fill-in. A "symbolic factorization" technique for predicting fill-in is developed in [G5]. When using a sparse matrix package like SPARSPAK (see Section 8-2), the user selects a reordering algorithm and inputs the data using a sparse storage structure. Then the package reallocates storage to make room for the fill-in and performs the factorization.

Exercise

2-4.20. Write a symbolic program to solve a factored bordered system where the factorization is generated by (64).

2-5. COMPACT ELIMINATION

There are two variations of Gaussian elimination, which we call the **compact row oriented method** and the **compact column oriented method**. Consider the following matrix:

$$\mathbf{A} = \begin{bmatrix} 3 & 2 & -2 \\ 9 & 7 & -9 \\ 6 & 8 & -8 \end{bmatrix}. \tag{67}$$

In the row method, we annihilate elements in the successive rows. The first step subtracts 3 times the first row from the second row giving

$$\begin{bmatrix} 3 & 2 & -2 \\ 0 & 1 & -3 \\ 6 & 8 & -8 \end{bmatrix}.$$

The multiplier l_{21} is 3. The next step processes row 2. We subtract 2 times the first row from the third row to get

$$\begin{bmatrix} 3 & 2 & -2 \\ 0 & 1 & -3 \\ 0 & 4 & -4 \end{bmatrix},$$

and we subtract 4 times the second row from the third row to obtain

$$\begin{bmatrix} 3 & 2 & -2 \\ 0 & 1 & -3 \\ 0 & 0 & 8 \end{bmatrix}.$$

The multipliers for the second step are $l_{31} = 2$ and $l_{32} = 4$. Each step of the row method processes one row of the coefficient matrix. Collecting multipliers, the factorization of **A** is

$$\underset{\mathbf{A}}{\begin{bmatrix} 3 & 2 & -2 \\ 9 & 7 & -9 \\ 6 & 8 & -8 \end{bmatrix}} = \underset{\mathbf{L}}{\begin{bmatrix} 1 & 0 & 0 \\ 3 & 1 & 0 \\ 2 & 4 & 1 \end{bmatrix}} \underset{\mathbf{U}}{\begin{bmatrix} 3 & 2 & -2 \\ 0 & 1 & -3 \\ 0 & 0 & 8 \end{bmatrix}}.$$

A symbolic program for the row method is

```
k = 2 to n
    i = 1 to k-1
        l_ki ← a_ki/a_ii
        j = i to n
            a_kj ← a_kj - l_ki a_ij
        next j
    next i
next k
```
(68)

Here k corresponds to the row being processed. Inside the i loop, we compute the multiplier l_{ki} and we subtract l_{ki} times row i from row k. Since rows $k+1$ through n are not needed when row k is processed, this scheme is called compact. In contrast, each step of Gaussian elimination processes every coefficient in the lower right corner of Figure 2-2.

In the compact row-oriented method, all the elimination steps pertaining to a single row are performed before we proceed to the next row. In the compact column-oriented method, all the elimination steps pertaining to a single column are performed before we proceed to the next column. For the matrix (67), the first step of the column method computes the multipliers in **L**'s column 1: $l_{21} = 9/3 = 3$, and $l_{31} = 6/3 = 2$. In the standard Gaussian elimination algorithm, the first row is multiplied by l_{21} and subtracted from the second row, and the first row is multiplied by l_{31} and subtracted from the third row. In the first step of the compact column method, we just perform that part of the elimination step which is relevant to columns 1 and 2. That is, zeros are inserted in column 1 beneath the diagonal, and the product between a_{12} (= 2) and the multipliers l_{21} (= 3) and l_{31} (= 2) is subtracted from a_{22} (= 7) and a_{32} (= 8), respectively. In other words, we subtract from column 2 of the matrix (67), a_{12} (= 2) times the first column of **L**, giving us the new column 2:

$$\begin{bmatrix} 2 \\ 7 \\ 8 \end{bmatrix} - 2\begin{bmatrix} 0 \\ 3 \\ 2 \end{bmatrix} = \begin{bmatrix} 2 \\ 1 \\ 4 \end{bmatrix}.$$

So after the first step of the compact column method, the partly processed matrix is

$$\begin{bmatrix} 3 & 2 & -2 \\ 0 & 1 & -9 \\ 0 & 4 & -8 \end{bmatrix}.$$

The second step uses the second column to compute the multipliers in **L**'s column 2: $l_{32} = 4/1 = 4$. Then zeros are inserted in column 2 of the coefficient matrix beneath the diagonal and column 3 is processed using the multipliers. We subtract from column 3, the first column of **L** times $a_{13} = -2$:

$$\begin{bmatrix} -2 \\ -9 \\ -8 \end{bmatrix} - (-2)\begin{bmatrix} 0 \\ 3 \\ 2 \end{bmatrix} = \begin{bmatrix} -2 \\ -3 \\ -4 \end{bmatrix}.$$

At this point, the partly processed coefficient matrix is

$$\begin{bmatrix} 3 & 2 & -2 \\ 0 & 1 & -3 \\ 0 & 0 & -4 \end{bmatrix}.$$

Then we subtract from the third column of the coefficient matrix, $a_{23} = -3$ times the second column of **L**:

$$\begin{bmatrix} -2 \\ -3 \\ -4 \end{bmatrix} - (-3)\begin{bmatrix} 0 \\ 0 \\ 4 \end{bmatrix} = \begin{bmatrix} -2 \\ -3 \\ 8 \end{bmatrix}.$$

This completes the second step, and the upper triangular factor is

$$\begin{bmatrix} 3 & 2 & -2 \\ 0 & 1 & -3 \\ 0 & 0 & 8 \end{bmatrix}.$$

In general, the column process can be stated:

```
k = 1 to n−1
    i = k+1 to n
        l_ik ← a_ik / a_kk
        a_ik ← 0
    next i
    j = 1 to k
        i = j+1 to n
            a_i,k+1 ← a_i,k+1 − l_ij a_j,k+1
        next i
    next j
next k
```

(69)

The kth step of this algorithm computes the multipliers in **L**'s column k and processes column $k + 1$ in the coefficient matrix. We subtract from column $k + 1$ of the coefficient matrix, $a_{j,k+1}$ times the jth column of **L** for $j = 1, 2, \cdots, k$.

These compact methods are essentially new ways to organize the computations in Gaussian elimination. Although the compact schemes involve the same multiplications, divisions, and subtractions as the standard elimination schemes, the compact schemes delay part of the computation, and in each step, just one column or just one row is updated. One attractive feature of the compact schemes is improved accuracy if double-precision arithmetic is used. Since each step of a compact scheme updates a single row or a single column of the coefficient matrix, the rounding error associated with this step is reduced by performing the arithmetic inside a double-precision array with the same length as that of a column or of a row. In contrast, Gaussian elimination processes the lower right corner of the matrix each step and the entire right corner must be stored in double precision to achieve the same accuracy.

Another attractive aspect of the compact schemes is convenient storage manipulation. For example, suppose a large matrix is stored on a magnetic tape and just s columns of the matrix fit in main memory at one time. Using the compact column-oriented scheme, the factorization is accomplished by the following procedure: Transferring $s - 1$ columns from tape to main memory, we process them using (69) where k is restricted to the range "1 to $s - 1$." Next, each of the remaining columns is transferred to main memory, it is processed using the first $s - 1$ columns of multipliers, and it is returned to tape. This phase can be stated as follows:

```
k = s to n
    j = 1 to s-1
        i = j+1 to n
            a_ik ← a_ik − a_jk l_ij
        next i
    next j
next k
```
(70)

At this point, the partly processed matrix resembles Figure 2-2. The first $s - 1$ columns of **L** and **U** are stored in main memory and the remaining columns of the partly processed matrix reside on the tape. Then these $s - 1$ columns are returned to tape, and the lower right corner of Figure 2-2 is processed using the procedure just outlined. That is, as many columns as possible are transferred to main memory where they are processed using the column method (69). In contrast, Gaussian elimination transfers the entire right corner between tape and main memory each elimination step (instead of each $s - 1$ elimination steps).

2-6. RUNNING TIME

Although a computer program that factors a matrix, solves a system, or multiplies a matrix by a vector can contain more than 100 statements, the running time of this program is often controlled by the execution speed of one statement. We first illustrate this phenomenon with a program to compute the product between a square matrix **A** and a vector **b** (see Table 2-10). The program in Table 2-10 multiplies each column of **A** by the corresponding component of **b** and accumulates these vector-scalar products in array X. To determine the execution time of a program, we multiply the statement execution time by the number of times that the statement is executed. In other words, if E_i is the execution time of the ith statement in Table 2-10 and N_i is the number of times that the ith statement is executed, then the execution time T_N of the program is

$$T_N = \sum E_i N_i .$$

The first step in determining T_N is to extract the loops from the code and count the number of times statements are executed. For the program in Table 2-10, there are three loops:

```
      DO 10 I = 1,N
10    CONTINUE
      DO 30 J = 1,N
         DO 20 I = 1,N
20       CONTINUE
30    CONTINUE
```

The number of times the ith statement is executed is counted by the following technique: Replace the statement by the assignment NI = NI + 1 where NI is 0 initially. After executing the code, NI stores the number of times that the statement is executed. In the following code, N1, N2, and N3 count the number of times the three assignment statements of Table 2-10 are executed.

```
      N1 = 0
      N2 = 0
      N3 = 0
      DO 10 I = 1,N
         N1 = N1 + 1
10    CONTINUE
      DO 30 J = 1,N
         N2 = N2 + 1
         DO 20 I = 1,N
            N3 = N3 + 1
20       CONTINUE
30    CONTINUE
```

TABLE 2-10 MULTIPLY A MATRIX BY A VECTOR

```
      DO 10 I = 1,N
         X(I) = 0.
10    CONTINUE
      DO 30 J = 1,N
         T = B(J)
         DO 20 I = 1,N
            X(I) = X(I) + A(I,J)*T
20       CONTINUE
30    CONTINUE
```

Of course, the number of executions can also be expressed using the summation notation. In particular,

$$N_1 = \sum_{j=1}^{N} 1, \qquad N_2 = \sum_{j=1}^{N} 1, \qquad N_3 = \sum_{j=1}^{N}\sum_{i=1}^{N} 1. \tag{71}$$

These expressions are obtained by summing over the index range of loops. For example, the N1 assignment is contained in a loop where index I sweeps from 1 to N. Similarly the N3 assignment is contained in two loops. I increments from 1 to N in the inner loop, and J sweeps from 1 to N in the outer loop giving us the double sum in (71). To evaluate N_1, we add the quantity to the right of the summation sign for each i between 1 and N. Adding the quantity 1 for each i between 1 and N just counts the number of integers between 1 and N inclusive so N_1 is N. Similarly, N_2 equals N. To evaluate N_3, we start at the right and work left. The sum over i adds together the quantity 1 for each i between 1 and N, and we just saw that this sum is N. Making this substitution, N_3 is expressed

$$N_3 = \sum_{j=1}^{N}\sum_{i=1}^{N} 1 = \sum_{j=1}^{N} N = N\sum_{j=1}^{N} 1 = NN = N^2.$$

If E_1, E_2, and E_3 denote the respective execution times for the three statements "X(I) = 0," "T = B(J)," and "X(I) = X(I) + A(I,J)*T," the running time for the program in Table 2-10 is

$$T_N = E_1N_1 + E_2N_2 + E_3N_3.$$

Since $N_1 = N_2 = N$ and $N_3 = N^2$, we have

$$T_N = NE_1 + NE_2 + N^2E_3. \tag{72}$$

From a practical viewpoint, the estimate (72) is only important when N is large. (The running time is insignificant when N is small.) Hence let us consider the asymptotic behavior of T_N as N grows. Factoring N^2 from (72) yields

$$T_N = N^2\{E_3 + N^{-1}(E_1 + E_2)\}.$$

As N grows, N^{-1} approaches zero. Neglecting the N^{-1} term, we see that $T_N \approx N^2E_3$. The expression N^2E_3 is the **asymptotic time** for the program in Table 2-10 and the factor N^2 is the **asymptotic parameter**.

The relation $T_N \approx N^2E_3$ slightly underestimates the actual running time since it neglects the loop overhead. A more precise estimate of running time is obtained by replacing the statement execution time E_3 with the loop cycle time. Consider the following loop:

```
      DO 10 K = 5,N
         A(K) = ...
   10 CONTINUE
```
(73)

Each cycle of this loop assigns a new value to A(K), increments K by one, checks that K is less than N, and branches to the start of the loop. The time for one cycle of a loop is called the **loop cycle time**. For the program in Table 2-10, a more precise estimate of the running time T_N is N^2C, where C denotes the cycle time for the loop given below.

```
      DO 10 I = 1,N
         X(I) = X(I) + A(I,J)*T
   10 CONTINUE
```

For the IBM 370 model 3090 computer and the Fortran IV-G compiler, this loop's cycle time is about .5 microseconds so that the asymptotic time to multiply a matrix by a vector using the code in Table 2-10 is about $.5N^2$ microseconds.

To measure the cycle time of the loop (73), one is tempted to replace N by 104, run the code, and divide the running time by 100. This estimate of the cycle time is poor for two reasons. First, a computer's timer is often inaccurate for short time measurements. Since 100 cycles of the loop (73) may take just .0001 second, the time measurement may be relatively inaccurate. Second, if the computer runs several programs simultaneously, the execution time of short programs can vary depending on the state of the machine when the program executes. For these reasons, it is better to embed the loop (73) inside another loop so that the augmented code runs for several seconds. For example, the cycle time of the loop (73) is measured with the following code:

```
      DO 20 L = 1,10000
         DO 10 K = 1,100
            A(K) = ...                                    (74)
10       CONTINUE
20    CONTINUE
```

Since the A(K) assignment is executed 10000 × 100 = 1,000,000 times, the cycle time of the loop (73) is the running time of (74) divided by 1 million. Typically, the loop cycle time does not depend on the index range. For the loop (73), the index range is 5 to N, but when the cycle time is measured in (74), we use the convenient range 1 to 100.

Now let us estimate the time of forward elimination for a general matrix (see Table 2-11). The asymptotic time to multiply a matrix by a vector reduces to the time consumed in the deepest loop. Almost always, the asymptotic time for nested loops reduces to the time consumed in the deepest loop. For the program in Table 2-11, there are two nested loops:

```
      DO 20 J = 1,N - 1
         DO 10 I = J+1,N
10       CONTINUE
20    CONTINUE
```

The deepest loop surrounds the B(I) assignment, and the code's asymptotic running time is the product between the asymptotic parameter and the cycle time for this loop. If N_1 denotes the number of times that B(I) is assigned a

TABLE 2-11 FORWARD ELIMINATION

```
      NM1 = N - 1
      DO 20 J = 1,NM1
         JP1 = J + 1
         T = B(J)
         DO 10 I = JP1,N
            B(I) = B(I) - L(I,J)*T
10       CONTINUE
20    CONTINUE
```

new value, then N_1 can be expressed in terms of the index range of loops containing the B(I) assignment:

$$N_1 = \sum_{j=1}^{N-1} \sum_{i=j+1}^{N} 1.$$

To evaluate this double sum, we start at the right and work left. The sum over i adds together the quantity 1 for each i between $j + 1$ and N. Thus the sum over i just counts the number of integers between $j + 1$ and N inclusive. Since the number of integers between a and b inclusive is b minus a plus one, it follows that

$$\sum_{i=j+1}^{N} 1 = N - (j + 1) + 1 = N - j.$$

Making this substitution, N_1 can be expressed

$$N_1 = \sum_{j=1}^{N-1} \sum_{i=j+1}^{N} 1 = \sum_{j=1}^{N-1} (N - j) = (N - 1) + (N - 2) + \cdots + 2 + 1.$$

A series whose successive terms differ by a constant, independent of the term, is called an **arithmetic series**, and its sum is the first term plus the last term all times the number of terms over 2. The series $1 + 2 + \cdots + N - 1$ is an arithmetic series since the difference in successive terms is one: $2 - 1 = 1$, $3 - 2 = 1$, and so on. The first term is 1 and the last term is $N - 1$, so the first term plus the last term is N. Since the number of terms is $N - 1$, the sum N_1 is

$$N_1 = 1 + 2 + \cdots + (N - 1) = N \times \frac{(N - 1)}{2}.$$

To evaluate the asymptotic limit of N_1, the product $N(N - 1)$ is computed, and every term is dropped except the term with the largest exponent. It follows that

$$N\frac{(N - 1)}{2} = \frac{N^2 - N}{2} \approx \frac{N^2}{2}.$$

Therefore, the asymptotic time for the code in Table 2-11 is $(N^2/2)C$, where C is the cycle time for the loop

```
      DO 10 I = 1,N
         B(I) = B(I) - L(I,J)*T
10    CONTINUE
```

Since this loop is almost identical to the loop associated with the program in Table 2-10, the asymptotic time for forward elimination is about half the asymptotic time to multiply a matrix by a vector.

Finally, let us estimate the time to factor a matrix using the program in Table 2-12. For the program in Table 2-12, the deepest loop surrounds the A(I,J) assignment and the code's asymptotic running time is the product between the asymptotic paramater and the cycle time for this deepest loop. Extracting the index ranges from Table 2-12, the number of times A(I,J) is assigned a new value is given by the triple sum

$$\sum_{k=1}^{N-1} \sum_{j=k+1}^{N} \sum_{i=k+1}^{N} 1. \tag{75}$$

TABLE 2-12 COLUMN ORIENTED ROW ELIMINATION

```
      NM1 = N - 1
      DO 20 K = 1,NM1
         KP1 = K + 1
         T = A(K,K)
         DO 10 I = KP1,N
10          A(I,K) = A(I,K)/T
         DO 20 J = KP1,N
            T = A(K,J)
            DO 20 I = KP1,N
               A(I,J) = A(I,J) - A(I,K)*T
20    CONTINUE
```

The sum over i just counts the number of integers between $k + 1$ and N. Replacing this sum by $N - k$, we have

$$\sum_{k=1}^{N-1} \sum_{j=k+1}^{N} \sum_{i=k+1}^{N} 1 = \sum_{k=1}^{N-1} \sum_{j=k+1}^{N} (N - k) = \sum_{k=1}^{N-1} (N - k) \sum_{j=k+1}^{N} 1$$

$$= \sum_{k=1}^{N-1} (N - k)^2 = (N - 1)^2 + (N - 2)^2 + \cdots + 2^2 + 1^2.$$

To sum this series, we use the formula for sums of squares:

$$1^2 + 2^2 + \cdots + m^2 = \frac{m(m + 1)(2m + 1)}{6}.$$

In our case, m is $N - 1$, so the sum (75) is equal to $N(N - 1)(2N - 1)/6$. To evaluate the asymptotic parameter, the product $N(N - 1)(2N - 1)$ is computed and every term is dropped except the term with the largest exponent, giving

$$\frac{N(N - 1)(2N - 1)}{6} = \frac{2N^3 - 3N^2 + N}{6} \approx \frac{N^3}{3}.$$

Therefore, the asymptotic time for the code in Table 2-12 is $(N^3/3)C$, where C is the cycle time for the loop

```
      DO 20 I = 1,N
         A(I,J) = A(I,J) - A(I,K)*T
20    CONTINUE
```

Each of the examples analyzed above possesses a unique deepest loop. The following code illustrates a program where asymptotic parameters associated with two different loops have the same exponent.

```
      DO 10 I = 1,N
        .
10    CONTINUE
      DO 20 I = 1,N+N                  (76)
        .
20    CONTINUE
```

The asymptotic parameters corresponding to the loops in (76) are N and $2N$, respectively, and the asymptotic time for the code is $T_N = NC_{10} + 2NC_{20}$, where C_{10} and C_{20} are the loop cycle times. Although the code (76) executes $3N$ cycles, the time for each of the first N cycles may be different from the time for each of the last $2N$ cycles. When different loops have different cycle times, we can define an effective cycle time in the following way: For the code (76), we factor $3N$ from the asymptotic time to obtain

$$T_N = 3N(\tfrac{1}{3}C_{10} + \tfrac{2}{3}C_{20}).$$

Defining the **effective cycle time** $C = \frac{1}{3}C_{10} + \frac{2}{3}C_{20}$, the execution time T_N equals the product $3NC$ between the asymptotic parameter $3N$ and the effective cycle time C.

Table 2-13 gives asymptotic parameters for the programs developed in Sections 2-3 and 2-4, and Table 2-14 gives loop cycle times for an IBM 370 model 3090 computer. The asymptotic running time is the product between the asymptotic parameter and the loop cycle time. The table heading "Solve" means forward and back solve a factored system. The .65-microsecond entries in Table 2-14 correspond to the cycle time for the loop

```
      DO 10 I = 1,N
         A(I) = A(I) - T*A(I+J)                        (77)
10    CONTINUE
```

which is the deepest loop in many programs. For programs that forward and back solve tridiagonal systems, the cycle times are really effective times since these programs have two different loops with the same asymptotic parameter. In factoring an unsymmetric tridiagonal matrix, the overhead associated with pivoting is significant, and the running time depends on the number of row interchanges. Without a pivot, the cycle time is 7.1 microseconds. With a pivot,

TABLE 2-13 ASYMPTOTIC PARAMETERS FOR A $N \times N$ MATRIX

Matrix type	Factor	Solve	Invert
General	$\frac{1}{3}n^3$	n^2	n^3
Tridiagonal (with pivots)	n	$2n$	$\frac{3}{2}n^2$
(l, u)-band (row elimination, row pivots)	$l(l + u)n$	$(u + 2l)n$	$\frac{1}{2}(2u + 3l)n^2$
Symmetric	$\frac{1}{6}n^3$	n^2	$\frac{1}{2}n^3$
Symmetric h-band	$\frac{1}{2}h^2n$	$2hn$	$\frac{1}{2}(h + 2n)hn$
Symmetric tridiagonal (no pivots)	n	$2n$	n^2

TABLE 2-14 CYCLE TIMES IN MICROSECONDS FOR THE IBM 370 MODEL 3090 COMPUTER AND THE FORTRAN IV-G COMPILER

Matrix type	Factor	Solve	Invert
General	.65	.65	.65
Tridiagonal	8.51	1.57	1.60
(l, u)-band	.65	.65	.65
Symmetric	.65	.65	.65
Symmetric h-band	.65	.65	.65
Symmetric tridiagonal	2.69	.79	.79

the cycle time is 9.9 microseconds. The entry 8.51 in Table 2-14 reflects the "average" tridiagonal matrix which requires a pivot every other elimination step.

Exercises

2-6.1. For each of the following codes, obtain a formula for the value of L in terms of N. What is the asymptotic parameter for the deepest loop in each code?

(a)

```
      L = 0
      M = 2*N
      DO 10 I = 1,N
      DO 10 J = I,M
         .
      L = L + 1
   10 CONTINUE
```

(b)

```
      L = 0
      DO 10 K = 1,N
      DO 10 J = 1,K
      DO 10 I = 1,J
         .
      L = L + 1
   10 CONTINUE
```

(c)

```
      L = 0
      DO 10 K = 1,N
      DO 10 J = K,N
      DO 10 I = 1,J
         .
      L = L + 1
   10 CONTINUE
```

2-6.2. Using a nonoptimizing compiler (such as a WATFIV compiler), evaluate the cycle time for each of the following two loops.

Loop 1

```
      DO 10 I = 2,N
         T = (I-1)*3.14159/(2.*FLOAT(N))
         U = (1+DT)*OLD(I+1)+OLD(I-1)
         S = U-(DT**2)*SIN(T)
         NEW(I) = S/(2.+DT-DT**2)
   10 CONTINUE
```

Loop 2

```
      DO 10 I = 2,N
         T = D*OLD(I+1)+OLD(I-1)
         NEW(I) = C*(T-B(I))
   10 CONTINUE
```

Of course, you must assign values to the variables OLD(I) and DT before running loop 1, and you must assign values to the variables OLD(I), B(I), C, and D before running loop 2. In general, what values for B(I), C, and D will make these loops equivalent? In other words, what values for B(I), C, and D will make the value of NEW(I) computed by loop 1 equal to the value of NEW(I) computed by loop 2. (To estimate the cycle time, set all the variables equal to zero.)

2-6.3. What is the asymptotic parameter corresponding to Aasen's algorithm (56)?

2-7. REPEATED SOLVES

In some applications, we wish to solve a linear system several times; each time the coefficient matrix **A** is the same, but the right side **b** is different. Since the solution to $\mathbf{Ax} = \mathbf{b}$ is $\mathbf{x} = \mathbf{A}^{-1}\mathbf{b}$, one is tempted to compute $\mathbf{A}^{-1}$ and multiply by each right side **b**. However, in almost every case, this is the wrong approach. It is better to factor **A** initially and then for each right side, solve the factored system. To contrast these two different approaches, we study a specific application. Let us return to the system (28) of Chapter 1. These $N - 1$ equations have the form

$$\begin{aligned}
2x_1 - x_2 \qquad\qquad &= (\Delta t)^2 e^{-x_1},\\
-x_1 + 2x_2 - x_3 \qquad &= (\Delta t)^2 e^{-x_2},\\
- x_2 + 2x_3 - x_4 &= (\Delta t)^2 e^{-x_3},\\
&\ \ \vdots\\
- x_{n-1} + 2x_n &= (\Delta t)^2 e^{-x_n},
\end{aligned}$$

where $n = N - 1$ and $\Delta t = 1/N$. Using matrix notation, this system is written

$$\begin{bmatrix} 2 & -1 & & & \\ -1 & 2 & \cdot & & \\ & \cdot & \cdot & \cdot & \\ & & \cdot & \cdot & -1 \\ & & & -1 & 2 \end{bmatrix} \begin{bmatrix} x_1 \\ x_2 \\ \vdots \\ x_n \end{bmatrix} = (\Delta t)^2 \begin{bmatrix} e^{-x_1} \\ e^{-x_2} \\ \vdots \\ e^{-x_n} \end{bmatrix}. \tag{78}$$

Letting **A** denote the coefficient matrix, **x** denote the unknown vector, and **b(x)** denote the right side, equation (78) can be expressed $\mathbf{Ax} = \mathbf{b}(\mathbf{x})$. This equation is nonlinear since the right side depends on the unknown vector **x**. We propose the following iterative technique for computing **x**: Starting from the initial guess $\mathbf{x}^{\text{old}} = \mathbf{0}$, generate iterations by the rule

$$\mathbf{Ax}^{\text{new}} = \mathbf{b}(\mathbf{x}^{\text{old}}). \tag{79}$$

For the first iteration, $\mathbf{x}^{\text{old}}$ is **0**, $\mathbf{b}(\mathbf{x}^{\text{old}})$ is $\mathbf{b}(\mathbf{0})$, and the new **x** is the solution to the tridiagonal system $\mathbf{Ax}^{\text{new}} = \mathbf{b}(\mathbf{0})$. For the next iteration, **b(x)** is evaluated at the current **x** and another tridiagonal system is solved to obtain the new **x**. Since $\mathbf{x}^{\text{old}}$ changes each iteration, the right side $\mathbf{b}(\mathbf{x}^{\text{old}})$ changes each iteration. Since **A** does not depend on **x**, the coefficient matrix is fixed during the iterations.

As stated above, we will compare two different approaches for implementing the iteration (79):

1. Invert **A**, then multiply each right side by the inverse.
2. Factor **A**, then for each right side, solve the factored system.

Referring to Section 2-6, the time to invert a $n \times n$ symmetric tridiagonal matrix is n^2C_1, while the time to multiply a matrix by a vector is n^2C_2, where C_1 and C_2 are the relevant loop cycle times. Therefore, using approach 1, the time for m iterations of (79) is $(C_1 + mC_2)n^2$. On the other hand, the time to factor a symmetric tridiagonal matrix is nC_3 and the time to solve the factored system is $2nC_4$, where C_3 and C_4 are the relevant loop cycle times. Using approach 2, the time for m iterations of (79) is $(C_3 + 2mC_4)n$. Since the time for approach 1 is proportional to n^2, while the time for approach 2 is proportional to n, approach 2 is much faster than approach 1 when n is large.

In almost every case, factoring the matrix and solving the factored system is faster than inverting the matrix and multiplying by the inverse. The reason for the superiority of the factorization approach is that the factorization process often preserves zeros while inversion does not. For illustration, we present the inverse of a specific tridiagonal matrix whose structure is similar to that of the coefficient matrix in (78):

$$\begin{bmatrix} 1 & -1 & 0 & 0 \\ -1 & 2 & -1 & 0 \\ 0 & -1 & 2 & -1 \\ 0 & 0 & -1 & 2 \end{bmatrix}^{-1} = \begin{bmatrix} 4 & 3 & 2 & 1 \\ 3 & 3 & 2 & 1 \\ 2 & 2 & 2 & 1 \\ 1 & 1 & 1 & 1 \end{bmatrix}.$$

Observe that the inverse matrix is completely nonzero even though the original matrix is tridiagonal. For a sparse matrix, the many zeros encountered in the factorization and the solution process can be skipped to speed up the algorithm. The inversion process, on the other hand, eliminates zeros.

As we have seen, solving several systems of equations using the inverse matrix is often inefficient. In fact, as we now show, the inverse matrix itself is often computed by solving factored systems. Let us define the vectors

$$\mathbf{e}_1 = \begin{bmatrix} 1 \\ 0 \\ 0 \\ \vdots \\ 0 \end{bmatrix}, \quad \mathbf{e}_2 = \begin{bmatrix} 0 \\ 1 \\ 0 \\ \vdots \\ 0 \end{bmatrix}, \quad \cdots.$$

That is, $\mathbf{e}_i$ is the vector with every component zero except for the ith component which is 1. The solution to $\mathbf{Ax} = \mathbf{e}_1$ is $\mathbf{x} = \mathbf{A}^{-1}\mathbf{e}_1$. Since multiplying $\mathbf{A}^{-1}$ by $\mathbf{e}_1$ yields the first column of $\mathbf{A}^{-1}$, the solution to $\mathbf{Ax} = \mathbf{e}_1$ is the first column of $\mathbf{A}^{-1}$. Similarly, the ith column of $\mathbf{A}^{-1}$ is the solution to the equation

$$\mathbf{Ax} = \mathbf{e}_i. \tag{80}$$

As i increments from 1 to n, the solution to (80) yields the n columns of $\mathbf{A}^{-1}$. As discussed earlier, an efficient way to solve these n systems is to factor $\mathbf{A}$ and solve a factored system corresponding to each right side $\mathbf{e}_1, \mathbf{e}_2, \cdots, \mathbf{e}_n$. The code in Table 2-15 uses subroutine FACT to factor $\mathbf{A}$ and subroutine SOLVE to replace the columns of array B with the columns of $\mathbf{A}^{-1}$.

Observe that the arguments of SOLVE in Table 2-15 are A and B(1,J). If the vector $\mathbf{b}$ is stored in B's column J, then the statement

```
CALL SOLVE(B(1,J),A,B(1,J))
```

TABLE 2-15 MATRIX INVERSION

```
      CALL FACT(A,N)
      DO 20 J = 1,N
         DO 10 I = 1,N
10          B(I,J) = 0.
         B(J,J) = 1.
         CALL SOLVE(B(1,J),A,B(1,J))
20    CONTINUE
```

overwrites column J with the solution to the linear system $\mathbf{Ax} = \mathbf{b}$. As J increments from 1 to N in Table 2-15, the columns of $\mathbf{A}^{-1}$ overwrite the columns of B. When **A** has special structure, the special subroutines developed in Section 2-4 can be used in place of FACT and SOLVE. For example, if **A** is a band matrix, it is more efficient to use BFACT and BSOLVE rather than FACT and SOLVE.

Since the code in Table 2-15 factors **A**, then solves n factored systems, it appears that the asymptotic parameter for inverting a matrix is equal to the parameter for factoring a matrix plus n times the parameter for solving a factored system. Since the asymptotic parameters for FACT and SOLVE are $\frac{1}{3}n^3$ and n^2, respectively, the parameter for inversion would be $\frac{1}{3}n^3 + nn^2 = \frac{4}{3}n^3$. This analysis overlooks an important point. Since the vectors $\mathbf{e}_i$ have many zeros, some forward elimination steps can be skipped. A careful analysis reveals that the asymptotic parameter for solving these n special systems is $\frac{2}{3}n^3$, not n^3, and the asymptotic parameter corresponding to the code in Table 2-15 is $\frac{1}{3}n^3 + \frac{2}{3}n^3 = n^3$.

The code in Table 2-15 is a little disappointing since two arrays A and B are required. For completeness, NAPACK includes a special set of subroutines (with names like VERT, TVERT, and BVERT) to invert a matrix. When possible, the original matrix is replaced by its inverse so that no additional storage is required.

Exercises

2-7.1. Implement the iteration (79). Start with the program developed in Section 1-6 and insert arrays D and U to store the diagonal and off-diagonal elements of the coefficient matrix in (78). Call subroutine TFACT in the initialization segment of your program and call subroutine TSOLVE during each iteration. Run your program taking N = 10.

2-7.2. Given a $n \times n$ matrix **A** and a $n \times m$ matrix **B**, explain how to compute the matrix-matrix product $\mathbf{A}^{-1}\mathbf{B}$ without computing the inverse of **A**. Using subroutines TFACT and TSOLVE, write a program to compute $\mathbf{A}^{-1}\mathbf{B}$ when **A** is tridiagonal.

2-7.3. Compare approach 1 and approach 2 for solving multiple systems when the coefficient matrix is general, with no special structure.

2-8. RANK 1 CHANGES

If $\mathbf{u}$ and $\mathbf{v}$ are column vectors, then $\mathbf{u}^T$ is a row vector and the product $\mathbf{u}^T\mathbf{v}$ between a row vector and a column vector is a scalar, the dot product. For example,

$$[1 \quad 2 \quad 3]\begin{bmatrix}4\\1\\2\end{bmatrix} = 1{\times}4 + 2{\times}1 + 3{\times}2 = 12.$$

On the other hand, the product $\mathbf{uv}^T$ between a column vector and a row vector is a matrix whose (i, j) element is $u_i v_j$. For example,

$$\begin{bmatrix}1\\2\\3\end{bmatrix}[4 \quad 1 \quad 2] = \begin{bmatrix}1\times4 & 1\times1 & 1\times2\\ 2\times4 & 2\times1 & 2\times2\\ 3\times4 & 3\times1 & 3\times2\end{bmatrix} = \begin{bmatrix}4 & 1 & 2\\ 8 & 2 & 4\\ 12 & 3 & 6\end{bmatrix}.$$

As this example illustrates, each column of the matrix $\mathbf{uv}^T$ is a multiple of the vector $\mathbf{u}$: The first column is v_1 times $\mathbf{u}$, the second column is v_2 times $\mathbf{u}$, and so on.

Suppose that $\mathbf{A}$ and $\mathbf{B}$ are square matrices, and $\mathbf{u}$ and $\mathbf{v}$ are vectors with the property that

$$\mathbf{A} = \mathbf{B} - \mathbf{uv}^T.$$

The **Sherman-Morrison formula** relates the inverse of $\mathbf{A}$ to the inverse of $\mathbf{B}$:

$$\mathbf{A}^{-1} = \mathbf{B}^{-1} + \alpha\mathbf{B}^{-1}\mathbf{uv}^T\mathbf{B}^{-1}, \tag{81}$$

where $\alpha = 1/(1 - \mathbf{v}^T\mathbf{B}^{-1}\mathbf{u})$. Thus if $\mathbf{B}$ has an inverse and $\mathbf{v}^T\mathbf{B}^{-1}\mathbf{u} \neq 1$, then $\mathbf{A}$ has an inverse which can be expressed in terms of $\mathbf{B}^{-1}$. You will verify (81) in Exercise 2-8.1.

The Sherman-Morrison formula is a helpful tool for solving systems with the property that a change in a row or column of the coefficient matrix generates a "nice structure." For example, consider the bordered system (60). The coefficient matrix can be expressed as the sum of an upper triangular matrix and a matrix that is completely zero except for the first column:

$$\begin{bmatrix}1 & 2 & 2 & 2\\ 1 & 1 & 0 & 0\\ 1 & 0 & -1 & 0\\ 1 & 0 & 0 & 1\end{bmatrix} = \begin{bmatrix}1 & 2 & 2 & 2\\ 0 & 1 & 0 & 0\\ 0 & 0 & -1 & 0\\ 0 & 0 & 0 & 1\end{bmatrix} + \begin{bmatrix}0 & 0 & 0 & 0\\ 1 & 0 & 0 & 0\\ 1 & 0 & 0 & 0\\ 1 & 0 & 0 & 0\end{bmatrix}.$$

The latter matrix can be expressed in the form $\mathbf{uv}^T$:

$$\begin{bmatrix}0 & 0 & 0 & 0\\ 1 & 0 & 0 & 0\\ 1 & 0 & 0 & 0\\ 1 & 0 & 0 & 0\end{bmatrix} = \begin{bmatrix}0\\1\\1\\1\end{bmatrix}[1 \quad 0 \quad 0 \quad 0].$$

Hence, the Sherman-Morrison formula can be applied and $\mathbf{B}$ is upper triangular.

For a second example, let us return to Problem 1-9, where a boundary-value problem with periodic boundary conditions is approximated using finite differences. The coefficient matrix for the finite difference approximation has the structure of the matrix in Figure 2-8 (although the numerical values of the nonzero coefficients in Problem 1-9 and in Figure 2-8 are different).

$$\mathbf{A} = \begin{bmatrix}3 & -1 & & & & -1\\ -1 & 4 & -1 & & & \\ & -1 & 4 & \cdot & & \\ & & \cdot & \cdot & \cdot & \\ & & & \cdot & 4 & -1\\ -1 & & & & -1 & 3\end{bmatrix}$$

Figure 2-8 Coefficient structure for periodic boundary conditions.

Notice that $\mathbf{A}$ is tridiagonal except for two corner elements. If $\mathbf{B}$ is the tridiagonal matrix with 4's on the diagonal and -1's off the diagonal, and

$$\mathbf{u} = \mathbf{v} = \begin{bmatrix} 1 \\ 0 \\ \cdot \\ \cdot \\ \cdot \\ 0 \\ 1 \end{bmatrix},$$

then we have $\mathbf{A} = \mathbf{B} - \mathbf{u}\mathbf{v}^T$. The Sherman-Morrison formula can be applied to $\mathbf{A}$, and $\mathbf{B}$ is a tridiagonal matrix.

To utilize (81), it appears that $\mathbf{B}^{-1}$ must be computed. We now show how to circumvent the computation of $\mathbf{B}^{-1}$ when solving $\mathbf{A}\mathbf{x} = \mathbf{b}$. Since $\mathbf{x} = \mathbf{A}^{-1}\mathbf{b}$, it follows from (81) that

$$\begin{aligned} \mathbf{x} = \mathbf{A}^{-1}\mathbf{b} &= (\mathbf{B}^{-1} + \alpha\mathbf{B}^{-1}\mathbf{u}\mathbf{v}^T\mathbf{B}^{-1})\mathbf{b} \\ &= \mathbf{B}^{-1}\mathbf{b} + \alpha\mathbf{B}^{-1}\mathbf{u}\mathbf{v}^T\mathbf{B}^{-1}\mathbf{b} \\ &= \mathbf{B}^{-1}\mathbf{b} + \alpha(\mathbf{v}^T\mathbf{B}^{-1}\mathbf{b})\mathbf{B}^{-1}\mathbf{u} \\ &= \mathbf{B}^{-1}\mathbf{b} + \beta\mathbf{B}^{-1}\mathbf{u}, \end{aligned}$$

where $\beta = \alpha(\mathbf{v}^T\mathbf{B}^{-1}\mathbf{b}) = \mathbf{v}^T\mathbf{B}^{-1}\mathbf{b}/(1 - \mathbf{v}^T\mathbf{B}^{-1}\mathbf{u})$. Defining $\mathbf{y} = \mathbf{B}^{-1}\mathbf{b}$ and $\mathbf{z} = \mathbf{B}^{-1}\mathbf{u}$, $\mathbf{x}$ is expressed as

$$\mathbf{x} = \mathbf{A}^{-1}\mathbf{b} = \mathbf{B}^{-1}\mathbf{b} + \beta\mathbf{B}^{-1}\mathbf{u} = \mathbf{y} + \beta\mathbf{z},$$

where

$$\beta = \frac{\mathbf{v}^T\mathbf{B}^{-1}\mathbf{b}}{1 - \mathbf{v}^T\mathbf{B}^{-1}\mathbf{u}} = \frac{\mathbf{v}^T\mathbf{y}}{1 - \mathbf{v}^T\mathbf{z}}.$$

Also, the definitions of $\mathbf{y}$ and $\mathbf{z}$ tell us that $\mathbf{B}\mathbf{y} = \mathbf{b}$ and $\mathbf{B}\mathbf{z} = \mathbf{u}$. Combining these observations, we conclude that $\mathbf{A}\mathbf{x} = \mathbf{b}$ is solved in four steps:

1. Factor $\mathbf{B}$.
2. Solve $\mathbf{B}\mathbf{y} = \mathbf{b}$ for the unknown $\mathbf{y}$ and solve $\mathbf{B}\mathbf{z} = \mathbf{u}$ for the unknown $\mathbf{z}$.
3. Compute the scalar $\beta = \mathbf{v}^T\mathbf{y}/(1 - \mathbf{v}^T\mathbf{z})$.
4. Set $\mathbf{x} = \mathbf{y} + \beta\mathbf{z}$.

When applying the Sherman-Morrison formula, we often think of $\mathbf{u}\mathbf{v}^T$ as a perturbation of the matrix $\mathbf{B}$. The Sherman-Morrison formula can be generalized to handle more complicated perturbations. If $\mathbf{A} = \mathbf{B} - \mathbf{U}\mathbf{V}^T$, where $\mathbf{B}$ is $n \times n$ and both $\mathbf{U}$ and $\mathbf{V}$ are $n \times m$, then the **Woodbury formula** states that

$$\mathbf{A}^{-1} = \mathbf{B}^{-1} + \mathbf{B}^{-1}\mathbf{U}(\mathbf{I} - \mathbf{V}^T\mathbf{B}^{-1}\mathbf{U})^{-1}\mathbf{V}^T\mathbf{B}^{-1}.$$

Here $\mathbf{I}$ denotes the **identity matrix**, the diagonal matrix whose diagonal elements are 1. If $\mathbf{U}$ and $\mathbf{V}$ are vectors, then the Woodbury formula reduces to the Sherman-Morrison formula since $(\mathbf{I} - \mathbf{V}^T\mathbf{B}^{-1}\mathbf{U})^{-1} = (1 - \mathbf{v}^T\mathbf{B}^{-1}\mathbf{u})^{-1} = \alpha$, which is a scalar that can be factored in front of $\mathbf{B}^{-1}\mathbf{U}\mathbf{V}^T\mathbf{B}^{-1} = \mathbf{B}^{-1}\mathbf{u}\mathbf{v}^T\mathbf{B}^{-1}$. The Woodbury formula is useful whenever a change in a few columns or rows of $\mathbf{A}$ produces a nice structure.

Exercises

2-8.1. Verify the Sherman-Morrison formula (80). [*Hint:* To prove that a matrix $\mathbf{X}$ is the inverse of $\mathbf{A}$, you have to show that $\mathbf{AX} = \mathbf{I}$. Hence, to verify the Sherman-Morrison formula, confirm that the product between $\mathbf{A} = \mathbf{B} - \mathbf{uv}^T$ and the right side of (80) equals $\mathbf{I}$. The product you get can be simplified since $\mathbf{v}^T\mathbf{B}^{-1}\mathbf{u}$ is a scalar.]

2-8.2. As noted above, the finite-difference system derived in Problem 1-9 can be written in the form $\mathbf{Ax} = \mathbf{b}$, where $\mathbf{A}$ is tridiagonal except for the two nonzero coefficients a_{1n} and a_{n1}. Solve this finite-difference system using the Sherman-Morrison formula and subroutines TFACT and TSOLVE.

2-8.3. Suppose that $\mathbf{A}$ is $l \times m$ and $\mathbf{B}$ is $m \times n$. If $\mathbf{a}_k$ denotes column k of $\mathbf{A}$ and $\mathbf{b}_k{}^T$ denotes row k of $\mathbf{B}$, then show that

$$\mathbf{AB} = \sum_{k=1}^{m} \mathbf{a}_k \mathbf{b}_k{}^T.$$

2-8.4. Verify the Woodbury formula.

2-8.5. In the special case where $\mathbf{A} = \mathbf{B} - \mathbf{UV}^T$, develop an algorithm to compute $\mathbf{A}^{-1}\mathbf{b}$ similar to the algorithm developed above for the Sherman-Morrison formula. In particular, explain how to evaluate the product $\mathbf{V}^T\mathbf{B}^{-1}\mathbf{b}$.

2-9. DETERMINANTS

Determinants play an important role in mathematical fields such as algebra and analysis. For example, if we make a change of variables in a multidimensional integral, a determinant associated with the transformation appears in the transformed integral. Also, the solution to a linear system $\mathbf{Ax} = \mathbf{b}$ can be expressed in terms of the ratio of two determinants: By Cramer's rule, x_i is the ratio between the determinant of the matrix obtained by substituting $\mathbf{b}$ for the ith column of $\mathbf{A}$ and the determinant of $\mathbf{A}$. Although determinants are very important from a theoretical viewpoint, they seem to be less significant from a computational viewpoint. For comparison, the computational effort involved in computing *one* determinant is roughly equivalent to the effort required to factor a matrix. Consequently, computing the solution to a linear system using Cramer's rule is about n times slower than Gaussian elimination when the coefficient matrix is $n \times n$.

We will discuss two different but equivalent techniques for evaluating a determinant. First, a determinant is a sum of signed products of coefficients. We form all possible products of coefficients with at most one coefficient taken from each column and from each row, then we sum these products after adjusting the sign in the following way: If $b_1 b_2 \cdots b_n$ denotes a typical product, where b_i is a coefficient from row i of $\mathbf{A}$, and if $c_1, c_2, \cdots, c_n$ are the columns of $\mathbf{A}$ corresponding to each factor in the product, then the sign is either plus or minus depending on whether the number of permutations of adjacent elements in the sequence $c_1, c_2, \cdots, c_n$ needed to obtain the ordering $1, 2, \cdots, n$ is even or odd, respectively.

For illustration, let us consider the 2×2 matrix

$$\mathbf{A} = \begin{bmatrix} 1 & 3 \\ 2 & 4 \end{bmatrix}. \tag{82}$$

There are two different products, (1)(4) and (3)(2), consisting of one coefficient from each column and one coefficient from each row. Since the column sequence corresponding to the product (1)(4) is {1, 2}, no permutations are needed to obtain the ordering {1, 2} and the sign is plus. Since the column sequence corresponding to the product (3)(2) is {2, 1}, one permutation is needed to obtain the ordering {1, 2} and the sign is minus. Hence we have det **A** = (1)(4) − (3)(2) = −2. For the general 2 × 2 matrix

$$\mathbf{A} = \begin{bmatrix} a & c \\ b & d \end{bmatrix},$$

det **A** = $ad - bc$, the product between the diagonal elements minus the product between the off-diagonal elements.

This procedure for evaluating a determinant can be time consuming when n is large since $n!$ products are computed altogether. A more efficient technique to compute a determinant is based on three observations: (a) the determinant of a triangular matrix is the product of its diagonal elements, (b) an elimination step does not change the value of a determinant, and (c) a row interchange or a column interchange simply reverses the sign of a determinant. The fact that the determinant of a triangular matrix is the product of its diagonal elements follows directly from the definition of a determinant; the determinant is a sum of signed products and each product contains the factor zero except possibly the product formed from the diagonal elements. Although we do not prove properties (b) and (c), a good reference is Strang's book [S15].

Based on these observations, a determinant can be evaluated in the following way: Reduce the coefficient matrix **A** to upper triangular form **U** using partial pivoting. Then the determinant of **A** is plus or minus the product of **U**'s diagonal elements:

$$\det \mathbf{A} = (-1)^p u_{11} u_{22} \cdots u_{nn},$$

where p is the number of times two different rows are interchanged during the elimination process. Functions DET, BDET, SDET, and so on use this procedure to evaluate the determinant of a matrix. The input for one of these subroutines is a factored matrix and the output is the determinant. The following program computes the determinant of the matrix (82):

```
REAL A(10)
DATA A/1.,2.,3.,4./
CALL FACT(A,2,2)
D = DET(I,A)
WRITE(6,*) D
END
```

In addition to the factored array A, note that the parameter list for DET above also includes the variable I. The determinant of a matrix is potentially very large or very small in absolute value. If the computation of the determinant is in danger of generating an overflow or underflow, the determinant is divided by a power of 10 to bring its magnitude closer to 1 and the exponent of 10 is stored in argument I of DET. Hence, when the magnitude of the determinant is exceedingly large or small, its actual value is DET*10**I. In the program above, the argument I of DET is ignored since I is zero. When the returned value of I is nonzero, be sure to print DET and I, not DET*10.**I, which will overflow or underflow.

Exercises

2-9.1. Compute the determinant of the following matrix using both the factorization method and the signed product method:

$$\begin{bmatrix} 1 & 4 & 7 \\ 2 & 5 & 8 \\ 3 & 6 & 9 \end{bmatrix}.$$

2-9.2. Use the signed product method to derive a formula for the determinant of a general 3×3 matrix.

2-9.3. Let **A** be the $n \times n$ tridiagonal matrix with every diagonal element equal to .5 and with every off-diagonal element equal to .25. Using subroutine TFACT and function TDET, compute the determinant of **A** when n is 40, 80, 160, and 320. As n increases, what does det **A** approach? Then let **A** be the tridiagonal matrix with every diagonal element equal to 8 and with every off-diagonal element equal to 4. Again compute the determinant of **A** when n is 40, 80, 160, and 320. What does det **A** approach as n increases?

2-10. VECTOR PROCESSING

The development of hardware to process arrays of numbers has led to substantial reductions in the cycle time associated with algorithms for matrix computations. Let us consider a code that adds the respective components of two arrays A and B and stores the result in a third array C:

```
          DO 10 I = 1,1000
   10        C(I) = A(I) + B(I)
```

In executing this code, a scalar computer loads the value of A(I) into a register, loads the value of B(I) into a register, performs the addition, stores the sum in C(I), increments I, and branches to the start of the loop if I is less than 1000. Also, additional arithmetic may be needed to determine the location in memory of the array elements. If the start of each array is stored in a separate register, then I is added to each of these registers to determine the memory locations for A(I), B(I), and C(I). On a scalar computer, each instruction is executed sequentially and the execution of an instruction does not start until the execution of the previous instruction is complete.

A computer with a vector processor has hardware which facilitates operations between vectors. Instead of waiting for one instruction to complete before starting to execute the next instruction, the vector hardware performs many tasks simultaneously. An important component of the vector hardware are "pipes" which quickly transport arrays through memory. Since data reach and leave the arithmetic unit quickly, the arithmetic unit is utilized efficiently and there is little idle time in which the arithmetic unit waits for data to process. For an array with 1000 elements, the vector processor may be 10 or 100 times faster than the corresponding scalar processor. For short vectors there may be no speed-up since there is a fixed overhead, independent of the array length, associated with the start up of an array processor. But for a long array, this fixed overhead is insignificant and the running time is proportional to the vector length. A vector processor should not be confused with a parallel processor. When the array A is added to B as in the example above, the array elements are added together sequentially, one after the other. The effective speed-up in the vector processor is related to the speed with which data reach and leave the adder.

When implementing the algorithms developed in the previous sections using a vector processor, we must be sure to invoke the vector hardware inside the deepest loops. For example, let us consider the code in Table 2-5 for row elimination by columns. A vector version of this code is the following:

```
      NM1 = N - 1
      DO 20 K = 1,NM1
        KP1 = K + 1
        T = 1./A(K,K)
C
C     *** COMPUTE MULTIPLIERS ***
C
          A(KP1:N,K) = T*A(KP1:N,K)
          DO 20 J = KP1,N
C
C     *** ELIMINATE BY COLUMNS ***
C
20              A(KP1:N,J) = A(KP1:N,J) - A(K,J)*A(KP1:N,K)
```

The statement "A(KP1:N,K) = T*A(KP1:N,K)" is equivalent to multiplying elements KP1 through N of A's column K by T. Similarly, the statement "A(KP1:N,J) = A(KP1:N,J) − A(K,J)*A(KP1:N,K)" is equivalent to multiplying elements KP1 through N of A's column K by A(K,J) and subtracting the result from the corresponding elements in column J. An optimizing compiler can accomplish some vectorization on its own. Note though that for some applications, a vectorized program is simpler and easier to read than the scalar version so there may be an advantage to coding with vector operations. In other applications, even a sophisticated optimizing compiler will not be able to vectorize the code and the programmer must arrange the storage structure and the arithmetic operations to take advantage of the vector processor. The article [D11] by Dongarra, Gustavson, and Karp is an excellent reference for vector processing of "dense" matrices.

Exercises

2-10.1. Code the following iteration using vector operations:

$$x_i^{\text{new}} = \tfrac{1}{2}(x_{i-1}^{\text{old}} + x_{i+1}^{\text{old}}), \qquad 1 < i < n.$$

2-10.2. Code the following iteration using vector operations:

$$x_{ij}^{\text{new}} = \tfrac{1}{4}(x_{i-1,j}^{\text{old}} + x_{i+1,j}^{\text{old}} + x_{i,j-1}^{\text{old}} + x_{i,j+1}^{\text{old}}), \qquad 1 < i < n,\ 1 < j < n.$$

When writing your code, assume that x_{ij} is stored in a single-subscripted array and employ vector operations which process vectors with length on the order of n^2 rather than vectors with length on the order of n. (By processing long vectors rather than short vectors, the cost associated with the array processor start-up is reduced.) Also, assume that there exists a vector subroutine

MASK(V1,V2,IV,U,N)

that sets U(I) = V1(I) if IV(I) = 0 and U(I) = V2(I) if IV(I) = 1 for I between 1 and N.

2-10.3. Suppose that the subdiagonal, diagonal, and superdiagonal of a tridiagonal matrix A are stored in three separate arrays L, D, and U, respectively. Write a code that uses vector operations to compute the matrix-vector product **Ax**. [*Note:* The statement "A(1:N) = B(1:N)*C(1:N)" stores in array A the product between respective elements of B and C.]

2-11. PARALLEL PROCESSING

Loosely speaking, a parallel computer has the ability to perform several different processes at the same time. The architecture of parallel computers falls into two classes: Single-instruction multiple data systems (SIMD), where each processor executes the same instruction simultaneously using different data, and multiple-instruction multiple data systems (MIMD), where each processor executes (possibly) different instructions using different data. Parallel computers are also classified according to their memory configuration. In a shared memory machine such as the Alliant FX/8, the Encore Multimax, or the Sequent Balance 21000, each processor can access a common memory, while in a distributed memory machine such as a hypercube, each processor has its own memory which cannot be accessed directly by the other processors. With distributed memory machines, data and instructions must be passed between the processors using communication links and the output from each processor must be combined to yield the final result. With a hypercube computer, for example, each of the n processors is linked to $\log_2 n$ neighbors.

To contrast parallel processing with vector processing, let us consider the elimination segment of the program in Table 2-5:

```
            DO 20 J = KP1,N
              DO 20 I = KP1,N
   20           A(I,J) = A(I,J) - A(K,J)*A(I,K)
```

When vectorizing this code, the deepest loop is replaced by a vector operation to obtain

```
      DO 20 J = KP1,N
 20     A(KP1:N,J) = A(KP1:N,J) - A(K,J)*A(KP1:N,K)
```

Since the index J in the DO loop above corresponds to the number of the column that is being updated in the elimination process, the vectorized code processes each column in the A array sequentially. With a parallel computer, on the other hand, different columns of A can be processed independently. For example, the loop

```
          DO 20 I = KP1,N
   20       A(I,J) = A(I,J) - A(K,J)*A(I,K)
```

corresponding to J = 1 can be executed by processor 1, while the loop corresponding to J = 2 can be executed by processor 2. If there are 64 separate processors, then 64 columns can be processed simultaneously.

In many applications that involve two nested loops (like the factorization program above), a vector processor can be applied to the inner loop while the outer loop can be partitioned among parallel processors. In some applications, however, completely new algorithms must be devised to make effective use of more than one processor. To illustrate this principle, we study a tridiagonal system of equations **Ax** = **b**. In Section 2-4.1 we present the following factorization routine:

$$
\begin{array}{l}
k = 1 \text{ to } n-1 \\
\quad l_k \leftarrow l_k/d_k \\
\quad d_{k+1} \leftarrow d_{k+1} - l_k u_k \\
\text{next } k
\end{array}
$$

This is a serial algorithm since d_{k+1} cannot be evaluated before d_k is evaluated. Given a parallel computer with two processors, the time to solve the tridiagonal system **Ax** = **b** can be cut in half by the following procedure: Using one processor, we start at the top of the matrix and successively annihilate subdiagonal coefficients by subtracting a multiple of one row from the next row. Using the other processor, we start at the bottom of the matrix and annihilate superdiagonal coefficients by subtracting a multiple of one row from the row above. If the matrix dimension is odd (say 5 × 5), then elimination generates a matrix with the following structure:

$$
\begin{bmatrix}
* & * & 0 & 0 & 0 \\
0 & * & * & 0 & 0 \\
0 & 0 & * & 0 & 0 \\
0 & 0 & * & * & 0 \\
0 & 0 & 0 & * & *
\end{bmatrix}. \tag{83}
$$

In the top half of the matrix, the subdiagonal coefficients are annihilated. In the bottom half of the matrix, the superdiagonal coefficients are annihilated. And in the middle row, all coefficients are annihilated except for the diagonal element. When n is odd, a symbolic program for this elimination process takes the form

$$
\begin{array}{l}
k = 1 \text{ to } (n-1)/2 \\
\quad l_k \leftarrow l_k/d_k \\
\quad d_{k+1} \leftarrow d_{k+1} - l_k u_k \\
\text{next } k \\
k = n-1 \text{ down to } (n+1)/2 \\
\quad u_k \leftarrow u_k/d_{k+1} \\
\quad d_k \leftarrow d_k - l_k u_k \\
\text{next } k
\end{array} \tag{84}
$$

The first loop, which annihilates subdiagonal coefficients, can be executed by one processor while the second loop, which annihilates superdiagonal coefficients, can be executed by the other processor.

Given a tridiagonal matrix factored in this way, the system $\mathbf{Ax} = \mathbf{b}$ can be solved in a parallel fashion. In the first phase, we apply each elimination step to the right-side vector to obtain

$$\boxed{\begin{array}{l} k = 1 \text{ to } (n-1)/2 \\ \quad b_{k+1} \leftarrow b_{k+1} - l_k b_k \\ \text{next } k \\ k = n-1 \text{ down to } (n+1)/2 \\ \quad b_k \leftarrow b_k - u_k b_{k+1} \\ \text{next } k \end{array}} \tag{85}$$

The first loop applies the elimination steps to the top half of the right-side vector while the second loop applies the elimination steps to the bottom half of the right-side vector. After executing these elimination steps, the coefficient matrix for the resulting tridiagonal system has the structure indicated in (83). To complete the solution process, we divide the middle equation by its diagonal element to evaluate the middle unknown. Then back substitution is applied to the top equations to obtain the top half of the unknown vector $\mathbf{x}$ while forward elimination is applied to the bottom equations to obtain the bottom half of $\mathbf{x}$. This second phase can be stated:

$$\boxed{\begin{array}{l} m \leftarrow (n+1)/2 \\ b_m \leftarrow b_m/d_m \\ k = m-1 \text{ down to } 1 \\ \quad b_k \leftarrow (b_k - u_k b_{k+1})/d_k \\ \text{next } k \\ k = m+1 \text{ to } n \\ \quad b_k \leftarrow (b_k - l_{k-1} b_{k-1})/d_k \\ \text{next } k \end{array}} \tag{86}$$

The factorization algorithm (84) and the solution algorithms (85) and (86) make effective use of two parallel processors. If more parallel processors are available, additional modifications in the algorithm are needed to fully utilize all the processors. An algorithm for solving a tridiagonal system which effectively utilizes a large number of parallel processors will be illustrated using a system of seven equations:

$$\begin{aligned} 2x_1 - x_2 \qquad\qquad\qquad\qquad\qquad &= 1, \\ -x_1 + 2x_2 - x_3 \qquad\qquad\qquad\qquad &= 1, \\ - x_2 + 2x_3 - x_4 \qquad\qquad\qquad &= 1, \\ - x_3 + 2x_4 - x_5 \qquad\qquad &= 1, \\ - x_4 + 2x_5 - x_6 \qquad &= 1, \\ - x_5 + 2x_6 - x_7 &= 1, \\ - x_6 + 2x_7 &= 1. \end{aligned} \tag{87}$$

Using groups of three equations, every other unknown can be eliminated. For example, we add to the second equation, $\frac{1}{2}$ times the first equation and $\frac{1}{2}$ times the third equation to get

$$x_2 - \tfrac{1}{2}x_4 = 2.$$

Then we add to the fourth equation, $\frac{1}{2}$ times the third equation and $\frac{1}{2}$ times the fifth equation to obtain

$$-\tfrac{1}{2}x_2 + x_4 - \tfrac{1}{2}x_6 = 2.$$

Finally, we add to the sixth equation, $\frac{1}{2}$ times the fifth equation and $\frac{1}{2}$ times the seventh equation, giving us

$$-\tfrac{1}{2}x_4 + x_6 = 2.$$

After these elimination steps, we are left with three equations in three unknowns:

$$\begin{aligned} x_2 - \tfrac{1}{2}x_4 \qquad &= 2, \\ -\tfrac{1}{2}x_2 + \ x_4 - \tfrac{1}{2}x_6 &= 2, \\ -\ \tfrac{1}{2}x_4 + \ x_6 &= 2. \end{aligned} \tag{88}$$

The elimination process now repeats. Using groups of three equations, every other unknown is eliminated. For the system (88) there is just one group of three equations. We add to the second equation, the first equation times $\frac{1}{2}$ and the third equation times $\frac{1}{2}$ to obtain

$$\tfrac{1}{2}x_4 = 4.$$

Therefore, $x_4 = 8$. Substituting $x_4 = 8$ into (88), we can solve for x_2 and x_6: $x_2 = 6$ and $x_6 = 6$. Finally, substituting $x_2 = 6$, $x_4 = 8$, and $x_6 = 6$ into equation (87), the remaining unknowns can be computed: $x_1 = 3.5$, $x_3 = 7.5$, $x_5 = 7.5$, and $x_7 = 3.5$.

This algorithm is well suited for parallel computation. Given p processors, the equations can be partitioned into p sets. Each processor eliminates every other unknown from its set of equations by forming multiples of groups of three equations. After every other unknown is eliminated, the process is repeated for the new set of equations. That is, the smaller set of equations are partitioned into p sets and each processor eliminates every other unknown for its set of equations. After elimination is complete, we divide by the diagonal element to evaluate the final unknown and then we successively substitute the known values into the previous equations to compute the remaining unknowns. These substitutions can also be partitioned among the processors. The computing time associated with this parallel algorithm will be proportional to the number of equations n divided by the number of processors p. That is, the computing time will be some constant times n/p. Note, though, that since the startup cost associated with a parallel processor can be large, n must be large before the computing time is proportional to n/p.

In this section we have given a brief introduction to parallel processing strategies. Ortega and Voigt provide a comprehensive survey of vector and parallel processing in [O5], while additional references appear at the end of the chapter. Some interesting numerical experiments for a computer (CRAY X-MP-4) with four parallel vector processors are reported by Dongarra and Hewitt in [D7]. They show that approaching 1,000,000,000 flops (a gigaflop) can be performed in a second, where a flop is one cycle of the following n-cycle loop:

for $i = 1$ to n
    $t \leftarrow t + a_{ij}b_{ik}$
next i

Exercises

2-11.1. Modify the elimination and substitution schemes (84), (85), and (86) to handle an even number of equations.

2-11.2. Consider the system of six equations obtained by deleting the last equation and the last unknown in (87). Solve this system of six equations using the reduction process developed above.

REVIEW PROBLEMS

2-1. A matrix that is zero beneath the subdiagonal is called **upper Hessenberg** (an example appears in part (d) of Exercise 2-1.2). An upper Hessenberg matrix can be factored in the following way: Starting at column n, we compare a_{nn} to $a_{n,n-1}$. If $|a_{nn}| < |a_{n,n-1}|$, then the last two columns are interchanged. After this pivot step, we multiply column n by $a_{n,n-1}/a_{nn}$ and subtract from column $n - 1$ to annihilate the coefficient $a_{n,n-1}$. In the next step, $a_{n-1,n-1}$ is compared to $a_{n-1,n-2}$, a pivot is performed if necessary, and $a_{n-1,n-2}$ is annihilated. Moving up the diagonal in this way, each subdiagonal coefficient is annihilated.

(a) Write a symbolic program to implement this factorization scheme.

(b) Why is the usual algorithm for column elimination with column pivoting a bad strategy when the coefficient matrix is upper Hessenberg?

(c) Why is the algorithm in part (a) preferable in Fortran to the usual algorithm for row elimination with row pivoting? (*Hint:* Consider the paging issue.)

2-2. A matrix that is zero above the superdiagonal is called **lower Hessenberg**. One of the best ways to factor a lower Hessenberg matrix is column-oriented column elimination. Modify the program (35) to process a lower Hessenberg matrix using column-oriented operations.

2-3. A $n \times n$ matrix which is entirely zero except for its diagonal elements a_{ii} and its cross-diagonal elements $a_{i,n-i+1}$, $i = 1, \cdots, n$, is called a **cross matrix**. Cross matrices appear in parts (e) and (f) of Exercise 2-1.2. The nonzero elements of a cross matrix can be stored in two arrays of length n. Write subroutines to factor a cross matrix without pivots and to solve a factored system. Test your subroutines using the coefficients and the right sides that appear in parts (e) and (f) of Exercises 2-1.2 and 2-2.1.

2-4. In Table 2-13, we see that for a (l, u)-band matrix, the asymptotic parameter for row elimination with row pivoting is $l(l + u)n$. Similarly, the asymptotic parameter for column elimination with column pivoting is $u(l + u)n$. For a (l, u)-band matrix, when is column elimination faster than row elimination?

2-5. Suppose that a matrix **A** is factored using the program in Table 2-5. Write a column-oriented program to solve $\mathbf{A}^T\mathbf{x} = \mathbf{b}$ using for input the array A generated by the program in Table 2-5. [Remember that the transpose of a product is the product of the transposes in the reverse order. Hence $\mathbf{A}^T = (\mathbf{LU})^T = \mathbf{U}^T\mathbf{L}^T$.]

2-6. Use the ideas of Exercise 2-3.2 to write a subroutine RPACK(A,LA,M,N) that pushes a M × N matrix stored in a double-subscripted array A with leading dimension LA to the beginning of the array so that the columns of the matrix are stored sequentially in computer memory. For example, the output of the following program is 1, 2, · · ·, 12 (subroutine LIST appears in Exercise 2-3.2).

```
      REAL A(8,10)
      DO 10 J = 1,4
      DO 10 I = 1,3
10       A(I,J) = I + 3*(J-1)
      CALL RPACK(A,8,3,4)
      CALL LIST(A,12)
      END
```

2-7. Write a column-oriented subroutine to compute the product **P** = **AB** between a $l \times m$ matrix **A** and a $m \times n$ matrix **B**. (*Hint:* Apply the program appearing in Table 2-10 to each column of B.) What is the asymptotic parameter for your program?

2-8. Write a column-oriented subroutine BMULT(P,B,A,LA,N,L,U) that computes the matrix-vector product **p** = **Ab** when **A** is a (l, u)-band matrix. Assume that **A** is stored using the structure described in Section 2-4.2. When multiplying a matrix **A** by a vector **b**, the matrix element a_{ij} is multiplied by the vector component b_j and the resulting products are added together for j between 1 and n to obtain the ith component of **p**. To compute the product **Ab** when **A** is a band matrix, first initialize array P to zero. Your program should start at the beginning of array A and process one element after the other, multiplying each element of A by the appropriate element of B and adding this product to the appropriate element of P.

2-9. This exercise explores one strategy for testing whether a program correctly solves a system of equations. We start with a vector **y** and a matrix **A** whose elements are randomly generated, and we compute the product **b** = **Ay**. Then our equation solver is applied to the system **Ax** = **b**. If the program is correct, then the computed solution is near the starting vector **y**. Letting RAND(X) denote a function that generates a random number between 0 and X, the following program creates a (3, 2)-band matrix with coefficients randomly distributed between +1 and −1. The coefficients are stored in array A using the band structure described in Section 2-4.2.

```
      INTEGER U
      REAL A(10,20),Y(15)
      N = 15
      L = 3
      U = 2
      DO 10 J = 1,N
         Y(J) = 1.
         DO 10 I = 1,6
10          A(I,J) = 1. - RAND(2.)
```

Inside the J loop, an array Y is initialized to 1. Complete this program using BMULT (see Problem 2-8) to multiply the band matrix by the vector **y**, using BFACT to factor the matrix and using BSOLVE to solve the factored system. Each component of the computed solution should be near 1.

2-10. Identify the deepest loops in the following program and compute the asymptotic parameters for these loops.

```
      DO 40 I = 1,N
      I1 = N - I + 2
      I2 = 2*N - I
      X = 0.
```

```
      DO 40 J = I1,I2
      I3 = (J-I1+1)*3
      IF ( I3 .GT. I2 ) GOTO 10
      I3 = I2 + 1
10    CONTINUE
      DO 20 K = I2,I3
      X = X + 1./FLOAT(I3) - 1./FLOAT(K*K)
20    CONTINUE
      K1 = I3/2 + I + J/3
      DO 30 M = I,K1
      X = X + 1./FLOAT(M*M)
30    CONTINUE
40    CONTINUE
```

To determine the asymptotic parameter CN^p, first strip the code to its basic loop structure by removing assignment statements unrelated to the loop limits. Insert counters as we did in Section 2-6 and run the stripped program taking N = 4, 8, 16, · · ·. Print the number of loop executions corresponding to each N. Using the technique developed in Section 1-7 [equations (30) and (31)], estimate the exponent p and the coefficient C of the asymptotic parameter.

2-11. Gaussian elimination annihilates coefficients below the diagonal each elimination step. **Jordan elimination** annihilates coefficients both above and below the diagonal each elimination step. For example, consider the following system of three equations:

$$2x_1 + 4x_2 + 2x_3 = 4, \tag{89a}$$

$$4x_1 + 7x_2 + 7x_3 = 13, \tag{89b}$$

$$-2x_1 - 7x_2 + 5x_3 = 7. \tag{89c}$$

Jordan elimination divides the first equation by the diagonal element 2 to obtain

$$x_1 + 2x_2 + x_3 = 2. \tag{90}$$

To eliminate x_1 from the second equation and from the third equation, 4 times (90) is subtracted from (89b) and -2 times (90) is subtracted from (89c), giving us

$$x_1 + 2x_2 + x_3 = 2, \tag{91a}$$

$$-x_2 + 3x_3 = 5, \tag{91b}$$

$$-3x_2 + 7x_3 = 11. \tag{91c}$$

Before the second elimination step, equation (91b) is divided by the diagonal element -1, yielding

$$x_2 - 3x_3 = -5. \tag{92}$$

Then x_2 is eliminated from the first equation and from the third equation by subtracting 2 times (92) from (91a) and by subtracting -3 times (92) from (91c), giving us

$$x_1 \quad + 7x_3 = 12,$$

$$x_2 - 3x_3 = -5,$$

$$-2x_3 = -4.$$

In the last elimination step, the last equation is divided by -2 to get $x_3 = -4/-2 = 2$, and x_3 is eliminated from the first and second equation to obtain $x_1 = -2$ and $x_2 = 1$. In general, the kth step of Jordan elimination divides the kth equation by the kth diagonal element and eliminates x_k from equations 1 through $k - 1$ and from equations $k + 1$ through n. Write a symbolic row-oriented program to perform Jordan elimination on a $n \times n$ matrix **A**. When a coefficient is annihilated, store the multiplier in place of zero, and when the kth row is divided by a_{kk}, store $1/a_{kk}$ in place of a_{kk}.

2-12. Write a computer program that implements the algorithm developed in Problem 2-11. What is the asymptotic parameter for your program? Which is faster, Gaussian elimination or Jordan elimination?

2-13. Write a symbolic program to solve the general system $\mathbf{Ax} = \mathbf{b}$, where the input to your program is a vector $\mathbf{b}$ and a matrix $\mathbf{A}$ which has been processed previously using the program developed in Problem 2-11.

2-14. The compact row version of Jordan elimination processes one row and its predecessors each elimination step. For illustration, we consider (89). In the first step, equation (89a) is divided by 2 and x_1 is eliminated from equation (89b) to get

$$x_1 + 2x_2 + x_3 = 2, \tag{93a}$$

$$-x_2 + 3x_3 = 5, \tag{93b}$$

$$-2x_1 - 7x_2 + 5x_3 = 7. \tag{93c}$$

In the second step, equation (93b) is divided by -1, x_2 is eliminated from (93a), and x_1 and x_2 are eliminated from (93c), yielding

$$x_1 + 7x_3 = 12, \tag{94a}$$

$$x_2 - 3x_3 = -5, \tag{94b}$$

$$-2x_3 = -4. \tag{94c}$$

In the third step, equation (94c) is divided by -2 and x_3 is eliminated from the first two equations to obtain the solution $x_1 = -2$, $x_2 = 1$, and $x_3 = 2$. In general, the kth step divides equation k by a_{kk}, eliminates x_k from equations 1 through $k - 1$, and eliminates x_1 through x_k from equation $k + 1$. Write a symbolic program similar to that of Problem 2-11 for the compact row version of Jordan elimination. What is the asymptotic parameter associated with the program?

2-15. Write a symbolic program to solve the system $\mathbf{Ax} = \mathbf{b}$, where the input for your program is the vector $\mathbf{b}$ and a matrix $\mathbf{A}$ which has been processed previously using the algorithm developed in Problem 2-14.

2-16. Suppose we wish to solve $\mathbf{Ax} = \mathbf{b}$ using the compact row version of Jordan elimination. We will apply the multipliers to the right side and then discard the multipliers as the elimination progresses. Assume that matrix $\mathbf{A}$ is stored on a tape and one row of $\mathbf{A}$ after another is transferred to main memory where the row is processed. What is the least amount of main memory needed to solve the system?

2-17. The finite-difference system derived in Problem 1-8 can be written in the form $\mathbf{Ax} = \mathbf{b}$, where $\mathbf{A}$ is tridiagonal. Determine the elements of $\mathbf{A}$ and $\mathbf{b}$ and solve the system of equations using subroutines TFACT and TSOLVE.

2-18. (by N. Ghosh) Suppose that $\mathbf{A}$ and $\mathbf{B}$ are $n \times n$ matrices.

(a) If $\mathbf{A}$ is a (l_1, u_1)-band matrix and $\mathbf{B}$ is a (l_2, u_2)-band matrix, what are the band widths of the product $\mathbf{P} = \mathbf{AB}$?

(b) Show that the asymptotic parameter associated with the computation of the product $\mathbf{P} = \mathbf{AB}$ is

$$(l_1 + u_1 + 1)(l_2 + u_2 + 1)n.$$

(In determining the asymptotic parameter, l_1, u_1, l_2, and u_2 are held fixed as n tends to infinity.)

(c) Suppose that $\mathbf{A}$ is tridiagonal. Decide which of the following two methods to compute $\mathbf{D} = \mathbf{A}^6$ is asymptotically faster.

Method 1: Compute $\mathbf{B} = \mathbf{A}^2$, compute $\mathbf{C} = \mathbf{AB}$, and compute $\mathbf{D} = \mathbf{C}^2$.

Method 2: Compute $\mathbf{B} = \mathbf{A}^2$, compute $\mathbf{C} = \mathbf{B}^2$, and compute $\mathbf{D} = \mathbf{BC}$.

REFERENCES

As pointed out in the LINPACK user's guide [D8], "Gaussian elimination was known long before Gauss. A 3 by 3 example occurs in a Chinese manuscript over 2000 years old and Gauss himself refers to the method as commonly known." Although the material on row elimination is fairly standard, the discussion of column elimination and its relative merits does not seem to appear elsewhere. Some excellent numerical algebra references include the book [G15] by Golub and Van Loan, the book [S14] by Stewart, and the book [S15] by Strang. Diagonal pivoting factorization techniques to factor a symmetric matrix with pivoting are developed in papers by Bunch [B16] and [B17], Bunch and Kaufman [B18], Bunch and Parlett [B20], and Bunch, Kaufman, and Parlett [B19]. An algorithm related to Aasen's algorithm appears in [P5]. Barwell and George compare various methods for solving symmetric linear systems in [B6]. The compact row-oriented elimination scheme in Section 2-5 is essentially the Doolittle or the Crout reduction scheme. The elimination scheme in Section 2-11 which uses many processors to solve a tridiagonal system is related to the cyclic reduction algorithm (see [B24], [H9], [S19], and [S21]). Other references for parallel processing techniques, besides the paper by Ortega and Voigt, include the bibliography [S4] by Satyanaranyanan, the tutorial [K8], the conference proceedings [W8], the book [S5] by Schendel, and Heller's survey article [H10]. The solution to Problem 2-16 and the algorithm developed in Problem 2-14 are contained in [W2].

Gaussian elimination for a general matrix is called an $O(n^3)$ algorithm since the computing time is proportional to n^3, the matrix dimension cubed. Researchers in complexity theory have developed $O(n^\alpha)$ algorithms where $\alpha < 3$. Strassen's [S18] $O(n^{2.81})$ algorithm was the first of this type. A series of improvements in Strassen's algorithm led to the $O(n^{2.495})$ scheme [C6] of Coppersmith and Winograd.† A survey of these new schemes is given by Pan [P2]. Although the exponent α is at least 2, the smallest α that can be achieved is currently unknown. It should be noted that with these new algorithms the proportionality constant in front of the n^α can be very large. Consequently, even though the new schemes are asymptotically faster than Gaussian elimination, n must be large before their asymptotic superiority is observed.

† More recently, Strassen has developed a method with $\alpha < 2.48$ and Coppersmith and Winograd have modified Strassen's method to obtain a scheme with $\alpha < 2.38$ (see the Journal für de reine und angewandte Mathematatik, Volume 375/376, 1987, pp. 406–443).

3

Conditioning

3-1. AN EXAMPLE

As stated in Chapter 1, three issues must be confronted when a problem is solved numerically: problem conditioning, algorithm stability, and cost. Chapter 2 develops efficient elimination schemes to solve a linear system and these schemes are stable when pivoting is employed. But alas, some ill-conditioned equations are tough to solve by any method. These ill-conditioned equations are identified in this chapter. We begin with an illustration involving two equations:

$$\begin{bmatrix} 600 & 800 \\ 30{,}001 & 40{,}002 \end{bmatrix} \begin{bmatrix} x_1 \\ x_2 \end{bmatrix} = \begin{bmatrix} 200 \\ 10{,}001 \end{bmatrix}. \tag{1}$$

If this equation is solved using an IBM 370 computer and subroutines FACT and SOLVE, the computed solution $\bar{\mathbf{x}}$ is

$$\begin{cases} \bar{x}_1 = -1.013887 \\ \bar{x}_2 = 1.010416. \end{cases} \tag{2}$$

On the other hand, it can be verified that the correct solution is $x_1 = -1$ and $x_2 = 1$. Even though FACT employs pivoting and the machine epsilon is less than 10^{-6}, the relative error in the computed solution is greater than 10^{-2}. This system is ill conditioned and the solution computed by any algorithm is probably inaccurate.

As stated in Chapter 1, problem conditioning is related to how the solution changes when parameters change. The natural parameters for a linear system are the coefficients and the right side. To investigate the conditioning of (1), let us change the right-side component 10,001 to 9999. The exact solution changes to $x_1 = 3$ and $x_2 = -2$. Changing the right side by 2 causes x_1 to change to 3. The relative change $|9999 - 10{,}001|/10{,}001 \approx .0002$ in the right side generates a relative change $(3 - (-1))/|-1| = 4$ in x_1. Since the relative change in the solution is much larger than the relative change in the right side, the equation is sensitive to perturbations in the right side.

The coefficients of equation (1) are integers that can be stored exactly in single precision on most computers and the inaccuracy in the computed solution (2) results from rounding errors that occur during Gaussian elimination. Now let us divide the equations in (1) by 2000 and by 100,000, respectively, to obtain the equivalent system

$$\begin{bmatrix} .3 & .4 \\ .30001 & .40002 \end{bmatrix} \begin{bmatrix} x_1 \\ x_2 \end{bmatrix} = \begin{bmatrix} .1 \\ .10001 \end{bmatrix}. \tag{3}$$

Although the fractions appearing in (3) are finite decimal fractions, they are infinite binary fractions. And when (3) is stored in the computer, these decimal fractions are replaced by binary approximations. Hence the computed solution $\bar{\mathbf{x}}$ to (3) is also polluted by the error that results from replacing the coefficients and the right side by their floating-point representations. Solving equation (3) on an IBM 370 computer using subroutines FACT and SOLVE, we obtain

$$\bar{x}_1 = -.9879876,$$

$$\bar{x}_2 = .9909909,$$

which differs significantly from the previous solution (2).

Pretending for the moment that the exact solution to (1) is unknown, we now describe a practical approach for estimating the error in the computed solution: Perform the computation both in single precision and in double precision. Typically, the double-precision answer is much more accurate than the single-precision solution. Comparing the double-precision solution to the single-precision solution, we can estimate the number of digits lost due to rounding errors. For the system (1), the solution is

$$\bar{x}_1 = -1.000000000007461,$$

$$\bar{x}_2 = 1.000000000005596$$

in double precision. Subtracting the double-precision solution from the single-precision solution, we see that the relative error in each component of the single-precision solution is about .01, which corresponds to two correct digits. Since single precision on an IBM 370 computer is equivalent to about seven decimal digits in the mantissa, $7 - 2 = 5$ digits were lost. Similarly, five digits are lost in double precision. Since double precision on an IBM 370 computer corresponds to nearly 17 decimal digits in the mantissa, the double-precision solution has nearly $17 - 5 = 12$ correct digits.

Our goal in this chapter is to present a parameter, the condition number, which quantitatively measures the conditioning of a linear system. The condition number is greater than or equal to one and as the equation becomes more ill conditioned, the condition number increases. After factoring a matrix, the condition number can be estimated in roughly the same time as it takes to solve a few factored systems $(\mathbf{LU})\mathbf{x} = \mathbf{b}$. Hence, after factoring a matrix, the extra computer time needed to estimate the condition number is usually insignificant.

3-2. VECTOR NORMS

When solving $\mathbf{Ax} = \mathbf{b}$ on a computer, $\mathbf{A}$ and $\mathbf{b}$ are replaced by their floating-point representations $\mathrm{fl}(\mathbf{A})$ and $\mathrm{fl}(\mathbf{b})$, respectively. Hence the computer works with the equation $\mathrm{fl}(\mathbf{A})\mathbf{y} = \mathrm{fl}(\mathbf{b})$. The conditioning of a linear system is related to the distance between $\mathbf{x}$ and $\mathbf{y}$ relative to the differences $\mathbf{A} - \mathrm{fl}(\mathbf{A})$ and $\mathbf{b} - \mathrm{fl}(\mathbf{b})$. If a relatively small change in the coefficients or in the right side causes a relatively large change in the solution, the system is ill conditioned. To investigate a system's conditioning, we must compare the length of the vector $\mathbf{x} - \mathbf{y}$ to the length of the matrix $\mathbf{A} - \mathrm{fl}(\mathbf{A})$ and to the length of the vector $\mathbf{b} - \mathrm{fl}(\mathbf{b})$. The most common notion of length, attributed to Euclid, is based on the Pythagorean theorem (Figure 3-1). The vector from the origin to the point (3, 4) in the xy-plane has length $\sqrt{3^2 + 4^2} = \sqrt{25} = 5$, and the vector from the origin to the point (a, b) has length $\sqrt{a^2 + b^2}$.

The length of a vector with two components is the square root of the sum of each component squared. Extending Euclid's concept of length to n dimensions, a vector $\mathbf{x}$ with n components has length

$$\sqrt{x_1^2 + x_2^2 + \cdots + x_n^2}.$$

The Euclidean length of a vector, often denoted $\|\mathbf{x}\|_2$ since the components of $\mathbf{x}$ are raised to the second power, has the following properties:

(a) $\|\mathbf{x}\|_2 \geqq 0$ and $\|\mathbf{x}\|_2 = 0$ only if $\mathbf{x} = \mathbf{0}$.

(b) $\|\alpha\mathbf{x}\|_2 = |\alpha|\,\|\mathbf{x}\|_2$ for all scalars α.

(c) $\|\mathbf{x} + \mathbf{y}\|_2 \leqq \|\mathbf{x}\|_2 + \|\mathbf{y}\|_2$.

Property (c) is the **triangle inequality**, which is equivalent to the observation that each side of a triangle is shorter than the other two sides combined (see Figure 3-2).

Mathematicians often use the terminology "norm" instead of length. Any function that associates a scalar with a vector and which obeys (a), (b), and (c) is a **norm**. Another common norm is the 1-norm defined by

$$\|\mathbf{x}\|_1 = |x_1| + |x_2| + \cdots + |x_n|.$$

That is, the 1-norm of a vector is the sum of the absolute vector components. Property (a) obviously holds since $\|\mathbf{x}\|_1 \geqq 0$ and $\|\mathbf{x}\|_1 = 0$ only if each component of $\mathbf{x}$ is zero. Properties (b) and (c) are verified below:

$$\begin{aligned} \|\alpha\mathbf{x}\|_1 &= |\alpha x_1| + |\alpha x_2| + \cdots + |\alpha x_n| \\ &= |\alpha||x_1| + |\alpha||x_2| + \cdots + |\alpha||x_n| \\ &= |\alpha|\{|x_1| + |x_2| + \cdots + |x_n|\} \\ &= |\alpha|\,\|\mathbf{x}\|_1, \end{aligned} \qquad \text{(b)}$$

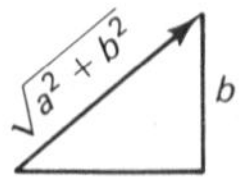

Figure 3-1 Pythagorean theorem.

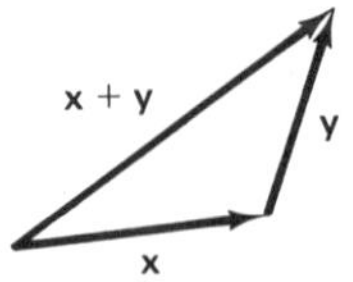

Figure 3-2 One side $\mathbf{x} + \mathbf{y}$ of a triangle is shorter than the other two sides $\mathbf{x}$ and $\mathbf{y}$ combined.

$$\begin{aligned}\|\mathbf{x} + \mathbf{y}\|_1 &= |x_1 + y_1| + |x_2 + y_2| + \cdots + |x_n + y_n| \\ &\leqq |x_1| + |y_1| + |x_2| + |y_2| + \cdots + |x_n| + |y_n| \qquad \text{(c)} \\ &= |x_1| + |x_2| + \cdots + |x_n| + |y_1| + |y_2| + \cdots + |y_n| \\ &= \|\mathbf{x}\|_1 + \|\mathbf{y}\|_1 .\end{aligned}$$

The max-norm of a vector is the largest absolute component. That is, the max-norm of **x**, denoted $\|\mathbf{x}\|_\infty$, is defined by

$$\|\mathbf{x}\|_\infty = \text{maximum } \{|x_1|, |x_2|, \cdots, |x_n|\}.$$

In Exercise 3-2.1 you will verify properties (a), (b), and (c) for the max-norm. In general, if p is a real number greater than or equal to 1, the p-norm is defined by

$$\|\mathbf{x}\|_p = (|x_1|^p + |x_2|^p + \cdots + |x_n|^p)^{1/p}.$$

Note that $p = 2$ corresponds to the Euclidean norm and $p = 1$ corresponds to the 1-norm. It can be shown that

$$\|\mathbf{x}\|_\infty \leqq \|\mathbf{x}\|_p \leqq n^{1/p}\|\mathbf{x}\|_\infty. \tag{4}$$

Since $n^{1/p}$ approaches 1 as p increases, it follows that $\|\mathbf{x}\|_p$ is sandwiched between $\|\mathbf{x}\|_\infty$ and a number close to $\|\mathbf{x}\|_\infty$ when p is large. In the limit as p tends to infinity, the p-norm of **x** approaches the max-norm of **x**.

To help visualize these norms, we examine the collection of vectors with unit length. In the plane, a unit vector satisfies

$$\|\mathbf{x}\|_2 = \sqrt{x_1^2 + x_2^2} = 1,$$

which is the equation of a circle. In general, the set of vectors for which $\|\mathbf{x}\| = 1$ is called the **unit sphere**, although this set resembles a sphere only in the Euclidean case. For the 1-norm, the vectors forming the unit sphere satisfy the equation

$$\|\mathbf{x}\|_1 = |x_1| + |x_2| = 1.$$

This equation describes a square whose diagonals coincide with the coordinate axes (see Figure 3-3). For example, when x_1 and x_2 are positive, the 1-norm of **x** is $x_1 + x_2$ and the equation $x_1 + x_2 = 1$ describes the line connecting the points (0, 1) and (1, 0). The unit sphere for the max-norm satisfies the relation

$$\|\mathbf{x}\|_\infty = \text{maximum } \{|x_1|, |x_2|\} = 1,$$

which is the equation of a square with sides parallel to the coordinate axes (see Figure 3-4). Along the right and left sides of the square, $|x_1|$ is 1. Along the top

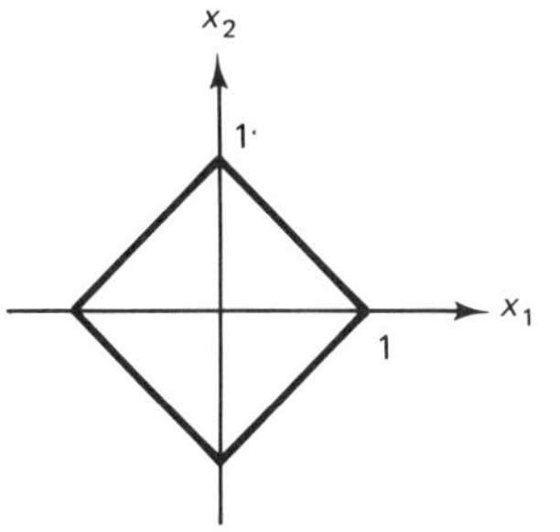

Figure 3-3 Unit sphere for the 1-norm.

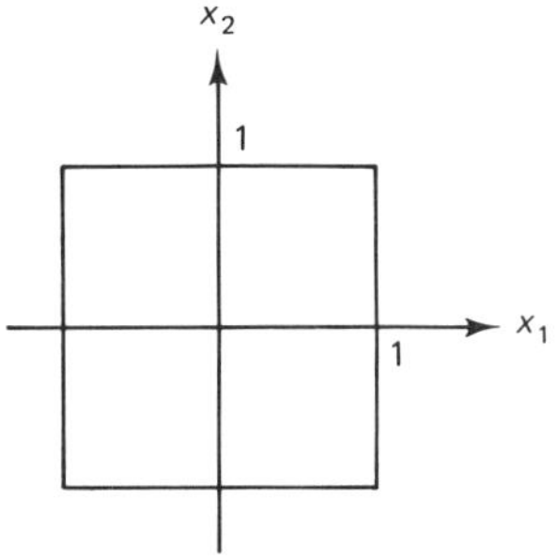

Figure 3-4 Unit sphere for the max-norm.

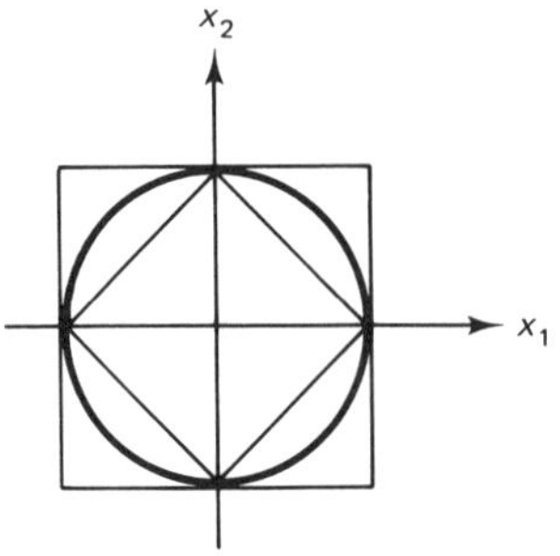

Figure 3-5 Unit spheres for the p-norm, $p = 1, 2, \infty$.

and bottom of the square, $|x_2| = 1$. Superimposing these three unit spheres, we get Figure 3-5. As p increases, the unit sphere in the p-norm deforms from the square inscribed in the circle to the square circumscribing the circle.

Figure 3-5 reveals that the shape of the unit sphere is similar for different values of p. In fact, it can be shown that any two vector norms are equivalent in the sense that the ratio between a vector's length in one norm and a vector's length in another norm is bound from above and from below by constants which do not depend upon the vector. For example, dividing (4) by $\|\mathbf{x}\|_\infty$, we see that the max-norm and the p-norm are equivalent since the norm ratio obeys the inequality

$$1 \leqq \frac{\|\mathbf{x}\|_p}{\|\mathbf{x}\|_\infty} \leqq n^{1/p} \tag{5}$$

for every nonzero vector $\mathbf{x}$. In other words, the p-norm of $\mathbf{x}$ is greater than the max-norm of $\mathbf{x}$, but less than $n^{1/p}$ times the max-norm of $\mathbf{x}$. In general, the p-norm and the q-norm are equivalent and the norm ratio satisfies the inequality

$$1 \leqq \frac{\|\mathbf{x}\|_p}{\|\mathbf{x}\|_q} \leqq n^{(q-p)/pq} \tag{6}$$

provided that $p \leqq q$. As q tends to infinity, $(q - p)/pq$ approaches $1/p$ and $\|\mathbf{x}\|_q$ approaches $\|\mathbf{x}\|_\infty$. Thus (6) reduces to (5) in the limit as q tends to infinity.

Exercises

3-2.1. Show that the max-norm satisfies properties (a), (b), and (c) above.

3-2.2. In Figure 3-5 we see that the unit sphere for the max-norm contains the unit sphere for the 2-norm, which contains the unit sphere for the 1-norm. These inclusions are equivalent to saying that $\|\mathbf{x}\|_\infty \leqq \|\mathbf{x}\|_2 \leqq \|\mathbf{x}\|_1$ for every vector $\mathbf{x}$. Prove that $\|\mathbf{x}\|_p$ is a decreasing function of $p > 0$ by computing the derivative of $\|\mathbf{x}\|_p$ with respect to p when $\|\mathbf{x}\|_p = 1$ and observing that the derivative is nonpositive.

3-2.3. Verify the inequality (4).

3-3. MATRIX NORMS

How do we measure the length of a matrix? One approach is to sum the square of each element and take the square root. This direct extension of Euclidean length is called the Frobenius norm. In this section we discuss another way to measure the length of a matrix which is also closely related to vector norms.

We can think of a matrix $\mathbf{A}$ as being an operator that transforms a vector $\mathbf{x}$ into another vector $\mathbf{Ax}$. For example, as $\mathbf{x}$ travels around the unit sphere in the plane, $\mathbf{Ax}$ might form the ellipse in Figure 3-6. The size of $\mathbf{A}$ is related to the size of $\mathbf{Ax}$ relative to $\mathbf{x}$. If the elements of $\mathbf{A}$ are large, then for some choice of $\mathbf{x}$, the components of $\mathbf{Ax}$ are large. To measure the amplification of $\mathbf{Ax}$ relative to $\mathbf{x}$, we form the ratio $\|\mathbf{Ax}\|/\|\mathbf{x}\|$, where $\|\cdot\|$ is a vector norm. By definition, the norm of $\mathbf{A}$ is the maximum amplification, or equivalently,

$$\|\mathbf{A}\| = \text{maximum}\left\{\frac{\|\mathbf{Ax}\|}{\|\mathbf{x}\|} : \mathbf{x} \neq 0\right\}. \tag{7}$$

Since the amplification ratio $\|\mathbf{Ax}\|/\|\mathbf{x}\|$ is expressed in terms of a vector norm, we say that the matrix norm (7) is induced by (or subordinate to) the corresponding vector norm.

This definition is now simplified. Observe that $\alpha\mathbf{x}$ is a unit vector when α is the scalar $1/\|\mathbf{x}\|$. That is, $\|\alpha\mathbf{x}\| = |\alpha|\,\|\mathbf{x}\| = \|\mathbf{x}\|/\|\mathbf{x}\| = 1$ when $\alpha = 1/\|\mathbf{x}\|$. Thus the amplification ratio can be written

$$\frac{\|\mathbf{Ax}\|}{\|\mathbf{x}\|} = \frac{\alpha\|\mathbf{Ax}\|}{\alpha\|\mathbf{x}\|} = \frac{\|\alpha(\mathbf{Ax})\|}{\|\alpha\mathbf{x}\|} = \|\alpha(\mathbf{Ax})\| = \|\mathbf{A}(\alpha\mathbf{x})\| = \|\mathbf{Ay}\|,$$

where $\mathbf{y} = \alpha\mathbf{x}$ has unit length. Hence, maximizing the ratio $\|\mathbf{Ax}\|/\|\mathbf{x}\|$ over nonzero $\mathbf{x}$ is equivalent to maximizing $\|\mathbf{Ay}\|$ over the unit sphere:

$$\|\mathbf{A}\| = \text{maximum}\,\{\|\mathbf{Ay}\| : \|\mathbf{y}\| = 1\}. \tag{8}$$

Let us return to Figure 3-6 and the ellipse representing the set of vectors $\mathbf{Ax}$ corresponding to $\mathbf{x}$ with unit length. Since $\|\mathbf{A}\|_2$ is the maximum of $\|\mathbf{Ax}\|_2$ over the unit sphere, we conclude that $\|\mathbf{A}\|_2$ is the maximum distance between the ellipse and the origin. In other words, $\|\mathbf{A}\|_2$ is half the length of the ellipse's major axis.

Six properties of matrix norms are summarized below:

(a) $\|\mathbf{A}\| \geqq 0$ and $\|\mathbf{A}\| = 0$ only if $\mathbf{A} = \mathbf{0}$.

(b) $\|\alpha\mathbf{A}\| = |\alpha|\,\|\mathbf{A}\|$ for all scalars α.

(c) $\|\mathbf{A} + \mathbf{B}\| \leqq \|\mathbf{A}\| + \|\mathbf{B}\|$.

(d) $\|\mathbf{Ax}\| \leqq \|\mathbf{A}\|\,\|\mathbf{x}\|$.

(e) $\|\mathbf{I}\| = 1$.

(f) $\|\mathbf{AB}\| \leqq \|\mathbf{A}\|\,\|\mathbf{B}\|$.

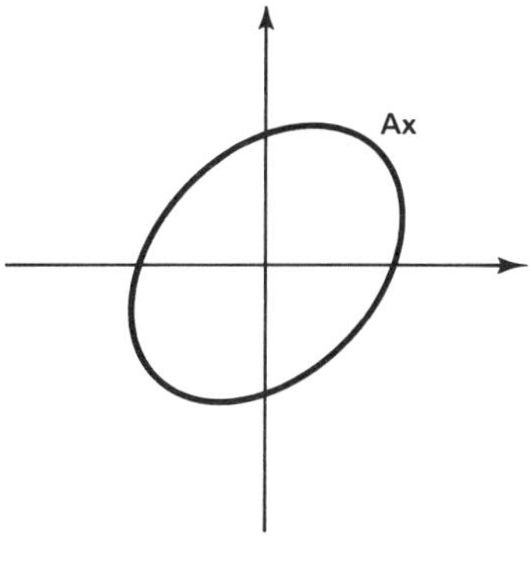

Figure 3-6 The collection of vectors $\mathbf{Ax}$ corresponding to unit vectors $\mathbf{x}$.

Property (b) is derived from (8):

$$\begin{aligned}\|\alpha\mathbf{A}\| &= \max\{\|(\alpha\mathbf{A}\mathbf{y})\| : \|\mathbf{y}\| = 1\} = \max\{\|\alpha(\mathbf{A}\mathbf{y})\| : \|\mathbf{y}\| = 1\}\\ &= \max\{|\alpha|\,\|\mathbf{A}\mathbf{y}\| : \|\mathbf{y}\| = 1\} = |\alpha| \max\{\|\mathbf{A}\mathbf{y}\| : \|\mathbf{y}\| = 1\}\\ &= |\alpha|\,\|\mathbf{A}\|.\end{aligned}$$

Property (d) follows directly from (7). Since $\|\mathbf{A}\|$ is the maximum ratio $\|\mathbf{Ax}\|/\|\mathbf{x}\|$, we conclude that

$$\|\mathbf{A}\| \geqq \frac{\|\mathbf{Ax}\|}{\|\mathbf{x}\|}$$

for any particular nonzero $\mathbf{x}$. Multiplying both sides by $\|\mathbf{x}\|$ gives $\|\mathbf{Ax}\| \leqq \|\mathbf{A}\|\,\|\mathbf{x}\|$. Properties (a), (c), (e), and (f) are established in Exercise 3-3.1.

Although the evaluation of $\|\mathbf{A}\|$ involves maximizing $\|\mathbf{Ay}\|$ over the unit sphere, there are two special cases, the 1-norm and the ∞-norm, where this maximum can be evaluated explicitly. If $\mathbf{A}$ is $m \times n$, then

$$\|\mathbf{A}\|_1 = \text{maximum}\left\{\sum_{i=1}^{m} |a_{ij}| : j = 1, 2, \cdots, n\right\},$$

$$\|\mathbf{A}\|_\infty = \text{maximum}\left\{\sum_{j=1}^{n} |a_{ij}| : i = 1, 2, \cdots, m\right\}.$$

That is, the 1-norm of a matrix is the maximum absolute column sum and the ∞-norm of a matrix is the maximum absolute row sum. For example, $\|\mathbf{A}\|_1 = 12$ and $\|\mathbf{A}\|_\infty = 17$ for the matrix

$$\mathbf{A} = \begin{bmatrix} 1 & -2 & 4 & -3 \\ 2 & 0 & 3 & -1 \\ 6 & 2 & -1 & 8 \end{bmatrix}.$$

The formula given above for the 1-norm of a matrix is now derived while the formula for the ∞-norm is established in Exercise 3-3.10. By property (d), we have the inequality $\|\mathbf{A}\|_1 \geqq \|\mathbf{Ax}\|_1/\|\mathbf{x}\|_1$ for every choice of $\mathbf{x}$. If $\mathbf{x}$ is the vector whose first component is one and whose remaining components are zero and if $\mathbf{a}_1$ denotes the first column of $\mathbf{A}$, then $\|\mathbf{x}\|_1 = 1$, $\mathbf{Ax} = \mathbf{a}_1$, and $\|\mathbf{Ax}\|_1 = \|\mathbf{a}_1\|_1$. Substituting $\|\mathbf{Ax}\|_1 = \|\mathbf{a}_1\|_1$ and $\|\mathbf{x}\|_1 = 1$ in the inequality $\|\mathbf{A}\|_1 \geqq \|\mathbf{Ax}\|_1/\|\mathbf{x}\|_1$ yields $\|\mathbf{A}\|_1 \geqq \|\mathbf{a}_1\|_1$. In other words, the 1-norm of $\mathbf{A}$ is at least as large as $\|\mathbf{a}_1\|_1$, the absolute sum of elements from the first column of $\mathbf{A}$. In the same fashion, if $\mathbf{x}$ is the vector whose jth component is one and whose remaining components are zero and $\mathbf{a}_j$ denotes the jth column of $\mathbf{A}$, the inequality $\|\mathbf{A}\|_1 \geqq \|\mathbf{Ax}\|_1/\|\mathbf{x}\|_1$ implies that the 1-norm of $\mathbf{A}$ is at least as large as $\|\mathbf{a}_j\|_1$, the absolute sum of elements from column j of $\mathbf{A}$. Thus the 1-norm of $\mathbf{A}$ is at least as large as the maximum absolute column sum.

An upper bound for the 1-norm of $\mathbf{A}$ is obtained from a bound for the quantity $\|\mathbf{Ay}\|$ in the definition (8). Applying the triangle inequality to the expression $\|\mathbf{Ay}\|_1$, where $\mathbf{y}$ is a unit vector, yields

$$\begin{aligned}\|\mathbf{Ay}\|_1 &= \left\|\sum_{j=1}^{n} y_j\,\mathbf{a}_j\right\|_1 \leqq \sum_{j=1}^{n} |y_j|\,\|\mathbf{a}_j\|_1\\ &\leqq \left(\underset{j=1,2,\cdots,n}{\text{maximum}} \|\mathbf{a}_j\|_1\right) \sum_{j=1}^{n} |y_j| = \underset{j=1,2,\cdots,n}{\text{maximum}} \|\mathbf{a}_j\|_1.\end{aligned}$$

Since $\|\mathbf{Ay}\|_1$ is bounded by the maximum absolute column sum when $\mathbf{y}$ is a unit vector, it follows from (8) that the 1-norm of $\mathbf{A}$ is bounded by the maximum absolute column sum. Since we already showed that the 1-norm of $\mathbf{A}$ is at least as large as the maximum absolute column sum, it follows that the 1-norm of $\mathbf{A}$ actually equals the maximum absolute column sum.

Although the 2-norm of a matrix cannot be expressed explicitly, definition (8) simplifies in the Euclidean case. First, observe that the square of the Euclidean norm of a vector is the dot product of the vector with itself:

$$(\|\mathbf{y}\|_2)^2 = y_1^2 + y_2^2 + \cdots + y_n^2 = \mathbf{y}^T\mathbf{y}.$$

Hence, the square of the 2-norm of $\mathbf{Ay}$ is the dot product of $\mathbf{Ay}$ with itself:

$$(\|\mathbf{Ay}\|_2)^2 = (\mathbf{Ay})^T(\mathbf{Ay}) = \mathbf{y}^T\mathbf{A}^T\mathbf{Ay} = \mathbf{y}^T(\mathbf{A}^T\mathbf{A})\mathbf{y}. \tag{9}$$

In (9) we utilize both the associative law and the formula for the transpose of a product: The transpose of the product $\mathbf{Ay}$ is equal to the product $\mathbf{y}^T\mathbf{A}^T$ of the transposes in the reverse order. Squaring (8) and referring to (9), we see that the 2-norm of a matrix maximizes the quadratic form $\mathbf{y}^T(\mathbf{A}^T\mathbf{A})\mathbf{y}$ over the unit sphere:

$$(\|\mathbf{A}\|_2)^2 = \text{maximum}\,\{\mathbf{y}^T(\mathbf{A}^T\mathbf{A})\mathbf{y} : \|\mathbf{y}\|_2 = 1\}. \tag{10}$$

Exercises 3-3.3 and 3-3.4 investigate two cases where this maximum can be evaluated explicitly.

Exercises

3-3.1. Verify matrix norm properties (a), (c), (e), and (f). [*Hint:* For property (a), note that $\|\mathbf{A}\| = 0$ if and only if $\mathbf{Ax} = \mathbf{0}$ for every $\mathbf{x}$. For property (c), apply the triangle inequality for vector norms and observe that the maximum of a sum is less than or equal to the sum of the maximums. Property (f) follows from (d).]

3-3.2. If $\mathbf{A}$ is a square matrix, then for each positive integer p, $\mathbf{A}^p$ is defined to be the p-fold product $\mathbf{AA} \cdots \mathbf{A}$. In the special case $p = 0$, $\mathbf{A}^0$ is defined to be the identity matrix $\mathbf{I}$.

(a) Show that $\|\mathbf{A}^p\| \leqq \|\mathbf{A}\|^p$ for each integer $p \geqq 0$. [*Hint:* Use property (f).]

(b) Consider the matrix

$$\mathbf{A} = \begin{bmatrix} .1 & .7 \\ .5 & .4 \end{bmatrix}.$$

Using matrix norms, show that the elements of $\mathbf{A}^p$ are bounded by $.9^p$ for each integer $p \geqq 0$. (*Hint:* $|a_{ij}| \leqq \|\mathbf{A}\|_q$ for each i and j.)

(c) Consider the matrix

$$\mathbf{A} = \begin{bmatrix} 0 & .5 & 0 \\ .5 & 0 & .5 \\ 0 & .5 & 0 \end{bmatrix}.$$

Show that the elements of $\mathbf{A}^p$ are bounded by $1/(\sqrt{2}\,)^p$ for each even integer $p \geqq 0$. [*Hint:* If n is even, then $\mathbf{A}^n = (\mathbf{A}^2)^{n/2}$.]

3-3.3. Consider the matrix

$$\mathbf{A} = \begin{bmatrix} 2 & 0 \\ 0 & 1 \end{bmatrix}.$$

(a) Using (10), evaluate the 2-norm of $\mathbf{A}$.

(b) If $\mathbf{A}$ is a $n \times n$ diagonal matrix, what is $\|\mathbf{A}\|_1$?

(c) If $\mathbf{A}$ is a $n \times n$ diagonal matrix, what is $\|\mathbf{A}\|_\infty$?

(d) If $\mathbf{A}$ is a $n \times n$ diagonal matrix, what is $\|\mathbf{A}\|_2$?
(e) What is your conjecture for the p-norm of a diagonal matrix?

3-3.4. Evaluate the 2-norm of the matrix

$$\mathbf{A} = \begin{bmatrix} .6 & -.8 \\ .8 & .6 \end{bmatrix}.$$

If the rows of $\mathbf{A}$ are unit vectors in the Euclidean norm and the dot product between different rows is zero, then $\mathbf{A}$ is an **orthogonal matrix** and its 2-norm is always the same.

3-3.5. A scalar λ is an eigenvalue of the matrix $\mathbf{A}$ if there exists a nonzero vector $\mathbf{x}$, called the eigenvector, such that $\mathbf{Ax} = \lambda\mathbf{x}$. In Chapter 6, we will see that a $n \times n$ matrix has n eigenvalues (counting multiplicities). Show that $\|\mathbf{A}\| \geqq |\lambda|$ for each eigenvalue λ.

3-3.6. For the 1-norm, what vector $\mathbf{y}$ attains the maximum in (8)? For the ∞-norm, what vector $\mathbf{y}$ attains the maximum in (8)?

3-3.7. Establish the following identity for any square, invertible matrix $\mathbf{A}$:

$$\text{minimum } \{\|\mathbf{Ay}\| : \|\mathbf{y}\| = 1\} = \frac{1}{\|\mathbf{A}^{-1}\|}.$$

3-3.8. For any symmetric matrix $\mathbf{S}$, the following relation holds:

$$\|\mathbf{S}\|_2 = \text{maximum } \{|\mathbf{y}^T\mathbf{S}\mathbf{y}| : \|\mathbf{y}\|_2 = 1\}. \tag{11}$$

Using this result, show that $\|\mathbf{A}\|_2^2 = \|\mathbf{A}^T\mathbf{A}\|_2$.

3-3.9. It can be shown that a vector which attains the maximum in (11) is an eigenvector of $\mathbf{S}$. Using this fact and Exercise 3-3.5, show that for any symmetric matrix $\mathbf{S}$ and for any matrix norm $\|\cdot\|$ subordinate to a vector norm, the inequality $\|\mathbf{S}\|_2 \leqq \|\mathbf{S}\|$ holds.

3-3.10. Show that the ∞-norm of a matrix is the maximum absolute row sum.

3-4. CONDITION NUMBER

Now let us consider the conditioning of the linear system $\mathbf{Ax} = \mathbf{b}$. First suppose that the $\mathbf{b}$ is replaced by $\mathbf{b} + \delta\mathbf{b}$ where $\delta\mathbf{b}$ might be the error from converting $\mathbf{b}$ into floating-point form. If $\mathbf{x} + \delta\mathbf{x}$ is the solution corresponding to the right side $\mathbf{b} + \delta\mathbf{b}$, then we have

$$\mathbf{A}(\mathbf{x} + \delta\mathbf{x}) = \mathbf{b} + \delta\mathbf{b}. \tag{12}$$

The conditioning of a linear system is related to the ratio between the relative change in the solution and the relative change in the right side. If $\delta\mathbf{x}$ is relatively large compared to $\delta\mathbf{b}$, the system is ill conditioned. To analyze the conditioning of (12), we first express $\delta\mathbf{x}$ in terms of $\delta\mathbf{b}$. Since $\mathbf{Ax} = \mathbf{b}$, the $\mathbf{Ax}$ term and the $\mathbf{b}$ term in (12) cancel to give

$$\mathbf{A}\delta\mathbf{x} = \delta\mathbf{b}. \tag{13}$$

Multiplying by $\mathbf{A}^{-1}$, we have

$$\delta\mathbf{x} = \mathbf{A}^{-1}\delta\mathbf{b}. \tag{14}$$

Recall that the product between a matrix and a vector is equal to the sum of the product between each component of the vector and the corresponding column of the matrix. If $\mathbf{c}_i$ denotes the ith column of $\mathbf{A}^{-1}$, then (14) implies that

$$\delta\mathbf{x} = \sum_{i=1}^{n} \delta b_i \mathbf{c}_i. \tag{15}$$

Hence, a change δb_i in the ith component of **b** produces a change $\delta b_i \mathbf{c}_i$ in the solution, and if the ith column of $\mathbf{A}^{-1}$ is large, then a small δb_i can generate a large $\boldsymbol{\delta}\mathbf{x}$.

Equation (15) gives us the microscopic dependence of $\boldsymbol{\delta}\mathbf{x}$ on each component of $\boldsymbol{\delta}\mathbf{b}$. Taking the norm of (15) and utilizing the triangle inequality, we obtain the macroscopic dependence of $\boldsymbol{\delta}\mathbf{x}$ on $\boldsymbol{\delta}\mathbf{b}$:

$$\|\boldsymbol{\delta}\mathbf{x}\| \leqq \sum_{i=1}^{n} |\delta b_i| \, \|\mathbf{c}_i\|.$$

This shows that $\boldsymbol{\delta}\mathbf{x}$ is small in norm if the product between $|\delta b_i|$ and the norm of the ith column of $\mathbf{A}^{-1}$ is small for each i. Another macroscopic bound for $\|\boldsymbol{\delta}\mathbf{x}\|$ is obtained by taking the norm of (14) and utilizing property (d) of Section 3-3:

$$\|\boldsymbol{\delta}\mathbf{x}\| = \|\mathbf{A}^{-1}\boldsymbol{\delta}\mathbf{b}\| \leqq \|\mathbf{A}^{-1}\| \, \|\boldsymbol{\delta}\mathbf{b}\|. \tag{16}$$

Thus the change $\|\boldsymbol{\delta}\mathbf{x}\|$ in the solution is bounded by $\|\mathbf{A}^{-1}\|$ times the change $\|\boldsymbol{\delta}\mathbf{b}\|$ in the right side.

Often we are concerned with the digits of agreement between $\mathbf{x}$ and $\mathbf{x} + \boldsymbol{\delta}\mathbf{x}$, not with the magnitude $\|\boldsymbol{\delta}\mathbf{x}\|$ of the error. Recall that the relative error $\|\boldsymbol{\delta}\mathbf{x}\|/\|\mathbf{x}\|$ measures the digits of agreement between $\mathbf{x}$ and $\mathbf{x} + \boldsymbol{\delta}\mathbf{x}$. The conditioning of a linear system is connected with the ratio between the relative error and the relative change $\|\boldsymbol{\delta}\mathbf{b}\|/\|\mathbf{b}\|$ in the right side. If this ratio is large, then a relatively small change in the right side can generate a relatively large change in the solution. To estimate the ratio between the relative change in the solution and the relative change in the right side, we substitute $\mathbf{b} = \mathbf{A}\mathbf{x}$ and $\boldsymbol{\delta}\mathbf{x} = \mathbf{A}^{-1}\boldsymbol{\delta}\mathbf{b}$, to obtain

$$\frac{\|\boldsymbol{\delta}\mathbf{x}\|/\|\mathbf{x}\|}{\|\boldsymbol{\delta}\mathbf{b}\|/\|\mathbf{b}\|} = \frac{\|\mathbf{A}^{-1}\boldsymbol{\delta}\mathbf{b}\|/\|\mathbf{x}\|}{\|\boldsymbol{\delta}\mathbf{b}\|/\|\mathbf{A}\mathbf{x}\|} = \frac{\|\mathbf{A}\mathbf{x}\|}{\|\mathbf{x}\|} \, \frac{\|\mathbf{A}^{-1}\boldsymbol{\delta}\mathbf{b}\|}{\|\boldsymbol{\delta}\mathbf{b}\|}. \tag{17}$$

Utilizing property (d) in Section 3-3, $\|\mathbf{A}\mathbf{x}\| \leqq \|\mathbf{A}\| \, \|\mathbf{x}\|$ and $\|\mathbf{A}^{-1}\boldsymbol{\delta}\mathbf{b}\| \leqq \|\mathbf{A}^{-1}\| \, \|\boldsymbol{\delta}\mathbf{b}\|$. Hence identity (17) implies that

$$\frac{\|\boldsymbol{\delta}\mathbf{x}\|/\|\mathbf{x}\|}{\|\boldsymbol{\delta}\mathbf{b}\|/\|\mathbf{b}\|} \leqq \|\mathbf{A}\| \, \|\mathbf{A}^{-1}\|. \tag{18}$$

The quantity $\|\mathbf{A}\| \, \|\mathbf{A}^{-1}\|$ is called the **condition number** of $\mathbf{A}$. Letting c denote the product $\|\mathbf{A}\| \, \|\mathbf{A}^{-1}\|$, (18) can be written

$$\frac{\|\boldsymbol{\delta}\mathbf{x}\|}{\|\mathbf{x}\|} \leqq c \, \frac{\|\boldsymbol{\delta}\mathbf{b}\|}{\|\mathbf{b}\|}. \tag{19}$$

Thus the relative change in the solution is bounded by the condition number times the relative change in the right side. When the product between the condition number and the relative change in the right side is small, the relative change in the solution is small.

The effect of replacing the coefficient matrix $\mathbf{A}$ by $\mathbf{A} + \boldsymbol{\delta}\mathbf{A}$ while retaining the right side $\mathbf{b}$ is analyzed in a similar manner. Here the perturbation $\boldsymbol{\delta}\mathbf{A}$ might be the error from converting the coefficients into floating-point form. If $\mathbf{x} + \boldsymbol{\delta}\mathbf{x}$ is the solution corresponding to the coefficient matrix $\mathbf{A} + \boldsymbol{\delta}\mathbf{A}$, we have

$$(\mathbf{A} + \boldsymbol{\delta}\mathbf{A})(\mathbf{x} + \boldsymbol{\delta}\mathbf{x}) = \mathbf{b}. \tag{20}$$

Since $(\mathbf{A} + \boldsymbol{\delta}\mathbf{A})(\mathbf{x} + \boldsymbol{\delta}\mathbf{x}) = \mathbf{A}\mathbf{x} + \mathbf{A}\boldsymbol{\delta}\mathbf{x} + \boldsymbol{\delta}\mathbf{A}(\mathbf{x} + \boldsymbol{\delta}\mathbf{x})$ and $\mathbf{A}\mathbf{x} = \mathbf{b}$, (20) can be rearranged to obtain

$$\mathbf{A}\boldsymbol{\delta}\mathbf{x} = -\boldsymbol{\delta}\mathbf{A}(\mathbf{x} + \boldsymbol{\delta}\mathbf{x}).$$

Multiplying by $\mathbf{A}^{-1}$ yields

$$\boldsymbol{\delta}\mathbf{x} = -\mathbf{A}^{-1}\boldsymbol{\delta}\mathbf{A}(\mathbf{x} + \boldsymbol{\delta}\mathbf{x}).$$

Taking norms and applying the triangle inequality, we have

$$\|\boldsymbol{\delta}\mathbf{x}\| = \|\mathbf{A}^{-1}\boldsymbol{\delta}\mathbf{A}(\mathbf{x} + \boldsymbol{\delta}\mathbf{x})\| \leqq \|\mathbf{A}^{-1}\|\,\|\boldsymbol{\delta}\mathbf{A}\|(\|\mathbf{x}\| + \|\boldsymbol{\delta}\mathbf{x}\|).$$

Moving the $\boldsymbol{\delta}\mathbf{x}$ term to the left side and dividing by $\|\mathbf{x}\|$ gives

$$\frac{\|\boldsymbol{\delta}\mathbf{x}\|}{\|\mathbf{x}\|} \leqq \frac{\|\boldsymbol{\delta}\mathbf{A}\|\,\|\mathbf{A}^{-1}\|}{1 - \|\boldsymbol{\delta}\mathbf{A}\|\,\|\mathbf{A}^{-1}\|} = \frac{c\,\|\boldsymbol{\delta}\mathbf{A}\|/\|\mathbf{A}\|}{1 - \|\boldsymbol{\delta}\mathbf{A}\|\,\|\mathbf{A}^{-1}\|}. \tag{21}$$

If the product $\|\boldsymbol{\delta}\mathbf{A}\|\,\|\mathbf{A}^{-1}\|$ is much smaller than 1, the denominator in (21) is near 1. Consequently, when $\|\boldsymbol{\delta}\mathbf{A}\|\,\|\mathbf{A}^{-1}\|$ is much smaller than 1, (21) implies that the relative change in the solution is bounded by the condition number $c = \|\mathbf{A}\|\,\|\mathbf{A}^{-1}\|$ times the relative change in the coefficient matrix.

Since the condition number is expressed in terms of a norm, the condition number c_1 associated with the 1-norm will be different from the condition number c_2 associated with the 2-norm. However, by inequality (6), different p-norms are closely related, and if the condition number is large in one norm, it is large in other norms. Moreover, one property of the condition number is valid for every norm: The condition number is at least 1. In fact, if $\mathbf{A}$ is a scalar, then $\mathbf{A}^{-1} = 1/\mathbf{A}$ and $c = \|\mathbf{A}\|\,\|\mathbf{A}^{-1}\| = |\mathbf{A}|/|\mathbf{A}| = 1$. Hence, c is exactly 1 for one equation. The analogous result $c \geqq 1$ for systems of equations is based on properties (e) and (f) (see Section 3-3) of a matrix norm. Since $\mathbf{I} = \mathbf{A}\mathbf{A}^{-1}$, (e) and (f) with $\mathbf{B} = \mathbf{A}^{-1}$ yield

$$1 = \|\mathbf{I}\| = \|\mathbf{A}\mathbf{A}^{-1}\| \leqq \|\mathbf{A}\|\,\|\mathbf{A}^{-1}\| = c.$$

That is, $c \geqq 1$.

The accuracy in the computed solution to a linear system can be estimated in terms of the condition number and the residual. Let $\bar{\mathbf{x}}$ denote the computed solution to $\mathbf{A}\mathbf{x} = \mathbf{b}$. Rearranging the definition $\mathbf{r} = \mathbf{b} - \mathbf{A}\bar{\mathbf{x}}$ of the residual, we have $\mathbf{A}\bar{\mathbf{x}} = \mathbf{b} - \mathbf{r}$, which is equivalent to

$$\mathbf{A}(\mathbf{x} + \bar{\mathbf{x}} - \mathbf{x}) = \mathbf{b} - \mathbf{r}. \tag{22}$$

If $\boldsymbol{\delta}\mathbf{x}$ denotes $\bar{\mathbf{x}} - \mathbf{x}$ and $\boldsymbol{\delta}\mathbf{b}$ denotes $-\mathbf{r}$, then (22) has the form

$$\mathbf{A}(\mathbf{x} + \boldsymbol{\delta}\mathbf{x}) = \mathbf{b} + \boldsymbol{\delta}\mathbf{b}.$$

Substituting $\boldsymbol{\delta}\mathbf{x} = \bar{\mathbf{x}} - \mathbf{x}$ and $\boldsymbol{\delta}\mathbf{b} = -\mathbf{r}$, (19) tells us that

$$\frac{\|\bar{\mathbf{x}} - \mathbf{x}\|}{\|\mathbf{x}\|} \leqq c\,\frac{\|\mathbf{r}\|}{\|\mathbf{b}\|}. \tag{23}$$

Thus the relative error in $\bar{\mathbf{x}}$ is less than or equal to the condition number times the relative residual $\|\mathbf{r}\|/\|\mathbf{b}\|$. *Observe that a relatively small residual does not ensure a relatively small error.* Only when the product between the condition number and the relative residual is small can we be certain that the relative error in $\bar{\mathbf{x}}$ is small. If the relative residual $\|\mathbf{r}\|/\|\mathbf{b}\|$ is near our machine's epsilon E while c is greater than $1/E$, then $c\|\mathbf{r}\|/\|\mathbf{b}\|$ is greater than 1 and the computed solution $\bar{\mathbf{x}}$ may disagree with $\mathbf{x}$ in the first significant digit. Loosely speaking, a linear system is well conditioned with respect to the computing precision if c is small relative to $1/E$.

Equation (1) at the beginning of the chapter is an example of an ill-conditioned system. For this equation, we have

$$\mathbf{A} = \begin{bmatrix} 600 & 800 \\ 30001 & 40002 \end{bmatrix} \quad \text{and} \quad \mathbf{A}^{-1} \approx \begin{bmatrix} 100 & -2.0 \\ -75 & 1.5 \end{bmatrix}.$$

Since $\|\mathbf{A}\|_1 = 40802$ and $\|\mathbf{A}^{-1}\|_1 \approx 175$, the 1-norm condition number for $\mathbf{A}$ is about 7×10^6, which is greater than the reciprocal of the machine epsilon $.95 \times 10^{-6}$ for an IBM 370 computer. The relatively large condition number for $\mathbf{A}$ and the relatively large error in the computed solution (2) go hand in hand.

Notice that relation (23) gives us an *upper bound* for the error in the computed solution $\bar{\mathbf{x}}$. A more precise expression for the error is obtained from equation (15). Identifying $\boldsymbol{\delta}\mathbf{x}$ with $\bar{\mathbf{x}} - \mathbf{x}$ and $\boldsymbol{\delta}\mathbf{b}$ with $-\mathbf{r}$, equation (15) yields:

$$\mathbf{x} - \bar{\mathbf{x}} = \sum_{i=1}^{n} r_i \mathbf{c}_i \tag{24}$$

where $\mathbf{c}_i$ is the ith column of $\mathbf{A}^{-1}$. Since (24) is an equality while (23) is an inequality, (24) estimates the error in $\bar{\mathbf{x}}$ more precisely than (23). On the other hand, (23) is cheaper to apply than (24) since the condition number can be approximated faster than a matrix can be inverted. An algorithm to estimate the condition number of a matrix appears at the end of this section.

Although the expression $c\|\boldsymbol{\delta}\mathbf{b}\|/\|\mathbf{b}\|$ may overestimate the relative error $\|\boldsymbol{\delta}\mathbf{x}\|/\|\mathbf{x}\|$, there are many cases where $\|\boldsymbol{\delta}\mathbf{x}\|/\|\mathbf{x}\|$ is close to $c\|\boldsymbol{\delta}\mathbf{b}\|/\|\mathbf{b}\|$. Referring to the derivation of (19), we see that $\|\boldsymbol{\delta}\mathbf{x}\|/\|\mathbf{x}\|$ equals $c\|\boldsymbol{\delta}\mathbf{b}\|/\|\mathbf{b}\|$ if and only if $\|\mathbf{A}\mathbf{x}\| = \|\mathbf{A}\|\,\|\mathbf{x}\|$ and $\|\mathbf{A}^{-1}\boldsymbol{\delta}\mathbf{b}\| = \|\mathbf{A}^{-1}\|\,\|\boldsymbol{\delta}\mathbf{b}\|$. Recall that the 1-norm of a matrix is the largest column sum. If j is a column index corresponding to the largest column sum for $\mathbf{A}$ and if every component of $\mathbf{x}$ is zero except for component j, then $\|\mathbf{A}\mathbf{x}\|_1 = \|\mathbf{A}\|_1\|\mathbf{x}\|_1$. Similarly, if k is a column index corresponding to the largest column sum for $\mathbf{A}^{-1}$ and if every component of $\boldsymbol{\delta}\mathbf{b}$ is zero except for component k, then $\|\mathbf{A}^{-1}\boldsymbol{\delta}\mathbf{b}\|_1 = \|\mathbf{A}^{-1}\|_1\|\boldsymbol{\delta}\mathbf{b}\|_1$. Therefore, the equality $\|\boldsymbol{\delta}\mathbf{x}\|_1/\|\mathbf{x}\|_1 = c_1\|\boldsymbol{\delta}\mathbf{b}\|_1/\|\mathbf{b}\|_1$ holds whenever each component of $\mathbf{x}$ and $\boldsymbol{\delta}\mathbf{b}$ is zero except for components j and k, respectively. Finally, note that if every component of $\mathbf{x}$ is zero except for component j, then $\mathbf{b} = \mathbf{A}\mathbf{x}$ is a multiple of column j from $\mathbf{A}$.

We now state an algorithm, developed in [H4], which gives an estimate for $\|\mathbf{A}^{-1}\|_1$ that averages just 3% smaller than the true $\|\mathbf{A}^{-1}\|_1$. Moreover, neglecting arithmetic errors, this scheme obtains the correct $\|\mathbf{A}^{-1}\|_1$ in over 80% of the cases. Since $\|\mathbf{A}\|_1$ can be evaluated explicitly (it is the maximum absolute column sum), an estimate for the 1-norm condition number is given by the product between the 1-norm of $\mathbf{A}$ and the estimated 1-norm of $\mathbf{A}^{-1}$.

1. $b_i \leftarrow 1/n$ for $i = 1$ to n and $\rho \leftarrow 0$.
2. Solve $\mathbf{A}\mathbf{x} = \mathbf{b}$ for the unknown $\mathbf{x}$. If $\|\mathbf{x}\| \le \rho$, go to step 7. Else $\rho \leftarrow \|\mathbf{x}\|_1$ and proceed to step 3.
3. $i = 1$ to n
   - $y_i \leftarrow 1$ if $x_i \geqq 0$
   - $y_i \leftarrow -1$ if $x_i < 0$

   next i
4. Solve $\mathbf{A}^T\mathbf{z} = \mathbf{y}$ for the unknown $\mathbf{z}$.
5. $j \leftarrow \arg\max \{|z_i| : i = 1 \text{ to } n\}$.
6. If $|z_j| > \mathbf{z}^T\mathbf{b}$, go to step 2 after making the assignments $b_i \leftarrow 0$ for $i = 1$ to n and $b_j \leftarrow 1$. Else go to step 7.
7. $\|\mathbf{A}^{-1}\|_1 \approx \rho$ and $c_1 \approx \rho\|\mathbf{A}\|_1$.

(25)

As observed in [H4], this algorithm usually stops after two iterations so that the running time is roughly the time to solve $\mathbf{Ax} = \mathbf{b}$ with two different choices of $\mathbf{b}$ plus the time to solve $\mathbf{A}^T\mathbf{z} = \mathbf{y}$ with two different choices of $\mathbf{y}$. Improvements in this algorithm which yield still sharper estimates for $\|\mathbf{A}^{-1}\|_1$ also appear in [H4]. These improved estimates, however, involve more computations than (25). Function routines in the program library use the solution subroutines of Chapter 2 to implement the condition estimator (25). Function CON estimates the condition number of a general factored matrix. Analogous routines for matrices with special structure have names of the form _CON. For example, BCON estimates the condition number for a system that has been factored previously using BFACT. The condition number for the coefficient matrix of equation (1) is determined by the following program:

```
REAL A(10),B(2)
DATA A/600.,30001.,800.,40002./
CALL FACT(A,2,2)
WRITE(6,*) CON(A,B)
END
```

Observe that the arguments of the condition estimation routine consist of the factored array followed by a working array (B above) with at least n elements, where n is the matrix dimension.

Exercises

3-4.1. Consider the matrix

$$\mathbf{A} = \begin{bmatrix} .6 & -.8 \\ .8 & .6 \end{bmatrix}.$$

(a) Evaluate the 2-norm condition number of $\mathbf{A}$. (*Hint:* $\mathbf{A}^{-1} = \mathbf{A}^T$.)

(b) Given a fixed parameter θ, evaluate the 2-norm condition number of the matrix

$$\mathbf{A} = \begin{bmatrix} \cos\theta & -\sin\theta \\ \sin\theta & \cos\theta \end{bmatrix}.$$

The condition number of a matrix is at least 1 and for the 2-norm, the only matrices with unit condition number are multiples of an orthogonal matrix.

3-4.2. If $\mathbf{A}$ and $\mathbf{B}$ are $n \times n$ invertible matrices and $c(\mathbf{A})$ denotes the condition number of $\mathbf{A}$, show that

$$c(\mathbf{AB}) \leqq c(\mathbf{A})c(\mathbf{B}).$$

In other words, show that the condition number for the product $\mathbf{AB}$ is less than or equal to the product between the condition number for $\mathbf{A}$ and the condition number for $\mathbf{B}$.

3-4.3. If λ is an eigenvalue of $\mathbf{A}$ and $\mathbf{A}$ has an inverse, show that $1/\lambda$ is an eigenvalue of $\mathbf{A}^{-1}$. Referring to Exercise 3-3.5, show that the condition number of $\mathbf{A}$ satisfies the inequality

$$c(\mathbf{A}) = \|\mathbf{A}\|\,\|\mathbf{A}^{-1}\| \geqq \frac{|\lambda_b|}{|\lambda_s|},$$

where λ_b and λ_s are, respectively, the absolute largest eigenvalue and the absolute smallest eigenvalue of $\mathbf{A}$.

3-4.4. Use inequality (23) to obtain an estimate for the 1-norm relative error in the computed solution (2) for equation (1). What is the ratio between the estimated relative error and the true relative error?

3-4.5. Consider the perturbed linear system $\mathbf{A}(\mathbf{x} + \boldsymbol{\delta}\mathbf{x}) = \mathbf{b} + \boldsymbol{\delta}\mathbf{b}$. What is the minimum value for the ratio $(\|\boldsymbol{\delta}\mathbf{x}\|/\|\mathbf{x}\|)/(\|\boldsymbol{\delta}\mathbf{b}\|/\|\mathbf{b}\|)$? (*Hint:* See Exercise 3-3.7.) For the 1-norm, determine the $\mathbf{x}$ and the $\boldsymbol{\delta}\mathbf{b}$ for which the ratio $(\|\boldsymbol{\delta}\mathbf{x}\|_1/\|\mathbf{x}\|_1)/(\|\boldsymbol{\delta}\mathbf{b}\|_1/\|\mathbf{b}\|_1)$ is a minimum? (*Hint:* See Exercise 3-3.6.) What is the $\mathbf{b}$ that corresponds to this minimizing $\mathbf{x}$? In some sense, this minimizing choice for $\mathbf{b}$ and $\boldsymbol{\delta}\mathbf{b}$ are the best right side and the best perturbation since the relative change in the solution is as small as possible.

3-4.6. The coefficients of the Hilbert matrix are defined by $a_{ij} = 1/(i + j - 1)$. Using subroutines SFACT and SCON, estimate the condition number of the $n \times n$ Hilbert matrix for n between 2 and 20. First compute the condition number using single precision and then compute the condition number using double precision. When a matrix is ill conditioned, the condition number of its single-precision representation and its double-precision representation can be quite different.

3-5. GEOMETRIC SERIES

In Section 3-4 we used norms to study the conditioning of a linear system. In this section, we use norms to study the convergence of a matrix geometric series. Some applications of these series appear in the exercises. For a scalar geometric series $1 + r^1 + r^2 + r^3 + \cdots$, each term approaches zero and the sum is $1/(1 - r)$ provided $|r| < 1$. Now consider the matrix geometric series $\mathbf{S}$ given by

$$\mathbf{S} = \mathbf{I} + \mathbf{A} + \mathbf{A}^2 + \mathbf{A}^3 + \cdots. \tag{26}$$

If $\|\mathbf{A}\| < 1$ for some matrix norm, then each term in the series (26) tends to zero since $\|\mathbf{A}^k\| \leqq \|\mathbf{A}\|^k$ (see Exercise 3-3.2). Since each term in the series (26) is bounded by $\|\mathbf{A}\|^k$ and since the scalar geometric series $1 + \|\mathbf{A}\|^1 + \|\mathbf{A}\|^2 + \cdots$ converges, it follows from the dominated convergence theorem that the matrix geometric series (26) also converges when $\|\mathbf{A}\| < 1$.

To sum the matrix geometric series, we multiply by $\mathbf{A}$ to obtain

$$\mathbf{AS} = \mathbf{A} + \mathbf{A}^2 + \mathbf{A}^3 + \cdots. \tag{27}$$

Subtracting (27) from (26) yields $\mathbf{S} - \mathbf{AS} = \mathbf{I}$ or $(\mathbf{I} - \mathbf{A})\mathbf{S} = \mathbf{I}$. Since the product $(\mathbf{I} - \mathbf{A})\mathbf{S}$ equals $\mathbf{I}$, we conclude that $\mathbf{S}$ is the inverse of $\mathbf{I} - \mathbf{A}$. Hence the sum of the matrix geometric series is $\mathbf{S} = (\mathbf{I} - \mathbf{A})^{-1}$. By truncating the series (26), we obtain an approximation to the inverse of $\mathbf{I} - \mathbf{A}$. Let $\mathbf{P}_k$ denote the partial sum defined by

$$\mathbf{P}_k = \mathbf{I} + \mathbf{A} + \mathbf{A}^2 + \cdots + \mathbf{A}^k.$$

Subtracting $\mathbf{P}_k$ from $\mathbf{S}$ yields

$$\mathbf{S} - \mathbf{P}_k = \mathbf{A}^{k+1} + \mathbf{A}^{k+2} + \cdots = \mathbf{A}^{k+1}(\mathbf{I} + \mathbf{A} + \mathbf{A}^2 + \cdots).$$

Taking norms and utilizing the triangle inequality and the fact that the norm of a product is bounded by the product of the norms, we have

$$\|\mathbf{S} - \mathbf{P}_k\| = \|(\mathbf{A} - \mathbf{I})^{-1} - \mathbf{P}_k\| \leqq \|\mathbf{A}\|^{k+1}[1 + \|\mathbf{A}\| + \|\mathbf{A}\|^2 + \cdots] = \frac{\|\mathbf{A}\|^{k+1}}{1 - \|\mathbf{A}\|}.$$

In other words, the norm of the difference between the partial sum $\mathbf{P}_k$ and the true sum $\mathbf{S} = (\mathbf{A} - \mathbf{I})^{-1}$ is at most $\|\mathbf{A}\|^{k+1}/(1 - \|\mathbf{A}\|)$.

Exercises

3-5.1. Let **S** be the symmetric tridiagonal matrix with 0's on the diagonal and 1's off the diagonal and let **D** be the diagonal matrix with 10's on the diagonal. Consider the problem of inverting the tridiagonal matrix $\mathbf{T} = \mathbf{D} - \mathbf{S}$. Since $\mathbf{T} = \mathbf{D} - \mathbf{S} = \mathbf{D}(\mathbf{I} - \mathbf{D}^{-1}\mathbf{S})$ and since the inverse of a product **BC** is the product $\mathbf{C}^{-1}\mathbf{B}^{-1}$ of the inverses in reverse order, the inverse of **T** can be written

$$\mathbf{T}^{-1} = (\mathbf{I} - \mathbf{D}^{-1}\mathbf{S})^{-1}\mathbf{D}^{-1}.$$

Letting **A** denote $\mathbf{D}^{-1}\mathbf{S}$, (26) implies that

$$\mathbf{T}^{-1} = (\mathbf{I} + \mathbf{A} + \mathbf{A}^2 + \cdots)\mathbf{D}^{-1}.$$

Consider the following approximation to the inverse of **T**:

$$\mathbf{T}^{-1} = (\mathbf{I} + \mathbf{A} + \mathbf{A}^2 + \cdots + \mathbf{A}^k)\mathbf{D}^{-1}.$$

For what value of k does the 1-norm of the approximate inverse differ by at most .001 from the true inverse? [*Hint:* $\mathbf{D}^{-1} = .1\ \mathbf{I}$ and $\mathbf{T}^{-1} = .1(\mathbf{I} - .1\mathbf{S})^{-1}$.]

3-5.2. Show that $\|(\mathbf{I} - \mathbf{A})^{-1}\| \leqq 1/(1 - \|\mathbf{A}\|)$ whenever **A** is a square matrix and $\|\mathbf{A}\| < 1$.

3-5.3. Suppose that **B** is an invertible matrix. Show that for $\delta\mathbf{B}$ sufficiently small, the perturbed matrix $\mathbf{B} + \delta\mathbf{B}$ is invertible and $(\mathbf{B} + \delta\mathbf{B})^{-1}$ approaches $\mathbf{B}^{-1}$ as $\delta\mathbf{B}$ tends to zero. [*Hint:* Note that $\mathbf{B} + \delta\mathbf{B} = \mathbf{B}(\mathbf{I} - \mathbf{A})$ where $\mathbf{A} = -\mathbf{B}^{-1}\delta\mathbf{B}$.]

3-5.4. Suppose that **A** is a square matrix, the elements of **A** are nonnegative, and $\|\mathbf{A}\| < 1$ for some matrix norm. Show that the elements of $(\mathbf{I} - \mathbf{A})^{-1}$ are nonnegative.

3-5.5. Suppose that **B** is a square invertible matrix, $b_{ii} > 0$ for every i, $b_{ij} \leqq 0$ for every $i \neq j$, and each row sum for **B** is nonnegative. Show that every element of $\mathbf{B}^{-1}$ is nonnegative. [*Hint:* Let **D** denote the diagonal matrix formed from **B**'s diagonal and define $\mathbf{E} = \mathbf{D} - \mathbf{B}$. Observe that the diagonal of **E** is zero while the off-diagonal elements are the negatives of the off-diagonal elements of **B**. Letting ε denote a small positive number, consider the matrix

$$\mathbf{C} = \mathbf{B} + \varepsilon\mathbf{I} = (\mathbf{D} + \varepsilon\mathbf{I}) - \mathbf{E} = (\mathbf{D} + \varepsilon\mathbf{I})(\mathbf{I} - (\mathbf{D} + \varepsilon\mathbf{I})^{-1}\mathbf{E}).$$

Apply Exercise 3-5.4 with $\mathbf{A} = (\mathbf{D} + \varepsilon\mathbf{I})^{-1}\mathbf{E}$; then let ε approach zero and utilize Exercise 3-5.3.]

REVIEW PROBLEMS

3-1. Given a $m \times n$ matrix **A**, let $\mathbf{a}_j$ denote its jth column. Establish the following relationship between the p-norm of the matrix **A** and the p-norm of each column:

$$\underset{j=1,\cdots,n}{\text{maximum}} \|\mathbf{a}_j\|_p \leqq \|\mathbf{A}\|_p \leqq n^{(p-1)/p} \underset{j=1,\cdots,n}{\text{maximum}} \|\mathbf{a}_j\|_p .$$

[*Hint:* To establish the upper bound for $\|\mathbf{A}\|_p$, observe that $\mathbf{Ay} = y_1\mathbf{a}_1 + \cdots + y_n\mathbf{a}_n$. Apply the triangle inequality to **Ay** and utilize the inequality (6).]

3-2. Given a $m \times n$ matrix **A**, use Problem 3-1 and inequality (5) to show that

$$\underset{i,j}{\text{maximum}} |a_{ij}| \leqq \|\mathbf{A}\|_p \leqq n^{(p-1)/p} m^{1/p} \underset{i,j}{\text{maximum}} |a_{ij}|.$$

3-3. Given a $m \times n$ matrix **A**, use Problem 3-1 and inequality (6) to show that

$$m^{(1-p)/p}\|\mathbf{A}\|_1 \leqq \|\mathbf{A}\|_p \leqq n^{(p-1)/p}\|\mathbf{A}\|_1.$$

3-4. If **x** and **y** are vectors, show that $|\mathbf{x}^T\mathbf{y}| \leqq \|\mathbf{x}\|_\infty\|\mathbf{y}\|_1$. For which unit vectors **y** does the equality $|\mathbf{x}^T\mathbf{y}| = \|\mathbf{x}\|_\infty$ hold?

3-5. The generalization of Problem 3-4, known as Hölder's inequality, can be stated

$$\text{maximum } \{|\mathbf{x}^T\mathbf{y}| : \|\mathbf{y}\|_q = 1\} = \|\mathbf{x}\|_p$$

whenever p and q are positive and $1/p + 1/q = 1$. Use Hölder's inequality and the definition (8) for a matrix norm to show that $\|\mathbf{A}\|_p = \|\mathbf{A}^T\|_q$.

3-6. Given a $m \times n$ matrix $\mathbf{A}$, show that

$$n^{-1/p}\|\mathbf{A}\|_\infty \leqq \|\mathbf{A}\|_p \leqq m^{1/p}\|\mathbf{A}\|_\infty .$$

(*Hint:* Combine Problems 3-3 and 3-5.)

3-7. Prove that for any matrix $\mathbf{A}$ and for any positive p and q for which $1/p + 1/q = 1$, we have

$$\|\mathbf{A}\|_2 \leqq \sqrt{\|\mathbf{A}\|_p\|\mathbf{A}\|_q} .$$

(*Hint:* By Exercise 3-3.9, $\|\mathbf{A}^T\mathbf{A}\|_2 \leqq \|\mathbf{A}^T\mathbf{A}\|_p$ since $\mathbf{A}^T\mathbf{A}$ is symmetric.)

3-8. Use relation (6) to establish the following inequality for any $m \times n$ matrix $\mathbf{A}$ and for any $q \geqq p \geqq 1$:

$$n^{(p-q)/pq}\|\mathbf{A}\|_q \leqq \|\mathbf{A}\|_p \leqq m^{(q-p)/pq}\|\mathbf{A}\|_q .$$

3-9. Given a $m \times n$ matrix $\mathbf{A}$, let $\|\mathbf{A}\|_F$ denote the Frobenius norm defined by

$$\|\mathbf{A}\|_F = \left\{\sum_{i=1}^{m}\sum_{j=1}^{n} a_{ij}^2\right\}^{\frac{1}{2}} .$$

Establish the following relationship between the 2-norm and the Frobenius norm:

$$\|\mathbf{A}\|_2 \leqq \|\mathbf{A}\|_F \leqq \sqrt{n}\ \|\mathbf{A}\|_2 .$$

(*Hint:* To establish the lower bound, first apply the triangle inequality to the sum $x_1\mathbf{a}_1 + \cdots + x_n\mathbf{a}_n$ and then apply the Cauchy-Schwarz inequality "$|\mathbf{x}^T\mathbf{y}| \leqq \|\mathbf{x}\|_2\|\mathbf{y}\|_2$." To establish the upper bound, utilize Problem 3-1.)

3-10. Suppose that $\mathbf{A} = \mathbf{u}\mathbf{v}^T$, where $\mathbf{u}$ is a vector with m components and $\mathbf{v}$ is a vector with n components. Use Hölder's inequality to show that $\|\mathbf{A}\|_p = \|\mathbf{u}\|_p\|\mathbf{v}\|_q$.

3-11. Given vectors $\mathbf{u}$ and $\mathbf{v}$, show that the matrix $\mathbf{A} = \mathbf{v}\mathbf{u}^T/\mathbf{u}^T\mathbf{u}$ is a matrix with smallest 2-norm that satisfies the equation $\mathbf{A}\mathbf{u} = \mathbf{v}$. [*Hint:* First show that $\mathbf{A} = \mathbf{v}\mathbf{u}^T/\mathbf{u}^T\mathbf{u}$ satisfies the equation $\mathbf{A}\mathbf{u} = \mathbf{v}$. Then use matrix norm property (d) to obtain a lower bound for $\|\mathbf{A}\|_2$. Finally, use Problem 3-10 to show that $\mathbf{A} = \mathbf{v}\mathbf{u}^T/\mathbf{u}^T\mathbf{u}$ achieves this lower bound.]

3-12. Given two matrices $\mathbf{B}$ and $\mathbf{C}$ with the same number of columns, let $\mathbf{A}$ denote the composite matrix formed by $\mathbf{B}$ above $\mathbf{C}$:

$$\mathbf{A} = \begin{bmatrix} \mathbf{B} \\ \mathbf{C} \end{bmatrix} .$$

Show that $\|\mathbf{A}\|_2^2 \leqq \|\mathbf{B}\|_2^2 + \|\mathbf{C}\|_2^2$.

3-13. Suppose that $\mathbf{A}$ is the symmetric tridiagonal matrix with 2's on the diagonal and with -1's off the diagonal. Investigate how the condition number depends on the matrix dimension n by using TFACT and TCON to compute the 1-norm condition number corresponding to $n = 10$, 20, 40, and 80. In this example, the condition number has the form $c_1 \approx CN^p$. Use the technique developed in Chapter 1, equations (30) and (31), to determine C and p.

3-14. Consider a system of 100 equations $\mathbf{A}\mathbf{x} = \mathbf{b}$, where $\mathbf{A}$ is the symmetric tridiagonal matrix with every diagonal element equal to 20,000 and with every off-diagonal element equal to $-10{,}000$, and $\mathbf{b}$ is the vector with components $b_i = i/100$. Solve $\mathbf{A}\mathbf{x} = \mathbf{b}$ using TFACT and TSOLVE, compute the residual $\mathbf{r} = \mathbf{b} - \mathbf{A}\mathbf{x}$, and determine the condition number using TCON. Utilizing relation (23), determine an upper bound for the 1-norm relative error in the computed solution. Finally, resolve $\mathbf{A}\mathbf{x} = \mathbf{b}$ using double precision and determine the actual error in the single-precision solution by comparing it to the double-precision solution.

3-15. When solving a boundary-value with periodic boundary conditions, suppose that one obtains a $n \times n$ matrix of the form

$$\mathbf{A} = \begin{bmatrix} a & -2 & & & & -1 \\ -2 & b & -2 & & & \\ & -2 & b & \cdot & & \\ & & \cdot & \cdot & \cdot & \\ & & & \cdot & b & -2 \\ -1 & & & & -2 & a \end{bmatrix},$$

where $a = 3 + n^{-2}$ and $b = 4 + n^{-2}$. Using algorithm (25), estimate the condition number of $\mathbf{A}$ when $n = 10$, $n = 20$, and $n = 40$. (*Hint:* To solve $\mathbf{Ax} = \mathbf{b}$, use the Sherman-Morrison formula as you did in Exercise 2-8.2. Also observe that $\mathbf{A}^T = \mathbf{A}$.)

3-16. Show that the p-norm of a diagonal matrix is the magnitude of the absolute largest diagonal element. That is, if $\mathbf{A}$ is $n \times n$ diagonal matrix, then

$$\|\mathbf{A}\|_p = \text{maximum } \{|a_{11}|, \cdots, |a_{nn}|\}$$

for each $p \geqq 1$.

3-17. Given a nonsingular matrix $\mathbf{W}$ and a vector norm $\|.\|$, define a new vector norm by $\|\mathbf{x}\|_W = \|\mathbf{Wx}\|$. Show that the induced matrix norm is given by $\|\mathbf{A}\|_W = \|\mathbf{WAW}^{-1}\|$.

REFERENCES

Other references for norms include Householder's classic [H18] as well as books by Golub and Van Loan [G15], Horn and Johnson [H16], Ortega [O2], and Stewart [S14]. The algorithm (25) is essentially a gradient algorithm for maximizing the convex function $f(\mathbf{x}) = \|\mathbf{A}^{-1}\mathbf{x}\|_1$ over $\mathbf{x}$ on the unit sphere $\|\mathbf{x}\|_1 = 1$. Starting at the center of a face for the unit sphere, we move to the vertex that yields the locally largest increase in the value of f. Then in each successive iteration, we move to the vertex that yields the locally largest increase in the value of f. The iterations stop when the value of f locally decreases in the direction of other vertices. Another scheme to estimate the condition number of a matrix is developed in [C2]. An algorithm to compute the condition number of a tridiagonal matrix appears in [H14]. With regard to Exercises 3-5.4 and 3-5.5, a wealth of results concerning matrices with nonnegative inverses appears in Varga's book [V2].

4

Nonlinear Systems

4-1. SUBSTITUTION TECHNIQUES

Linear equations can be solved directly by Gaussian elimination using a finite number of additions, subtractions, multiplications, and divisions. Nonlinear equations are distinctly different. Even a simple equation like

$$x^3 = \sin x \tag{1}$$

can never be solved by algebraic manipulation, so our algorithms must be iterative. Graphing the function $f(x) = x^3 - \sin x$, we observe three points where f vanishes. Each of these points is a **root** of equation (1) or a **zero** of function f. Let us compute the zero that lies near $x = 1$ in Figure 4-1 using the following procedure: We rewrite equation (1) in the form $x = g(x)$ and we perform the substitution $x_{k+1} = g(x_k)$. For example, taking cube roots, (1) reduces to

$$x = \sqrt[3]{\sin x}. \tag{2}$$

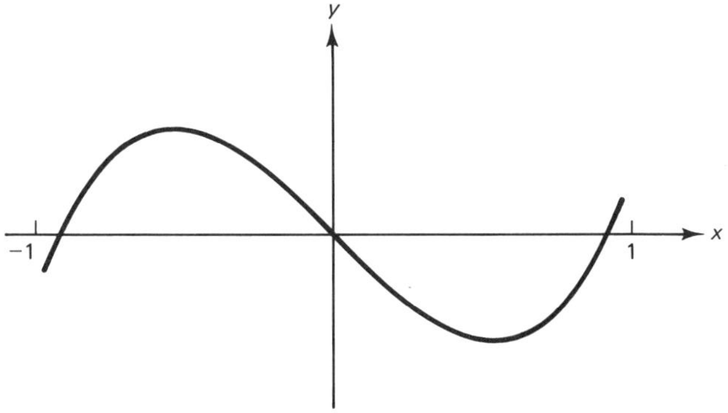

Figure 4-1 The graph of $y = x^3 - \sin x$.

TABLE 4-1
ITERATION (3)
STARTING FROM
$x_1 = 1$

k	x_k
1	1.000
2	.944
3	.932
4	.929
5	.929

Letting $g(x)$ be defined by $g(x) = \sqrt[3]{\sin x}$, the iteration $x_{k+1} = g(x_k)$ is

$$x_{k+1} = \sqrt[3]{\sin x_k}\,. \tag{3}$$

If the starting guess is $x_1 = 1$, then

$$x_2 = \sqrt[3]{\sin x_1} = \sqrt[3]{\sin 1.00} \approx .944 \text{ and } x_3 = \sqrt[3]{\sin x_2} = \sqrt[3]{\sin .944} \approx .932.$$

The first five iterations appear in Table 4-1. Notice that the iterations are converging to the root $x \approx .929$ of equation (1). When an equation is written in the form $x = g(x)$, a root is also called a **fixed point** of g. If $x_k = x$, a fixed point of g, then $x_{k+1} = g(x_k) = g(x) = x = x_k$ and the iterations repeat.

Of course, there are many different ways to transform equation (1) into the form $x = g(x)$. Another possibility is to divide by x^2, giving

$$x = \frac{\sin x}{x^2}.$$

In this case, $g(x) = \sin x/x^2$ and the iteration $x_{k+1} = g(x_k)$ is

$$x_{k+1} = \frac{\sin x_k}{{x_k}^2}. \tag{4}$$

Starting from $x_1 = 1$, we obtain the iterations in Table 4-2. Notice that iteration (4) diverges from the root $x \approx .929$.

To understand why (3) converges while (4) diverges, we return to the iteration cobwebs of Chapter 1. Recall that the iteration $x_{k+1} = e^{-x_k}$ is generated by a cobweb spun between the line $y = x$ and the curve $y = e^{-x}$. Similarly, the iteration $x_{k+1} = g(x_k)$ is generated by a cobweb spun between the line $y = x$

TABLE 4-2
ITERATION (4)
STARTING FROM
$x_1 = 1$

k	x_k
1	1.000
2	.841
3	1.053
4	.783
5	1.149

and the curve $y = g(x)$. The cobweb corresponding to iteration (3) appears in Figure 4-2a while the cobweb corresponding to iteration (4) appears in Figure 4-2b. At the intersection of the line $y = x$ and the curve $y = g(x)$, we have $x = y = g(x)$. Hence the intersection point corresponds to the root of the equation $x = g(x)$. The cobweb approaches the root in Figure 4-2a, while the cobweb diverges from the root in Figure 4-2b. Experimenting with various g's, we discover that the iteration behavior is controlled by the slope of g near the root. Letting α denote the root, the iterations converge if $|g'(\alpha)| < 1$ and the starting guess x_1 is near α. Conversely, the iterations diverge if $|g'(\alpha)| > 1$. The derivative of the first function $g(x) = (\sin x)^{1/3}$ is

$$g'(x) = \frac{(\sin x)^{-2/3}}{3} \cos x.$$

At the root $\alpha \approx .929$, we have $g'(.929) \approx .23$. Since .23 is less than one, iteration (3) converges when the starting guess is near the root. The derivative of the second function $g(x) = \sin x/x^2$ is

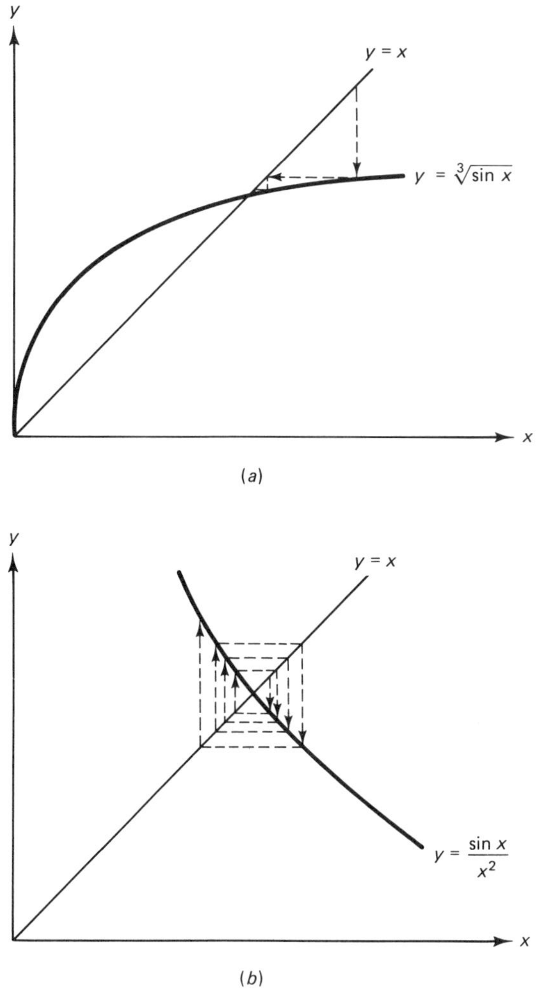

Figure 4-2 Iteration cobwebs: (a) for $g(x) = \sqrt[3]{\sin x}$; (b) for $g(x) = \sin x/x^2$.

$$g'(x) = \frac{\cos x}{x^2} - 2\frac{\sin x}{x^3}.$$

At the root $\alpha \approx .929$, we have $g'(.929) \approx -1.3$. Since the absolute value of $g'(\alpha)$ is greater than 1, iteration (4) must diverge.

The techniques developed above to transform the equation $x^3 = \sin x$ into the form $x = g(x)$ are very specialized. Now we have to find a systematic way to construct an equivalent form $x = g(x)$ starting from any given equation $f(x) = 0$. One way is to add x to both sides of the equation $f(x) = 0$. The new equation is $x = f(x) + x$, and $g(x)$ is equal to $f(x) + x$. But Newton found a better way to obtain $x = g(x)$. His idea is the fundamental principle: *Linearize the problem*. As always, we start from an initial guess x_1 for the solution. Then we replace f by the tangent line at x_1 (see Figure 4-3). The tangent line in the figure has the same slope and height at $x = x_1$ as that of f. The equation of the tangent is

$$y = f(x_1) + f'(x_1)(x - x_1). \tag{5}$$

At $x = x_1$, we see that $y = f(x_1)$. Thus the height of the line (5) at $x = x_1$ is the same as the height of f at x_1. Moreover, the slope of the line is $f'(x_1)$, the same as the slope of f at x_1. Newton's second approximation x_2 to the root is the point where the tangent (5) crosses the x-axis. Since the tangent crosses the x-axis when y is zero, x_2 satisfies the equation

$$0 = f(x_1) + f'(x_1)(x_2 - x_1).$$

Solving for x_2, we obtain

$$x_2 = x_1 - \frac{f(x_1)}{f'(x_1)}.$$

To generate x_3, we linearize about x_2 and again compute the zero of the tangent:

$$x_3 = x_2 - \frac{f(x_2)}{f'(x_2)}.$$

The iteration continues in the same way. It amounts to the substitution $x_{k+1} = g(x_k)$ for a particular g, which we denote g_N:

$$g_N(x) = x - \frac{f(x)}{f'(x)}.$$

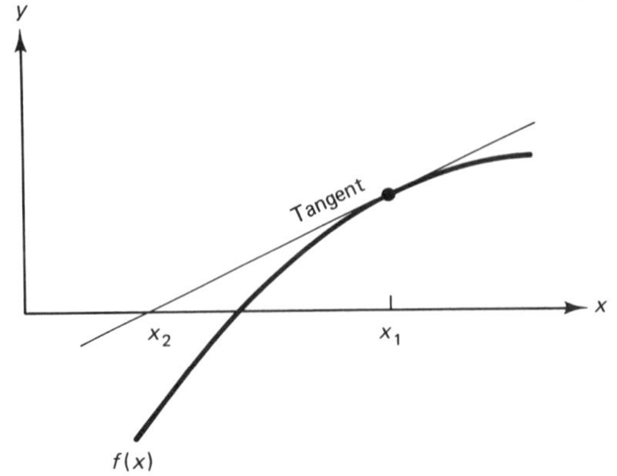

Figure 4-3 The linear approximation at $x = x_1$.

You can see that the first step was $x_2 = g_N(x_1)$ and the second was $x_3 = g_N(x_2)$. We could have constructed g_N by starting from $0 = f(x)$, dividing by $f'(x)$, and adding x to both sides, giving us the equivalent relation

$$x = x - \frac{f(x)}{f'(x)}.$$

Since the right side is g_N, the original equation $f(x) = 0$ is equivalent to $x = g_N(x)$, which suggests the iteration $x_{k+1} = g_N(x_k)$. These blind algebraic manipulations, however, hide the geometry of Newton's method depicted in Figure 4-4. For the right triangle we equate the change in y over the change in x to the slope to obtain

$$f'(x_k) = \frac{f(x_k)}{x_k - x_{k+1}}.$$

Solving for x_{k+1} yields Newton's iteration:

$$x_{k+1} = x_k - \frac{f(x_k)}{f'(x_k)} = g_N(x_k). \tag{6}$$

Let us apply Newton's method to the equation $x^3 = \sin x$. To obtain f, we rewrite the equation so that the right side is zero:

$$f(x) = x^3 - \sin x = 0.$$

Hence, $g_N(x) = x - f(x)/f'(x)$ is given by

$$g_N(x) = x - \frac{x^3 - \sin x}{3x^2 - \cos x}.$$

Table 4-3 (on page 150) shows the iteration $x_{k+1} = g_N(x_k)$ starting from $x_1 = 1$.

Convergence is fast! The number of correct digits nearly doubles each step. Earlier we observed that the iteration $x_{k+1} = g(x_k)$ converges to a root α if $|g'(\alpha)| < 1$ and x_1 is near α. As we will see in the next section, the convergence speed depends on the absolute value of $g'(\alpha)$ and the error at each step is nearly multiplied by the factor $|g'(\alpha)|$. That is,

$$|x_{k+1} - \alpha| \approx |g'(\alpha)||x_k - \alpha|.$$

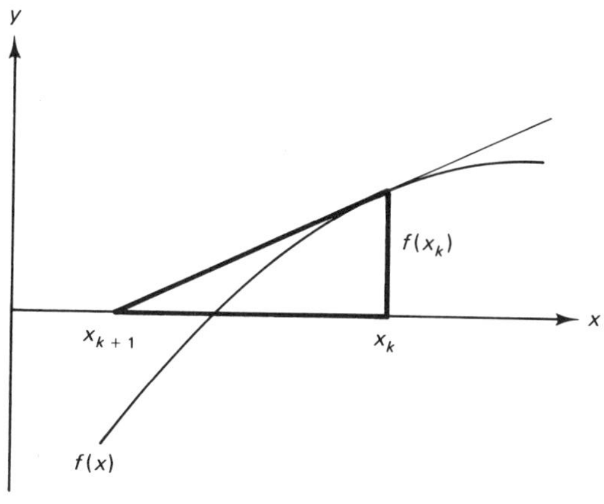

Figure 4-4 Newton's method.

TABLE 4-3 NEWTON'S ITERATION FOR $x^3 - \text{SIN } x = 0$

k	x_k
1	1.000000000000000
2	.935549390654669
3	.928701808803824
4	.928626317865637
5	.928626308731734
6	.928626308731734

The derivative of Newton's choice $g_N(x) = x - f(x)/f'(x)$ is

$$\frac{dg_N}{dx}(x) = 1 - \frac{f'(x)^2 - f(x)f''(x)}{f'(x)^2} = \frac{f(x)f''(x)}{f'(x)^2}.$$

Since $f(\alpha) = 0$, it follows that the convergence factor $g_N'(\alpha)$ is zero for Newton's method. The iterations converge rapidly in Table 4-3 because the error is multiplied by a factor near zero in each iteration.

Newton's method approximates f by its tangent at each step. This suggests a possibility that looks even better: Approximate f by a parabola instead of a straight line. If we apply it to simple examples, the convergence is fantastic. But like many good ideas in numerical analysis (and in life), there is a place to stop and Newton found it. Often, approximations utilizing high-degree polynomials are awkward to implement—for systems of equations the calculations can be horrendous and a straightforward linearization is better. In fact, even for a single equation, there are situations where it is not easy to evaluate the derivative $f'(x)$ and Newton's method itself is either too awkward or too time consuming.

One strategy for reducing the cost of each Newton iteration is to employ a previous derivative in the current iteration. This strategy, called **modified Newton's method**, is described by the iteration

$$x_{k+1} = x_k - \frac{f(x_k)}{f'(x_1)}.$$

In Newton's method (6), we evaluate the derivative $f'(x_k)$ in each iteration. In modified Newton's method, the derivative $f'(x_1)$ is utilized in each iteration. Thus the derivative of f is evaluated only once, during the first iteration.

The modified Newton scheme is illustrated in Figure 4-5. In each iteration,

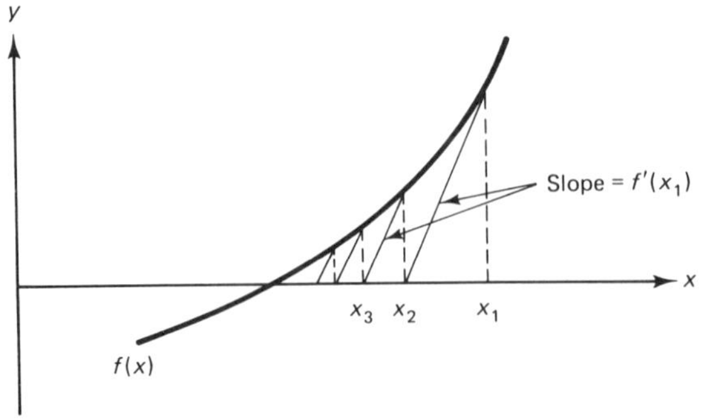

Figure 4-5 Modified Newton's method.

TABLE 4-4
MODIFIED NEWTON ITERATION (7).

k	x_k
1	1.000000
2	.935549
3	.929891
4	.928866
5	.928672
6	.928635
7	.928628
8	.928626

we start at a point on the curve $y = f(x)$ and we move along a line with slope $f'(x_1)$ until reaching the x-axis. If the slope of f at x_1 differs significantly from the slope of f at the root, the modified Newton scheme is slow. For this reason, it is a good idea to reevaluate the derivative of f at the current iteration and restart the iterations occasionally. For the equation $f(x) = x^3 - \sin x = 0$ and the starting guess $x_1 = 1$, we have $f'(x) = 3x^2 - \cos x$ and $f'(1) \approx 2.4597$. Hence the modified Newton scheme is

$$x_{k+1} = x_k - \frac{x_k^{\,3} - \sin x_k}{2.4597}. \tag{7}$$

With the starting guess $x_1 = 1$, we have the iterations in Table 4-4. In this example, the convergence is reasonably fast, but not as fast as the original Newton iteration in Table 4-3.

The **secant method** is a compromise between Newton's method, where x_{k+1} is constructed using a line of slope $f'(x_k)$ (see Figure 4-4), and the modified Newton scheme, where x_{k+1} is constructed using a line of slope $f'(x_1)$ (see Figure 4-5). In the secant method x_{k+1} is determined using a line whose slope coincides with the slope of the chord that subtends the graph of f at $x = x_k$ and $x = x_{k-1}$ (see Figure 4-6). The slope of the chord that subtends the graph of f at $x = x_k$ and $x = x_{k-1}$ is

$$\frac{f(x_k) - f(x_{k-1})}{x_k - x_{k-1}}. \tag{8}$$

Replacing $f'(x_k)$ in (6) by the approximation (8) gives us the secant iteration:

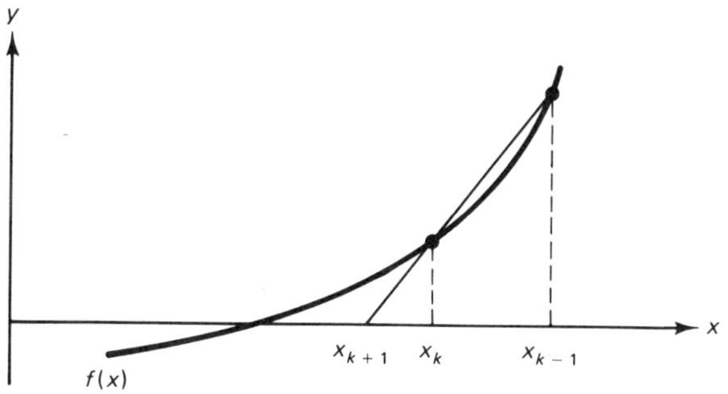

Figure 4-6 The secant method.

TABLE 4-5 SECANT ITERATION FOR $x^3 - \sin x = 0$

k	x_k
1	1.100000000000000
2	1.000000000000000
3	.943636858361971
4	.930174607737405
5	.928662795531490
6	.928626399090439
7	.928626308737018
8	.928626308731734
9	.928626308731734

$$x_{k+1} = x_k - \frac{x_k - x_{k-1}}{f(x_k) - f(x_{k-1})} f(x_k). \tag{9}$$

Since the secant iteration expresses x_{k+1} in terms of x_k and x_{k-1}, x_3 is determined from x_2 and x_1. Unlike Newton's method, which just requires one starting guess x_1, the secant method requires two starting points x_1 and x_2. In practice, x_2 is usually any approximation to the root and x_1 is any nearby point. Table 4-5 gives the secant iterations (9) corresponding to the equation $x^3 - \sin x = 0$ and the starting guess $x_2 = 1.0$ and $x_1 = 1.1$. Comparing Tables 4-3, 4-4, and 4-5, we see that the secant method is faster than the modified Newton iteration but slower than Newton's method.

Newton's method and related schemes converge when the starting guess is near the root. But for a poor starting guess, the iterations can diverge. In contrast, the **bisection method** converges even when the starting guess is far from the root. The bisection method is based on the following principle: If f has opposite signs at L and R and f is continuous† on the interval $[L, R]$, then f vanishes at least once between L and R (see Figure 4-7).

In stating the bisection method, we employ the "sign" function, a function that provides a numerical representation of a number's sign. If x is positive, then sign(x) is $+1$ and if x is negative, then sign(x) is -1. Although the sign

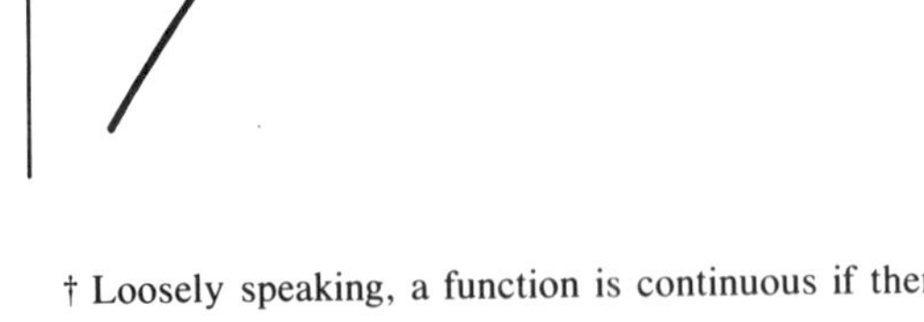

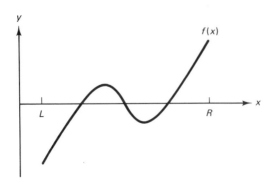

Figure 4-7 A function with opposite signs at L and R.

† Loosely speaking, a function is continuous if there are no gaps in the graph.

of x is undefined when x is zero, it will be convenient to set $\text{sign}(x) = 1$ if $x = 0$. In other words, we define $\text{sign}(x)$ in the following way:

$$\text{sign}(x) = \begin{cases} +1 & \text{if } x \geqq 0, \\ -1 & \text{if } x < 0. \end{cases}$$

Suppose that f has opposite signs at L and R and let s_L denote $\text{sign}(f(L))$. As its name suggests, the bisection method repeatedly bisects an interval that contains a zero of f. Let M denote the midpoint of the interval $[L, R]$:

$$M = \frac{L + R}{2}.$$

Inspecting the sign of f at M, we determine a smaller interval that contains a zero of f. If $\text{sign}(f(M)) = s_L$, then f has opposite signs at M and at R and a zero lies on the interval $[M, R]$. If $\text{sign}(f(M)) = -s_L$, then f has opposite signs at L and at M and a zero lies on the interval $[L, M]$. In either case, a smaller interval is determined which contains a zero. Then this smaller interval is bisected and the iteration continues in the same way:

1. $s \leftarrow \text{sign}(f(L))$.
2. $M \leftarrow (L+R)/2$.
3. If $\text{sign}(f(M)) = s$, then $L \leftarrow M$. Else $R \leftarrow M$.
4. Return to step 2 unless $R - L$ is sufficiently small.

(10)

The bisection method

Table 4-6 shows the successive intervals $[L, R]$ generated by the bisection method for the equation $x^3 - \sin x = 0$ and for the starting points $L = .5$ and $R = 1$.

Comparing Tables 4-3, 4-5, and 4-6, we see that both Newton's method and the secant method are faster than the bisection method. In particular, each iteration of Newton's method roughly doubles the number of correct digits in the approximation x_k to the root while each bisection step divides the bracketing interval in half. For the bisection method, the error in our estimate of the root is typically proportional to the width of the bracketing interval. To improve the accuracy by one digit, we must reduce the width of the bracketing interval by

TABLE 4-6 BISECTION METHOD FOR $x^3 - \text{SIN } x = 0$

Iteration	L	R
1	.500	1.000
2	.750	1.000
3	.875	1.000
4	.875	.938
5	.906	.938
6	.922	.938
7	.922	.930
8	.926	.930
9	.928	.930
10	.928	.929

the factor $\frac{1}{10}$. One bisection reduces the width by the factor $\frac{1}{2}$, two bisections reduces the width by the factor $\frac{1}{4}$, and roughly 3.3 bisections reduces the width by the factor $\frac{1}{10}$. (Actually, $1/\log_{10}2$ bisections reduces the width by the factor $\frac{1}{10}$). To summarize, each Newton iteration roughly doubles the number of correct digits while 3.3 bisections increases the number of correct digits by one. Even though the bisection method is slower than either Newton's method or the secant method, it has the advantage of guaranteed convergence. This observation suggests that a hybrid bisection-secant scheme is better than either method by itself. We can use the slow but sure bisection method until we are so close to the root that a scheme like the secant method converges rapidly. Function routine ROOT uses a strategy like this to solve $f(x) = 0$ starting from two points where f has opposite sign. ROOT has four arguments:

L: left side of bracketing interval
R: right side of bracketing interval
T: computing tolerance
F: name of routine to evaluate function

The iteration continues until the error in the root is less than or equal to the computing tolerance T.

Let us briefly describe the algorithm employed by ROOT. The first iteration is a secant step. In the second iteration, a zero of the quadratic that agrees with the function at the three previous points is used to approximate the root. Thereafter, Newton's method is employed except that the derivative appearing in g_N is approximated by the derivative of an interpolating cubic polynomial. Whenever the convergence appears to be slow in the sense that the successive iteration differences do not decrease by at least the factor $\frac{1}{2}$, a bisection step is performed and the entire process is restarted. To illustrate the use of ROOT, we present a program that computes a zero for two different functions: $f(x) = x^3 - \sin x$ and $h(x) = x - e^{-x}$. Each of these functions has a zero between $\frac{1}{2}$ and 1. The following program determines an approximation to each zero which deviates by at most 10^{-5} from the true zero.

```
EXTERNAL F,H
X = ROOT(.5,1.,1.E-5,F)
WRITE(6,*)X
X = ROOT(.5,1.,1.E-5,H)
WRITE(6,*)X
END

FUNCTION F(X)
F = X**3 - SIN(X)
RETURN
END

FUNCTION H(X)
H = X - EXP(-X)
RETURN
END
```

The final argument of ROOT is the name of the function evaluation routine. Note that a function name can be employed as a subroutine argument only if that name is declared external at the start of the program.

Exercises

4-1.1. Using a pocket calculator, compute the roots of the following equations with three digit accuracy:

$$\text{(a) } x^4 = \sin x \quad \text{(b) } x/2 = \sin x \quad \text{(c) } 1 + x = \tan x,\ 0 \leqq x \leqq 3$$

4-1.2. Consider the equation $x^2 - 2 = 0$.

(a) Show graphically how Newton's method behaves when the starting guess x_1 is near zero.

(b) Show graphically how the secant method behaves when $x_1 \approx -.3$ and $x_2 \approx .3$.

(c) If the secant iteration is converging to a zero of f, then due to rounding errors, it is possible to have $f(x_k) = f(x_{k-1}) \approx 0$. In this case, what is x_{k+1}? When programming the secant method, what action should be taken when $f(x_k) = f(x_{k-1}) \approx 0$?

4-1.3. Newton's method can be used to find a complex zero of a polynomial, however, the starting guess must be a complex number and the program variables must be declared "COMPLEX." Use Newton's method to find the zeros of the polynomial

$$p(z) = z^4 + 4z^3 + 3z^2 + 10z + 50.$$

(*Hint:* For different starting guesses, the Newton iterations converge to different zeros. Also note that the zeros occur in conjugate pairs since the polynomial coefficients are real numbers. That is, if α is a zero of p, so is the complex conjugate of α.)

4-2. CONVERGENCE THEORY

Let us now explain rigorously why the iteration $x_{k+1} = g(x_k)$ converges to a root α of the equation $x = g(x)$ if $|g'(\alpha)| < 1$ and x_1 is near the root. Subtracting the identity $\alpha = g(\alpha)$ from $x_{k+1} = g(x_k)$ gives us the relation

$$x_{k+1} - \alpha = g(x_k) - g(\alpha). \tag{11}$$

Letting e_k denote the error $x_k - \alpha$ at step k, (11) is written

$$e_{k+1} = g(x_k) - g(\alpha). \tag{12}$$

We apply the mean value theorem to the right side of (12). Recall that if $g'(x)$ depends continuously on x for x between a and b, then by the mean value theorem, there exists a point c between a and b such that

$$g'(c) = \frac{g(b) - g(a)}{b - a}. \tag{13}$$

The right side of (13) is the slope of the chord subtending the graph of g at $x = a$ and $x = b$. The left side of (13) is the slope of the tangent at $x = c$. By the mean value theorem, there exists a point c between a and b such that the slope of the tangent is equal to the slope of the chord. The construction of c is illustrated in Figure 4-8 (on page 156).

Rearranging (13), we have

$$g(b) - g(a) = g'(c)(b - a). \tag{14}$$

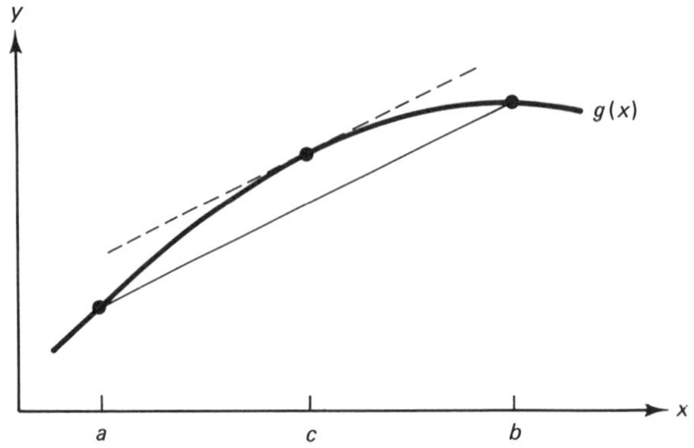

Figure 4-8 The mean value theorem.

Identifying a with α and b with x_k, (12) and (14) combine to give

$$e_{k+1} = g'(c_k)(x_k - \alpha) = g'(c_k)e_k, \tag{15}$$

where c_k lies between x_k and α. Let us assume that for some number $\lambda < 1$ and for some number $r > 0$, we have

$$|g'(x)| \leqq \lambda < 1 \tag{16}$$

whenever the distance from x to α is less than or equal to r. In other words, we assume that $|g'(x)| \leqq \lambda < 1$ whenever x lies on the interval $[\alpha - r, \alpha + r]$. If x_0 lies on the interval $[\alpha - r, \alpha + r]$, then since c_0 lies between x_0 and α, c_0 also lies on the interval $[\alpha - r, \alpha + r]$. Putting $k = 0$ in (15) and utilizing (16), we have

$$|e_1| = |g'(c_0)e_0| \leqq \lambda|e_0| \leqq r. \tag{17}$$

This shows that the distance from x_1 to α is less than or equal to both r and $\lambda|e_0|$. Since c_1 lies between x_1 and α, the distance from c_1 to α is less than or equal to r. Putting $k = 1$ in (15) and utilizing (16) and (17) yields

$$|e_2| = |g'(c_1)||e_1| \leqq \lambda|e_1| \leqq \lambda^2|e_0|.$$

Observe that the error is multiplied by λ each iteration: $|e_1| \leqq \lambda|e_0|$ and $|e_2| \leqq \lambda|e_1| \leqq \lambda^2|e_0|$. Therefore, the error at step k is at most λ^k times the initial error:

$$|e_k| \leqq \lambda^k|e_0|. \tag{18}$$

When the error contracts by a factor $\lambda < 1$ each iteration, the convergence is called **contracting convergence.**

If (16) holds, the iterations not only converge to α starting from any x_0 in the interval $[\alpha - r, \alpha + r]$, but α is the unique root of $x = g(x)$ on the interval $[\alpha - r, \alpha + r]$. To verify this uniqueness property, suppose that β is a second root of $x = g(x)$ on the interval $[\alpha - r, \alpha + r]$. Subtracting the relation $\beta = g(\beta)$ from $\alpha = g(\alpha)$ gives us

$$\alpha - \beta = g(\alpha) - g(\beta).$$

By the mean value theorem and (16), we have

$$|\alpha - \beta| = |g(\alpha) - g(\beta)| \leqq \lambda|\alpha - \beta|.$$

Rearranging this inequality yields

$$(1 - \lambda)|\alpha - \beta| \leqq 0. \tag{19}$$

Since $1 - \lambda > 0$ and $|\alpha - \beta| \geqq 0$, (19) implies that $|\alpha - \beta| = 0$ and $\alpha = \beta$.

In our analysis of the iteration $x_{k+1} = g(x_k)$, we have assumed that the root α is known. In most cases, however, we wish to determine whether an iteration will converge even though the precise location of the root is not known. Let us now examine some convergence results that do not require precise knowledge of α. First, let us examine the case where g' is positive on an interval $[a, b]$ which contains a root of the equation $x = g(x)$. Equation (15) and the inequality $g'(c_k) > 0$ imply that the sign of the error at x_{k+1} is the same as the sign of the error at x_k. Assuming that there exists a scalar $\lambda < 1$ such that $g'(x) \leqq \lambda$ for every x between a and b, (15) implies that x_{k+1} is closer to α than x_k. When the iterations converge to a root and the error has the same sign in each iteration, the convergence is **monotonic**.

Monotone convergence is illustrated using the equation

$$x = 3 - \frac{1}{1 + x} \tag{20}$$

and the iteration $x_{k+1} = 3 - 1/(1 + x_k)$. The g corresponding to this iteration is given by $g(x) = 3 - 1/(1 + x)$. Moving all terms to the left, we see that (20) can be expressed $f(x) = 0$, where $f(x) = x - 3 + 1/(1 + x)$. Since $f(1) = -1.5 < 0$ and $f(3) = \frac{1}{4} > 0$, f vanishes between $x = 1$ and $x = 3$, or equivalently, (20) has a root on the interval $[1, 3]$. Observe that $g'(x) = 1/(1 + x)^2$. For x between 1 and 3, $g'(x)$ is less than 1. Since g' is positive for $1 \leqq x \leqq 3$, the iterations $x_{k+1} = g(x_k)$ converge to the root whenever the starting guess is between 1 and 3. The value of λ corresponding to the interval $[1, 3]$ is $\frac{1}{4}$, the largest value for g' on $[1, 3]$. By (18), the error satisfies the relation $|e_{k+1}| \leqq (\frac{1}{4})^k|e_0|$; furthermore, when x_0 lies between 1 and 3, the error in x_0 is at most 2, the width of the interval that contains the root. Combining these observations, we have $|e_{k+1}| \leqq 2(\frac{1}{4})^k$. The iterations corresponding to (20) appear in Table 4-7.

Next, we analyze the case where g' is negative on an interval that contains a root. In other words, there exists $\lambda < 1$ such that $-\lambda \leqq g'(x) < 0$. Since $g'(c_k)$ is negative, (15) implies that the sign of the error at x_{k+1} is opposite to the sign of the error at x_k. That is, x_{k+1} and x_k lie on opposite sides of the root. In particular, the starting guess x_0 and the first iteration x_1 lie on opposite sides of the root. If $-\lambda \leqq g'(x) \leqq 0$ for each x between x_0 and x_1, then by (15), the iterations oscillate back and forth about the root, and in each iteration, the error contracts by the factor λ at least.

TABLE 4-7 ITERATION $x_{k+1} = 3 - 1/(1 + x_k)$

k	x_k
0	1.00000
1	2.50000
2	2.71428
3	2.73077
4	2.73196
5	2.73204
6	2.73205
7	2.73205

Oscillating convergence is illustrated using the equation

$$x = 3 - \log x \tag{21}$$

and the iteration $x_{k+1} = 3 - \log x_k$. Moving each term to the left, equation (21) is written $f(x) = 0$, where $f(x) = x - 3 + \log x$. Since $f(2) \approx -.3$ and $f(3) \approx 1.1$, f vanishes at a point between $x = 2$ and $x = 3$, or equivalently, (21) has a root on the interval [2, 3]. Since $g(x) = 3 - \log x$, we have $g'(x) = -1/x$ and $-\frac{1}{2} \leqq g'(x) \leqq -\frac{1}{3}$ for x between 2 and 3. Since the maximum magnitude of g's derivative on the interval [2, 3] is less than 1, the iterations converge when the starting guess is sufficiently close to the root. Taking $x_0 = 2$, we have $x_1 \approx 2.3$. Since $g'(x) < 0$ and $|g'(x)| = 1/x \leqq \frac{1}{2}$ for x between $x_0 = 2$ and $x_1 \approx 2.3$, it follows from (18) that $|e_k| \leqq (\frac{1}{2})^k|e_0|$. Of course, the error e_0 in x_0 is at most .3 since the root lies on the interval [2, 2.3]. Combining these observations, we have $|e_k| \leqq .3(\frac{1}{2})^k$. The actual iterations corresponding to (21) appear in Table 4-8.

Finally, let us consider the case where g' vanishes at a point near a root. Initially, we assume for convenience that there exists $\lambda < 1$ such that $|g'(x)| \leqq \lambda$ for every x. Subtracting each side of the relation $x_k = g(x_{k-1})$ from the corresponding side of the relation $x_{k+1} = g(x_k)$ yields $x_{k+1} - x_k = g(x_k) - g(x_{k-1})$. Applying the mean value theorem as in (15), we have $d_{k+1} = g'(c_k)d_k$, where $d_k = x_k - x_{k-1}$. Taking absolute values and replacing $g'(c_k)$ by λ gives us an inequality: $|d_{k+1}| \leqq \lambda|d_k|$. Since this inequality holds for each k, it follows [as in (18)] that $|d_k| \leqq \lambda^k|d_0|$. As a consequence, the iterations not only converge, but they never stray too far from x_0. In particular, using the triangle inequality, we have the following estimate for the distance from x_0 to x_k:

$$\begin{aligned}|x_k - x_0| &= |x_k - x_{k-1} + x_{k-1} - x_{k-2} + x_{k-2} + \cdots + x_2 - x_1 + x_1 - x_0| \\ &\leqq |d_k| + |d_{k-1}| + \cdots + |d_0| \\ &\leqq |d_0|(\lambda^k + \lambda^{k-1} + \cdots + \lambda^0) \\ &\leqq |d_0|(1 + \lambda^1 + \lambda^2 + \cdots) \\ &= \frac{|d_0|}{1 - \lambda} = \frac{|x_1 - x_0|}{1 - \lambda}.\end{aligned}$$

Since the iterations never stray more than the distance $|x_1 - x_0|/(1 - \lambda)$ away from x_0, the initial assumption "$|g'(x)| \leqq \lambda$ for every x" can be weakened to "$|g'(x)| \leqq \lambda$ for every x in the interval $[x_0 - r, x_0 + r]$" where

$$r = \frac{|x_1 - x_0|}{1 - \lambda}.$$

TABLE 4-8 ITERATION $x_{k+1} = 3 - \text{LOG}\ x_k$

k	x_k	k	x_k
0	2.000	5	2.212
1	2.307	6	2.206
2	2.164	7	2.209
3	2.228	8	2.208
4	2.199	9	2.208

After expressing λ in terms of r, our convergence result can be stated in the following way:

If there exists $r > 0$ such that $|g'(x)| \leqq 1 - |x_1 - x_0|/r$ for every x in the interval $[x_0 - r, x_0 + r]$, then the iteration $x_{k+1} = g(x_k)$ converges to a root of the equation $x = g(x)$ and the error satisfies the inequality (18) with $\lambda = 1 - |x_1 - x_0|/r$.

You apply this convergence result to a numerical example in Exercise 4-2.2.

There is one special case, namely Newton's method, which exhibits monotone convergence near a root even though the derivative of g changes sign. Often the derivative of g_N vanishes at the root, with g_N' positive on one side of the root, and g_N' negative on the other side of the root. Since $g_N'(\alpha) = 0$, $|g_N'(x)| < 1$ for x in a neighborhood of α (assuming that the derivative is continuous). By our previous analysis, we know that the iteration converges monotonically to the root whenever the starting guess is in the region where the derivative is both nonnegative and less than 1. On the other hand, if the starting guess lies in the region where the derivative is negative, the first iteration takes us to the other side of the root, where the derivative is positive and the convergence is monotone thereafter.

For Newton's method, the region associated with monotone convergence is actually larger than the previous analysis predicts. In particular, for Newton's method monotone convergence occurs whenever the starting guess lies in the region where g_N' is positive, even though $|g_N'(x)| > 1$. Figure 4-9 illustrates monotone convergence for Newton's method. In Section 4-1 we found that the derivative of g_N is given by

$$\frac{dg_N}{dx}(x) = \frac{f(x)f''(x)}{f'(x)^2}. \tag{22}$$

Since denominator of (22) is nonnegative, the product between f and f'' is positive

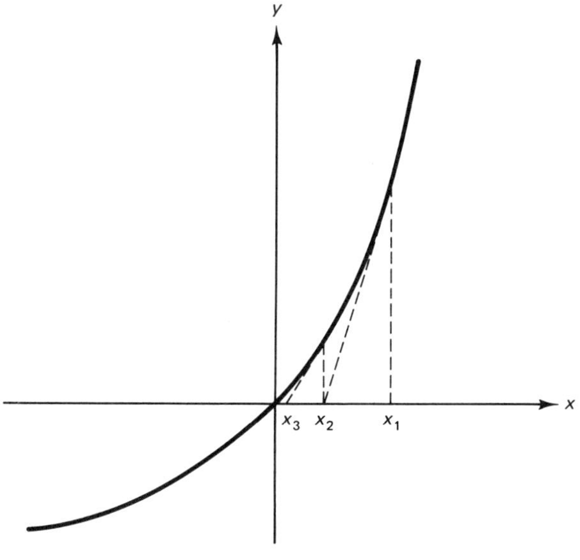

Figure 4-9 Monotone convergence for Newton's method.

in the region where the derivative of g_N is positive. If the product between f and f'' is positive, then either both f and its second derivative are positive and the graph of f is concave upward as in Figure 4-9 or both f and its second derivative are negative and the graph of f is concave downward. In either case, the iterations converge monotonically. We emphasize that in general Newton's method is not guaranteed to converge. Monotone convergence is only guaranteed when the starting guess lies in an open interval (a, b) with the following property: Either a or b is the root α and $g_N'(x)$ is positive [or equivalently, the product $f(x)f''(x)$ is positive] for every x between a and b.

The factor λ in (18) is related to convergence speed since the error contracts by at least the factor λ in each iteration. If the iterations $x_{k+1} = g(x_k)$ lie near a root α, then $g'(x) \approx g'(\alpha)$ for x near α and the iteration error contracts by about the factor $|g'(\alpha)|$ each iteration. On the other hand, (18) grossly underestimates the convergence speed of Newton's method. Relation (15) implies that

$$e_{k+1} \leqq \lambda_k e_k, \tag{23}$$

where λ_k is the maximum absolute value for g' on the interval $[x_k, \alpha]$. Since the derivative of Newton's g is zero at the root, $g_N'(x)$ approaches zero as x approaches α, and λ_k in (23) typically approaches zero as k increases. An inequality like (18) that involves a fixed parameter λ will always underestimate the speed of Newton's method. This leads us to a better measure of convergence speed: We say that the sequence $x_0, x_1, \cdots$ converges to α at **order** p if there exists a positive constant C such that

$$\lim_{k\to\infty} \frac{\|x_{k+1} - \alpha\|}{\|x_k - \alpha\|^p} = C. \tag{24}$$

If $p = 1$, then the convergence is called **linear** and we also require that $C < 1$. It can be shown that $C = |g'(\alpha)|$ when the convergence is linear. If $p = 2$, the convergence is called **quadratic**.

To investigate the meaning of convergence order, let us suppose that $C = 1$, $p = 2$, $\|x_0 - \alpha\| = .1$, and the ratio $\|x_{k+1} - \alpha\|/\|x_k - \alpha\|^p$ actually equals C every iteration. The equality $\|x_{k+1} - \alpha\| = C\|x_k - \alpha\|^2$ implies that

$$\|x_1 - \alpha\| = \|x_0 - \alpha\|^2 = (.1)^2 = .01,$$

$$\|x_2 - \alpha\| = \|x_1 - \alpha\|^2 = (.01)^2 = .0001,$$

$$\|x_3 - \alpha\| = \|x_2 - \alpha\|^2 = (.0001)^2 = .00000001.$$

The iteration error decreases like the sequence $10^{-1}, 10^{-2}, 10^{-4}, 10^{-8}, \cdots$. Loosely speaking, the number of correct digits doubles each iteration when p is 2. For a pth-order method, the number of correct digits increases by the factor p each iteration. To illustrate the importance of convergence order, let us perform the following experiment: Assuming that $C = .8$, $\|x_0 - \alpha\| = .1$, and the ratio $\|x_{k+1} - \alpha\|/\|x_k - \alpha\|^p$ actually equals C each iteration, we determine the first iteration for which $\|x_k - \alpha\|$ is less than 10^{-6}. Our results are displayed in Table 4-9 for several choices of p. Observe that convergence is very rapid near a root when p is greater than or equal to 2.

The convergence order depends on both the algorithm and the equation, and in some cases p is not an integer. For example, the convergence order of the secant method is $(1 + \sqrt{5})/2 \approx 1.6$. On the other hand, when p is an

TABLE 4-9
CONVERGENCE ORDER VERSUS NUMBER OF ITERATIONS

p	Iterations
1	52
2	3
3	2
4	2

integer, there is a simple procedure for finding its value. The iteration $x_{k+1} = g(x_k)$ converges at order p to the root α of the equation $x = g(x)$ if the first through the $(p - 1)$st derivatives of g vanish at α, the pth derivative is nonzero at α, and the starting guess is sufficiently close to α. The convergence is linear if $g'(\alpha)$ is nonzero and $|g'(\alpha)| < 1$. Examining the iterations in Table 4-3, we suspect that Newton's method is quadratic. To determine the convergence order, we compute the derivatives of $g_N(x) = x - f(x)/f'(x)$ at a root α of $f(x) = 0$. The first derivative of g_N appears in (22). Observe that the first derivative is zero at $x = \alpha$ provided that $f'(\alpha) \neq 0$. To compute the second derivative, we differentiate (22). Recall that the derivative of a product like $f(x)$ times $(f''(x)/f'(x)^2)$ is the first factor times the derivative of the second plus the second factor times the derivative of the first. Therefore, we have

$$\frac{d^2 g_N}{dx^2}(x) = f(x)\left(\frac{f''(x)}{f'(x)^2}\right)' + \left(\frac{f''(x)}{f'(x)^2}\right)f'(x) = f(x)\left(\frac{f''(x)}{f'(x)^2}\right)' + \frac{f''(x)}{f'(x)}.$$

The first term is zero at $x = \alpha$ since $f(\alpha)$ is 0, while the second term is nonzero if $f''(\alpha) \neq 0$. Therefore, Newton's method is second order when both $f'(\alpha)$ and $f''(\alpha)$ are nonzero. But if $f''(\alpha)$ is zero, the second derivative of g_N vanishes at the root and Newton's method is at least third order.

As stated above, the convergence speed depends on both the algorithm and the equation. Let us examine the effect of root multiplicity on the convergence rate. We say that the zero $x = 1$ of the polynomial $f(x) = x^3 - 3x + 2 = (x - 1)^2(x + 2)$ is a zero with multiplicity 2 since the factor $(x - 1)$ is squared. Notice that $f(1) = f'(1) = 0$ while $f''(1) = 6$. In general a zero α of f has **multiplicity** m if the derivatives of f through order $m - 1$ vanish at α while the mth derivative is nonzero. That is, $f(\alpha) = f'(\alpha) = \cdots = f^{(m-1)}(\alpha) = 0$ and $f^{(m)}(\alpha) \neq 0$. A zero of multiplicity 1 is a **simple zero**. Almost any algorithm converges slowly at a multiple zero. To illustrate this property, we display Newton iterations for two similar equations in Table 4-10.

For both equations in Table 4-10 (on page 162), $x = 1$ is a root. In the first example, $x = 1$ is a simple zero of f and the number of correct digits doubles each iteration. In the second example, $x = 1$ is a zero with multiplicity 2 and the number of correct digits increases by 1 every 3.3 iterations. The reason for slow convergence in the second example is that Newton's method is just a first-order scheme at a multiple root. Observe that the derivative of g_N at $x = \alpha$ involves the ratio $f(\alpha)/f'(\alpha)^2$, which is 0/0 at a multiple root:

TABLE 4-10 NEWTON'S METHOD AT A SIMPLE ROOT AND AT A MULTIPLE ROOT

$f(x) = (x-1)(x+2) = 0$		$f(x) = (x-1)^2(x+2) = 0$	
k	x_k	k	x_k
1	2.000000000	1	2.000000000
2	1.199999999	2	1.555555555
3	1.011764705	3	1.297906602
4	1.000045777	4	1.155390199
5	1.000000000	5	1.079562210

$$\frac{dg_{\mathrm{N}}}{dx}(\alpha) = \frac{f(\alpha)f''(\alpha)}{f'(\alpha)^2}. \tag{25}$$

To evaluate the indeterminate ratio in (25), we must take the limit of $g_{\mathrm{N}}'(x)$ as x approaches α. It can be shown (see Exercise 4-2.6) that

$$\lim_{x\to\alpha} \frac{dg_{\mathrm{N}}}{dx}(x) = 1 - \frac{1}{m} \tag{26}$$

at a zero of multiplicity m. Newton's method is linear if $m > 1$ since $g_{\mathrm{N}}'(\alpha) \neq 0$.

At the beginning of this section we saw that if the iteration $x_{k+1} = g(x_k)$ converges to a root α of $x = g(x)$, then the error $|x_k - \alpha|$ contracts by the factor $\lambda \approx |g'(\alpha)|$ every iteration. Hence, at a zero of multiplicity m, the error in Newton's method decays by about the factor $1 - 1/m$ each step. In particular, when the multiplicity is 2, the error decays by about the factor $\frac{1}{2}$ each step. As the multiplicity increases, the contraction factor $1 - 1/m$ in (26) approaches 1 and the convergence speed decreases.

Besides slow convergence, another problem encountered at a multiple zero of f is that the function value is often inaccurate (see the discussion in Section 1-4 concerning multiple zeros of polynomials). To improve the accuracy in f's value near a multiple zero, we can evaluate f using either double- or quadruple-precision arithmetic. But to accelerate the convergence of an algorithm near a multiple zero, we must modify the algorithm. Newton's method converges slowly at a multiple zero because the step $f(x_k)/f'(x_k)$ is too small. The convergence rate is improved if the step is multiplied by the multiplicity m, giving us the iteration

$$x_{k+1} = x_k - \frac{mf(x_k)}{f'(x_k)}.$$

This iteration can also be written $x_{k+1} = g_m(x_k)$, where g_m is defined by

$$g_m(x) = x - \frac{mf(x)}{f'(x)}. \tag{27}$$

To illustrate the dramatic impact of the factor m in (27), consider the equation $x^2 = 0$ and the root $x = 0$ with multiplicity 2. Newton's g is

$$g_{\mathrm{N}}(x) = x - \frac{f(x)}{f'(x)} = x - \frac{x^2}{2x} = \frac{x}{2},$$

and the iterations $x_{k+1} = g_{\mathrm{N}}(x_k) = x_k/2$ converge to the root $x = 0$. For example, if $x_0 = 1$, then $x_1 = \frac{1}{2}$, $x_2 = \frac{1}{4}$, $x_3 = \frac{1}{8}$, and so on. Now, inserting $m = 2$ in (27) yields

$$g_2(x) = x - \frac{2f(x)}{f'(x)} = x - \frac{2x^2}{2x} = 0.$$

The iteration $x_{k+1} = g_2(x_k)$ convergences in one step since x_1 is equal to the root for any choice of x_0: $x_1 = g_2(x_0) = 0$. In general, the scheme $x_{k+1} = g_m(x_k)$ is quadratically convergent at a root of multiplicity m.

The multiplicity of a zero can be estimated using the following identity (verified in Exercise 4-2.7): If α is a zero of f with multiplicity m, then

$$m = \lim_{x \to \alpha} \frac{f'(x)^2}{f'(x)^2 - f(x)f''(x)}. \tag{28}$$

Therefore, if the iterations $x_0, x_1, x_2, \cdots$ converge to the root α, we have

$$m \approx \frac{f'(x_k)^2}{f'(x_k)^2 - f(x_k)f''(x_k)}.$$

We have seen that Newton's method generally converges to a root α of $f(x) = 0$ when the starting guess is sufficiently close to α. The derivative of Newton's g is

$$\frac{dg_N}{dx}(x) = \frac{f(x)f''(x)}{f'(x)^2}$$

and $g_N'(\alpha) = 0$ provided that $f'(\alpha) \neq 0$ and $f''(\alpha)$ is finite. Since $|g_N'(x)| < 1$ for x near α, the iterations contract if the starting guess is near α. There are a few weird functions, however, where the iterations always diverge. An example is $f(x) = \sqrt[3]{x}$. At the root $x = 0$ of the equation $f(x) = 0$, both the first derivative and the second derivative of f are infinite (see Figure 4-10). Newton's g corresponding to $f(x) = x^{1/3}$ is

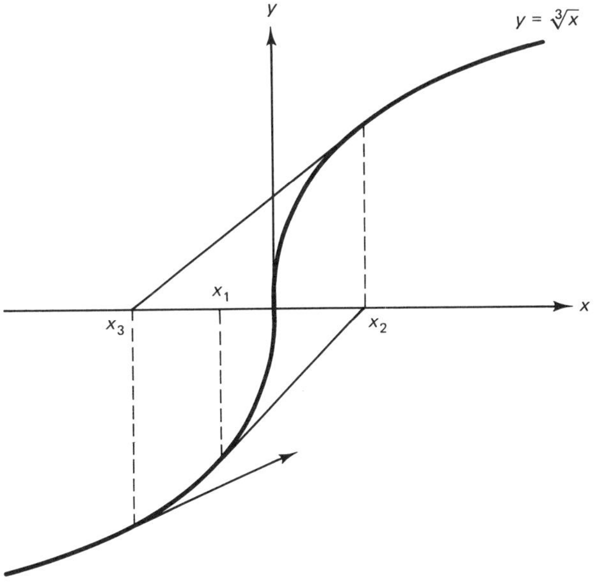

Figure 4-10 Divergence of Newton's method for $f(x) = \sqrt[3]{x}$.

$$g_N(x) = x - \frac{f(x)}{f'(x)} = x - \frac{x^{1/3}}{\frac{1}{3}x^{-2/3}} = x - 3x = -2x.$$

The iteration reduces to $x_{k+1} = g_N(x_k) = -2x_k = (-2)(-2)x_{k-1} = \cdots = (-2)^k x_1$ and the x_k grow proportionally to $(-2)^k$ for any starting guess.

Exercises

4-2.1. In each case (a)–(c), evaluate g' near the positive root to determine whether the iteration will exhibit monotone or oscillating convergence (see Exercise 4-1.1).

(a) Solve $x^4 = \sin x$ using the iteration $x_{k+1} = \sqrt[4]{\sin x_k}$.

(b) Solve $x/2 = \sin x$ using the iteration $x_{k+1} = 2 \sin x_k$.

(c) Solve $1 + x = \tan x$ for $0 \leqq x \leqq 3$ using the iteration $x_{k+1} = \arctan (1 + x_k)$.

4-2.2. Let us consider the function $g(x) = \arctan (1 + x)$ and the iteration $x_{k+1} = g(x_k)$. Taking $x_0 = 1$, determine a r such that $|g'(x)| \leqq 1 - |x_1 - x_0|/r$ for every x in the interval $[x_0 - r, x_0 + r]$. [By the theory developed earlier, the iterations converge to a root of the equation $x = \arctan (1 + x)$ and the error satisfies (18) with $\lambda = 1 - |x_1 - x_0|/r$.]

4-2.3. Suppose that each of the following equations is solved using Newton's method and a starting guess near the indicated root. What is the convergence order in each case?

(a) $\cos x = 0$, $x = \pi/2$.

(b) $1 + \cos 2x = 0$, $x = \pi/2$.

(c) $x^2 - 4 = 0$, $x = 2$.

4-2.4. If $|g'(\alpha)| > 1$, then the iteration $x_{k+1} = g(x_k)$ diverges from the root α of the equation $x = g(x)$. Explain why the iteration $x_{k+1} = g^{-1}(x_k)$ converges when the starting guess is sufficiently close to α. (*Moral:* If the original iteration diverges, the inverse iteration converges.)

4-2.5. Let us solve the equation $x = 3 - \log x$ using the iteration $x_{k+1} = 3 - \log x_k$ with the starting guess $x_0 = 2$. Estimate mathematically (without performing more than one iteration) the number of iterations that are required before the relative error is at most 10^{-8}.

4-2.6. Suppose that $f(x) = (x - \alpha)^m h(x)$, where $h(\alpha) \neq 0$ and h is twice continuously differentiable at α. Verify the identity (26).

4-2.7. Suppose that $f(x) = (x - \alpha)^m h(x)$, where $h(\alpha) \neq 0$ and h is twice continuously differentiable at α. Verify the identity (28).

4-2.8. Suppose that f has the form $f(x) = \sqrt[3]{(x - \alpha)}\, h(x)$, where α is the desired (but unknown) root. We saw above that Newton's method can diverge for a function with this structure. How can the original equation $f(x) = 0$ be converted to a new equation, say $F(x) = 0$, to which Newton's method can be applied?

4-3. NEWTON'S METHOD FOR SYSTEMS

Newton's idea for solving a single equation also applies to a system of equations. First, let us consider a special system of two equations in two unknowns:

$$\begin{aligned} 2x_1 - x_2 &= e^{-x_1}, \\ -x_1 + 2x_2 &= e^{-x_2}. \end{aligned} \tag{29}$$

Defining the two functions

$$\begin{aligned} f_1(x_1, x_2) &= 2x_1 - x_2 - e^{-x_1}, \\ f_2(x_1, x_2) &= 2x_2 - x_1 - e^{-x_2}, \end{aligned} \tag{30}$$

(29) is rewritten $\mathbf{f}(\mathbf{x}) = \mathbf{0}$, where $\mathbf{x}$ is a vector with two components x_1 and x_2 and $\mathbf{f}$ is a vector with two components f_1 and f_2, each component expressed in terms of x_1 and x_2:

$$\mathbf{f}(\mathbf{x}) = \begin{bmatrix} f_1(\mathbf{x}) \\ f_2(\mathbf{x}) \end{bmatrix} = \begin{bmatrix} f_1(x_1, x_2) \\ f_2(x_1, x_2) \end{bmatrix}, \qquad \mathbf{x} = \begin{bmatrix} x_1 \\ x_2 \end{bmatrix}. \tag{31}$$

For a system of equations, Newton's principle remains the same: Each iteration, linearize the problem and solve the linear system. For a function of two variables such as $f_1(x_1, x_2)$, the equation of the plane tangent to the surface $y_1 = f_1(x_1, x_2)$ at $x_1 = a_1$ and $x_2 = a_2$ is

$$y_1 = f_1(a_1, a_2) + \frac{\partial f_1}{\partial x_1}(a_1, a_2)(x_1 - a_1) + \frac{\partial f_1}{\partial x_2}(a_1, a_2)(x_2 - a_2). \tag{32}$$

The surface $y_1 = f_1(x_1, x_2)$ and the plane (32) are depicted in Figure 4-11.

Similarly, the equation of the plane tangent to the surface $y_2 = f_2(x_1, x_2)$ at $x_1 = a_1$ and $x_2 = a_2$ is

$$y_2 = f_2(a_1, a_2) + \frac{\partial f_2}{\partial x_1}(a_1, a_2)(x_1 - a_1) + \frac{\partial f_2}{\partial x_2}(a_1, a_2)(x_2 - a_2). \tag{33}$$

Equations (32) and (33) are equivalent to one matrix-vector equation

$$\mathbf{y} = \mathbf{f}(\mathbf{a}) + \mathbf{J}(\mathbf{a})(\mathbf{x} - \mathbf{a}) \tag{34}$$

involving the 2×2 Jacobian $\mathbf{J}$ of the function $\mathbf{f}$ in (31):

$$\mathbf{J}(\mathbf{a}) = \begin{bmatrix} \dfrac{\partial f_1}{\partial x_1}(\mathbf{a}) & \dfrac{\partial f_1}{\partial x_2}(\mathbf{a}) \\ \dfrac{\partial f_2}{\partial x_1}(\mathbf{a}) & \dfrac{\partial f_2}{\partial x_2}(\mathbf{a}) \end{bmatrix}.$$

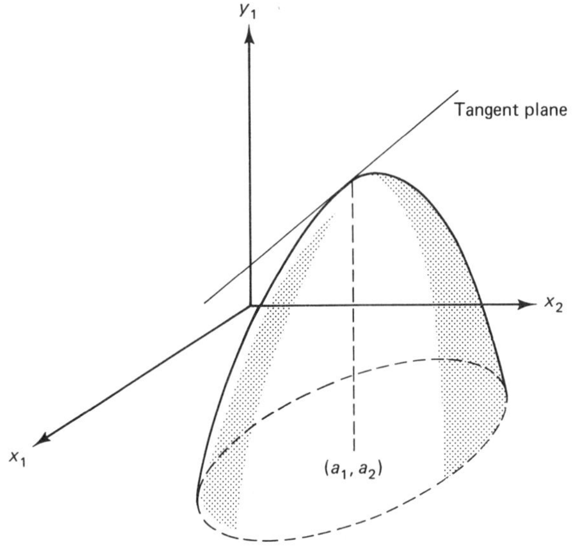

Figure 4-11 The plane tangent to $y_1 = f_1(x_1, x_2)$ at $\mathbf{x} = \mathbf{a}$.

That is, equation (32) is equivalent to equating the first component of $\mathbf{y}$ to the first component of $\mathbf{f}(\mathbf{a})$ plus the first row of $\mathbf{J}(\mathbf{a})$ times $\mathbf{x} - \mathbf{a}$. And equation (33) is equivalent to equating the second component of $\mathbf{y}$ to the second component of $\mathbf{f}(\mathbf{a})$ plus the second row of $\mathbf{J}(\mathbf{a})$ times $\mathbf{x} - \mathbf{a}$. The vector-matrix relation (34) includes both equation (32) and equation (33).

In Newton's method for a system of equations, we start from an initial guess $\mathbf{x}_1$ and we replace each component of $\mathbf{f}$ by its tangent plane approximation at $\mathbf{x} = \mathbf{x}_1$:

$$\mathbf{y} = \mathbf{f}(\mathbf{x}_1) + \mathbf{J}(\mathbf{x}_1)(\mathbf{x} - \mathbf{x}_1).$$

(Note that $\mathbf{x}_1$ is a vector of starting guesses not to be confused with x_1, the first component of $\mathbf{x}$.) The next approximation $\mathbf{x}_2$ to the root is the point where each tangent vanishes or equivalently where $\mathbf{y}$ is zero. Replacing $\mathbf{y}$ by zero gives us an equation for $\mathbf{x}_2$:

$$\mathbf{0} = \mathbf{f}(\mathbf{x}_1) + \mathbf{J}(\mathbf{x}_1)(\mathbf{x}_2 - \mathbf{x}_1).$$

Solving for $\mathbf{x}_2$, we have

$$\mathbf{x}_2 = \mathbf{x}_1 - \mathbf{J}(\mathbf{x}_1)^{-1}\mathbf{f}(\mathbf{x}_1).$$

The iteration continues in the same way. Linearizing next about $\mathbf{x}_2$, we obtain $\mathbf{x}_3 = \mathbf{g}_N(\mathbf{x}_2)$, where Newton's $\mathbf{g}$ is defined by

$$\mathbf{g}_N(\mathbf{x}) = \mathbf{x} - \mathbf{J}(\mathbf{x})^{-1}\mathbf{f}(\mathbf{x}). \tag{35}$$

Newton's method is again the successive substitution $\mathbf{x}_{k+1} = \mathbf{g}_N(\mathbf{x}_k)$.

These ideas also extend to a system of n equations in n unknowns:

$$\begin{aligned} f_1(x_1, x_2, \cdots, x_n) &= 0, \\ f_2(x_1, x_2, \cdots, x_n) &= 0, \\ &\vdots \\ f_n(x_1, x_2, \cdots, x_n) &= 0. \end{aligned}$$

This system can be written $\mathbf{f}(\mathbf{x}) = \mathbf{0}$, where $\mathbf{x}$ is the vector with components x_1 through x_n and $\mathbf{f}$ is the vector with components f_1 through f_n. Newton's iteration still has the form

$$\mathbf{x}_{k+1} = \mathbf{x}_k - \mathbf{J}(\mathbf{x}_k)^{-1}\mathbf{f}(\mathbf{x}_k),$$

but now the Jacobian is a $n \times n$ matrix:

$$\mathbf{J}(\mathbf{x}) = \begin{bmatrix} \dfrac{\partial f_1}{\partial x_1}(\mathbf{x}) & \cdots & \dfrac{\partial f_1}{\partial x_n}(\mathbf{x}) \\ \vdots & & \vdots \\ \dfrac{\partial f_n}{\partial x_1}(\mathbf{x}) & \cdots & \dfrac{\partial f_n}{\partial x_n}(\mathbf{x}) \end{bmatrix}.$$

To apply Newton's method, the Jacobian must be evaluated. This computation will be illustrated with some examples. For one equation in one unknown, n is 1, $\mathbf{J}(\mathbf{x})$ is $f'(x)$, and $\mathbf{J}(\mathbf{x})^{-1} = 1/f'(x)$. Thus Newton's $\mathbf{g}$ in (35) reduces to

$$\mathbf{g}_N(\mathbf{x}) = x - \frac{f(x)}{f'(x)},$$

the same expression derived in Section 4-1. Now consider a function $\mathbf{f}$ with two components that depend linearly on $\mathbf{x}$:

$$\mathbf{f}(\mathbf{x}) = \begin{bmatrix} f_1(x_1, x_2) \\ f_2(x_1, x_2) \end{bmatrix} = \begin{bmatrix} 2x_1 - x_2 + 5 \\ -x_1 + 2x_2 + 1 \end{bmatrix}. \tag{36}$$

In this case, n is 2, the Jacobian is a 2×2 matrix, and the four elements of the Jacobian are

$$\frac{\partial f_1}{\partial x_1} = 2, \qquad \frac{\partial f_1}{\partial x_2} = -1,$$

$$\frac{\partial f_2}{\partial x_1} = -1, \qquad \frac{\partial f_2}{\partial x_2} = 2.$$

So the Jacobian of $\mathbf{f}$, sometimes denoted $\nabla\mathbf{f}$, is

$$\nabla\mathbf{f} = \begin{bmatrix} 2 & -1 \\ -1 & 2 \end{bmatrix}.$$

Since (36) can be written

$$\mathbf{f}(\mathbf{x}) = \begin{bmatrix} 2 & -1 \\ -1 & 2 \end{bmatrix}\begin{bmatrix} x_1 \\ x_2 \end{bmatrix} + \begin{bmatrix} 5 \\ 1 \end{bmatrix},$$

we see that the Jacobian of $\mathbf{f}$ is the coefficient matrix. This property is true in general. The Jacobian of $\mathbf{f}(\mathbf{x}) = \mathbf{Ax} - \mathbf{b}$ is the coefficient matrix $\mathbf{A}$ or equivalently $\nabla(\mathbf{Ax} - \mathbf{b}) = \mathbf{A}$. Hence, for a linear system, the Newton function (35) is

$$\mathbf{g}_N(\mathbf{x}) = \mathbf{x} - \nabla\mathbf{f}(\mathbf{x})^{-1}\mathbf{f}(\mathbf{x}) = \mathbf{x} - \mathbf{A}^{-1}(\mathbf{Ax} - \mathbf{b}) = \mathbf{A}^{-1}\mathbf{b}.$$

Since $\mathbf{x}_2 = \mathbf{g}_N(\mathbf{x}_1) = \mathbf{A}^{-1}\mathbf{b}$ for any starting guess $\mathbf{x}_1$, Newton's method produces the solution to a linear system in one step. Observe that this step (multiplying $\mathbf{A}^{-1}$ by $\mathbf{b}$) amounts to solving the system. Since Newton's method is obtained by linearizing the system and since a linear system is its own linearization, one step of Newton's method applied to a linear system is equivalent to solving the system. Clearly, Newton's method should only be applied to a nonlinear system.

For another example, suppose that $\mathbf{f}$ is the sum of two functions $\mathbf{p}$ and $\mathbf{q}$: $\mathbf{f} = \mathbf{p} + \mathbf{q}$. Since the element in row i and column j of $\nabla\mathbf{f}$ is the partial derivative of f_i with respect to x_j, we have

$$\mathbf{J}_{ij}(\mathbf{x}) = \frac{\partial f_i}{\partial x_j}(\mathbf{x}) = \frac{\partial(p_i + q_i)}{\partial x_j}(\mathbf{x}) = \frac{\partial p_i}{\partial x_j}(\mathbf{x}) + \frac{\partial q_i}{\partial x_j}(\mathbf{x}). \tag{37}$$

Since the last two terms on the right side of (37) are the elements in row i and column j of the Jacobians for $\mathbf{p}$ and $\mathbf{q}$, it follows that the Jacobian of the sum of functions is the sum of the individual Jacobians. That is, if $\mathbf{f} = \mathbf{p} + \mathbf{q}$, then $\nabla\mathbf{f} = \nabla\mathbf{p} + \nabla\mathbf{q}$.

Finally, let us consider the following system of n equations in n unknowns:

$$\begin{aligned}
2x_1 - x_2 \qquad\qquad & - e^{-x_1} = 0, \\
-x_1 + 2x_2 - x_3 \qquad & - e^{-x_2} = 0, \\
- x_2 + 2x_3 - x_4 \qquad & - e^{-x_3} = 0, \\
& \;\;\vdots \\
- x_{n-1} + 2x_n & - e^{-x_n} = 0.
\end{aligned}$$

As we saw in Section 2-7, this system can be expressed $\mathbf{f}(\mathbf{x}) = \mathbf{Ax} - \mathbf{b}(\mathbf{x}) = \mathbf{0}$, where $\mathbf{x}$ is the vector with components x_1 through x_n, $\mathbf{A}$ is a $n \times n$ tridiagonal matrix:

$$\mathbf{A} = \begin{bmatrix} 2 & -1 & & & & \\ -1 & 2 & \cdot & & & \\ & \cdot & \cdot & \cdot & & \\ & & \cdot & \cdot & \cdot & \\ & & & \cdot & \cdot & -1 \\ & & & & -1 & 2 \end{bmatrix}, \tag{38}$$

and $\mathbf{b}$ is a vector whose components depend on $\mathbf{x}$:

$$\mathbf{b}(\mathbf{x}) = \begin{bmatrix} e^{-x_1} \\ e^{-x_2} \\ \cdot \\ \cdot \\ \cdot \\ e^{-x_n} \end{bmatrix}. \tag{39}$$

Since $\mathbf{f}(\mathbf{x}) = \mathbf{Ax} - \mathbf{b}(\mathbf{x})$, it follows that the Jacobian of $\mathbf{f}(\mathbf{x})$ is the Jacobian of $\mathbf{Ax}$ minus the Jacobian of $\mathbf{b}(\mathbf{x})$. Since the Jacobian of the linear term $\mathbf{Ax}$ is the coefficient matrix $\mathbf{A}$, we have $\nabla\mathbf{f}(\mathbf{x}) = \mathbf{A} - \nabla\mathbf{b}(\mathbf{x})$. Now let us evaluate the Jacobian of $\mathbf{b}$. The first row of this Jacobian contains the partial derivatives of $b_1(\mathbf{x}) = e^{-x_1}$ with respect to x_1 through x_n:

$$\frac{\partial b_1}{\partial x_1}(\mathbf{x}) = -e^{-x_1}, \quad \frac{\partial b_1}{\partial x_2}(\mathbf{x}) = 0, \quad \frac{\partial b_1}{\partial x_3}(\mathbf{x}) = 0, \quad \cdots.$$

The second row of the Jacobian contains the partial derivatives of $b_2(\mathbf{x}) = e^{-x_2}$ with respect to x_1 through x_n:

$$\frac{\partial b_2}{\partial x_1}(\mathbf{x}) = 0, \quad \frac{\partial b_2}{\partial x_2}(\mathbf{x}) = -e^{-x_2}, \quad \frac{\partial b_2}{\partial x_3}(\mathbf{x}) = 0, \quad \cdots.$$

The general rule is

$$(\nabla\mathbf{b}(\mathbf{x}))_{ij} = \frac{\partial b_i}{\partial x_j}(\mathbf{x}) = \begin{cases} -e^{-x_i} & \text{if } i = j, \\ 0 & \text{if } i \neq j. \end{cases}$$

Thus the Jacobian of $\mathbf{b}$ is a diagonal matrix:

$$\nabla\mathbf{b}(\mathbf{x}) = -\begin{bmatrix} e^{-x_1} & & & & & \\ & e^{-x_2} & & & & \\ & & \cdot & & & \\ & & & \cdot & & \\ & & & & \cdot & \\ & & & & & e^{-x_n} \end{bmatrix}.$$

And the Jacobian of $\mathbf{f}$ is a tridiagonal matrix:

$$\nabla\mathbf{f}(\mathbf{x}) = \mathbf{A} - \nabla\mathbf{b}(\mathbf{x}) = \begin{bmatrix} 2 + e^{-x_1} & -1 & & & \\ -1 & 2 + e^{-x_2} & -1 & & \\ & -1 & \cdot & \cdot & \\ & & \cdot & \cdot & -1 \\ & & & -1 & 2 + e^{-x_n} \end{bmatrix}.$$

Up to here, we have derived Newton's method and we have studied how to compute the Jacobian. Now let us consider the implementation of Newton's method. By the definition (35) of Newton's $\mathbf{g}$, the new iteration $\mathbf{x}^{\text{new}}$ is related to the previous iteration $\mathbf{x}^{\text{old}}$ by the formula

$$\mathbf{x}^{\text{new}} = \mathbf{x}^{\text{old}} - \mathbf{J}(\mathbf{x}^{\text{old}})^{-1}\mathbf{f}(\mathbf{x}^{\text{old}}). \tag{40}$$

Since matrix inversion is more costly than matrix factoring (see Section 2-7), (40) will be expressed in a way that involves the computation of $\mathbf{J}$'s factors rather than $\mathbf{J}$'s inverse. Multiplying (40) by $\mathbf{J}(\mathbf{x}^{\text{old}})$ gives us the relation

$$\mathbf{J}(\mathbf{x}^{\text{old}})(\mathbf{x}^{\text{new}} - \mathbf{x}^{\text{old}}) = -\mathbf{f}(\mathbf{x}^{\text{old}}).$$

Writing $\mathbf{y}$ for $\mathbf{x}^{\text{new}} - \mathbf{x}^{\text{old}}$, we have

$$\mathbf{J}(\mathbf{x}^{\text{old}})\mathbf{y} = -\mathbf{f}(\mathbf{x}^{\text{old}}) \quad \text{where} \quad \mathbf{y} = \mathbf{x}^{\text{new}} - \mathbf{x}^{\text{old}}.$$

This leads us to the following outline for Newton's method:

1. Evaluate $\mathbf{J}(\mathbf{x}^{\text{old}})$ and call it $\mathbf{A}$; then factor $\mathbf{A}$. 2. Evaluate $\mathbf{f}(\mathbf{x}^{\text{old}})$ and let $\mathbf{b}$ denote $-\mathbf{f}(\mathbf{x}^{\text{old}})$. 3. Forward and back solve $\mathbf{Ay} = \mathbf{b}$. 4. Set $\mathbf{x}^{\text{new}} = \mathbf{x}^{\text{old}} + \mathbf{y}$. 5. Return to step 1 unless the iterations have converged.	(41)

Newton's method for a system of equations

In steps 1 and 3 we use the special routines developed in Chapter 2. For example, if the Jacobian is a tridiagonal matrix, it is factored using either TFACT or PFACT.

As is true for the scalar Newton iteration $x_{k+1} = g_{\text{N}}(x_k)$, the vector Newton iteration $\mathbf{x}_{k+1} = \mathbf{g}_{\text{N}}(\mathbf{x}_k)$ typically converges whenever the starting guess is sufficiently close to a root. In general, the analysis of a vector iteration $\mathbf{x}_{k+1} = \mathbf{g}(\mathbf{x}_k)$ is similar to the analysis of a scalar iteration except that the derivative g' appearing in the scalar analysis is replaced by the Jacobian $\nabla\mathbf{g}$ and the absolute value of the derivative is replaced by the norm of the Jacobian. In particular, the following result applies to the vector iteration:

> *Suppose that $\boldsymbol{\alpha}$ is a root of the equation $\mathbf{x} = \mathbf{g}(\mathbf{x})$ and $\mathbf{g}$ is continuously differentiable. If there exists a norm $\|\cdot\|$ and scalars $\lambda < 1$ and $r > 0$ such that $\|\nabla\mathbf{g}(\mathbf{x})\| \leqq \lambda$ whenever the distance from $\mathbf{x}$ to $\boldsymbol{\alpha}$ is less than or equal to r, then the error $\mathbf{e}_k = \mathbf{x}_k - \boldsymbol{\alpha}$ satisfies the inequality $\|\mathbf{e}_k\| \leqq \lambda^k\|\mathbf{e}_0\|$ whenever the distance between the starting guess $\mathbf{x}_0$ and the root $\boldsymbol{\alpha}$ is less than or equal to r.*

Usually, the Jacobian of $\mathbf{g}_{\text{N}}$ vanishes at the root for Newton's method. Consequently, $\|\nabla\mathbf{g}_{\text{N}}\|$ is less than 1 in a neighborhood of the root and the iterations converge when the starting guess is sufficiently close to the root. In addition, since $\nabla\mathbf{g}_{\text{N}}(\boldsymbol{\alpha}) = \mathbf{0}$, the convergence order for Newton's method is usually quadratic; that is, there exists a constant C such that $\|\mathbf{e}_{k+1}\| \leqq C\|\mathbf{e}_k\|^2$ for every k.

When the starting guess is far from the root, there is one case where convergence of Newton's method (40) can be very slow: The inverse of the

Jacobian does not exist either at the starting point or near the starting point. If this inverse does not exist, then the iteration (40) cannot be performed. If the starting guess is near a point where the inverse does not exist, then the computation of the new **x** typically involves a division by a small number so that the new **x** is large and far from the desired root. For example, consider the function $f(x) = x^2 - 4$ and the zeros $x = \pm 2$. Using the starting guess $x_0 = 0$, we have $f'(x_0) = 2x_0 = 0$, $f(x_0) = -4$, and $x_1 = x_0 - f(x_0)/f'(x_0) = 4/0$ which is undefined. Furthermore, if x_0 is near zero, then $f'(x_0)$ is near zero and x_1 is near plus or minus infinity. In summary, the starting guess in Newton's method should never be close to a point where the inverse of the Jacobian does not exist.

Exercises

4-3.1. Compute the Jacobian of each of the following functions:

(a)
$$\mathbf{f}(\mathbf{x}) = \begin{bmatrix} f_1(x_1, x_2) \\ f_2(x_1, x_2) \end{bmatrix} = \begin{bmatrix} x_1 + \cos x_2 + x_1 x_2 \\ x_2^2 + x_1 \end{bmatrix}$$

(b)
$$\mathbf{f}(\mathbf{x}) = \begin{bmatrix} f_1(x_1, x_2, x_3) \\ f_2(x_1, x_2, x_3) \\ f_3(x_1, x_2, x_3) \end{bmatrix} = \begin{bmatrix} x_1^2 + 2x_1x_2 + \sin x_3 \\ x_1 + x_2 + x_3 \\ \cos x_1 + e^{x_3} \end{bmatrix}$$

For both functions, evaluate the Jacobian at $x_1 = x_2 = x_3 = 0$. Can Newton's method be applied to either of these functions starting from $x_1 = x_2 = x_3 = 0$? Explain.

4-3.2. Solve the system of equations (29) using Newton's method with subroutines FACT and SOLVE.

4-3.3. Taking $N = 10$, solve equation (78) of Chapter 2 using Newton's method with subroutines TFACT and TSOLVE.

4-3.4. Use Newton's method to find the roots of the following system of three equations in three unknowns:

$$\begin{aligned} f_1(x_1, x_2, x_3) &= 3x_1 - \cos x_2x_3 - .5 = 0, \\ f_2(x_1, x_2, x_3) &= x_1^2 - 625x_2^2 \qquad = 0, \\ f_3(x_1, x_2, x_3) &= e^{-x_1x_2} + 20x_3 + 9 \quad = 0. \end{aligned}$$

This system has more than one solution. Choose your starting guess carefully. If x_2 is zero, what is the zero pattern of **f**'s Jacobian? Why must the x_2 component of the starting guess be nonzero?

4-4. ALTERATIONS TO NEWTON'S METHOD

In some applications, evaluating and factoring the Jacobian every iteration is too costly. We now discuss some alterations of Newton's method that reduce the time for each iteration. Although these alterations decrease the iteration time, they increase the number of iterations. Whether any of these schemes is practical depends on the actual equation and the initial guess. In each of these methods, we replace the Jacobian by a simple approximation, say $\mathbf{A}(\mathbf{x})$, giving us the iteration

$$\mathbf{x}^{\text{new}} = \mathbf{x}^{\text{old}} - \mathbf{A}(\mathbf{x}^{\text{old}})^{-1}\mathbf{f}(\mathbf{x}^{\text{old}}).$$

In the **modified Newton** scheme, $\mathbf{A}(\mathbf{x})$ is $\mathbf{J}(\mathbf{x}_1)$. In other words, we evaluate the

Jacobian at the initial point $\mathbf{x}_1$ and we use this Jacobian in place of $\mathbf{J}(\mathbf{x}_k)$ for the succeeding iterations. Since $\mathbf{A}$ does not depend on $\mathbf{x}$ in this iteration, $\mathbf{A}$ is factored once before the first iteration. Hence the outline for the modified Newton scheme is identical to the outline for Newton's method except that we branch back to step 2 instead of step 1 after each iteration:

1. Evaluate $\mathbf{J}(\mathbf{x}^{\text{old}})$ and call it $\mathbf{A}$; then factor $\mathbf{A}$.
2. Evaluate $\mathbf{f}(\mathbf{x}^{\text{old}})$ and let $\mathbf{b}$ denote $-\mathbf{f}(\mathbf{x}^{\text{old}})$.
3. Forward and back solve $\mathbf{A}\mathbf{y} = \mathbf{b}$. (42)
4. Set $\mathbf{x}^{\text{new}} = \mathbf{x}^{\text{old}} + \mathbf{y}$.
5. Return to step 2 unless the iterations have converged.

Modified Newton's method for a system of equations

For one equation in one unknown, $\mathbf{J}(x_1) = f'(x_1)$ and the modified Newton iteration (42) reduces to $x_{k+1} = x_k - f(x_k)/f'(x_1)$, which is identical to the modified Newton iteration in Section 4-1. As noted in Section 4-1, the convergence speed for the modified Newton scheme can be slow if the slope of f at the starting point differs significantly from the slope of f at the root. Similarly, for a system of equations, the convergence speed can be slow if the Jacobian at the starting point differs significantly from the Jacobian at the root. For this reason, it is a good idea to reevaluate $\mathbf{J}$ and its factorization at the current iteration and restart the iterations occasionally, especially when the starting guess is far from the root.

We now discuss three schemes where $\mathbf{A}$ is a lower triangular approximation to $\mathbf{J}$. Since $\mathbf{A}$ is lower triangular, it is already factored and the factorization cost associated with Newton's method is eliminated. On the other hand, these methods usually converge more slowly than Newton's method and in some cases, they diverge no matter how close the starting guess is to the root. To decide whether any of these methods is practical, we must balance the time saved in each iteration against the increased number of iterations. In the **Jacobi scheme**, $\mathbf{A}$ is the diagonal matrix $\mathbf{D}$ whose diagonal is the same as the diagonal of $\mathbf{J}$. In other words, the Jacobi iteration is

$$\mathbf{x}^{\text{new}} = \mathbf{x}^{\text{old}} - \mathbf{D}(\mathbf{x}^{\text{old}})^{-1}\mathbf{f}(\mathbf{x}^{\text{old}}), \tag{43}$$

where $\mathbf{D}$ is a diagonal matrix with diagonal elements

$$d_{ii}(\mathbf{x}) = \frac{\partial f_i}{\partial x_i}(\mathbf{x}).$$

If the off-diagonal elements of the Jacobian are small compared to the diagonal elements, then $\mathbf{D}$ is a good approximation to $\mathbf{J}$ and the iteration (43) is a good approximation to Newton's iteration (40). For equation (29) and the functions defined in (30), the Jacobi iteration takes the form

$$\mathbf{x}^{\text{new}} = \mathbf{x}^{\text{old}} - \begin{bmatrix} 2 + e^{-x_1^{\text{old}}} & 0 \\ 0 & 2 + e^{-x_2^{\text{old}}} \end{bmatrix}^{-1} \begin{bmatrix} 2x_1^{\text{old}} - x_2^{\text{old}} - e^{-x_1^{\text{old}}} \\ 2x_2^{\text{old}} - x_1^{\text{old}} - e^{-x_2^{\text{old}}} \end{bmatrix}.$$

Since inverting a diagonal matrix is equivalent to inverting the diagonal elements, this iteration can be expressed

$$x_1^{\text{new}} = x_1^{\text{old}} - \frac{2x_1^{\text{old}} - x_2^{\text{old}} - e^{-x_1^{\text{old}}}}{2 + e^{-x_1^{\text{old}}}},$$
$$x_2^{\text{new}} = x_2^{\text{old}} - \frac{2x_2^{\text{old}} - x_1^{\text{old}} - e^{-x_2^{\text{old}}}}{2 + e^{-x_2^{\text{old}}}}. \tag{44}$$

Comparing the Jacobi iteration (43) to the Newton iteration (40), the only difference is that **J** is replaced by **D**. Hence the outline for the Jacobi scheme is identical to the outline for Newton's method except that the factorization step is omitted since a diagonal matrix is already factored:

1. Evaluate $\mathbf{f}(\mathbf{x}^{\text{old}})$ and let $\mathbf{b}$ denote $-\mathbf{f}(\mathbf{x}^{\text{old}})$.
2. Solve $\mathbf{D}(\mathbf{x}^{\text{old}})\mathbf{y} = \mathbf{b}$.
3. Set $\mathbf{x}^{\text{new}} = \mathbf{x}^{\text{old}} + \mathbf{y}$.
4. Return to step 1 unless the iterations have converged.

(45)

Jacobi's method

Since **A** is a diagonal matrix in the Jacobi scheme, step 2 is equivalent to dividing each component of **b** by the corresponding diagonal element of **D**. Each iteration of the Jacobi scheme is cheap compared to the Newton scheme since we only evaluate the diagonal of the Jacobian and the factorization step has been eliminated. Since a cheap iteration still yields an expensive algorithm when convergence is slow, the decreased iteration time must be balanced against the increased number of iterations to determine whether Jacobi's method is practical.

The **Gauss-Seidel scheme** for a nonlinear system of equations is given by the iteration

$$\mathbf{x}^{\text{new}} = \mathbf{x}^{\text{old}} - \mathbf{L}(\mathbf{x}^{\text{old}})^{-1}\mathbf{f}(\mathbf{x}^{\text{old}}),$$

where **L** is the lower triangular matrix which is identical to **J** except for the zeros above the diagonal. In some sense, the Gauss-Seidel scheme is a better approximation to Newton's method than the Jacobi scheme since **L** contains more coefficients from the actual Jacobian matrix than **D**. For a function **f** with two components, **L** is given by

$$\mathbf{L}(\mathbf{x}) = \begin{bmatrix} \frac{\partial f_1}{\partial x_1}(\mathbf{x}) & 0 \\ \frac{\partial f_2}{\partial x_1}(\mathbf{x}) & \frac{\partial f_2}{\partial x_2}(\mathbf{x}) \end{bmatrix}.$$

Since a lower triangular matrix is already factored, the outline for the Gauss-Seidel scheme is the same as the outline for the Jacobi scheme except that **D** is replaced by **L**.

Finally, the **successive overrelaxation** or SOR scheme for a nonlinear system is

$$\mathbf{x}^{\text{new}} = \mathbf{x}^{\text{old}} - (\rho\mathbf{D}(\mathbf{x}^{\text{old}}) + \mathbf{L}(\mathbf{x}^{\text{old}}))^{-1}\mathbf{f}(\mathbf{x}^{\text{old}}), \tag{46}$$

where ρ is a fixed scalar. The quantity $\omega = 1/(\rho + 1)$ is called the **relaxation parameter** associated with the iteration (46). Since the matrix $\rho\mathbf{D}(\mathbf{x}) + \mathbf{L}(\mathbf{x})$ is lower triangular, the outline for SOR is identical to the outline for the Jacobi scheme except that $\mathbf{D}$ is replaced by $\rho\mathbf{D} + \mathbf{L}$. A detailed discussion of Jacobi, Gauss-Seidel, and SOR iterations for a linear system of equations appears in Chapter 7.

As indicated above, Jacobi, Gauss-Seidel, and SOR iterations are not guaranteed to converge, even if the starting guess is near the root. There are two different strategies for studying the convergence properties of an iteration. To establish contracting convergence, we must find a neighborhood of the root where $\|\nabla\mathbf{g}\|$ is less than one. For systems of equations, however, it is often difficult to guess the location of a root, let alone analyze the norm of the Jacobian. A more specialized but often more amenable strategy is to try to apply monotone convergence results. A comprehensive reference for monotone convergence results is the book [O4] by Ortega and Rheinboldt. Examples of monotonically convergent iterations include iteration (29) of Chapter 1 and iteration (79) of Chapter 2.

Let us examine a particular monotone convergence result which applies to iterations of the form

$$\mathbf{x}_{k+1} = \mathbf{x}_k - \mathbf{P}(\mathbf{x}_k)\mathbf{f}(\mathbf{x}_k). \tag{47}$$

Note that all the alterations to Newton's method discussed above have this form. For the modified Newton scheme, $\mathbf{P}(\mathbf{x}) = \mathbf{J}(\mathbf{x}_1)^{-1}$; for the Jacobi scheme, $\mathbf{P}(\mathbf{x}) = \mathbf{D}(\mathbf{x})^{-1}$; for the Gauss-Seidel scheme, $\mathbf{P}(\mathbf{x}) = \mathbf{L}(\mathbf{x})^{-1}$; and for the SOR scheme, $\mathbf{P}(\mathbf{x}) = (\rho\mathbf{D}(\mathbf{x}) + \mathbf{L}(\mathbf{x}))^{-1}$. Many iterations can be written in the form (47). For another example, let us consider the equation $\mathbf{Ax} = \mathbf{b}(\mathbf{x})$, where $\mathbf{A}$ and $\mathbf{b}$ are defined in (38) and (39), respectively, and let us apply the iteration proposed in Chapter 1 for the semiconductor equation. Using vector notation, this iteration can be expressed

$$\mathbf{x}_{k+1} = \mathbf{D}^{-1}[(\mathbf{D} - \mathbf{A})\mathbf{x}_k + \mathbf{b}(\mathbf{x}_k)], \tag{48}$$

where $\mathbf{D}$ is the diagonal matrix with every diagonal element equal to 2. After minor rearrangement, we have

$$\mathbf{x}_{k+1} = \mathbf{x}_k - \mathbf{D}^{-1}(\mathbf{Ax}_k - \mathbf{b}(\mathbf{x}_k)). \tag{49}$$

Comparing this to (47), we see that $\mathbf{P}(\mathbf{x}) = \mathbf{D}^{-1}$ and $\mathbf{f}(\mathbf{x}) = \mathbf{Ax} - \mathbf{b}(\mathbf{x})$. In Exercise 4-4.1 you will express iteration (79) of Chapter 2 in the form (47).

To establish monotone convergence for an iteration, we must show that $\mathbf{f}$ satisfies various inequalities. An analog of a result due to Ortega and Rheinboldt [O3] will be stated, but first some notation: The vector inequality $\mathbf{x} \leqq \mathbf{y}$ means that $x_i \leqq y_i$ for every i. Similarly, the matrix inequality $\mathbf{P} \leqq \mathbf{Q}$ means that $p_{ij} \leqq q_{ij}$ for every i and j. Suppose that there exist two vectors $\mathbf{x}_1$ and $\mathbf{y}_1$ and a matrix $\mathbf{Q}(\mathbf{x})$ with the following properties:

(a) $\mathbf{x}_1 \leqq \mathbf{y}_1$ and $\mathbf{f}(\mathbf{x}_1) \leqq \mathbf{0} \leqq \mathbf{f}(\mathbf{y}_1)$.
(b) $\mathbf{0} \leqq \mathbf{P}(\mathbf{x})$, $\mathbf{P}(\mathbf{x})\mathbf{Q}(\mathbf{x}_1) \leqq \mathbf{I}$, and $\mathbf{Q}(\mathbf{x})\mathbf{P}(\mathbf{x}) \leqq \mathbf{I}$ whenever $\mathbf{x}_1 \leqq \mathbf{x} \leqq \mathbf{y}_1$.
(c) $\mathbf{f}(\mathbf{y}) - \mathbf{f}(\mathbf{x}) \geqq \mathbf{Q}(\mathbf{y})(\mathbf{y} - \mathbf{x})$ whenever $\mathbf{x}_1 \leqq \mathbf{x} \leqq \mathbf{y} \leqq \mathbf{y}_1$.

Then the iterations generated by (47) approach a limit $\mathbf{x}_\infty$ and the convergence is monotone in the sense that $\mathbf{x}_{k+1} \geqq \mathbf{x}_k$ for every k. If $\mathbf{f}$ is continuous at $\mathbf{x}_\infty$ and there exists an invertible matrix $\mathbf{R}$ such that $\mathbf{P}(\mathbf{x}_k) \geqq \mathbf{R} \geqq \mathbf{0}$ for every k, then $\mathbf{x}_\infty$ is a solution to the equation $\mathbf{f}(\mathbf{x}) = \mathbf{0}$. There are many different monotone convergence results corresponding to slightly different hypotheses and slightly different conclusions. For example, if the inequality "$\mathbf{f}(\mathbf{x}_1) \leqq \mathbf{0} \leqq \mathbf{f}(\mathbf{y}_1)$" in assumption (a) is changed to "$\mathbf{f}(\mathbf{x}_1) \geqq \mathbf{0} \geqq \mathbf{f}(\mathbf{y}_1)$," then the condition "$\mathbf{P}(\mathbf{x})\mathbf{Q}(\mathbf{x}_1) \leqq \mathbf{I}$" in part (b) is changed to "$\mathbf{P}(\mathbf{x})\mathbf{Q}(\mathbf{x}) \leqq \mathbf{I}$."

For a rigorous derivation of monotone convergence results, the reader is referred to the book [O4]. Loosely speaking, conditions (a)–(c) can be visualized in the following way: The inequality $\mathbf{f}(\mathbf{x}_1) \leqq \mathbf{0} \leqq \mathbf{f}(\mathbf{y}_1)$ in part (a) implies that the components of $\mathbf{f}$ at $\mathbf{x}_1$ are nonpositive while the components of $\mathbf{f}$ at $\mathbf{y}_1$ are nonnegative. The remaining conditions guarantee that in successive Newton iterations, the components of $\mathbf{x}_k$ increase while being bounded from above by $\mathbf{y}_1$. Since the components are bounded from above, the $\mathbf{x}_k$ approach a limit $\mathbf{x}_\infty$. Since the difference $\mathbf{x}_{k+1} - \mathbf{x}_k$ tends to zero, it follows from (47) that the product $\mathbf{P}(\mathbf{x}_k)\mathbf{f}(\mathbf{x}_k)$ approaches zero. If $\mathbf{P}(\mathbf{x}_k)$ is bounded away from singularity, then the $\mathbf{f}(\mathbf{x}_k)$ must approach zero, and by continuity of $\mathbf{f}$, $\mathbf{f}(\mathbf{x}_\infty)$ is equal to zero. In conditions (b) and (c), think of $\mathbf{Q}$ as an approximation to the Jacobian matrix and think of $\mathbf{P}$ as an approximation to the inverse of $\mathbf{Q}$. Condition (b) requires that $\mathbf{P}$ is nonnegative and that the product between $\mathbf{P}$ and $\mathbf{Q}$ is in some sense near the identity matrix. If $\mathbf{f}$ is linear, then $\mathbf{f}(\mathbf{x}) = \mathbf{A}\mathbf{x} - \mathbf{b}$, $\mathbf{J}(\mathbf{x}) = \mathbf{A}$, and $\mathbf{f}(\mathbf{y}) - \mathbf{f}(\mathbf{x}) = \mathbf{A}(\mathbf{y} - \mathbf{x}) = \mathbf{J}(\mathbf{y})(\mathbf{y} - \mathbf{x})$. On the other hand, when $\mathbf{f}$ is not linear, we have the approximation $\mathbf{f}(\mathbf{y}) - \mathbf{f}(\mathbf{x}) \approx \mathbf{J}(\mathbf{y})(\mathbf{y} - \mathbf{x})$. In condition (c), $\mathbf{Q}$ is required to satisfy the related inequality $\mathbf{f}(\mathbf{y}) - \mathbf{f}(\mathbf{x}) \geqq \mathbf{Q}(\mathbf{y})(\mathbf{y} - \mathbf{x})$.

Let us now show that assumptions (a)–(c) are satisfied by the iteration (49) when $\mathbf{x}_1 = \mathbf{0}$. Since $\mathbf{f}(\mathbf{x}) = \mathbf{A}\mathbf{x} - \mathbf{b}(\mathbf{x})$, we have $\mathbf{f}(\mathbf{0}) = -\mathbf{b}(\mathbf{0})$, where $\mathbf{b}$ is the vector of exponentials in (39). Since $e^0 = 1$, it follows that $\mathbf{f}(\mathbf{0}) = -\mathbf{1} \leqq \mathbf{0}$, where $\mathbf{1}$ denotes the vector with every component equal to 1. For the exponential function, $e^{-x} - e^{-y} \geqq 0$ whenever $y \geqq x$. Consequently, $\mathbf{b}(\mathbf{x}) - \mathbf{b}(\mathbf{y}) \geqq \mathbf{0}$ whenever $\mathbf{y} \geqq \mathbf{x}$ and we have $\mathbf{f}(\mathbf{y}) - \mathbf{f}(\mathbf{x}) = \mathbf{A}(\mathbf{y} - \mathbf{x}) + \mathbf{b}(\mathbf{x}) - \mathbf{b}(\mathbf{y}) \geqq \mathbf{A}(\mathbf{y} - \mathbf{x})$. Thus hypothesis (c) is satisfied with $\mathbf{Q} = \mathbf{A}$. Since $\mathbf{D}^{-1} = .5\mathbf{I}$, the condition $\mathbf{P}(\mathbf{x}) \geqq \mathbf{0}$ in (b) holds trivially. Since $\mathbf{D}^{-1}\mathbf{A} = \mathbf{A}\mathbf{D}^{-1}$ is the tridiagonal matrix with 1's on the diagonal and with $-.5$'s off the diagonal, the remaining parts of hypothesis (b) hold. Finally, in Exercise 4-4.2 you construct $\mathbf{y}_1$ such that $\mathbf{0} \leqq \mathbf{y}_1$ and $\mathbf{0} \leqq \mathbf{f}(\mathbf{y}_1)$. Since hypotheses (a)–(c) are satisfied, the iteration proposed in Chapter 1 for the semiconductor equation converges monotonically.

Exercises

4-4.1. Express iteration (79) of Chapter 2 in the form (47). What is the $\mathbf{P}$ and the $\mathbf{f}$ corresponding to this iteration?

4-4.2. Since e^{-x} approaches zero as x approaches infinity, the vector $\mathbf{b}(\mathbf{x})$ defined in (39) approaches zero as each component of $\mathbf{x}$ approaches infinity. If $\mathbf{p}$ is a vector with positive components such that each component of $\mathbf{A}\mathbf{p}$ is positive, then $\mathbf{b}(s\mathbf{p})$ approaches zero and $s\mathbf{A}\mathbf{p}$ approaches $+\infty$ as s increases. Hence, for s sufficiently large, each component of $\mathbf{f}(s\mathbf{p}) = s\mathbf{A}\mathbf{p} - \mathbf{b}(s\mathbf{p})$ is positive and the vector $\mathbf{y}_1 = s\mathbf{p}$ satisfies assumption (a) above. Letting $\mathbf{p}$ be the vector with components $p_i = (n + 1)i - i^2$ for i between 1 and n, show that the components of $\mathbf{p}$ and the components of $\mathbf{A}\mathbf{p}$ are positive, where $\mathbf{A}$ is defined in (38).

4-4.3. Show that iteration (79) of Chapter 2 converges monotonically. (*Hint:* See Exercises 3-5.5, 4-4.1, and 4-4.2.)

4-4.4. Show that the iteration (44) converges monotonically when the starting guess is $\mathbf{x}_1 = \mathbf{0}$.

4-4.5. What is the Jacobi iteration corresponding to the system (28) of Chapter 1? Express this iteration in component form similar to equation (29) of Chapter 1. Show that this iteration converges monotonically when the starting guess is $\mathbf{x}_1 = \mathbf{0}$.

4-4.6. What is the Gauss-Seidel iteration corresponding to the system (28) of Chapter 1? Express this iteration in component form similar to equation (29) of Chapter 1. Show that this iteration converges monotonically when the starting guess is $\mathbf{x}_1 = \mathbf{0}$. (*Hint:* See Exercise 3-5.5.)

4-4.7. Consider the system (29) of two equations and the iteration

$$\begin{bmatrix} 2 & -1 \\ -1 & 2 \end{bmatrix} \begin{bmatrix} x_1^{\text{new}} \\ x_2^{\text{new}} \end{bmatrix} = \begin{bmatrix} e^{-x_1^{\text{old}}} \\ e^{-x_2^{\text{old}}} \end{bmatrix}.$$

If this iteration is expressed in the form $\mathbf{x}^{\text{new}} = \mathbf{g}(\mathbf{x}^{\text{old}})$, what is $\mathbf{g}$? Use the identity

$$\begin{bmatrix} 2 & -1 \\ -1 & 2 \end{bmatrix}^{-1} = \tfrac{1}{3}\begin{bmatrix} 2 & 1 \\ 1 & 2 \end{bmatrix}$$

to simplify $\mathbf{g}$. By the monotone convergence theory developed above, equation (29) has a solution with both x_1 and x_2 positive. Is the 1-norm of $\nabla \mathbf{g}(\mathbf{x})$ greater than one or less than one when the components of $\mathbf{x}$ are positive?

4-5. QUASI-NEWTON METHODS

Each of the alterations to Newton's method discussed in Section 4-4 requires the computation of partial derivatives. Another strategy is to approximate partial derivatives using finite differences as we did in the secant method of Section 4-1. The element in row i and column j of the Jacobian is the partial derivative of f_i with respect to x_j. By the definition of the partial derivative, we have

$$\frac{\partial f_i}{\partial x_j}(\mathbf{x}) = \lim_{\delta \to 0} \frac{f_i(\mathbf{x} + \delta \mathbf{e}_j) - f_i(\mathbf{x})}{\delta},$$

where $\mathbf{e}_j$ is the vector with every component equal to zero except for the jth component which is 1, and δ is a scalar which tends to zero. Hence, if δ is near zero, we can approximate the Jacobian $\mathbf{J}$ by the matrix $\mathbf{A}$ with ij element defined by†

$$\mathbf{A}(\mathbf{x})_{ij} = \frac{f_i(\mathbf{x} - \delta \mathbf{e}_j) - f_i(\mathbf{x})}{\delta}. \tag{50}$$

For a function $\mathbf{f}$ with n components, (50) estimates the n^2 elements of the Jacobian using n^2 finite differences. Replacing the Jacobian in Newton's method by the finite difference approximation $\mathbf{A}$ defined in (50), we are led to the iteration

$$\mathbf{x}_{k+1} = \mathbf{x}_k - \mathbf{A}(\mathbf{x}_k)^{-1}\mathbf{f}(\mathbf{x}_k). \tag{51}$$

† In Section 1-4, we observe that as δ tends to zero, (50) involves the difference of nearly equal numbers and the relative error in the computed difference tends to infinity. We must take δ small in order that the finite difference approximation (50) is near the actual partial derivative, but not so small that the numerical errors involved in computing the finite-difference dominate the error in the finite-difference approximation.

In some applications, iteration (51) is too expensive since each step requires the evaluation of $\mathbf{f}$ at $n + 1$ points. To construct an approximation to the Jacobian using fewer than $n + 1$ function evaluations, Broyden devised a family of **quasi-Newton schemes** that successively build better approximations to the Jacobian using function values from previous iterations. Since it is actually the inverse Jacobian $\mathbf{J}(\mathbf{x}_k)^{-1}$ that appears in Newton's method, Broyden constructs a sequence of approximations $\mathbf{H}_2, \mathbf{H}_3, \cdots$ to the inverse Jacobian starting from a given approximation $\mathbf{H}_1$ to the inverse Jacobian evaluated at the starting guess $\mathbf{x}_1$. Defining $\mathbf{s}_1 = -\mathbf{H}_1\mathbf{f}(\mathbf{x}_1)$, one iteration in Broyden's first method [B15] appears below.

$$\begin{aligned}
\mathbf{x}_{k+1} &= \mathbf{x}_k + \mathbf{s}_k \\
\mathbf{u}_k &= -\mathbf{H}_k\mathbf{f}(\mathbf{x}_{k+1}) \\
\mathbf{v}_k &= \gamma_k\mathbf{H}_k{}^T\mathbf{s}_k, \quad \gamma_k = 1/\mathbf{s}_k{}^T(\mathbf{s}_k - \mathbf{u}_k) \\
\mathbf{H}_{k+1} &= \mathbf{H}_k + \mathbf{u}_k\mathbf{v}_k{}^T \\
\mathbf{s}_{k+1} &= \gamma_k(\mathbf{s}_k{}^T\mathbf{s}_k)\mathbf{u}_k
\end{aligned} \tag{52}$$

If the denominator in γ_k's definition is zero, then we set $\mathbf{H}_{k+1} = \mathbf{H}_k$.

Briefly, quasi-Newton schemes such as (52) are derived in the following way: Suppose that $\mathbf{f}(\mathbf{x}) = \mathbf{0}$ is a linear equation or equivalently, $\mathbf{f}(\mathbf{x}) = \mathbf{Ax} - \mathbf{b}$. Subtracting $\mathbf{f}(\mathbf{x}_{k-1})$ from $\mathbf{f}(\mathbf{x}_k)$, we see that $\mathbf{f}(\mathbf{x}_k) - \mathbf{f}(\mathbf{x}_{k-1}) = \mathbf{A}(\mathbf{x}_k - \mathbf{x}_{k-1})$. Multiplying by $\mathbf{A}^{-1}$ yields $\mathbf{x}_k - \mathbf{x}_{k-1} = \mathbf{A}^{-1}(\mathbf{f}(\mathbf{x}_k) - \mathbf{f}(\mathbf{x}_{k-1}))$. Since the Jacobian of a linear function is the coefficient matrix, $\mathbf{J}(\mathbf{x}_k) = \mathbf{A}$ and

$$\mathbf{x}_k - \mathbf{x}_{k-1} = \mathbf{J}(\mathbf{x}_k)^{-1}(\mathbf{f}(\mathbf{x}_k) - \mathbf{f}(\mathbf{x}_{k-1})).$$

Now consider a nonlinear function $\mathbf{f}$. Even though the equality

$$\mathbf{x}_k - \mathbf{x}_{k-1} = \mathbf{J}(\mathbf{x}_k)^{-1}(\mathbf{f}(\mathbf{x}_k) - \mathbf{f}(\mathbf{x}_{k-1}))$$

typically does not hold, we still have the approximation

$$\mathbf{x}_k - \mathbf{x}_{k-1} \approx \mathbf{J}(\mathbf{x}_k)^{-1}(\mathbf{f}(\mathbf{x}_k) - \mathbf{f}(\mathbf{x}_{k-1})). \tag{53}$$

In Broyden's method, the approximation $\mathbf{H}_k$ to $\mathbf{J}(\mathbf{x}_k)^{-1}$ is chosen to satisfy the condition

$$\mathbf{H}_k(\mathbf{f}(\mathbf{x}_k) - \mathbf{f}(\mathbf{x}_{k-1})) = \mathbf{x}_k - \mathbf{x}_{k-1} \tag{54}$$

every iteration. In equation (54), $\mathbf{x}_k$ and $\mathbf{x}_{k-1}$ and the corresponding function values are known while the coefficient matrix $\mathbf{H}_k$ is unknown. If $\mathbf{x}$ and $\mathbf{f}$ have n components, then (54) is essentially n equations in n^2 unknowns (the elements of $\mathbf{H}_k$). In general, there are an infinite number of solutions, or equivalently, there are an infinite number of $\mathbf{H}_k$ that satisfy (54). Broyden utilizes one of the simplest solutions. The $\mathbf{H}_k$ are constructed using the recurrence $\mathbf{H}_{k+1} = \mathbf{H}_k + \mathbf{u}_k\mathbf{v}_k{}^T$, where the vectors $\mathbf{u}_k$ and $\mathbf{v}_k$ are chosen so that (54) holds in each iteration. In Exercise 4-5.1 you will verify that (54) holds for the $\mathbf{u}_k$ and $\mathbf{v}_k$ given in (52).

Having computed an approximation to the inverse Jacobian, the new iterate $\mathbf{x}_{k+1}$ is constructed from the previous iterate $\mathbf{x}_k$ exactly as in Newton's method. Recall that for Newton's method, we have $\mathbf{x}_{k+1} = \mathbf{x}_k - \mathbf{J}(\mathbf{x}_k)^{-1}\mathbf{f}(\mathbf{x}_k)$. For the quasi-Newton method, the inverse Jacobian $\mathbf{J}(\mathbf{x}_k)^{-1}$ is replaced by its approximation

$\mathbf{H}_k$ to obtain the iteration

$$\mathbf{x}_{k+1} = \mathbf{x}_k - \mathbf{H}_k\mathbf{f}(\mathbf{x}_k). \tag{55}$$

In Exercise 4-5.1 you also show that the formula for $\mathbf{x}_{k+1}$ in (52) is equivalent to (55).

Broyden's scheme converges to a root whenever $\mathbf{x}_1$ is sufficiently close to the root and $\mathbf{H}_1$ is sufficiently close to $\mathbf{J}(\mathbf{x}_1)^{-1}$. For one equation in one unknown, (52) reduces to the secant method (see Exercise 4-5.2). For a linear system $\mathbf{Ax} - \mathbf{b} = \mathbf{0}$ of n equations, Gay [G2] shows that the iterations converge to the root $\mathbf{x} = \mathbf{A}^{-1}\mathbf{b}$ in at most $2n$ steps. In some sense, $2n$ quasi-Newton steps are equivalent to one Newton step since Newton's method solves a linear system in one iteration. Subroutine QUASI implements the quasi-Newton scheme (52). The subroutine arguments are the following:

X:	starting guess(input) and solution(output)
H:	starting guess for inverse Jacobian
LH:	leading (row) dimension of array H
N:	number of equations
DIF:	input for subroutine WHATIS
SIZE:	input for subroutine WHATIS
NDIGIT:	desired number of correct digits
LIMIT:	maximum number of iterations
SUB:	name of subroutine to evaluate **f**
W:	work array (length at least 3N)

When a subroutine name is an argument of a subroutine, this name must be declared external at the start of the program. Thus the subroutine that evaluates **f** must be declared external. The following program solves equation (29) starting from $\mathbf{x}_1 = \mathbf{0}$. The initial approximation to the inverse Jacobian is

$$\mathbf{H}_1 = \begin{bmatrix} 3 & -1 \\ -1 & 3 \end{bmatrix}^{-1},$$

the inverse of the Jacobian evaluated at $\mathbf{x} = \mathbf{x}_1 = \mathbf{0}$.

```
EXTERNAL FUNC
REAL H(6),X(2),W(6)
DATA X/0.,0./
DATA H/3.,-1.,-1.,3./
CALL VERT(H,2,2)
CALL QUASI(X,H,2,2,D,S,4,100,FUNC,W)
CALL WHATIS(D,S)
WRITE(6,*)X(1),X(2)
END

SUBROUTINE FUNC(F,X)
REAL F(1),X(1)
F(1) = X(1) + X(1) - X(2) - EXP(-X(1))
F(2) = X(2) + X(2) - X(1) - EXP(-X(2))
RETURN
END
```

Exercises

4-5.1. Letting $\mathbf{f}_k$ stand for $\mathbf{f}(\mathbf{x}_k)$, we define $\mathbf{y}_k = \mathbf{f}_{k+1} - \mathbf{f}_k$. Assuming that $\mathbf{s}_k = -\mathbf{H}_k\mathbf{f}_k$, establish the following identities for algorithm (52): (a) $\mathbf{v}_k{}^T\mathbf{y}_k = 1$, (b) $\mathbf{u}_k = \mathbf{s}_k - \mathbf{H}_k\mathbf{y}_k$, (c) $\mathbf{H}_{k+1}\mathbf{y}_k = \mathbf{s}_k$, and (d) $\mathbf{s}_{k+1} = -\mathbf{H}_{k+1}\mathbf{f}_{k+1}$.

4-5.2. Show that for one equation in one unknown, iteration (52) reduces to the secant method (9). [*Hint:* First show that for a scalar equation, the following relation holds:

$$\mathbf{H}_{k+1} = \frac{x_{k+1} - x_k}{f_{k+1} - f_k}.]$$

4-6. GLOBAL TECHNIQUES

Except for special circumstances such as monotonically convergent iterations, most algorithms converge only when the starting guess is sufficiently close to the root. For the equation arc tan $x = 0$, we see in Figure 4-12 that Newton's method diverges from the root $x = 0$ when the starting guess is far from the root while the iterations converge when the starting guess is near the root. And for a special "intermediate" starting guess (computed in Exercise 4-6.1), the iterations cycle.

There is a simple modification of Newton's method that prevents divergence in many cases where the starting guess is poor. The standard Newton iteration is

$$\mathbf{x}_{k+1} = \mathbf{x}_k - \mathbf{J}(\mathbf{x}_k)^{-1}\mathbf{f}(\mathbf{x}_k).$$

The new point $\mathbf{x}_{k+1}$ is equal to the previous point $\mathbf{x}_k$ minus the increment $\mathbf{J}(\mathbf{x}_k)^{-1}\mathbf{f}(\mathbf{x}_k)$. When Newton's method diverges, the increment (also called the step) is usually too big. Let s be a positive parameter and let $\mathbf{y}(s)$ be defined by $\mathbf{y}(s) = \mathbf{x}_k - s\mathbf{z}$ where $\mathbf{z} = \mathbf{J}(\mathbf{x}_k)^{-1}\mathbf{f}(\mathbf{x}_k)$. Observe that $\mathbf{y}(0) = \mathbf{x}_k$ and $\mathbf{y}(1) = \mathbf{x}_{k+1}$, the point generated by a Newton step. Typically, the norm of $\mathbf{f}(\mathbf{y}(s))$ as a function of s resembles Figure 4-13. If $\mathbf{x}_k$ is near a root, then the minimum in Figure 4-13 is attained near $s = 1$. Since $s = 1$ corresponds to $\mathbf{x}_{k+1}$, it follows that $\|\mathbf{f}(\mathbf{x}_{k+1})\| < \|\mathbf{f}(\mathbf{x}_k)\|$. Since $\mathbf{f}$ vanishes at the root, $\|\mathbf{f}(\mathbf{x}_k)\|$ typically approaches zero monotonically when the starting guess is sufficiently close to the root.

When $\mathbf{x}_k$ is far from a root, the minimum in Figure 4-13 is often achieved

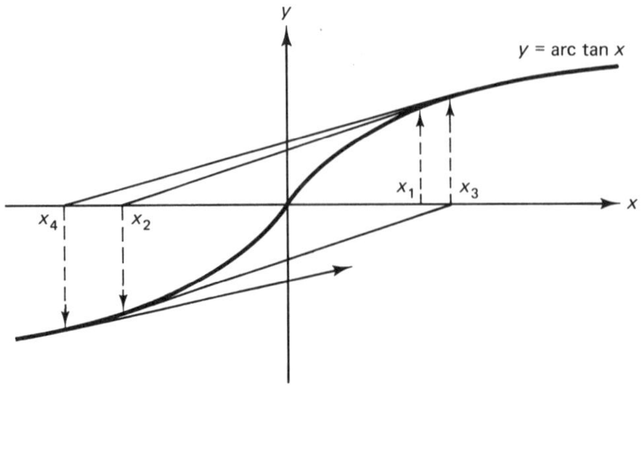

Figure 4-12 Divergence of Newton's method for $f(x)$ = arc tan x.

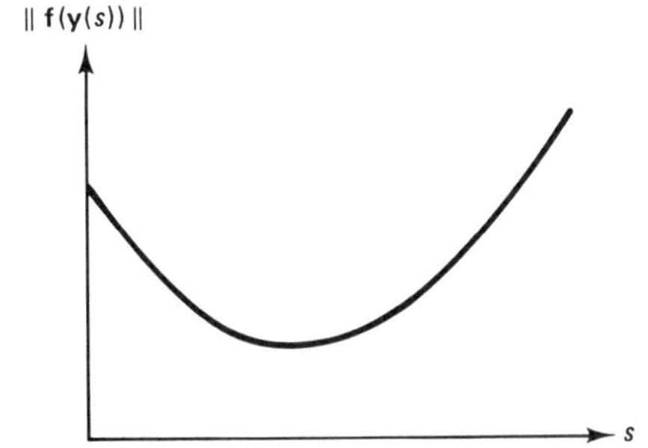

Figure 4-13 The norm of $\mathbf{f}(\mathbf{y}(s))$ versus s.

at a value of s less than 1 and it is conceivable that $\|\mathbf{f}(\mathbf{y}(1))\| > \|\mathbf{f}(\mathbf{y}(0))\|$, or equivalently, $\|\mathbf{f}(\mathbf{x}_{k+1})\| > \|\mathbf{f}(\mathbf{x}_k)\|$. The inequality $\|\mathbf{f}(\mathbf{x}_{k+1})\| < \|\mathbf{f}(\mathbf{x}_k)\|$ will be preserved if the Newton step is reduced. Armijo's rule for determining s is the following: Evaluate $\|\mathbf{f}(\mathbf{y}(s))\|$ at $s = 1, \frac{1}{2}, \frac{1}{4}, \cdots$, stopping when

$$\|\mathbf{f}(\mathbf{y}(s))\| \leqq \left(1 - \frac{s}{2}\right) \|\mathbf{f}(\mathbf{x}_k)\|. \tag{56}$$

If t denotes the first s which satisfies (56), then $\mathbf{x}_{k+1}$ is given by

$$\mathbf{x}_{k+1} = \mathbf{x}_k - t\mathbf{J}(\mathbf{x}_k)^{-1}\mathbf{f}(\mathbf{x}_k). \tag{57}$$

Usually, the Euclidean norm is used in (56) and the sum of the squares of **f**'s components are reduced at each iteration. To illustrate the iteration (57) corresponding to the Armijo strategy (56), we solve the equation arc tan $x = 0$ starting from $x_1 = 10$. The iterations in Table 4-11 converge to the root $x = 0$ even though the standard Newton iteration diverges when $x_1 > 1.39 \cdots$.

Armijo's adjustment to the Newton step emanates from optimization theory. Instead of trying to solve the equation $\mathbf{f}(\mathbf{x}) = \mathbf{0}$, let us try to minimize the 2-norm of **f**, or equivalently, let us consider the problem

$$\text{minimize} \sum_{i=1}^{n} f_i(\mathbf{x})^2. \tag{58}$$

At the minimum, the partial derivative with respect to each of the unknowns vanishes:

$$\sum_{i=1}^{n} \frac{\partial f_i}{\partial x_j}(\mathbf{x}) f_i(\mathbf{x}) = 0 \tag{59}$$

for $j = 1, 2, \cdots, n$. Using matrix-vector notation, (59) is written $\mathbf{J}(\mathbf{x})^T\mathbf{f}(\mathbf{x}) = \mathbf{0}$. If $\mathbf{J}(\mathbf{x})$ is invertible, then the equation $\mathbf{J}(\mathbf{x})^T\mathbf{f}(\mathbf{x}) = \mathbf{0}$ implies that any minimizer for (58) satisfies $\mathbf{f}(\mathbf{x}) = \mathbf{0}$.

TABLE 4-11 ITERATION (57) FOR THE EQUATION ARC TAN $x = 0$

k	t	x_k
1	.0625	10.000000
2	1.0000	.713507
3	1.0000	−.221728
4	1.0000	.007197
5	1.0000	.000000

We will draw a picture of Armijo's rule (56) using the 2-norm. Let $r(s)$ denote the 2-norm of $\mathbf{f}(\mathbf{y}(s))$; that is,

$$r(s) = \sqrt{\sum_{i=1}^{n} f_i(\mathbf{y}(s))^2}.$$

The derivative of $r(s)$ can be computed using the chain rule. Omitting the arithmetic, we have

$$\left.\frac{dr}{ds}\right|_{s=0} = -\|\mathbf{f}(\mathbf{x}_k)\|_2. \tag{60}$$

This shows that the slope of r is negative at $s = 0$. Since $r(s) \geqq 0$, the graph eventually bends upward as indicated in Figure 4-14. Rearranging (56) and utilizing (60) and the identity

$$r(0) = \|\mathbf{f}(\mathbf{y}(0))\|_2 = \|\mathbf{f}(\mathbf{x}_k)\|_2 = -r'(0),$$

yields the following equivalent formulation of (56):

$$\frac{r(s) - r(0)}{s} \leqq -\frac{1}{2}\|\mathbf{f}(\mathbf{x}_k)\| = \frac{1}{2}r'(0). \tag{61}$$

Since the left side of (61) is the slope of the chord that subtends the graph of r on the interval $[0,s]$, Armijo's rule is equivalent to saying that the slope of the chord is at most half the slope at $s = 0$. In Figure 4-14, those s's which satisfy (61) correspond to the interval $[0,a]$ where the slope of the chord is at most $.5r'(0)$. In Armijo's rule, we decrease s until it reaches $[0,a]$. In the vicinity of a root, the point $s = 1$ is typically contained in the interval $[0,a]$ and Armijo's rule terminates with $s = 1$. Hence (57) reduces to Newton's method in the vicinity of a root. But far from a root, the s generated by Armijo's rule is often less than 1. It can be shown that if the region where $\|\mathbf{f}(\mathbf{x})\| \leqq \|\mathbf{f}(\mathbf{x}_1)\|$ is bounded and both $\mathbf{f}$ is twice continuously differentiable and its Jacobian is nonsingular throughout this region, then iteration (57) converges to a root.

Another strategy for transforming a locally convergent iteration into a globally convergent algorithm involves the continuation method. Unlike Armijo's rule, which just applies to Newton's method, the continuation method can be used in conjunction with almost any algorithm. Abstractly, the idea is to embed a parameter s in the original equation $\mathbf{f}(\mathbf{x}) = \mathbf{0}$ giving us a new equation

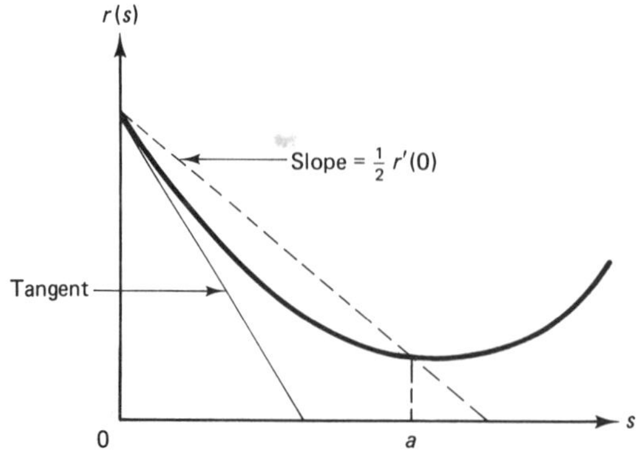

Figure 4-14 Armijo's rule.

$\mathbf{f}(\mathbf{x},s) = \mathbf{0}$ involving the parameter s. This parameter is chosen so that for $s = 0$, $\mathbf{f}(\mathbf{x},0) = \mathbf{f}(\mathbf{x})$, the original function, while for $s = 1$, $\mathbf{f}(\mathbf{x},1) = \mathbf{0}$ is easy to solve. For example, the equation $\mathbf{f}(\mathbf{x}) = \mathbf{A}\mathbf{x} - \mathbf{b}(\mathbf{x}) = \mathbf{0}$ can be cast in the form

$$\mathbf{f}(\mathbf{x},s) = \mathbf{A}\mathbf{x} - (1 - s)\mathbf{b}(\mathbf{x}) = \mathbf{0}.$$

Putting $s = 0$, we get the original equation $\mathbf{A}\mathbf{x} - \mathbf{b}(\mathbf{x}) = \mathbf{0}$ while $s = 1$ yields the trivial system $\mathbf{A}\mathbf{x} = \mathbf{0}$ with the solution $\mathbf{x} = \mathbf{0}$ (assuming that $\mathbf{A}$ is invertible).

For fixed s, let $\mathbf{x} = \mathbf{x}(s)$ be a solution to $\mathbf{f}(\mathbf{x},s) = \mathbf{0}$. Since $\mathbf{f}(\mathbf{x},0) = \mathbf{f}(\mathbf{x})$, $\mathbf{x}(0)$ is a solution to the original equation. In many applications, $\mathbf{x}(s)$ depends continuously on s. That is, small changes in s produce small changes in $\mathbf{x}$. By assumption, $\mathbf{f}(\mathbf{x},1) = \mathbf{0}$ is easy to solve and the solution is $\mathbf{x}(1)$. Given a small number Δ, let us solve $\mathbf{f}(\mathbf{x}, 1-\Delta) = \mathbf{0}$ using the starting guess $\mathbf{x}_1 = \mathbf{x}(1)$ [which is near the actual solution $\mathbf{x}(1-\Delta)$ since Δ is small and $\mathbf{x}(1-\Delta)$ is near $\mathbf{x}(1)$]. After computing $\mathbf{x}(1-\Delta)$, $\mathbf{x}(1-2\Delta)$ is computed using the starting guess $\mathbf{x}_1 = \mathbf{x}(1-\Delta)$. Continuing in this fashion, we trace the curve $\mathbf{x}(s)$ and we finally reach $\mathbf{x}(0)$, a solution to the original equation $\mathbf{f}(\mathbf{x}) = \mathbf{0}$. In implementing the continuation method, we initially use a large Δ and if the iterations diverge, Δ is reduced. For example, we could start with $\Delta = \frac{1}{2}$ and successively divide Δ by 2 until it is so small that $\mathbf{x}_1 = \mathbf{x}(1)$ is a good guess for a root of the equation $\mathbf{f}(\mathbf{x},1-\Delta) = \mathbf{0}$ and the iterations converge. When the iterations converge quickly, the step size may be increased. If there is no natural way to embed s in the equation $\mathbf{f}(\mathbf{x}) = \mathbf{0}$, try the following approach: Let $\mathbf{y}$ be an arbitrary guess for the solution and define

$$\mathbf{f}(\mathbf{x},s) = \mathbf{f}(\mathbf{x}) - s\mathbf{f}(\mathbf{y}).$$

Observe that $\mathbf{f}(\mathbf{x},0) = \mathbf{f}(\mathbf{x})$ while $\mathbf{f}(\mathbf{x},1) = \mathbf{0}$ has the trivial solution $\mathbf{x} = \mathbf{y}$.

There are three problems to watch for when applying the continuation method:

1. **No root**. For some s, $\mathbf{f}(\mathbf{x},s) = \mathbf{0}$ has no solution.
2. **Bifurcation**. For some s, $\mathbf{f}(\mathbf{x},s) = \mathbf{0}$ has many solutions, and we are soon lost in the solution branches (see Figure 4-15).
3. **Wrong root**. The equation $\mathbf{f}(\mathbf{x}) = \mathbf{0}$ has several solutions and the continuation method does not take us to the desired root.

Despite these pitfalls, the continuation method is something to try when an

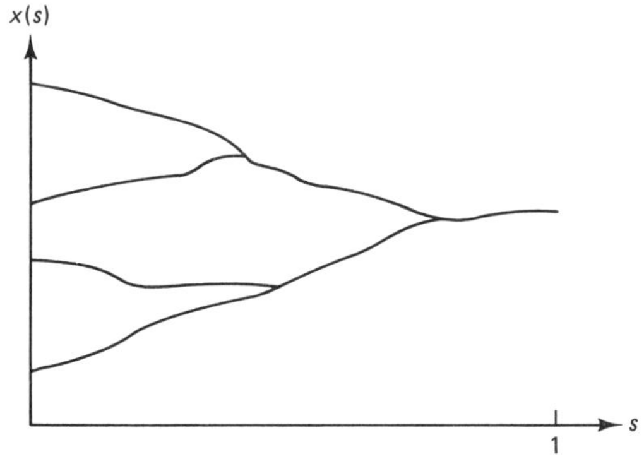

Figure 4-15 Bifurcation during the continuation method.

algorithm diverges due to a poor starting guess. References for the continuation method include the surveys [A3] and [A4] by Allgower and Georg and the book [R6] by Rheinboldt. Additional references appear at the end of the chapter.

To illustrate the continuation method, let us consider the equation arc tan $x = 0$ with the root $x = 0$ and with the starting guess $x = 3$, for which Newton's method diverges. The parametric equation we will use is

$$f(x,s) = \arctan x - (s)\arctan 3 = 0.$$

When $s = 1$, the root is $x = 3$ and when $s = 0$, $f(x,s) = 0$ reduces to the original equation arc tan $x = 0$. As s decreases from 1 down to 0, the root of $f(x,s) = 0$ travels from 3 to the desired "unknown" root 0. Taking $s = .5$, we apply Newton's method to the equation

$$f(x,.5) = \arctan x - (.5)\arctan 3 = 0$$

with the starting guess $x = 3$. Since the iterations diverge, we increase s. Taking $s = .75$ and applying Newton's method to

$$f(x,.75) = \arctan x - (.75)\arctan 3 = 0$$

with the starting guess $x = 3$, the iterations converge to 1.3600. Returning to $s = .5$, Newton's method is applied to $f(x,.5) = 0$ with the starting guess $x = 1.3600$. This time the iterations converge to .7208. Then taking $s = .25$ and applying Newton's method to $f(x,.25) = 0$ with the starting guess $x = .7208$, the iterations converge to .3228. Finally, taking $s = 0$, we apply Newton's method to $f(x,0) = 0$ with the starting guess $x = .3228$ to obtain the desired root $x = 0$.

Exercises

4-6.1. For the function $f(x) = \arctan x$, Newton's method diverges when $|x_1| > 1.39\cdots$ while Newton's method converges when $|x_1| < 1.39\cdots$. Obtain an equation whose solution is the smallest positive x_1 for which Newton's method diverges. Using any algorithm, compute the solution to this equation with six-place accuracy.

4-6.2. Explain how to evaluate the vector $\mathbf{z} = \mathbf{J}(\mathbf{x}_k)^{-1}\mathbf{f}(\mathbf{x}_k)$ in (57) without inverting $\mathbf{J}(\mathbf{x}_k)$.

4-6.3. Consider the following system of three equations in three unknowns:

$$6 \arctan (x_1 - 10) - 2e^{-x_2} - 2e^{-x_3} + 2x_2 + 2x_3 - 9 = 0,$$
$$2 \arctan (x_1 - 10) - 4e^{-x_2} - e^{-x_3} + 7x_2 - 2x_3 - 3 = 0,$$
$$2 \arctan (x_1 - 10) - e^{-x_2} - 3e^{-x_3} - x_2 + 5x_3 - 3 = 0.$$

(a) Using Newton's method and the starting guess $\mathbf{x}_1 = 0$, try to solve this system.

(b) Now incorporate Armijo's rule in the program for part (a) and rerun your program.

4-7. CONDITIONING

For the linear system $\mathbf{Ax} = \mathbf{b}$ and its perturbation $\mathbf{A}(\mathbf{x} + \boldsymbol{\delta}\mathbf{x}) = \mathbf{b} + \boldsymbol{\delta}\mathbf{b}$, we saw that $\boldsymbol{\delta}\mathbf{x}$ satisfies the inequality

$$\frac{\|\boldsymbol{\delta}\mathbf{x}\|}{\|\mathbf{x}\|} \leqq c\,\frac{\|\boldsymbol{\delta}\mathbf{b}\|}{\|\mathbf{b}\|},$$

where $c = \|\mathbf{A}\|\|\mathbf{A}^{-1}\|$ is $\mathbf{A}$'s condition number. Analogous sensitivity results for nonlinear equations are now developed. For a linear system, it is natural to study the solution's sensitivity with respect to changes in the right side $\mathbf{b}$ or in the coefficient matrix $\mathbf{A}$. For a nonlinear equation, the natural sensitivity parameter depends on the nature of the equation. For example, let us consider the equation $f(x) = x - e^{-x} = 0$. Due to rounding errors associated with the computer's approximation to the exponential function, the computed value of e^{-x} is equal to $(1 + \varepsilon)e^{-x}$ where ε is related to the machine precision. Thus the computer's zero of f really satisfies the equation $x - (1 + \varepsilon)e^{-x} = 0$. The sensitivity of the solution to $x - (1 + \varepsilon)e^{-x} = 0$ with respect to changes in ε is related to the conditioning of the original equation $x = e^{-x}$. For a second example, let us consider the polynomial

$$f(x) = x^5 + \tfrac{4}{3}x^4 + 1.$$

In this case, it is natural to study a zero's sensitivity with respect to changes in the coefficients—since the computer replaces the coefficient $\frac{4}{3}$ by a binary approximation, the computer's zero of f is really a zero of the polynomial $x^5 + (1 + \varepsilon)\frac{4}{3}x^4 + 1$ where ε is related to the machine precision. Again, the sensitivity of the solution to $x^5 + (1 + \varepsilon)\frac{4}{3}x^4 + 1 = 0$ with respect to changes in ε is related to the conditioning of the original equation $x^5 + \frac{4}{3}x^4 + 1 = 0$.

In both of these examples, we have an equation $f(x,\varepsilon) = 0$, which involves x and a parameter ε, and we are interested in how the solution $x = x(\varepsilon)$ depends upon ε. In the first example, $f(x,\varepsilon) = x - (1 + \varepsilon)e^{-x}$ and in the second example, $f(x,\varepsilon) = x^5 + (1 + \varepsilon)\frac{4}{3}x^4 + 1$. When ε is zero, $f(x, \varepsilon)$ reduces to the original function $f(x)$ and the solution to $f(x) = 0$ is $x(0)$. If a small ε produces a relatively small change $x(\varepsilon) - x(0)$ in the solution, then the equation $f(x) = 0$ is well conditioned. If a small ε produces a relatively large change in the solution, then the equation is ill conditioned. The derivative of $x(\varepsilon)$ with respect to ε is related to the absolute change $|x(\varepsilon) - x(0)|$ since

$$\frac{dx}{d\varepsilon}(0) = \lim_{\varepsilon\to 0} \frac{x(\varepsilon) - x(0)}{\varepsilon}.$$

For small ε, we have the approximation

$$\frac{dx}{d\varepsilon}(0) \approx \frac{x(\varepsilon) - x(0)}{\varepsilon}.$$

Multiplying both sides by ε yields

$$x(\varepsilon) - x(0) \approx \varepsilon \frac{dx}{d\varepsilon}(0). \tag{62}$$

Thus the change $x(\varepsilon) - x(0)$ in the solution is roughly ε times the derivative $x'(0)$. The equation's conditioning is related to the relative change in the solution. Dividing (62) by $x(0)$, the relative change satisfies

$$\frac{x(\varepsilon) - x(0)}{x(0)} \approx \varepsilon \frac{1}{x(0)} \frac{dx}{d\varepsilon}(0). \tag{63}$$

If the ratio $|x'(0)|/|x(0)|$ on the right side of (63) is large, then a small ε can generate a relatively large change in the solution and the equation is ill conditioned. To evaluate $x'(\varepsilon)$, the identity $f(x(\varepsilon), \varepsilon) = 0$ is differentiated with respect to ε.

By the chain rule, we have

$$\frac{df}{d\varepsilon}(x(\varepsilon), \varepsilon) = \left.\frac{\partial f}{\partial x}\right|_{x=x(\varepsilon)} \frac{dx}{d\varepsilon} + \left.\frac{\partial f}{\partial \varepsilon}\right|_{x=x(\varepsilon)} = 0.$$

Solving for $x'(\varepsilon)$ and dividing by $x(0)$ gives

$$\frac{1}{x(0)}\frac{dx}{d\varepsilon}(0) = -\frac{1}{x(0)}\left(\frac{\partial f}{\partial x}\right)^{-1}\left.\frac{\partial f}{\partial \varepsilon}\right|_{x=x(0)}. \tag{64}$$

The right side of (64) measures the equation's conditioning. If the absolute value of the right side is near 1 or smaller than 1, the equation is well conditioned. If the absolute value of the right side is much larger than 1, the equation is ill conditioned.

Let us use (64) to study the conditioning of a polynomial's zeros with respect to a change in a coefficient. In particular, if p is a polynomial of degree n:

$$p(x) = x^n + a_1x^{n-1} + a_2x^{n-2} + \cdots + a_{n-1}x + a_n,$$

we will consider the sensitivity of the zeros with respect to changes in a_1. The perturbed function is

$$f(x, \varepsilon) = x^n + (1 + \varepsilon)a_1x^{n-1} + \cdots + a_n = p(x) + \varepsilon a_1x^{n-1}.$$

To apply (64), we need the derivative of f with respect to x and ε:

$$\frac{\partial f}{\partial x} = \frac{\partial}{\partial x}(p(x) + \varepsilon a_1x^{n-1}) = \frac{dp}{dx}(x) + (n - 1)\varepsilon a_1x^{n-2}$$

and

$$\frac{\partial f}{\partial \varepsilon} = \frac{\partial}{\partial \varepsilon}(p(x) + \varepsilon a_1x^{n-1}) = a_1x^{n-1}.$$

Putting $\varepsilon = 0$ and utilizing (64), it follows that

$$\frac{1}{x(0)}\frac{dx}{d\varepsilon}(0) = -\left(\left.\frac{dp}{dx}\right|_{x=x(0)}\right)^{-1} a_1x(0)^{n-2}. \tag{65}$$

As these calculations indicate, a zero of a polynomial is a differentiable function of the coefficients provided the derivative of p does not vanish at the zero.

For the polynomial

$$p(x) = (x - 1)(x - 2) = x^2 - 3x + 2$$

and the zero $x = 2$, n is 2 and

$$f(x, \varepsilon) = x^2 - (1 + \varepsilon)3x + 2 = p(x) - 3\varepsilon x.$$

If $\varepsilon = 0$, then $x = 2$ is a zero and as ε changes, this zero wanders around. Of course, we can calculate the zero nearest 2 using the quadratic formula:

$$x(\varepsilon) = \frac{3(1 + \varepsilon) + \sqrt{9(1 + \varepsilon)^2 - 8}}{2}. \tag{66}$$

You can see that when $\varepsilon = 0$, $x(0)$ is 2. Although the ratio $x'(0)/x(0)$ can be computed by differentiating (66) and dividing by $x(0) = 2$, it is easier to apply (65). Since $p'(x) = 2x - 3$, $p'(x(0)) = p'(2) = 1$ and by (65), we have

$$\frac{1}{x(0)}\frac{dx}{d\varepsilon}(0) = -(1^{-1})(-3)(2^0) = 3.$$

Thus the relative change $(x(\varepsilon) - x(0))/x(0)$ is about 3ε. Since the relative change in the zero is small when ε is small, the zero $x = 2$ of the polynomial $x^2 - 3x + 2$ is well conditioned.

Next, let us apply (65) to the polynomial

$$p(x) = (x-1)(x-2)\cdots(x-20) = x^{20} - 210x^{19} + \cdots + 2432902008176640000.$$

In this case, n is 20, $f(x,\varepsilon) = p(x) - 210\varepsilon x^{19}$, and the zeros of p are $x = 1$, $x = 2, \cdots, x = 20$. To apply (65), we need the derivative of p. It can be shown that

$$\frac{dp}{dx}(x) = \sum_{i=1}^{20} \prod_{\substack{j=1 \\ j \neq i}}^{20} (x - j). \tag{67}$$

Evaluating (67) at a zero, say $x = k$, where k is an integer between 1 and 20, every term in the sum vanishes except for the kth term. Hence (65) reduces to the following expression at the zero $x = k$:

$$\frac{1}{x(0)}\frac{dx}{d\varepsilon} = \left(\prod_{\substack{j=1 \\ j \neq k}}^{20} (k - j)\right)^{-1} 210\, k^{18}. \tag{68}$$

At the zero $x = k = 1$, (68) is approximately -1.7×10^{-15} so the smallest zero is very well conditioned. But at the zero $x = k = 20$, (68) is approximately 4.5×10^8, which implies that the largest zero is ill conditioned. For example, taking $\varepsilon = 2^{-23}/210$, Wilkinson [W5] obtains the following zeros for the perturbed polynomial $f(x, \varepsilon) = p(x) - 210\varepsilon x^{19}$:

1.000	4.000	7.000	$10.095 \pm .643\mathbf{i}$	$16.731 \pm 2.813\mathbf{i}$
2.000	5.000	8.000	$11.794 \pm 1.652\mathbf{i}$	$19.502 \pm 1.940\mathbf{i}$
3.000	6.000	8.917	$13.992 \pm 2.519\mathbf{i}$	20.847

As expected, the change in the smallest zero is insignificant while the change in the largest zero is enormous. By equation (63), the estimated relative change in a zero is the magnitude of (68) times ε. For Wilkinson's ε, the estimated relative change in the zero $x = 20$ is $(4.5 \times 10^8) \times (2^{-23}/210) \approx .26$, while the actual relative change is

$$\frac{20.847 - 20}{20} \approx .04.$$

Although formula (63) overestimates the relative change in the zero $x = 20$ when $\varepsilon = 2^{-23}/210$, (63) becomes a better estimate for the actual relative change as ε tends to zero.

The results developed above for a single equation also apply to a system of equations, but partial derivatives with respect to x [see equation (64)] are replaced by the gradient with respect to $\mathbf{x}$. For example, if $\mathbf{x} = \mathbf{x}(\varepsilon)$ denotes a solution to $\mathbf{f}(\mathbf{x},\varepsilon) = 0$, then the derivative of $\mathbf{x}$ with respect to ε is

$$\frac{d\mathbf{x}}{d\varepsilon} = -(\nabla\mathbf{f})^{-1} \left.\frac{\partial\mathbf{f}}{\partial\varepsilon}\right|_{\mathbf{x}=\mathbf{x}(\varepsilon)}.$$

The analog of (64) is

$$\frac{1}{\|\mathbf{x}(0)\|}\left\|\frac{d\mathbf{x}}{d\varepsilon}(0)\right\| \le \frac{1}{\|\mathbf{x}(0)\|}\|\nabla\mathbf{f}^{-1}\| \left\|\frac{\partial\mathbf{f}}{\partial\varepsilon}\right\|_{\mathbf{x}=\mathbf{x}(0)}.$$

On the other hand, (64) does not extend to multiple zeros since the derivative $x'(0)$ may not exist. For example, the equation $x^2 = 0$ has the root $x = 0$ with multiplicity 2 while the perturbed equation $x^2 = \varepsilon$ has the root $x = \sqrt{\varepsilon}$. The derivative $x'(\varepsilon)$ is $1/(2\sqrt{\varepsilon})$, which is infinite at $\varepsilon = 0$. For a simple zero, the change $x(\varepsilon) - x(0)$ is proportional to ε. For a root of multiplicity 2, the change in the root is proportional to $\sqrt{\varepsilon}$. In general, if the root has multiplicity m, the change is proportional to $\sqrt[m]{\varepsilon}$. Since $\sqrt[m]{\varepsilon}$ is much larger than ε when ε is small, equations with multiple roots are almost always ill conditioned.

Exercises

4-7.1. Determine the conditioning of each of the zeros of the polynomial $p(x) = (x - 1)(x - 2) \cdots (x - 20)$ with respect to perturbations in the coefficient of x^{19}. In other words, evaluate the expression (68) at each of the zeros. Which zero has the worst condition number?

4-7.2. Determine the conditioning of each of the zeros of the polynomial $p(x) = (x - 1)(x - 2) \cdots (x - 20)$ with respect to perturbations in the coefficient of x^0. Which zero has the worst condition number?

4-7.3. When ε is zero, $x = 3$ is a root of the equation

$$x^3 - 6x^2 + 11x - (6 + \varepsilon) = 0. \tag{69}$$

Using equations (62) and (64), estimate the root of (69) near $x = 3$ that corresponds to $\varepsilon = .1$.

4-8. POLYNOMIAL EQUATIONS

Many special algorithms have been devised for finding the zeros of a polynomial. Typically, a polynomial will have both real and complex zeros. Let z denote the complex number $x + y\mathbf{i}$, where x is the real part of z, y is the imaginary part of z, and $\mathbf{i}$ is the square root of -1. A nth degree **monic** polynomial p has the form

$$p(z) = z^n + a_1 z^{n-1} + a_2 z^{n-2} + \cdots + a_{n-1}z + a_n.$$

A polynomial is monic if its **leading coefficient** (the coefficient of z^n above) is 1. Counting multiplicities, a nth-degree polynomial has n zeros. If a good estimate for a zero, say z_1 of p, is known, we can use Newton's method to determine z_1 precisely. Dividing p by $z - z_1$ gives us a polynomial of degree $n - 1$ to which Newton's method or some other scheme can be applied to obtain a new zero z_2. Removing a zero z_i from p by dividing out the factor $z - z_i$ to obtain a new polynomial of degree $n - 1$ is called **deflation**. The new polynomial has the same zeros as the original polynomial except that the zero z_i is missing. Although the strategy outlined above can be used to find a few zeros, we encounter a problem when searching for many zeros of a polynomial. The standard deflation process is unstable unless the zeros are arranged in increasing magnitude. That is, unless $|z_1| \leqq |z_2| \leqq \cdots$, the zeros of the deflated polynomials will soon differ significantly from the zeros of the original polynomial p due to the propagation of rounding errors. Consequently, when computing many or all the zeros of a polynomial, the zeros must be generated roughly in the order of increasing magnitude.

One of the best algorithms for finding many or all the zeros of a polynomial is based on the Jenkins-Traub scheme. In its simplest form, the Jenkins-Traub scheme constructs a sequence $q_0, q_1, \cdots$ of polynomials of degree $n - 1$, where the starting polynomial q_0 is the derivative of p. Thereafter, q_{k+1} is determined from q_k by the formula

$$q_{k+1}(z) = \frac{q_k(z) - \dfrac{q_k(s_k)}{p(s_k)}\, p(z)}{z - s_k}. \tag{70}$$

Since the numerator in (70) vanishes at $z = s_k$, the denominator $z - s_k$ evenly divides the numerator giving us a polynomial whose degree is one lower than that of the numerator. Since the degree of p is n and the degree of q_k is $n - 1$, the degree of the numerator in (70) is n and the degree of q_{k+1} is $n - 1$. The Jenkins-Traub algorithm has three phases. The value of s_k in each phase is summarized as follows:

$$\begin{aligned} &\text{Phase 1}: s_k = 0 \\ &\text{Phase 2}: s_k = \sigma \\ &\text{Phase 3}: s_{k+1} = s_k - l_{k+1}p(s_k)/q_{k+1}(s_k) \end{aligned}$$

where σ is an approximation to a zero of p and l_k is the leading coefficient of q_k. The starting value for s_k in phase 3 is σ. If σ is sufficiently close to a zero of p, the s_k converge to it and the convergence is faster than quadratic. After a zero, say z_1, of p is computed, we divide p by $z - z_1$ to obtain a polynomial of degree $n - 1$. The Jenkins-Traub algorithm is applied to this lower-degree polynomial, giving us a new zero, and so on. When q_{k+1} is computed using (70), we divide $q_k(s_k)$ by $p(s_k)$. Obviously, this division cannot be performed when $p(s_k)$ is zero. But if $p(s_k)$ is zero, a zero of p has been found and the iteration should stop.

We propose the following strategy for determining σ. During phase one, compute the parameter r_k defined by

$$r_0 = |a_0|^{1/n} \quad \text{and} \quad r_k = \left(\frac{n|p(0)|}{|q_{k-1}(0)|}\right)^{1/k} \quad \text{for} \quad k = 1, 2, \cdots,$$

and let m_k be the minimum of $r_0, r_1, \cdots, r_k$. It can be shown that m_k approaches the radius of the smallest circle centered at the origin in the complex plane which contains a zero of p. After performing several iterations of phase 1 and obtaining an estimate for this circle, we evaluate p at evenly spaced points on the estimated circle. That point that yields the smallest magnitude for p is the σ for phase 2. After performing a few phase 2 iterations, we proceed to phase 3. If phase 3 fails to converge quickly, we return to phase 1 and obtain a sharper estimate for the radius of the smallest circle centered at the origin that contains a zero of p, we evaluate p at more points on this circle, and we proceed to phase 2, using for σ that point where the magnitude of p is smallest. Eventually, a suitable starting guess for phase 3 is generated. The fast Fourier transform ([C5] and [S11]) can be used to evaluate p at evenly spaced points on a circle in the complex plane. A detailed statement of this starting procedure for the Jenkins-Traub algorithm appears in [H5]. Subroutine CZERO in NAPACK implements this scheme.

REVIEW PROBLEMS

4-1. Taking $N = 10$, solve equation (78) of Chapter 2 using the nonlinear Jacobi scheme, using the nonlinear Gauss-Seidel scheme, and using Newton's method (with subroutines TFACT and TSOLVE). Which of these methods is more efficient than the others for this equation?

4-2. Consider the boundary-value problem

$$\frac{d^4x}{dt^4}(t) = e^{-x(t)}, \qquad 0 \leqq t \leqq 1,$$

with the boundary condition

$$x(0) = x(1) = x'(0) = x'(1) = 0.$$

Using the finite-difference approximations

$$\frac{d^4x}{dt^4}(t) \approx \frac{x(t-2\Delta t) - 4x(t-\Delta t) + 6x(t) - 4x(t+\Delta t) + x(t+2\Delta t)}{\Delta t^4}$$

and

$$\frac{d^4x}{dt^4}(0) \approx \frac{12x(\Delta t) - 6x(2\Delta t) + \frac{4}{3}x(3\Delta t)}{\Delta t^4},$$

we obtain (as in Chapter 1) a system of equations for $x_1, x_2, \cdots, x_{N-1}$:

$$N^4 \begin{bmatrix} 36 & -18 & 4 & & & & & \\ -12 & 18 & -12 & 3 & & & & \\ 3 & -12 & 18 & \cdot & \cdot & & & \\ & 3 & -12 & \cdot & \cdot & 3 & & \\ & & \cdot & \cdot & \cdot & -12 & 3 \\ & & & 3 & -12 & 18 & -12 \\ & & & & 4 & -18 & 36 \end{bmatrix} \begin{bmatrix} x_1 \\ x_2 \\ x_3 \\ \cdot \\ \cdot \\ \cdot \\ x_{N-1} \end{bmatrix} = \begin{bmatrix} 3e^{-x_1} \\ 3e^{-x_2} \\ 3e^{-x_3} \\ \cdot \\ \cdot \\ \cdot \\ 3e^{-x_{N-2}} \\ 3e^{-x_{N-1}} \end{bmatrix}. \qquad (71)$$

Here x_i is the approximation to $x(i/N)$. Solve system (71) using Newton's method, subroutines BFACT and BSOLVE, and $N = 10$.

4-3. Observe that system (71) has the form $\mathbf{Ax} = \mathbf{b(x)}$, where $\mathbf{A}$ is a pentadiagonal matrix and $\mathbf{b}$ is the right side of (71). Taking $N = 10$, solve this system using the iteration $\mathbf{Ax}^{\text{new}} = \mathbf{b(x}^{\text{old}})$. Which subroutine should be used to factor $\mathbf{A}$?

4-4. (by G. Strang) Consider the factorization (to four decimal places) of a large tridiagonal matrix for which every diagonal element is equal to 4 and every off-diagonal element is equal to 1:

$$\begin{bmatrix} 4 & 1 & & & \\ 1 & 4 & 1 & & \\ & 1 & 4 & 1 & \\ & & 1 & 4 & \cdot \\ & & & \cdot & \cdot & \cdot \\ & & & & \cdot & \cdot \end{bmatrix} =$$

$$\begin{bmatrix} 1 & & & & & \\ .2500 & 1 & & & & \\ & .2667 & 1 & & & \\ & & .2679 & 1 & & \\ & & & .2679 & \cdot & \\ & & & & \cdot & \cdot \end{bmatrix} \begin{bmatrix} 4. & 1 & & & & \\ & 3.7500 & 1 & & & \\ & & 3.7333 & 1 & & \\ & & & 3.7321 & 1 & \\ & & & & 3.7320 & \cdot \\ & & & & & \cdot \end{bmatrix}$$

Observe that the diagonal elements d_k of the $\mathbf{U}$ factor approach $3.7320 \cdots$, while the multipliers l_k approach $.2679 \cdots$. If this factorization is performed on a computer, we can stop the factorization process when the relative difference between two successive d_k and two successive l_k is around the machine epsilon—the subsequent diagonal elements and multipliers are essentially equal to the last computed d_k and l_k. Now consider a general tridiagonal matrix **A** with every diagonal element equal to a, with every subdiagonal element equal to b, and with every superdiagonal element equal to c:

$$\mathbf{A} = \begin{bmatrix} a & c & & \\ b & a & c & \\ & b & \cdot & \cdot \\ & & \cdot & \cdot \end{bmatrix}.$$

In this case, the diagonal elements of **A**'s upper triangular factor satisfy the recurrence

$$d_{k+1} = a - bc/d_k, \qquad \text{where } d_1 = a,$$

and the multipliers are $l_k = b/d_k$. It can be shown that if $a^2 \geqq 4bc$, the d_k approach a limit d_∞. Obtain a formula for d_∞ that involves a, b, and c. Evaluate d_∞ when $a = 4$ and $b = 1 = c$. What is the function g with the property that $d_{k+1} = g(d_k)$? Show that if $a^2 \geqq 4bc$, the derivative of g evaluated at d_∞ is

$$g'(d_\infty) = \frac{a - \operatorname{sign}(a)\sqrt{a^2 - 4bc}}{a + \operatorname{sign}(a)\sqrt{a^2 - 4bc}}.$$

4-5. Consider the iteration $x_{k+1} = g(x_k)$ for the following choices of g. Determine the convergence order when the iterations start in a neighborhood of the specified fixed point.

(a) $g(x) = (4 + 4x - x^2)^{1/3}$ and the fixed point $x = 2$.
(b) $g(x) = (6 - x^3)/5$ and the fixed point $x = 1$.
(c) $g(x) = 2x^{-1} + x/2$ and the fixed point $x = 2$.
(d) $g(x) = \frac{1}{3} + \frac{2}{3}\left(x + \dfrac{x - 1}{3x^2 - 6x + 2}\right)$ and the fixed point $x = 1$.

4-6. Consider the function g defined by

$$g(x) = x - \frac{f(x)}{f'(x)} - \frac{f''(x)}{2f'(x)}\left(\frac{f(x)}{f'(x)}\right)^2.$$

If α is a zero of f and $f'(\alpha) \neq 0$, show that the convergence order of the iteration $x_{k+1} = g(x_k)$ is at least cubic when the starting guess is sufficiently close to α.

4-7. (a) Graphing the function $f(x) = x^3 - x - 2$, we see that a zero lies between $x = 1$ and $x = 2$. Using the monotone convergence theory of Section 4-2, show that Newton's method converges monotonically to this zero whenever the starting guess x_1 is greater than 2.

(b) Graphing the function $f(x) = x - e^{-x}$, we see that a zero lies between $x = 0$ and $x = 1$. Show that Newton's method converges monotonically to this zero whenever $x_1 \leqq 0$.

(c) Graphing the function $f(x) = x^3 - \cos x$, we see that a zero lies between $x = 0$ and $x = 1$. Give a starting guess for which Newton's method is monotonically convergent.

(d) Graphing the function $f(x) = 5x + \sin x - e^x$, we see that a zero lies between $x = 0$ and $x = 1$. Give a starting guess for which Newton's method is monotonically convergent.

4-8. Suppose that z is a complex variable and f is a complex-valued function of z. That is, $z = x + y\mathbf{i}$ and $f(z) = u(z) + v(z)\mathbf{i}$, where x, y, u, and v are real and $\mathbf{i} = \sqrt{-1}$. The complex equation $f(z) = 0$ is equivalent to two real equations $u(x,y) = 0$ and $v(x,y) = 0$. Show that the Newton step associated with the complex equation $f(z) = 0$ is the same as the Newton step associated with the real system $u(x,y) = 0$ and $v(x,y) = 0$. More precisely, let $J(x,y)$ denote the Jacobian corresponding to the real system. Assuming that f is differentiable, show that

$$\frac{f(z)}{f'(z)} = J(x,y)^{-1}\begin{bmatrix} u(x,y) \\ v(x,y) \end{bmatrix} \tag{72}$$

where the identity (72) is interpreted in the following sense: The real part of the left side is equal to the first component of the right side and the imaginary part of the left side is equal to the second component of the right side. You will need the following result from complex variables: At any point where a complex function is differentiable, the (Cauchy-Riemann) equations

$$\frac{\partial u}{\partial x} = \frac{\partial v}{\partial y} \quad \text{and} \quad \frac{\partial u}{\partial y} = -\frac{\partial v}{\partial x}$$

are satisfied. Moreover, the derivative f' is given by

$$f'(z) = \frac{\partial u}{\partial x}(z) + \frac{\partial v}{\partial x}(z)\mathbf{i}.$$

Also, note that the inverse of a 2×2 matrix is given by the rule:

$$\begin{bmatrix} a & c \\ b & d \end{bmatrix}^{-1} = \frac{1}{ad - bc}\begin{bmatrix} d & -c \\ -b & a \end{bmatrix}.$$

4-9. Let us use Newton's method to solve an equation $f(x) = 0$ which has several roots. After computing one root, say α_1, we would like to remove this root from f so that Newton's method will not converge to the previously computed root α_1. To accomplish this deflation process, we divide out the previously computed root, or equivalently, we apply Newton's method to the function $F(x) = f(x)/(x - \alpha_1)$. Show that Newton's method applied to F reduces to the iteration $x_{k+1} = g(x_k)$ where

$$g(x) = x - \frac{f(x)}{f'(x) - [f(x)/(x - \alpha_1)]}.$$

After computing the roots $\alpha_1, \alpha_2, \cdots, \alpha_m$, the deflated function is given by

$$F(x) = \frac{f(x)}{(x - \alpha_1)(x - \alpha_2)\cdots(x - \alpha_m)}.$$

Show that Newton's method applied to F reduces to the iteration $x_{k+1} = g(x_k)$ where

$$g(x) = x - \frac{f(x)}{f'(x) - \sum_{i=1}^{m} [f(x)/(x - \alpha_i)]}.$$

4-10. Using the deflation procedure developed in Problem 4-9, compute the zeros of the polynomial

$$p(x) = x^4 - 10x^3 + 35x^2 - 50x + 24.$$

Start the Newton iterations with a large initial guess like $x_1 = 50$.

REFERENCES

General references for techniques to solve nonlinear systems include the book [O4] by Ortega and Rheinboldt and the book [O7] by Ostrowski. Newton's method not only applies to systems of equations, but also to equations in abstract spaces (see the book [K3] by Kantorovich and Akilov). Scalar equations are analyzed thoroughly by Traub in [T1]. There are many excellent root finders for a scalar equation. Some references include [B14], [B22], [D3], and [S10]. A reference for secant methods is Dennis and Schnabel's book [D4]. Armijo's rule appears in [A6]. Some additional references for the continuation method include [A5], [C1], [H15], [K4], [M6], [P8], [R5], [W1], and [W4]. The Jenkin-Traub algorithm is developed in [J2] and [J3]. Note that although the standard deflation process is unstable unless the zeros are arranged in increasing magnitude, Peters and Wilkinson [P10] show that the standard deflation algorithm can be modified to handle more general orderings of the zeros.

5

Least Squares

5-1. INTRODUCTION

So far we have discussed techniques for computing the solution $\mathbf{x}$ to a linear system $\mathbf{Ax} = \mathbf{b}$ when the coefficient matrix is square. In this case, the number of equations is equal to the number of unknowns and usually there exists a unique solution. Now we consider linear systems where the coefficient matrix is rectangular. If $\mathbf{A}$ has m rows and n columns, then $\mathbf{x}$ is a vector with n components and $\mathbf{b}$ is a vector with m components. If m is greater than n, there are more equations than unknowns and the system $\mathbf{Ax} = \mathbf{b}$ is usually overdetermined. Typically, an overdetermined system has no solution. Conversely, if n is greater than m, there are more unknowns than equations and the system is underdetermined. Typically, an underdetermined system has an infinite number of solutions. Even though $m > n$ or $n > m$, the linear system $\mathbf{Ax} = \mathbf{b}$ still has a natural unique solution, its "least squares" solution. This chapter studies techniques for computing the least squares solution to a linear system.

First, let us consider a system of two equations in two unknowns:

$$x_1 - x_2 = 2,$$

$$x_1 + x_2 = 4.$$

The set of points which satisfy the first equation $x_1 - x_2 = 2$ is a line with slope 1 and x_2-intercept -2. The set of points which satisfy the second equation $x_1 + x_2 = 4$ is a line with slope -1 and x_2-intercept 4. These lines intersect at the point $x_1 = 3$ and $x_2 = 1$, where both equations are satisfied (see Figure 5-1).

Solving two equations in two unknowns is equivalent to finding the intersection point of two lines. As long as these lines are not parallel, there exists a unique intersection point. Similarly, solving three equations in three unknowns is equivalent to finding the intersection point for three planes; and except for special geometric configurations, there exists a unique intersection point.

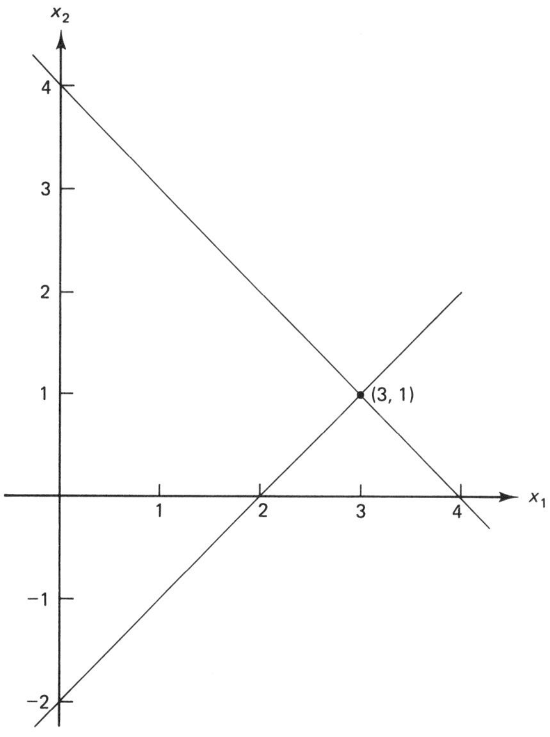

Figure 5-1 Two equations in two unknowns.

Now consider the following overdetermined linear system consisting of two equations in one unknown:

$$\begin{aligned} 2x_1 &= 3 \\ 4x_1 &= 1 \end{aligned} \tag{1}$$

To solve this system by Gaussian elimination, the first equation is multiplied by 2 and subtracted from the second equation to obtain the suspicious relation "$0 = -5$." Since 0 does not equal -5, we say that the two equations are inconsistent; that is, both equations cannot be satisfied since they imply that 0 equals -5, which is impossible.

A picture helps to clarify the inconsistency. Using vector notation, (1) is expressed

$$\begin{bmatrix} 2 \\ 4 \end{bmatrix} x_1 = \begin{bmatrix} 3 \\ 1 \end{bmatrix}. \tag{2}$$

When x_1 is zero, the left side of (2) is $[0, 0]^T$. When x_1 is 1, the left side of (2) is $[2, 4]^T$. As x_1 takes on all possible values, the left side of (2) generates the line in Figure 5-2 connecting the origin and the point (2, 4). On the other hand, the right side of (2) is the vector $[3, 1]^T$. Since the point (3, 1) does not lie on the line, the left side and the right side of (2) are never equal. Two equations in one unknown are only consistent (and have a solution) when the point corresponding to the right side is contained in the line corresponding to the left

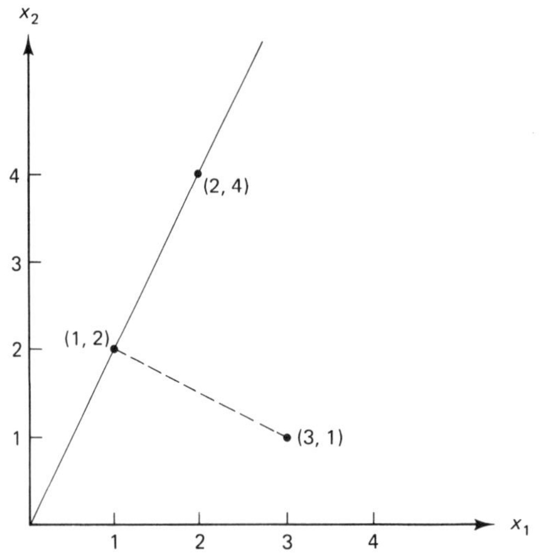

Figure 5-2 Least squares solution to an overdetermined system.

side. The least squares solution to (2) is the value of x_1 for which the point on the line is closest to (3, 1). In Figure 5-2, we see that the point (1, 2) on the line is closest to (3, 1). Since the left side of (2) reduces to $[1, 2]^T$ when $x_1 = \frac{1}{2}$, the least squares solution to (2) is $x_1 = \frac{1}{2}$.

For a general overdetermined system $\mathbf{Ax} = \mathbf{b}$, we search for the $\mathbf{x}$ for which $\mathbf{Ax}$ is closest to $\mathbf{b}$. In other words, we search for the $\mathbf{x}$ that makes the 2-norm of the residual $\mathbf{r} = \mathbf{b} - \mathbf{Ax}$ as small as possible. Thus finding the least squares solution to an overdetermined system is equivalent to solving the problem

$$\underset{\mathbf{x}}{\text{minimize}} \; \|\mathbf{b} - \mathbf{Ax}\|_2 . \tag{3}$$

When there are more unknowns than equations, we say that the system is underdetermined. The following single equation in two unknowns is an example of an underdetermined system:

$$x_1 + x_2 = 2. \tag{4}$$

Observe that there are an infinite number of solutions to this equation. Some solutions are $x_1 = 2$ and $x_2 = 0$, $x_1 = 1$ and $x_2 = 1$, and $x_1 = 0$ and $x_2 = 2$. The complete set of solutions forms the line depicted in Figure 5-3. The least squares (or minimum norm) solution to (4) is that solution which is closest to the origin. We see in Figure 5-3 that the point on the line which is closest to the origin is $x_1 = 1$ and $x_2 = 1$. Typically, an underdetermined linear system $\mathbf{Ax} = \mathbf{b}$ has an infinite number of solutions and the least squares solution is the one closest to the origin. In other words, the least squares solution to an underdetermined system $\mathbf{Ax} = \mathbf{b}$ is the solution $\mathbf{x}$ that makes $\|\mathbf{x}\|_2$ as small as possible, or equivalently, the least squares solution solves the problem:

$$\text{minimize} \; \{\|\mathbf{x}\|_2 : \mathbf{Ax} = \mathbf{b}\} \tag{5}$$

In this chapter we develop systematic techniques to solve both (3) and (5).

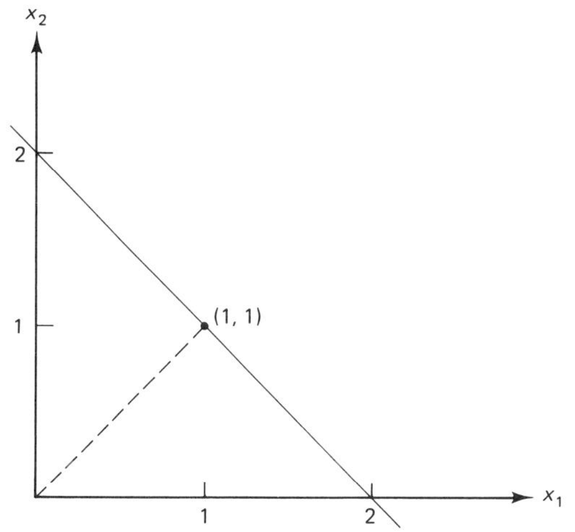

Figure 5-3 Least squares solution to an underdetermined system.

Exercises

5-1.1. Consider the two equations $ax_1 = b$ and $cx_1 = d$. Assuming that a is not zero, what relationship must be satisfied so that these two equations are consistent?

5-1.2. Consider the single equation $ax_1 + bx_2 = c$, where a and b are not both zero.
  (a) What is the equation of the line through the origin which is perpendicular to the line $ax_1 + bx_2 = c$?
  (b) What is the least squares (minimum norm) solution to $ax_1 + bx_2 = c$?

5-1.3. When is the single equation $ax_1 + bx_2 = c$ inconsistent?

5-2. OVERDETERMINED SYSTEMS

Let us begin with an application that leads to an overdetermined system of equations. Suppose that a ball is thrown into the air. If $h(t)$ denotes the height of the ball at time t, the graph of h is a parabola. Normalizing the height so that $h(0) = 0$, the equation for h is

$$h(t) = -\tfrac{1}{2}rt^2 + vt,$$

where r is the ratio between the gravitational force and the mass of the ball and v is the ball's initial vertical velocity. To evaluate r, we simply weigh the ball. Let us suppose that r is 10, in which case the height is given by

$$h(t) = -5t^2 + vt. \tag{6}$$

One way to determine the unknown initial velocity is to estimate the ball's height at some particular time t and solve for v:

$$v = \frac{h(t) + 5t^2}{t}. \tag{7}$$

However, due to experimental errors in measuring height and time, the computed velocity v is in error. One strategy for determining v more precisely is the following: Measure the ball's height at several different times and for each height-time measurement, compute the velocity using (7). The final estimate of the velocity is the average of the velocities determined from each height-time mea-

TABLE 5-1 ESTIMATED HEIGHT AND VELOCITY OF BALL

Time	Estimated height	Computed velocity
1	16	21.0
2	19	19.5
3	15	20.0
4	2	20.5

surement. Table 5-1 gives the estimated height of the ball at various times followed by the velocity computed from equation (7). To average the velocities in Table 5-1, we add the last column and divide by 4 to obtain

$$v \approx \frac{21 + 19.5 + 20 + 20.5}{4} = 20.25.$$

When there is one unknown such as the velocity v, this averaging technique is easily implemented. But for m equations in n unknowns, the computations can be unwieldy since the number of solutions

$$\frac{m(m-1)\cdots(m-n+1)}{(1)(2)\cdots(n)}$$

that are averaged together is so large. For example, if m is 10 and n is 5, 252 different solutions must be computed and averaged together. The least squares approach is a more practical method for estimating the solution to an overdetermined linear system. Referring to Table 5-1, the height is 16 at $t = 1$. Inserting $t = 1$ in equation (6) gives us the relation

$$-5 + v = 16.$$

Similarly, since the height is 19 at $t = 2$, equation (6) gives us the relation

$$-20 + 2v = 19.$$

Each time-height measurement in Table 5-1 provides one equation and there are four equations altogether:

$$\begin{aligned} -5 + v &= 16, \\ -20 + 2v &= 19, \\ -45 + 3v &= 15, \\ -80 + 4v &= 2. \end{aligned} \tag{8}$$

This is an overdetermined system of four equations in one unknown v.

In the least squares approach, v is chosen so that the Euclidean norm of the residual is as small as possible. Since the residual is the difference between the right side and the left side of the equation, the residual corresponding to system (8) is

$$\mathbf{r} = \begin{bmatrix} 16 + 5 - v \\ 19 + 20 - 2v \\ 15 + 45 - 3v \\ 2 + 80 - 4v \end{bmatrix} = \begin{bmatrix} 21 - v \\ 39 - 2v \\ 60 - 3v \\ 82 - 4v \end{bmatrix}.$$

The Euclidean norm of the residual is the square root of the sum of each component squared:

$$\|\mathbf{r}\|_2 = \sqrt{r_1^2 + r_2^2 + r_3^2 + r_4^2}.$$

Since minimizing $\|\mathbf{r}\|_2$ is equivalent to minimizing $(\|\mathbf{r}\|_2)^2$, the least squares solution to (8) is that value for v which minimizes the expression

$$(21 - v)^2 + (39 - 2v)^2 + (60 - 3v)^2 + (82 - 4v)^2. \tag{9}$$

The minimizing v is found by differentiating (9) with respect to v and setting the derivative to zero. Differentiating (9) leads us to the equation

$$-2[(21 - v) + 2(39 - 2v) + 3(60 - 3v) + 4(82 - 4v)] = 0.$$

Solving for the velocity, we obtain $v = 607/30 \approx 20.23$.

For a general overdetermined linear system $\mathbf{Ax} = \mathbf{b}$ of m equations in n unknowns, the residual is $\mathbf{r} = \mathbf{b} - \mathbf{Ax}$ and the Euclidean norm of the residual is the square root of $\mathbf{r}^T\mathbf{r}$. The least squares solution to $\mathbf{Ax} = \mathbf{b}$ minimizes

$$\mathbf{r}^T\mathbf{r} = (\mathbf{b} - \mathbf{Ax})^T(\mathbf{b} - \mathbf{Ax}). \tag{10}$$

At (10)'s minimum, the partial derivative with respect to each of the variables $x_1, x_2, \cdots, x_n$ is zero. We will differentiate (10) with respect to each variable and set the derivative to zero. First observe that (10) can be written

$$\mathbf{r}^T\mathbf{r} = r_1^2 + r_2^2 + \cdots + r_m^2, \tag{11}$$

where r_i is the ith component of $\mathbf{r}$. Differentiating (11) with respect to x_1, we have

$$\frac{\partial}{\partial x_1}\mathbf{r}^T\mathbf{r} = 2r_1\frac{\partial}{\partial x_1}r_1 + 2r_2\frac{\partial}{\partial x_1}r_2 + \cdots + 2r_m\frac{\partial}{\partial x_1}r_m. \tag{12}$$

Of course, the ith component of $\mathbf{r} = \mathbf{b} - \mathbf{Ax}$ is equal to b_i minus the ith row of $\mathbf{A}$ times $\mathbf{x}$:

$$r_i = b_i - a_{i1}x_1 - a_{i2}x_2 - \cdots - a_{in}x_n.$$

Since the partial derivative of r_i with respect to x_1 is $-a_{i1}$, (12) can be expressed

$$\frac{\partial}{\partial x_1}\mathbf{r}^T\mathbf{r} = -2r_1a_{11} - 2r_2a_{21} - \cdots - 2r_m a_{m1}.$$

In general, the partial derivative of $\mathbf{r}^T\mathbf{r}$ with respect to x_j is given by

$$\frac{\partial}{\partial x_j}\mathbf{r}^T\mathbf{r} = -2r_1a_{1j} - 2r_2a_{2j} - \cdots - 2r_m a_{mj}. \tag{13}$$

Examining the right side of (13), we see that the partial derivative of $\mathbf{r}^T\mathbf{r}$ with respect to x_j is -2 times the product between the jth column of $\mathbf{A}$ and $\mathbf{r}$. Recall that the jth column of $\mathbf{A}$ is the jth row of $\mathbf{A}^T$. Since the jth component of $\mathbf{A}^T\mathbf{r}$ is equal to the jth column of $\mathbf{A}$ times $\mathbf{r}$, the partial derivative of $\mathbf{r}^T\mathbf{r}$ with respect to x_j is the jth component of the vector $-2\mathbf{A}^T\mathbf{r}$. The Euclidean norm of the residual is minimized at that point $\mathbf{x}$ where all the partial derivatives vanish:

$$\frac{\partial}{\partial x_1}\mathbf{r}^T\mathbf{r} = \frac{\partial}{\partial x_2}\mathbf{r}^T\mathbf{r} = \cdots = \frac{\partial}{\partial x_n}\mathbf{r}^T\mathbf{r} = 0.$$

Since each of these partial derivatives is -2 times the corresponding component of $\mathbf{A}^T\mathbf{r}$, we conclude that $\mathbf{A}^T\mathbf{r} = 0$. Replacing $\mathbf{r}$ by $\mathbf{b} - \mathbf{Ax}$ yields the relation

$$\mathbf{A}^T(\mathbf{b} - \mathbf{A}\mathbf{x}) = \mathbf{0}, \tag{14}$$

and rearranging (14), we obtain the **normal equation**:

$$\mathbf{A}^T\mathbf{A}\mathbf{x} = \mathbf{A}^T\mathbf{b}. \tag{15}$$

Any $\mathbf{x}$ that minimizes the Euclidean norm of the residual $\mathbf{r} = \mathbf{b} - \mathbf{A}\mathbf{x}$ is a solution to the normal equation. Conversely, any solution to (15) is a least squares solution to the overdetermined linear system $\mathbf{A}\mathbf{x} = \mathbf{b}$.

Since this algebraic derivation of (15) hides some of the underlying geometry, we now give another derivation of the normal equation which may be easier to visualize. The set of vectors $\mathbf{A}\mathbf{x}$ corresponding to various choices of $\mathbf{x}$ is called the range space of $\mathbf{A}$. In Figure 5-4 the range space is depicted as a line while the vector $\mathbf{b}$ is a point. The collection of residuals $\mathbf{b} - \mathbf{A}\mathbf{x}$ is the set of vectors which point from the range space to $\mathbf{b}$. As we see in Figure 5-4, the shortest residual is perpendicular to the range space or equivalently, the dot product between each vector in the range space and the shortest residual is zero. Since each column of $\mathbf{A}$ is contained in its range space, the dot product between each column of $\mathbf{A}$ and the least squares residual must be zero. Using vector notation, the dot product between each column of $\mathbf{A}$ and the residual $\mathbf{b} - \mathbf{A}\mathbf{x}$ is given by the matrix-vector product $\mathbf{A}^T(\mathbf{b} - \mathbf{A}\mathbf{x})$. Hence the least squares solution to $\mathbf{A}\mathbf{x} = \mathbf{b}$ satisfies (14) as well as the equivalent equation (15).

To obtain the normal equation corresponding to (8), we move the terms that do not depend on v to the right side giving us the system

$$\begin{bmatrix} 1 \\ 2 \\ 3 \\ 4 \end{bmatrix} v = \begin{bmatrix} 21 \\ 39 \\ 60 \\ 82 \end{bmatrix}. \tag{16}$$

The coefficient matrix $\mathbf{A}$ for this system has 4 rows and 1 column:

$$\mathbf{A} = \begin{bmatrix} 1 \\ 2 \\ 3 \\ 4 \end{bmatrix},$$

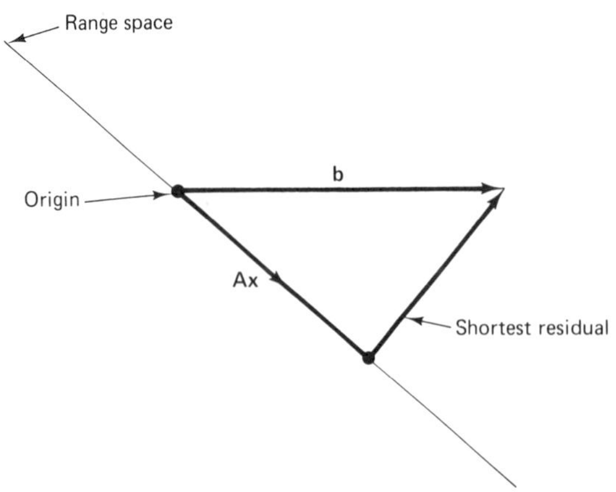

Figure 5-4 The range space of $\mathbf{A}$ and the least squares solution to $\mathbf{A}\mathbf{x} = \mathbf{b}$.

the unknown vector **x** has one component v, and the right side **b** has four components:

$$\mathbf{b} = \begin{bmatrix} 21 \\ 39 \\ 60 \\ 82 \end{bmatrix}.$$

The normal equation $\mathbf{A}^T\mathbf{A}\mathbf{x} = \mathbf{A}^T\mathbf{b}$ corresponding to (16) is

$$\underset{\mathbf{A}^T}{[1 \quad 2 \quad 3 \quad 4]} \underset{\mathbf{A}}{\begin{bmatrix} 1 \\ 2 \\ 3 \\ 4 \end{bmatrix}} \underset{\mathbf{x}}{v} = \underset{\mathbf{A}^T}{[1 \quad 2 \quad 3 \quad 4]} \underset{\mathbf{b}}{\begin{bmatrix} 21 \\ 39 \\ 60 \\ 82 \end{bmatrix}}. \tag{17}$$

Since $\mathbf{A}^T\mathbf{A} = 30$ and $\mathbf{A}^T\mathbf{b} = 607$, (17) reduces to $30v = 607$ and $v = 607/30 \approx 20.23$.

For a second example, we consider an overdetermined system of three equations in two unknowns:

$$\begin{aligned} 3x_1 &= 6, \\ 4x_1 + 5x_2 &= 12, \\ 4x_2 &= 4. \end{aligned} \tag{18}$$

Using matrix-vector notation, this system is written

$$\begin{bmatrix} 3 & 0 \\ 4 & 5 \\ 0 & 4 \end{bmatrix} \begin{bmatrix} x_1 \\ x_2 \end{bmatrix} = \begin{bmatrix} 6 \\ 12 \\ 4 \end{bmatrix}. \tag{19}$$

The corresponding normal equation $\mathbf{A}^T\mathbf{A}\mathbf{x} = \mathbf{A}^T\mathbf{b}$ is

$$\underset{\mathbf{A}^T}{\begin{bmatrix} 3 & 4 & 0 \\ 0 & 5 & 4 \end{bmatrix}} \underset{\mathbf{A}}{\begin{bmatrix} 3 & 0 \\ 4 & 5 \\ 0 & 4 \end{bmatrix}} \underset{\mathbf{x}}{\begin{bmatrix} x_1 \\ x_2 \end{bmatrix}} = \underset{\mathbf{A}^T}{\begin{bmatrix} 3 & 4 & 0 \\ 0 & 5 & 4 \end{bmatrix}} \underset{\mathbf{b}}{\begin{bmatrix} 6 \\ 12 \\ 4 \end{bmatrix}},$$

which reduces to

$$\begin{bmatrix} 25 & 20 \\ 20 & 41 \end{bmatrix} \begin{bmatrix} x_1 \\ x_2 \end{bmatrix} = \begin{bmatrix} 66 \\ 76 \end{bmatrix}. \tag{20}$$

The solution $x_1 = 1186/625 = 1.8976$ and $x_2 = 580/625 = .928$ to this system of two equations in two unknowns is the least squares solution to (18).

An especially important overdetermined system arises when we search for the best linear fit to a function $y = f(x)$. For example, consider the problem of determining the a and b such that the relationship $y = ax + b$ is the least squares fit to the data in Table 5-2. The data of Table 5-2 and a linear fit are plotted in Figure 5-5. Corresponding to each pair (x_i, y_i) in Table 5-2, the equation $ax_i + b = y_i$ should hold. In general, for m data points, there are m equations, which can be written

TABLE 5-2 COLLECTION OF EXPERIMENTAL DATA

i	x_i	y_i
1	1	2
2	2	5
3	3	6
4	4	8

$$\begin{bmatrix} x_1 & 1 \\ x_2 & 1 \\ \cdot & \cdot \\ \cdot & \cdot \\ x_m & 1 \end{bmatrix} \begin{bmatrix} a \\ b \end{bmatrix} = \begin{bmatrix} y_1 \\ y_2 \\ \cdot \\ \cdot \\ y_m \end{bmatrix}.$$

The normal equation $\mathbf{A}^T\mathbf{A}\mathbf{x} = \mathbf{A}^T\mathbf{b}$ reduces to

$$\begin{bmatrix} \sum_{i=1}^{m} x_i^2 & \sum_{i=1}^{m} x_i \\ \sum_{i=1}^{m} x_i & m \end{bmatrix} \begin{bmatrix} a \\ b \end{bmatrix} = \begin{bmatrix} \sum_{i=1}^{m} x_i y_i \\ \sum_{i=1}^{m} y_i \end{bmatrix}. \tag{21}$$

Since the solution to a system of two equations in two unknowns

$$\begin{bmatrix} d & f \\ e & g \end{bmatrix} \begin{bmatrix} x_1 \\ x_2 \end{bmatrix} = \begin{bmatrix} b_1 \\ b_2 \end{bmatrix}$$

is

$$x_1 = \frac{gb_1 - fb_2}{dg - ef} \quad \text{and} \quad x_2 = \frac{db_2 - eb_1}{dg - ef},$$

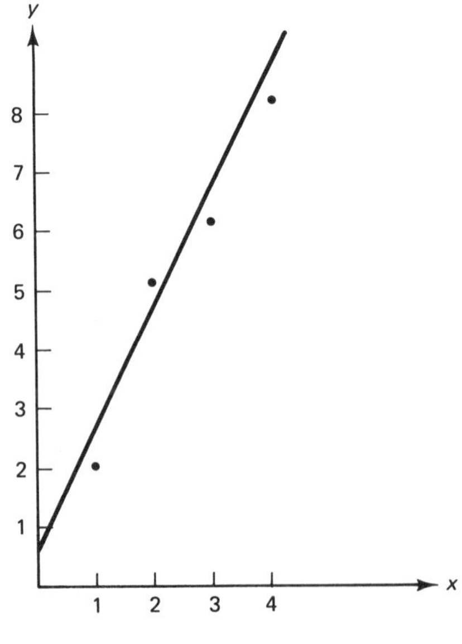

Figure 5-5 The data of Table 5-2 and a linear approximation.

the solution to (21) can be expressed

$$a = \frac{m\sum_{i=1}^{m} x_i y_i - \left(\sum_{i=1}^{m} x_i\right)\left(\sum_{i=1}^{m} y_i\right)}{m\sum_{i=1}^{m} x_i^2 - \left(\sum_{i=1}^{m} x_i\right)^2}$$

and

$$b = \frac{\left(\sum_{i=1}^{m} x_i^2\right)\left(\sum_{i=1}^{m} y_i\right) - \left(\sum_{i=1}^{m} x_i\right)\left(\sum_{i=1}^{m} x_i y_i\right)}{m\sum_{i=1}^{m} x_i^2 - \left(\sum_{i=1}^{m} x_i\right)^2}.$$

For a general overdetermined linear system $\mathbf{Ax} = \mathbf{b}$, the coefficient matrix for the normal equation consists of various dot products between columns of $\mathbf{A}$. For the normal equation (20) and for the $\mathbf{A}$ in (19), the coefficient 25 in the normal equation is column 1 dot column 1 of $\mathbf{A}$, the coefficient 20 is column 2 dot column 1, and the coefficient 41 is column 2 dot column 2.

In general, the coefficient in row i and column j for the matrix $\mathbf{A}^T\mathbf{A}$ *is the dot product between column i of* $\mathbf{A}$ *and column j of* $\mathbf{A}$.

Observe that the coefficient matrix in (20) is symmetric. *The coefficient matrix in the normal equation is always symmetric* since the coefficient in row i and column j is equal to the coefficient in row j and column i. In other words, for any matrix, column i dot column j is equal to column j dot column i. Since the coefficient matrix in the normal equation is symmetric, only the coefficients that are on the diagonal or above the diagonal must be evaluated. The remaining coefficients are determined from the symmetry property: $(\mathbf{A}^T\mathbf{A})_{ij} = (\mathbf{A}^T\mathbf{A})_{ji}$. When solving the normal equation, we can employ the special "symmetric" algorithms of Chapter 2. Moreover, by Exercise 2-4.16, pivoting is not required when the coefficient matrix $\mathbf{A}^T\mathbf{A}$ is factored. In particular, we can use HFACT and HSOLVE if $\mathbf{A}^T\mathbf{A}$ is a band matrix and we can use SFACT and SSOLVE if $\mathbf{A}^T\mathbf{A}$ has no special structure (other than its symmetry).

If $\mathbf{A}$ is a $m \times n$ matrix with no special structure, each column of $\mathbf{A}$ has m elements and the asymptotic parameter associated with computing the dot product between two columns is m. Since the $n \times n$ matrix $\mathbf{A}^T\mathbf{A}$ has $\frac{1}{2}n(n+1)$ elements on the diagonal or above the diagonal, the asymptotic parameter associated with computing the elements of $\mathbf{A}^T\mathbf{A}$ is $\frac{1}{2}mn^2$. Since the asymptotic parameter associated with solving a general symmetric system of n equations is $\frac{1}{6}n^3$, the asymptotic time to solve the normal equation (15) is

$$n^2\left(\frac{n}{6}C_s + \frac{m}{2}C_d\right),$$

where C_s is the cycle time for solving a symmetric system and C_d is the cycle time for computing a dot product.

Exercises

5-2.1. What is the a and b corresponding to the best linear least squares fit $y = ax + b$ to the data in Table 5-2.

5-2.2. Find the value of a such that the relationship $y = ax$ is the least squares fit to the data in Table 5-2. In other words, find the least squares solution a to the following system of four equations in one unknown:

$$ax_k = y_k, \qquad k = 1, 2, 3, 4.$$

In general, given m data points, what is the value of a for which the relationship $y = ax$ best fits the data in the least squares sense?

5-2.3. Solve the following overdetermined system of three equations in two unknowns:

$$\begin{bmatrix} 1 & 5 \\ 2 & -1 \\ -3 & 1 \end{bmatrix} \begin{bmatrix} x_1 \\ x_2 \end{bmatrix} = \begin{bmatrix} 36 \\ 45 \\ 0 \end{bmatrix}$$

5-2.4. Write a program to obtain the least squares solution to an overdetermined system $\mathbf{Ax} = \mathbf{b}$ when $\mathbf{A}$ has no special structure. Your program can call SFACT and SSOLVE. Test your program using equations (16) and (19).

5-3. UNDERDETERMINED SYSTEMS

Underdetermined systems of equations often arise in optimization theory and in economic modeling. In these applications, there are many ways to achieve some objective and the goal is to find the "best" way. For example, suppose that our doctor tells us to consume 2000 units of vitamin Z per day where vitamin Z can be obtained by eating either of two different foods. If food 1 contains 30 units of vitamin Z per gram and food 2 contains 10 units of vitamin Z per gram, then our daily diet must satisfy the equation $30a_1 + 10a_2 = 2000$, where a_1 and a_2 are the amounts of each food that we consume. Suppose that our daily diet includes 40 grams of food 1 and 80 grams of food 2. Since $a_1 = 40$ and $a_2 = 80$, $30(40) + 10(80) = 2000$ and we are complying with the doctor's orders. After some time, the doctor informs us that we should increase our daily consumption of vitamin Z to 2400 units. To fulfill this request, we will eat more of both foods. Let x_1 be the increase in consumption of food 1 and let x_2 be the increase in consumption of food 2. Since we must consume 400 additional units of vitamin Z, x_1 and x_2 must satisfy the equation

$$30x_1 + 10x_2 = 400. \tag{22}$$

The set of points (x_1, x_2) that satisfy (22) forms a line with slope $-30/10 = -3$ (see Figure 5-6). Since x_1 and x_2 correspond to the change in our current diet (the new diet is the old diet (a_1, a_2) plus the change (x_1, x_2)) the distance $(x_1^2 + x_2^2)^{1/2}$ from the origin to the point (x_1, x_2) measures in some sense the diet disruption associated with the consumption changes x_1 and x_2. To disrupt our diet as little as possible, we choose the consumption levels such that the point (x_1, x_2) is as close to the origin as possible. The point in Figure 5-6 which is closest to the origin is the least squares solution to (22). If s_1 and s_2 are the coordinates of the least squares solution, we see in Figure 5-6 that the vector from the origin to (s_1, s_2) is orthogonal to the line $30x_1 + 10x_2 = 400$. The

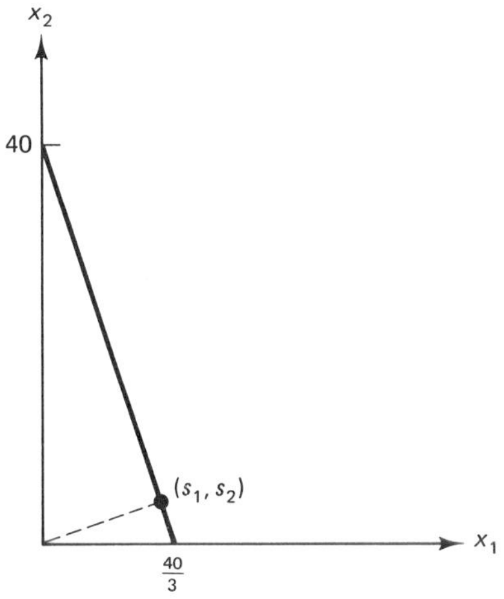

Figure 5-6 The graph of $x_2 = 40 - 3x_1$.

collection of points forming this perpendicular have the form

$$t\begin{bmatrix} 30 \\ 10 \end{bmatrix},$$

where t is an arbitrary scalar. Since (s_1, s_2) lies on this perpendicular, there exists a value of t such that

$$\begin{bmatrix} s_1 \\ s_2 \end{bmatrix} = t\begin{bmatrix} 30 \\ 10 \end{bmatrix}. \tag{23}$$

This equation provides two relations: $s_1 = 30t$ and $s_2 = 10t$. Since $x_1 = s_1$ and $x_2 = s_2$ must satisfy (22), we also have the relation

$$30s_1 + 10s_2 = 400. \tag{24}$$

Substituting $s_1 = 30t$ and $s_2 = 10t$ into (24) yields an equation for t:

$$(30^2 + 10^2)t = 400.$$

The solution is $t = 400/1000 = .4$. Inserting $t = .4$ into equation (23) gives us the least squares solution:

$$\begin{bmatrix} s_1 \\ s_2 \end{bmatrix} = (.4)\begin{bmatrix} 30 \\ 10 \end{bmatrix} = \begin{bmatrix} 12 \\ 4 \end{bmatrix}.$$

Now let us consider a general underdetermined linear system $\mathbf{Ax} = \mathbf{b}$. Suppose that $\mathbf{y}$ is any solution to $\mathbf{Ax} = \mathbf{b}$ and $\mathbf{z}$ is any vector for which $\mathbf{Az} = \mathbf{0}$. Since $\mathbf{A}(\mathbf{y} + \mathbf{z}) = \mathbf{Ay} + \mathbf{Az} = \mathbf{Ay} = \mathbf{b}$, we see that $\mathbf{y} + \mathbf{z}$ is a solution to $\mathbf{Ax} = \mathbf{b}$ whenever $\mathbf{Az} = \mathbf{0}$. Conversely, if $\mathbf{Ax} = \mathbf{b}$, then $\mathbf{x} = \mathbf{y} + (\mathbf{x} - \mathbf{y}) = \mathbf{y} + \mathbf{z}$ with $\mathbf{z} = \mathbf{x} - \mathbf{y}$. Since $\mathbf{A}(\mathbf{x} - \mathbf{y}) = \mathbf{b} - \mathbf{b} = \mathbf{0}$, it follows that $\mathbf{x}$ can be expressed as the sum $\mathbf{y} + \mathbf{z}$, where $\mathbf{z} = \mathbf{x} - \mathbf{y}$ and $\mathbf{Az} = \mathbf{0}$. Thus all solutions to $\mathbf{Ax} = \mathbf{b}$ have the form $\mathbf{y} + \mathbf{z}$, where $\mathbf{Az} = \mathbf{0}$. The set of all $\mathbf{z}$ such that $\mathbf{Az} = \mathbf{0}$ is called the **null space** of $\mathbf{A}$. Letting $\mathbf{N}$ denote the null space of $\mathbf{A}$, the set of solutions to an underdetermined system $\mathbf{Ax} = \mathbf{b}$ are sketched in Figure 5-7 on page 204. Each solution is the sum $\mathbf{y}$ plus $\mathbf{z}$, where $\mathbf{z}$ lies in $\mathbf{N}$.

Observe in Figure 5-7 that the solution closest to the origin is perpendicular to the null space. In linear algebra it is shown that the set of vectors perpendicular

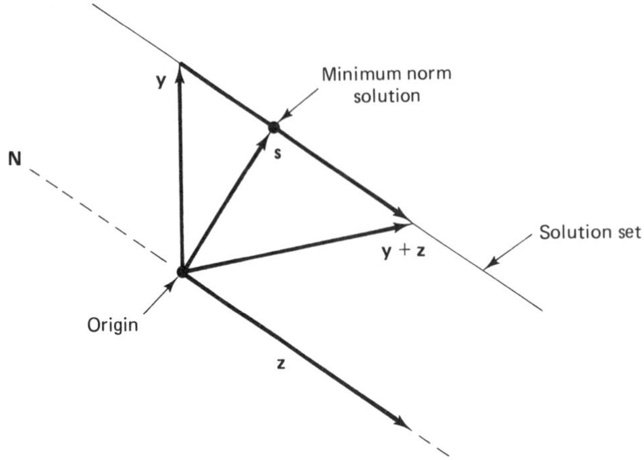

Figure 5-7 The collection of solutions to an underdetermined linear system.

to the null space of **A** are linear combinations of the rows of **A**. In other words, the set of vectors perpendicular to the null space of **A** have the form $\mathbf{A}^T\mathbf{t}$, where **t** is an arbitrary vector. In particular, if **s** denotes the least squares (minimum norm) solution to $\mathbf{Ax} = \mathbf{b}$ depicted in Figure 5-7, then $\mathbf{s} = \mathbf{A}^T\mathbf{t}$ for some **t**. Substituting $\mathbf{x} = \mathbf{s} = \mathbf{A}^T\mathbf{t}$ into the equation $\mathbf{Ax} = \mathbf{b}$, we have

$$\mathbf{AA}^T\mathbf{t} = \mathbf{b}. \tag{25}$$

Solving this equation yields **t**, while the least squares solution **s** to the underdetermined system $\mathbf{Ax} = \mathbf{b}$ is $\mathbf{s} = \mathbf{A}^T\mathbf{t}$.

For example, the underdetermined equation (22) can be expressed

$$[30 \quad 10]\begin{bmatrix} x_1 \\ x_2 \end{bmatrix} = 400.$$

In this case, the coefficient matrix **A** has one row and two columns:

$$\mathbf{A} = [30 \quad 10],$$

the unknown vector **x** has two components x_1 and x_2, the right side **b** has one component 400, and the equation $\mathbf{AA}^T\mathbf{t} = \mathbf{b}$ is

$$\underset{\mathbf{A}}{[30 \quad 10]}\underset{\mathbf{A}^T}{\begin{bmatrix} 30 \\ 10 \end{bmatrix}}\underset{\mathbf{t}}{t} = \underset{\mathbf{b}}{400}. \tag{26}$$

Observe that **t** has one component. Since $\mathbf{AA}^T$ is 1000, (26) reduces to $1000t = 400$ and $t = .4$. Finally, the least squares solution to (22) is $\mathbf{s} = \mathbf{A}^T\mathbf{t}$:

$$\mathbf{s} = \begin{bmatrix} 30 \\ 10 \end{bmatrix}(.4) = \begin{bmatrix} 12 \\ 4 \end{bmatrix}.$$

For a second example, let us consider an underdetermined system of two equations in three unknowns:

$$\begin{aligned} 3x_1 + 4x_2 \qquad &= 70, \\ 5x_2 + 4x_3 &= 81. \end{aligned} \tag{27}$$

Using matrix-vector notation, this system is written

$$\begin{bmatrix} 3 & 4 & 0 \\ 0 & 5 & 4 \end{bmatrix} \begin{bmatrix} x_1 \\ x_2 \\ x_3 \end{bmatrix} = \begin{bmatrix} 70 \\ 81 \end{bmatrix}.$$

For this system, equation (25) takes the form

$$\underset{\mathbf{A}}{\begin{bmatrix} 3 & 4 & 0 \\ 0 & 5 & 4 \end{bmatrix}} \underset{\mathbf{A}^T}{\begin{bmatrix} 3 & 0 \\ 4 & 5 \\ 0 & 4 \end{bmatrix}} \underset{\mathbf{t}}{\begin{bmatrix} t_1 \\ t_2 \end{bmatrix}} = \underset{\mathbf{b}}{\begin{bmatrix} 70 \\ 81 \end{bmatrix}},$$

which reduces to

$$\begin{bmatrix} 25 & 20 \\ 20 & 41 \end{bmatrix} \begin{bmatrix} t_1 \\ t_2 \end{bmatrix} = \begin{bmatrix} 70 \\ 81 \end{bmatrix}. \tag{28}$$

Solving for $\mathbf{t}$, we obtain $t_1 = 2$ and $t_2 = 1$, and the least squares solution $\mathbf{s}$ to (27) is $\mathbf{s} = \mathbf{A}^T\mathbf{t}$:

$$\underset{\mathbf{s}}{\begin{bmatrix} s_1 \\ s_2 \\ s_3 \end{bmatrix}} = \underset{\mathbf{A}^T}{\begin{bmatrix} 3 & 0 \\ 4 & 5 \\ 0 & 4 \end{bmatrix}} \underset{\mathbf{t}}{\begin{bmatrix} 2 \\ 1 \end{bmatrix}} = \begin{bmatrix} 6 \\ 13 \\ 4 \end{bmatrix}.$$

To summarize, the least squares solution to an underdetermined system $\mathbf{Ax} = \mathbf{b}$ is

$$\mathbf{x} = \mathbf{A}^T(\mathbf{AA}^T)^{-1}\mathbf{b},$$

while the least squares solution to an overdetermined system is

$$\mathbf{x} = (\mathbf{A}^T\mathbf{A})^{-1}\mathbf{A}^T\mathbf{b}.$$

Each coefficient in (28) is the dot product between two rows of $\mathbf{A}$: The coefficient 25 is row 1 dot row 1, the coefficient 20 is row 1 dot row 2, and the coefficient 41 is row 2 dot row 2.

> *In general, the coefficient in row i and column j for the matrix* $\mathbf{AA}^T$ *is the dot product between row i and row j from* $\mathbf{A}$.

Observe that the coefficient matrix in (28) is symmetric. *The coefficient matrix for the system* $(\mathbf{AA}^T)\mathbf{t} = \mathbf{b}$ *is always symmetric* since the coefficient in row i and column j is equal to the coefficient in row j and column i. In other words, row i dot row j of $\mathbf{A}$ is equal to row j dot row i of $\mathbf{A}$. When forming the matrix $\mathbf{AA}^T$, we just evaluate the coefficients that are on the diagonal or above the diagonal—the coefficients below the diagonal are determined from the symmetry property $(\mathbf{AA}^T)_{ij} = (\mathbf{AA}^T)_{ji}$. Due to this symmetry, (25) can be solved using the programs of Chapter 2 that process a symmetric system.

If $\mathbf{A}$ is a $m \times n$ matrix with no special structure, each row of $\mathbf{A}$ has n elements and the asymptotic parameter associated with computing the dot product between two rows of $\mathbf{A}$ is n. Since the $m \times m$ matrix $\mathbf{AA}^T$ has $\frac{1}{2}m(m+1)$ elements on the diagonal or above the diagonal, the asymptotic parameter associated with computing the elements of $\mathbf{AA}^T$ is $\frac{1}{2}nm^2$. Since the asymptotic parameter associated with solving a general symmetric system of m equations is $\frac{1}{6}m^3$, the asymptotic time to compute the least squares solution to a general underdetermined system is

$$m^2\left(\frac{n}{2}C_d + \frac{m}{6}C_s\right),$$

where C_s is the cycle time for solving a symmetric system and C_d is the cycle time for computing a dot product. Note that this formula for an underdetermined system is the same as the corresponding formula for an overdetermined system except that m and n are interchanged.

Exercises

5-3.1. Compute the least squares (minimum norm) solution to the following underdetermined linear system:

$$\begin{bmatrix} 1 & 2 & -3 \\ 5 & -1 & 1 \end{bmatrix}\begin{bmatrix} x_1 \\ x_2 \\ x_3 \end{bmatrix} = \begin{bmatrix} 42 \\ 54 \end{bmatrix}.$$

5-3.2. Solve the following minimization problem:

$$\text{minimize} \quad 4x_1^2 + 9x_2^2$$

$$\text{subject to} \quad 8x_1 + 9x_2 = 15$$

5-3.3. What is the point on the plane $y = 2x_1 + x_2 - 12$ which is closest to the origin?

5-3.4. Write a program to obtain the least squares (minimum norm) solution to an underdetermined linear system $\mathbf{Ax} = \mathbf{b}$ when $\mathbf{A}$ has no special structure. Assume that the rows of the coefficient matrix are stored in the columns of the input array. Test your program using equations (22) and (27).

5-4. ORTHOGONAL TECHNIQUES

Another way to solve least squares problems utilizes an **orthogonal transformation**. This approach may be slightly slower than the normal equation approach, but it is more stable numerically. An orthogonal transformation is a linear change of variables that preserves length. Either a rotation of the coordinate system about an axis or a reflection of the coordinate system across a plane are two examples of orthogonal transformations. For the following transformation

$$\begin{aligned} x_1 &= .6y_1 + .8y_2, \\ x_2 &= .8y_1 - .6y_2, \end{aligned} \tag{29}$$

the identity $\|\mathbf{x}\|_2 = \|\mathbf{y}\|_2$ is verified below:

$$(\|\mathbf{x}\|_2)^2 = x_1^2 + x_2^2 = (.6y_1 + .8y_2)^2 + (.8y_1 - .6y_2)^2 = y_1^2 + y_2^2 = (\|\mathbf{y}\|_2)^2$$

Thus the linear change of variables (29) is an orthogonal transformation.

For an underdetermined linear system, an orthogonal transformation is utilized in the following way: We change variables to obtain a new system for which the least squares solution is easily computed, then we convert back to the original variables. This procedure is illustrated with the underdetermined system

$$3x_1 + 4x_2 = 5.$$

Changing from the variable $\mathbf{x}$ to the variable $\mathbf{y}$ using the transformation (29), we obtain the equivalent system

$$5y_1 + 0y_2 = 5. \tag{30}$$

Since $\|\mathbf{x}\|_2 = \|\mathbf{y}\|_2$, minimizing the Euclidean length of $\mathbf{y}$ is equivalent to minimizing the Euclidean length of $\mathbf{x}$. Also, note that after the transformation (29), the coefficient of y_2 in (30) is zero. Since the coefficient of y_2 is zero, the least squares problem in the $\mathbf{y}$ variable

$$\text{minimize } \{y_1^2 + y_2^2 : 5y_1 = 5\}$$

is easily solved. Clearly, the value for y_2 which minimizes y_2^2 is $y_2 = 0$ while the equation $5y_1 = 5$ implies that $y_1 = 1$. Substituting $y_1 = 1$ and $y_2 = 0$ in (29) yields the least squares solution $x_1 = .6$ and $x_2 = .8$ to the original equation $3x_1 + 4x_2 = 5$.

For an overdetermined system, we apply the orthogonal transformation to the residual $\mathbf{r}$ (rather than to $\mathbf{x}$) in order to simplify the residual. To help us understand this procedure, let us first examine an orthogonal transformation from the matrix viewpoint. A linear change of variables such as (29) can be expressed $\mathbf{x} = \mathbf{Q}\mathbf{y}$. The coefficient matrix $\mathbf{Q}$ corresponding to an orthogonal transformation is called an **orthogonal matrix**. Since $\|\mathbf{x}\|_2 = \|\mathbf{y}\|_2$, it follows that multiplying a vector by an orthogonal matrix does not change the length of the vector. For an overdetermined system, we multiply the residual $\mathbf{r} = \mathbf{b} - \mathbf{A}\mathbf{x}$ by an orthogonal matrix $\mathbf{Q}$ to simplify the residual. Since the length of the transformed residual $\mathbf{Q}\mathbf{r}$ is the same as the length of the original residual $\mathbf{r}$, minimizing $\|\mathbf{b} - \mathbf{A}\mathbf{x}\|_2$ over $\mathbf{x}$ is equivalent to minimizing $\|\mathbf{Q}(\mathbf{b} - \mathbf{A}\mathbf{x})\|_2$ over $\mathbf{x}$. By an appropriate choice of $\mathbf{Q}$, this second minimum is easily evaluated.

For example, consider the overdetermined system

$$3x_1 = 5,$$

$$4x_1 = 10.$$

The corresponding residual is

$$\mathbf{r} = \begin{bmatrix} 5 - 3x_1 \\ 10 - 4x_2 \end{bmatrix}.$$

The coefficient matrix $\mathbf{Q}$ corresponding to the orthogonal transformation (29) is

$$\mathbf{Q} = \begin{bmatrix} .6 & .8 \\ .8 & -.6 \end{bmatrix}.$$

Multiplying $\mathbf{r}$ by $\mathbf{Q}$ gives us the transformed residual:

$$\mathbf{Q}\mathbf{r} = \begin{bmatrix} 11 - 5x_1 \\ -2 \end{bmatrix}.$$

It is easy to minimize the length of the transformed residual. The second component of $\mathbf{Q}\mathbf{r}$ is -2, regardless of our choice for x_1. The first component of $\mathbf{Q}\mathbf{r}$ is zero if $11 - 5x_1 = 0$, or equivalently, $x_1 = 11/5$. Thus the length of the residual is minimized when $x_1 = 11/5$.

To summarize, for an underdetermined system $\mathbf{A}\mathbf{x} = \mathbf{b}$, an orthogonal transformation $\mathbf{x} = \mathbf{Q}\mathbf{y}$ is employed to obtain an equivalent system $\mathbf{A}\mathbf{Q}\mathbf{y} = \mathbf{b}$ for which the minimum norm solution is computed easily. For an overdetermined linear system, the residual $\mathbf{r} = \mathbf{b} - \mathbf{A}\mathbf{x}$ is multiplied by an orthogonal matrix

$\mathbf{Q}$ to obtain a simplified residual $\mathbf{Qr} = \mathbf{Q}(\mathbf{b} - \mathbf{Ax})$ for which the $\mathbf{x}$ that minimizes the length of $\mathbf{Qr}$ is computed easily. In Sections 5-5 and 5-6 we explain how to generate orthogonal matrices while Section 5-7 gives a complete description of the algorithms introduced above.

Exercises

5-4.1. For the orthogonal transformation (29), sketch lines in the (x_1, x_2)-plane corresponding to the horizontal axis $y_2 = 0$ and the vertical axis $y_1 = 0$ in the (y_1, y_2)-plane. Is the linear transformation (29) a rotation or a reflection?

5-4.2. Suppose that $\mathbf{A}$ is a symmetric matrix and $\mathbf{x}^T\mathbf{Ax} = 0$ for every $\mathbf{x}$. What can you conclude about the $\mathbf{A}$ matrix? (*Hint:* See Exercise 3-3.8.)

5-4.3. For an orthogonal transformation $\mathbf{x} = \mathbf{Qy}$, the identity $\|\mathbf{x}\|_2 = \|\mathbf{y}\|_2$ is equivalent to

$$\mathbf{x}^T\mathbf{x} = (\mathbf{Qy})^T\mathbf{Qy} = \mathbf{y}^T\mathbf{Q}^T\mathbf{Qy} = \mathbf{y}^T\mathbf{y}.$$

Show that $\mathbf{Q}^T\mathbf{Q} = \mathbf{I}$. (*Hint:* See Exercise 5-4.2.) Thus for an orthogonal matrix, $\mathbf{Q}^T$ is the inverse of $\mathbf{Q}$. Since $\mathbf{AA}^{-1} = \mathbf{A}^{-1}\mathbf{A} = \mathbf{I}$, an orthogonal matrix has the property that $\mathbf{Q}^T\mathbf{Q} = \mathbf{QQ}^T = \mathbf{I}$.

5-4.4. Suppose that $\mathbf{Q}$ is a matrix for which $\mathbf{Q}^T\mathbf{Q} = \mathbf{I}$. Show that the transformation $\mathbf{x} = \mathbf{Qy}$ is an orthogonal transformation. (Combining Exercises 5-4.3 and 5-4.4, it follows that $\mathbf{Q}$ is an orthogonal matrix if and only if $\mathbf{Q}^T\mathbf{Q} = \mathbf{I}$.)

5-4.5. If $\mathbf{Q}_1$ and $\mathbf{Q}_2$ are orthogonal matrices, show that the product $\mathbf{Q}_1\mathbf{Q}_2$ is an orthogonal matrix.

5-4.6. If $\mathbf{Q}$ is an orthogonal matrix, show that $\mathbf{Q}^{-1}$ is an orthogonal matrix.

5-5. THE QR FACTORIZATION

In the **LU** factorization, a matrix is written as the product of a lower triangular matrix **L** containing the multipliers and an upper triangular matrix **U**. In the **QR** factorization, we express a matrix as the product of an orthogonal matrix **Q** and an upper triangular matrix **R** (actually **R** stands for right triangular). The **LU** factorization is a key step in algorithms to solve a linear system. Likewise, the **QR** factorization is an important step in algorithms to solve a linear system (see Section 5-7), to determine an orthonormal basis for a collection of vectors (see Section 5-8), and to compute the eigenvalues of a matrix (see Section 6-5). Although more computer time is needed to **QR** factor a matrix than to **LU** factor a matrix, algorithms based upon the **QR** factorization tend to be more stable.

The **QR** factorization of a matrix can be computed using either Householder reflections (discussed in this section), or Givens rotations (discussed in the next section). A **Householder reflection** is any matrix of the form

$$\mathbf{I} - 2\mathbf{ww}^T, \tag{31}$$

where $\mathbf{w}$ is a vector with Euclidean length one. For example, the vector

$$\mathbf{w} = \frac{1}{5}\begin{bmatrix} 3 \\ 4 \end{bmatrix}$$

is a unit vector since $(\frac{3}{5})^2 + (\frac{4}{5})^2 = 1$ and the corresponding matrix (31) is

$$\mathbf{I} - 2\mathbf{ww}^T = \begin{bmatrix} 1 & 0 \\ 0 & 1 \end{bmatrix} - \frac{2}{25}\begin{bmatrix} 9 & 12 \\ 12 & 16 \end{bmatrix} = \frac{1}{25}\begin{bmatrix} 7 & -24 \\ -24 & -7 \end{bmatrix}.$$

By Exercise 5-4.4, a matrix is orthogonal if and only if $\mathbf{Q}^T\mathbf{Q} = \mathbf{Q}\mathbf{Q}^T = \mathbf{I}$. A Householder matrix $\mathbf{H} = \mathbf{I} - 2\mathbf{w}\mathbf{w}^T$ is both symmetric and orthogonal: $\mathbf{H}^T = \mathbf{H}$ and $\mathbf{H}^T\mathbf{H} = \mathbf{I}$. Combining symmetry with orthogonality, we have

$$\mathbf{I} = \mathbf{H}^T\mathbf{H} = \mathbf{H}\mathbf{H} = \mathbf{H}^2. \tag{32}$$

Since $\mathbf{H}\mathbf{H} = \mathbf{I}$, a Householder matrix is its own inverse: $\mathbf{H}^{-1} = \mathbf{H}$. To verify (32), let us form the product $\mathbf{H}\mathbf{H}$:

$$\mathbf{H}\mathbf{H} = (\mathbf{I} - 2\mathbf{w}\mathbf{w}^T)(\mathbf{I} - 2\mathbf{w}\mathbf{w}^T) = \mathbf{I} - 4\mathbf{w}\mathbf{w}^T + 4\mathbf{w}\mathbf{w}^T\mathbf{w}\mathbf{w}^T \tag{33}$$

By the associative law for matrix multiplication, $\mathbf{w}\mathbf{w}^T\mathbf{w}\mathbf{w}^T = \mathbf{w}(\mathbf{w}^T\mathbf{w})\mathbf{w}^T$. Since $\mathbf{w}$ is a unit vector, $\mathbf{w}^T\mathbf{w} = 1$ and $\mathbf{w}(\mathbf{w}^T\mathbf{w})\mathbf{w}^T = \mathbf{w}\mathbf{w}^T$. Hence the last two terms in (33) cancel and $\mathbf{H}\mathbf{H} = \mathbf{I}$.

Multiplying a Householder matrix by a vector $\mathbf{x}$ is equivalent to reflecting $\mathbf{x}$ across the plane perpendicular to $\mathbf{w}$. To verify this reflection property, let us form the product $\mathbf{H}\mathbf{x}$:

$$\mathbf{H}\mathbf{x} = (\mathbf{I} - 2\mathbf{w}\mathbf{w}^T)\mathbf{x} = \mathbf{x} - 2\mathbf{w}\mathbf{w}^T\mathbf{x} = \mathbf{x} - 2\mathbf{w}(\mathbf{w}^T\mathbf{x}) = \mathbf{x} - 2(\mathbf{w}^T\mathbf{x})\mathbf{w}. \tag{34}$$

As depicted in Figure 5-8, $(\mathbf{w}^T\mathbf{x})\mathbf{w}$ is the projection of $\mathbf{x}$ onto $\mathbf{w}$. It follows from (34) that $\mathbf{H}\mathbf{x}$ is equal to $\mathbf{x}$ minus twice the projection of $\mathbf{x}$ onto $\mathbf{w}$. In Figure 5-9 we see that $\mathbf{x}$ minus twice its projection onto $\mathbf{w}$ is the same as the reflection of $\mathbf{x}$ across the plane perpendicular to $\mathbf{w}$. Due to this reflection property, Householder matrices are called **elementary reflectors**.

Another property of Householder matrices is the following: If $\mathbf{x}$ and $\mathbf{y}$ are nonzero vectors of equal Euclidean length and $\mathbf{w}$ is defined by

$$\mathbf{w} = \frac{1}{\|\mathbf{x} - \mathbf{y}\|_2}(\mathbf{x} - \mathbf{y}), \tag{35}$$

then

$$(\mathbf{I} - 2\mathbf{w}\mathbf{w}^T)\mathbf{x} = \mathbf{y}. \tag{36}$$

This result follows from the reflection property. By (35) $\mathbf{w}$ points in the direction of the vector $\mathbf{x} - \mathbf{y}$. In Figure 5-10 (on page 210) we see that $\mathbf{x} - \mathbf{y}$ is the vector connecting the tip of $\mathbf{y}$ to the tip of $\mathbf{x}$. Since $\mathbf{x}$ and $\mathbf{y}$ have the same length, the reflection of $\mathbf{x}$ across the plane perpendicular to $\mathbf{x} - \mathbf{y}$ is $\mathbf{y}$. In other words, $(\mathbf{I} - 2\mathbf{w}\mathbf{w}^T)\mathbf{x} = \mathbf{y}$.

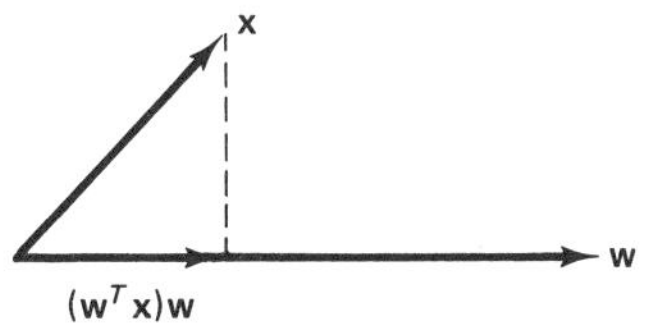

Figure 5-8 The projection of **x** onto **w**.

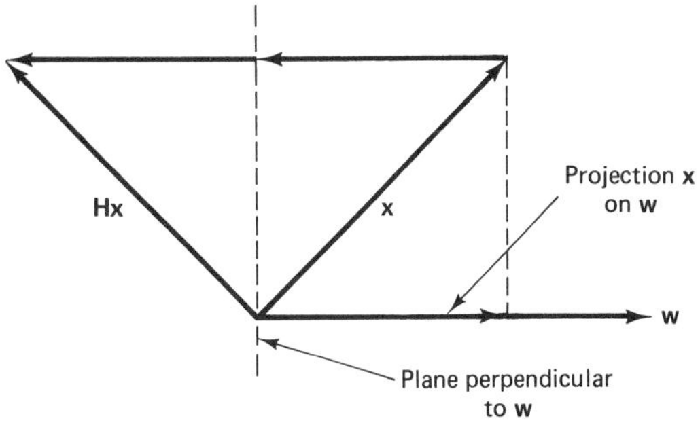

Figure 5-9 **Hx** is the reflection of **x** across the plane perpendicular to **w**.

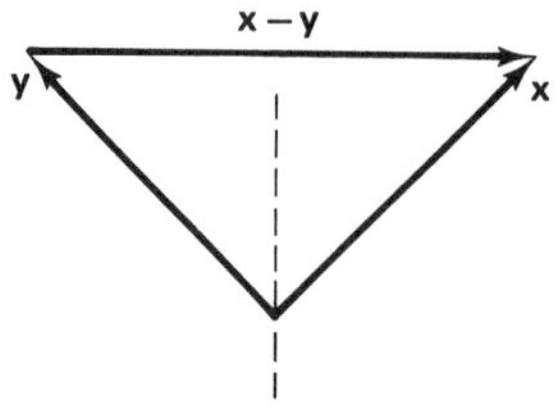

Figure 5-10 The reflection of **x** across the plane perpendicular to **x** − **y**.

Utilizing the special choice (35) for **w**, we can build a Householder matrix that annihilates some components of a given vector **x** while leaving all but one of the remaining components intact. In particular, given an integer $k \geqq 1$, let us construct a Householder matrix **H** with the property that the first $k - 1$ components of **Hx** are equal to the first $k - 1$ components of **x** while the last $n - k$ components of **Hx** are zero:

$$(\mathbf{Hx})_i = \begin{cases} x_i & \text{for } i = 1, 2, \cdots, k-1, \\ 0 & \text{for } i = k+1, k+2, \cdots, n. \end{cases}$$

Referring to (35), we make the following choice for **y**:

$$\begin{aligned} y_1 &= x_1, y_2 = x_2, \cdots, y_{k-1} = x_{k-1}, \\ y_k &= \pm\sqrt{x_k^2 + x_{k+1}^2 + \cdots + x_n^2}, \\ y_{k+1} &= y_{k+2} = \cdots = y_n = 0. \end{aligned} \tag{37}$$

Since $y_1^2 + y_2^2 + \cdots + y_n^2 = x_1^2 + x_2^2 + \cdots + x_n^2$, both **x** and **y** have the same length. If **w** is given by (35), then (36) tells us that $\mathbf{Hx} = (\mathbf{I} - 2\mathbf{ww}^T)\mathbf{x} = \mathbf{y}$. Since the first $k - 1$ components of **y** equal the corresponding components of **x** while the last $n - k$ components of **y** are zero, the desired Householder matrix has been constructed. Observe that there are two possible signs for the kth component of **y** in (37). Since the computation of **w** in (35) involves subtracting **y** from **x**, the relative error in the computed difference is smaller if the kth components of **x** and **y** have opposite signs. (Remember that subtracting nearly equal numbers is dangerous. If x_k and y_k have opposite signs, then $x_k - y_k$ is never the difference between nearly equal numbers.) If x_k and y_k have opposite signs and **y** is defined by (37), the vector **w** in (35) is

$$\mathbf{w} = \frac{1}{\sqrt{2s(s + |x_k|)}} \begin{bmatrix} 0 \\ \cdot \\ \cdot \\ \cdot \\ 0 \\ x_k + \operatorname{sign}(x_k)s \\ x_{k+1} \\ \cdot \\ \cdot \\ \cdot \\ x_n \end{bmatrix}, \tag{38}$$

where

$$s = \sqrt{x_k^2 + x_{k+1}^2 + \cdots + x_n^2}.$$

For illustration, let us construct **H** to annihilate the second component of the vector

$$\mathbf{x} = \begin{bmatrix} 4 \\ 3 \end{bmatrix}.$$

In this case, $k = 1$ and by (38), we have

$$\mathbf{w} = \frac{1}{\sqrt{90}}\begin{bmatrix} 9 \\ 3 \end{bmatrix} \quad \text{and} \quad \mathbf{H} = \mathbf{I} - 2\mathbf{w}\mathbf{w}^T = -\frac{1}{5}\begin{bmatrix} 4 & 3 \\ 3 & -4 \end{bmatrix}.$$

The sign rule formulated above specifies that the sign of y_k should be the opposite of x_k's sign. Since $k = 1$ and $x_k = x_1 = 4$ is positive, the first component of **y** is negative and (37) yields

$$\mathbf{y} = \begin{bmatrix} -5 \\ 0 \end{bmatrix}.$$

Finally, let us check that $\mathbf{Hx} = \mathbf{y}$:

$$-\frac{1}{5}\begin{bmatrix} 4 & 3 \\ 3 & -4 \end{bmatrix}\begin{bmatrix} 4 \\ 3 \end{bmatrix} = \begin{bmatrix} -5 \\ 0 \end{bmatrix}.$$

For a second example, we annihilate the third component of the vector

$$\mathbf{x} = \begin{bmatrix} 7 \\ 4 \\ 3 \end{bmatrix},$$

while leaving the first component intact. In this case, k is 2 and by (38), we have

$$\mathbf{w} = \frac{1}{\sqrt{90}}\begin{bmatrix} 0 \\ 9 \\ 3 \end{bmatrix} \quad \text{and} \quad \mathbf{H} = \mathbf{I} - 2\mathbf{w}\mathbf{w}^T = \frac{1}{5}\begin{bmatrix} 5 & 0 & 0 \\ 0 & -4 & -3 \\ 0 & -3 & 4 \end{bmatrix}. \tag{39}$$

Since $k = 2$ and $x_k = x_2 = 4$ is positive, the second component of **y** is negative and (37) yields

$$\mathbf{y} = \begin{bmatrix} 7 \\ -5 \\ 0 \end{bmatrix}.$$

Again, we see that $\mathbf{Hx} = \mathbf{y}$:

$$\underset{\mathbf{H}}{\frac{1}{5}\begin{bmatrix} 5 & 0 & 0 \\ 0 & -4 & -3 \\ 0 & -3 & 4 \end{bmatrix}} \underset{\mathbf{x}}{\begin{bmatrix} 7 \\ 4 \\ 3 \end{bmatrix}} = \underset{\mathbf{y}}{\begin{bmatrix} 7 \\ -5 \\ 0 \end{bmatrix}}. \tag{40}$$

The first component of **x** is equal to the first component of **y** and the last component of **x** is annihilated.

The **QR** factorization of a matrix **A** is computed using a sequence of Householder matrices to reduce **A** to upper triangular form. The resulting upper triangular matrix is **R** and the product of the Householder matrices is **Q**. To illustrate this process, let us consider the matrix

$$\mathbf{A} = \begin{bmatrix} 10 & 9 & 18 \\ 20 & -15 & -15 \\ 20 & -12 & 51 \end{bmatrix}.$$

The first column of $\mathbf{A}$ is

$$\begin{bmatrix} 10 \\ 20 \\ 20 \end{bmatrix}.$$

Using (38) with $k = 1$, we construct the Householder matrix $\mathbf{H}_1$ that annihilates components 2 and 3 in the first column:

$$\mathbf{H}_1 = -\frac{1}{3}\begin{bmatrix} 1 & 2 & 2 \\ 2 & -2 & 1 \\ 2 & 1 & -2 \end{bmatrix}.$$

The product $\mathbf{H}_1\mathbf{A}$ is a matrix that we denote $\mathbf{A}_1$:

$$\mathbf{A}_1 = \mathbf{H}_1\mathbf{A} = -\frac{1}{3}\begin{bmatrix} 1 & 2 & 2 \\ 2 & -2 & 1 \\ 2 & 1 & -2 \end{bmatrix}\begin{bmatrix} 10 & 9 & 18 \\ 20 & -15 & -15 \\ 20 & -12 & 51 \end{bmatrix} = \begin{bmatrix} -30 & 15 & -30 \\ 0 & -12 & -39 \\ 0 & -9 & 27 \end{bmatrix}.$$

Observe that each coefficient in the first column of $\mathbf{A}_1$ beneath the diagonal is zero. The second column of $\mathbf{A}_1$ is

$$\mathbf{x} = \begin{bmatrix} 15 \\ -12 \\ -9 \end{bmatrix}.$$

Using (38) with $k = 2$, we construct the Householder matrix $\mathbf{H}_2$ that annihilates the third component of the second column but leaves the first component intact:

$$\mathbf{H}_2 = \begin{bmatrix} 1 & 0 & 0 \\ 0 & -.8 & -.6 \\ 0 & -.6 & .8 \end{bmatrix}.$$

Premultiplying $\mathbf{A}_1$ by $\mathbf{H}_2$ gives us a matrix that we denote $\mathbf{A}_2$:

$$\mathbf{A}_2 = \mathbf{H}_2\mathbf{A}_1 = \begin{bmatrix} 1 & 0 & 0 \\ 0 & -.8 & -.6 \\ 0 & -.6 & .8 \end{bmatrix}\begin{bmatrix} -30 & 15 & -30 \\ 0 & -12 & -39 \\ 0 & -9 & 27 \end{bmatrix} = \begin{bmatrix} -30 & 15 & -30 \\ 0 & 15 & 15 \\ 0 & 0 & 45 \end{bmatrix}.$$

Observe that $\mathbf{A}_2$ is upper triangular. In fact, $\mathbf{A}_2$ is the matrix $\mathbf{R}$ in the $\mathbf{QR}$ factorization. By the structure of $\mathbf{H}_2$, the first column of $\mathbf{A}_1$ and the first column of $\mathbf{A}_2$ are identical. Thus the work done in the first step (multiplying $\mathbf{A}$ by $\mathbf{H}_1$ to annihilate the second and the third components of column 1) is not disturbed by the second step. At this point, we have constructed the upper triangular matrix $\mathbf{R} = \mathbf{A}_2 = \mathbf{H}_2\mathbf{A}_1 = \mathbf{H}_2\mathbf{H}_1\mathbf{A}$. Multiplying $\mathbf{R}$ by $\mathbf{H}_1\mathbf{H}_2$ gives us the relation

$$\mathbf{H}_1\mathbf{H}_2\mathbf{R} = \mathbf{H}_1\mathbf{H}_2\mathbf{H}_2\mathbf{H}_1\mathbf{A}. \tag{41}$$

Since a Householder matrix is its own inverse, $\mathbf{H}_2\mathbf{H}_2 = \mathbf{I}$ and $\mathbf{H}_1\mathbf{H}_1 = \mathbf{I}$. Hence (41) tells us that $\mathbf{A} = \mathbf{H}_1\mathbf{H}_2\mathbf{R}$. Letting $\mathbf{Q}$ denote the product $\mathbf{H}_1\mathbf{H}_2$, it follows that $\mathbf{A} = \mathbf{QR}$. To check that $\mathbf{Q}$ is orthogonal, we form the product $\mathbf{Q}^T\mathbf{Q}$:

$$\mathbf{Q}^T\mathbf{Q} = (\mathbf{H}_1\mathbf{H}_2)^T\,\mathbf{H}_1\mathbf{H}_2. \tag{42}$$

Since the transpose of a matrix product $\mathbf{H}_1\mathbf{H}_2$ is the product $\mathbf{H}_2{}^T\mathbf{H}_1{}^T$ of the transposes in the reverse order, (42) can be written

$$\mathbf{Q}^T\mathbf{Q} = \mathbf{H}_2{}^T\,\mathbf{H}_1{}^T\,\mathbf{H}_1\mathbf{H}_2.$$

Since each Householder matrix is orthogonal, we conclude that $\mathbf{Q}^T\mathbf{Q} = \mathbf{I}$ and $\mathbf{Q}$ is orthogonal. In summary, the $\mathbf{QR}$ factorization of the starting matrix $\mathbf{A}$ can be expressed

$$\mathbf{A} = \begin{bmatrix} 10 & 9 & 18 \\ 20 & -15 & -15 \\ 20 & -12 & 51 \end{bmatrix} = \left(-\frac{1}{15} \begin{bmatrix} 5 & -14 & 2 \\ 10 & 5 & 10 \\ 10 & 2 & -11 \end{bmatrix} \right) \begin{bmatrix} -30 & 15 & -30 \\ 0 & 15 & 15 \\ 0 & 0 & 45 \end{bmatrix},$$

where the first factor on the right side is $\mathbf{Q} = \mathbf{H}_1\mathbf{H}_2$ while the second factor is $\mathbf{R}$.

The **QR** factorization of a general $n \times n$ matrix $\mathbf{A}$ proceeds in the same way. In the first step we premultiply $\mathbf{A}$ by the Householder matrix $\mathbf{H}_1$ that annihilates elements 2 through n in the first column of $\mathbf{A}$, giving $\mathbf{A}_1 = \mathbf{H}_1\mathbf{A}$. In the second step we premultiply $\mathbf{A}_1$ by the Householder matrix $\mathbf{H}_2$ that annihilates elements 3 through n in the second column of $\mathbf{A}_1$. The second step produces $\mathbf{A}_2 = \mathbf{H}_2\mathbf{A}_1$. In the kth step, we premultiply $\mathbf{A}_{k-1}$ by the Householder matrix $\mathbf{H}_k$ that annihilates elements $k + 1$ through n in the kth column of $\mathbf{A}_{k-1}$. Step k generates $\mathbf{A}_k = \mathbf{H}_k\mathbf{A}_{k-1}$. After $n - 1$ steps, we arrive at $\mathbf{R}$:

$$\mathbf{R} = \mathbf{A}_{n-1} = \mathbf{H}_{n-1}\mathbf{A}_{n-2} = \mathbf{H}_{n-1}\mathbf{H}_{n-2}\mathbf{A}_{n-3} = \cdots = \mathbf{H}_{n-1}\mathbf{H}_{n-2} \cdots \mathbf{H}_1\mathbf{A}.$$

Finally, $\mathbf{A} = \mathbf{QR}$, where

$$\mathbf{Q} = \mathbf{H}_1\mathbf{H}_2 \cdots \mathbf{H}_{n-1}.$$

When a matrix is **LU** factored, pivoting is needed to alleviate the devastating interaction of rounding errors. As we noted in Chapter 2, the error in the computed solution to a linear system is related to the growth of coefficients during the elimination process. Pivoting helps to reduce the coefficient growth during elimination. On the other hand, when **QR** factoring a matrix, no pivoting is required† and there is essentially no growth in the coefficients. To show that there is no growth, let us compute the Euclidean norm of the matrix $\mathbf{A}_1 = \mathbf{H}_1\mathbf{A}$ generated by the first step in the **QR** factorization. In Chapter 3 we saw that the square of the Euclidean norm of a matrix $\mathbf{B}$ is the maximum of the quadratic form $\mathbf{y}^T(\mathbf{B}^T\mathbf{B})\mathbf{y}$ over the unit sphere:

$$(\|\mathbf{B}\|_2)^2 = \text{maximum } \{\mathbf{y}^T(\mathbf{B}^T\mathbf{B})\mathbf{y} : \|\mathbf{y}\|_2 = 1\}.$$

Identifying $\mathbf{B}$ with $\mathbf{A}_1$, it follows that the square of the Euclidean norm of $\mathbf{A}_1$ is the maximum of the quadratic form

$$\mathbf{y}^T(\mathbf{A}_1{}^T\mathbf{A}_1)\mathbf{y} = \mathbf{y}^T((\mathbf{H}_1\mathbf{A})^T\mathbf{H}_1\mathbf{A})\mathbf{y} = \mathbf{y}^T(\mathbf{A}^T\mathbf{H}_1{}^T\mathbf{H}_1\mathbf{A})\mathbf{y}$$

over the unit sphere. Since a Householder matrix is orthogonal, $\mathbf{H}_1{}^T\mathbf{H}_1 = \mathbf{I}$ and

$$\mathbf{y}^T(\mathbf{A}_1{}^T\mathbf{A}_1)\mathbf{y} = \mathbf{y}^T(\mathbf{A}^T\mathbf{A})\mathbf{y}.$$

Since the quadratic form $\mathbf{y}^T(\mathbf{A}_1{}^T\mathbf{A}_1)\mathbf{y}$ is equal to the quadratic form $\mathbf{y}^T(\mathbf{A}^T\mathbf{A})\mathbf{y}$, the Euclidean norm of $\mathbf{A}_1$ is equal to the Euclidean norm of $\mathbf{A}$. In the same manner, the Euclidean norm of $\mathbf{A}_2$ is equal to the Euclidean norm of $\mathbf{A}_1$. Proceeding in this way, we conclude that the Euclidean norm of each of the matrices $\mathbf{A}_k$ generated during the factorization process is equal to the Euclidean norm of the starting matrix $\mathbf{A}$. Loosely speaking, the coefficients do not grow during the **QR** factorization of a matrix. More precisely, the Euclidean norm of each intermediate matrix $\mathbf{A}_k$ is equal to the Euclidean norm of the starting matrix $\mathbf{A}$.

For some special matrices, the factorization algorithm must be modified to

† Although pivoting is not required, it still improves the accuracy in the **QR** factorization.

avoid division by zero. In particular, if elements k through n in column k of $\mathbf{A}_{k-1}$ are zero, then equation (38) involves zero over zero. For example, suppose that after the first step in the factorization of $\mathbf{A}$, we obtain the matrix

$$\mathbf{A}_1 = \begin{bmatrix} 1 & 2 & 3 \\ 0 & 0 & 3 \\ 0 & 0 & 4 \end{bmatrix}. \tag{43}$$

Blindly applying (38) to construct the Householder matrix that annihilates the third element in the second column of $\mathbf{A}_1$ while leaving the first element intact, we find that

$$\mathbf{w} = \frac{1}{0}\begin{bmatrix} 0 \\ 0 \\ 0 \end{bmatrix},$$

and the Householder matrix is undefined. To **QR** factor the matrix (43), simply skip the second column of $\mathbf{A}_1$ since it already has the proper structure and proceed to the third column, constructing the Householder matrix $\mathbf{H}_2$ that annihilates the third element of the third column while leaving the first element intact:

$$\mathbf{H}_2 = \begin{bmatrix} 1 & 0 & 0 \\ 0 & -.6 & -.8 \\ 0 & -.8 & .6 \end{bmatrix}.$$

The product $\mathbf{H}_2\mathbf{A}_1$ is $\mathbf{A}_2$:

$$\mathbf{A}_2 = \mathbf{H}_2\mathbf{A}_1 = \begin{bmatrix} 1 & 0 & 0 \\ 0 & -.6 & -.8 \\ 0 & -.8 & .6 \end{bmatrix}\begin{bmatrix} 1 & 2 & 3 \\ 0 & 0 & 3 \\ 0 & 0 & 4 \end{bmatrix} = \begin{bmatrix} 1 & 2 & 3 \\ 0 & 0 & -5 \\ 0 & 0 & 0 \end{bmatrix}.$$

In this example, the last row of $\mathbf{R} = \mathbf{A}_2$ is completely zero. Now consider the factorization of a general matrix. Suppose that column 1 through column $c - 1$ of matrix $\mathbf{A}$ have been processed before executing step k. If elements k through n in column c of $\mathbf{A}_{k-1}$ are zero at step k, the computer program should inspect columns $c + 1$ through n searching for a column of $\mathbf{A}_{k-1}$, where at least one element in rows k through n is nonzero. Applying (38) to this column, we construct the Householder matrix $\mathbf{H}_k$ which annihilates components $k + 1$ through n and we set $\mathbf{A}_k = \mathbf{H}_k\mathbf{A}_{k-1}$. If elements k through n in every column are zero, then the factorization is complete. For a square matrix, the number of columns that we skip during the factorization is equal to the number of rows of zeros at the bottom of $\mathbf{R}$.

Rectangular matrices can be factored just as easily as square matrices. Consider the 3×2 matrix

$$\mathbf{A} = \begin{bmatrix} 3 & 0 \\ 4 & 5 \\ 0 & 4 \end{bmatrix},$$

the coefficient matrix in equation (19). The first step in the **QR** factorization of $\mathbf{A}$ constructs the Householder matrix $\mathbf{H}_1$ that annihilates the second and third elements in column 1:

$$\mathbf{H}_1 = \frac{1}{5}\begin{bmatrix} -3 & -4 & 0 \\ -4 & 3 & 0 \\ 0 & 0 & 5 \end{bmatrix}.$$

Multiplying $\mathbf{A}$ by $\mathbf{H}_1$ yields $\mathbf{A}_1$:

$$\mathbf{A}_1 = \mathbf{H}_1\mathbf{A} = \frac{1}{5}\begin{bmatrix} -3 & -4 & 0 \\ -4 & 3 & 0 \\ 0 & 0 & 5 \end{bmatrix}\begin{bmatrix} 3 & 0 \\ 4 & 5 \\ 0 & 4 \end{bmatrix} = \begin{bmatrix} -5 & -4 \\ 0 & 3 \\ 0 & 4 \end{bmatrix}.$$

In the second step, we construct the Householder matrix $\mathbf{H}_2$ that annihilates the third element in the second column of $\mathbf{A}_1$ while leaving the first element intact:

$$\mathbf{H}_2 = \frac{1}{5}\begin{bmatrix} 5 & 0 & 0 \\ 0 & -3 & -4 \\ 0 & -4 & 3 \end{bmatrix}.$$

Multiplying $\mathbf{A}_1$ by $\mathbf{H}_2$ yields the upper triangular matrix $\mathbf{A}_2$:

$$\mathbf{R} = \mathbf{A}_2 = \mathbf{H}_2\mathbf{A}_1 = \frac{1}{5}\begin{bmatrix} 5 & 0 & 0 \\ 0 & -3 & -4 \\ 0 & -4 & 3 \end{bmatrix}\begin{bmatrix} -5 & -4 \\ 0 & 3 \\ 0 & 4 \end{bmatrix} = \begin{bmatrix} -5 & -4 \\ 0 & -5 \\ 0 & 0 \end{bmatrix}. \tag{44}$$

Again, $\mathbf{R} = \mathbf{A}_2$ and $\mathbf{A} = \mathbf{QR}$, where $\mathbf{Q} = \mathbf{H}_1\mathbf{H}_2$:

$$\mathbf{A} = \begin{bmatrix} 3 & 0 \\ 4 & 5 \\ 0 & 4 \end{bmatrix} = \left(\frac{1}{25}\begin{bmatrix} -15 & 12 & 16 \\ -20 & -9 & -12 \\ 0 & -20 & 15 \end{bmatrix}\right)\begin{bmatrix} -5 & -4 \\ 0 & -5 \\ 0 & 0 \end{bmatrix}.$$

To summarize, the kth step of the factorization process multiplies $\mathbf{A}_{k-1}$ by a Householder matrix $\mathbf{H}_k$ to obtain $\mathbf{A}_k$. When implementing the **QR** factorization, the matrix $\mathbf{H}_k$ is never actually constructed since the product $\mathbf{H}_k\mathbf{A}_{k-1}$ can be computed using the identity (34). Referring to (34), the product between a Householder matrix $\mathbf{H}$ and a vector $\mathbf{x}$ can be expressed in terms of $\mathbf{x}$ and the vector $\mathbf{w}$ which characterizes the Householder matrix: $\mathbf{Hx} = \mathbf{x} - 2(\mathbf{w}^T\mathbf{x})\mathbf{w}$. For illustration, let us compute the matrix-vector product (40) using the vector $\mathbf{w}$ given in (39):

$$\mathbf{Hx} = \mathbf{x} - 2(\mathbf{w}^T\mathbf{x})\mathbf{w} = \begin{bmatrix} 7 \\ 4 \\ 3 \end{bmatrix} - \frac{2}{\sqrt{90}}\left([0 \quad 9 \quad 3]\begin{bmatrix} 7 \\ 4 \\ 3 \end{bmatrix}\right)\frac{1}{\sqrt{90}}\begin{bmatrix} 0 \\ 9 \\ 3 \end{bmatrix}$$

$$= \begin{bmatrix} 7 \\ 4 \\ 3 \end{bmatrix} - \begin{bmatrix} 0 \\ 9 \\ 3 \end{bmatrix} = \begin{bmatrix} 7 \\ -5 \\ 0 \end{bmatrix}.$$

This agrees with $\mathbf{y}$ in (40). Now consider the product $\mathbf{H}_k\mathbf{A}_{k-1}$ between the Householder matrix $\mathbf{H}_k$ and the partly factored matrix $\mathbf{A}_{k-1}$. Since column j of the product is equal to $\mathbf{H}_k$ times column j of $\mathbf{A}_{k-1}$, relation (34) can be applied to compute each column of the product.

The following algorithm **QR** factors a $m \times n$ matrix $\mathbf{A}$, overwriting the coefficient matrix with both $\mathbf{R}$ and the vectors characterizing each Householder matrix. During the kth step, we annihilate $m - k$ elements in the coefficient matrix using a Householder matrix whose corresponding vector $\mathbf{w}$ has $m - k + 1$ nonzero components. Since the nonzero components of $\mathbf{w}$ do not fit within the space created by the annihilated coefficients, we store the diagonal of $\mathbf{R}$ in a separate array $\mathbf{d}$ to make room for the first nonzero component of $\mathbf{w}$. After executing the following algorithm, the diagonal of $\mathbf{R}$ is stored in $\mathbf{d}$, the remaining elements above $\mathbf{R}$'s diagonal are stored above the diagonal in $\mathbf{A}$, and the vectors $\mathbf{v} = \sqrt{2}\mathbf{w}$ related to each Householder matrix are stored on the diagonal and beneath the diagonal of **A**.

$$
\begin{array}{l}
k \leftarrow 0 \\
l = 1 \text{ to } n \\
\quad k \leftarrow k + 1, \text{ if } k = m \text{ then } d_l \leftarrow a_{kl} \text{ and exit } l \text{ loop} \\
\quad s \leftarrow \left(\sum_{i=k}^{m} a_{il}^2 \right)^{1/2}, \\
\quad \text{if } s = 0 \text{ then } d_l \leftarrow 0 \text{ and go to next } l \\
\quad t \leftarrow a_{kl},\ r \leftarrow 1/(s(s+|t|))^{1/2}, \text{ if } t < 0 \text{ then } s \leftarrow -s, \\
\quad d_l \leftarrow -s,\ a_{kk} \leftarrow r(t+s) \\
\quad a_{ik} \leftarrow ra_{il} \text{ for } i = k+1 \text{ to } m \\
\quad j = l+1 \text{ to } n \\
\qquad t \leftarrow 0 \\
\qquad t \leftarrow t + a_{ik}a_{ij} \text{ for } i = k \text{ to } m \\
\qquad a_{ij} \leftarrow a_{ij} - ta_{ik} \text{ for } i = k \text{ to } m \\
\quad \text{next } j \\
\text{next } l
\end{array}
\tag{45}
$$

Subroutine QR implements algorithm (45). The subroutine arguments are

A: array containing matrix
LA: leading (row) dimension of array A
M: number of rows in the coefficient matrix
N: number of columns in the coefficient matrix

For a $n \times n$ matrix, the asymptotic running time for this program is $\frac{2}{3}n^3$ times .62 microsecond on an IBM 370 model 3090 computer with the Fortran IV-G compiler. Notice that subroutine QR stores the **d** of (45) within the A array and the elements of the coefficient matrix are rearranged to make room for **d**. The following program **QR** factors the coefficient matrix contained in equation (19):

```
REAL A(20)
DATA A/3.,4.,0.,0.,5.,4./
CALL QR(A,3,3,2)
END
```

Exercises

5-5.1. Construct the Householder matrix **H** which annihilates the second component of the vector

$$\mathbf{x} = \begin{bmatrix} 5 \\ 12 \end{bmatrix}.$$

For this particular **H** and **x**, use a sketch to show how $\mathbf{y} = \mathbf{Hx}$ is generated from **x**.

5-5.2. Compute the **QR** factorization of the matrix

$$\begin{bmatrix} 5 & -13 \\ 12 & 26 \end{bmatrix}.$$

5-5.3. The product **Hx** between a Householder matrix $\mathbf{H} = \mathbf{I} - 2\mathbf{w}\mathbf{w}^T$ and a vector **x** can be expressed $\mathbf{x} - a\mathbf{w}$. What is the formula for a?

5-5.4. Using the formula for a derived in Exercise 5-5.3, write a symbolic program to compute the product **HA** between a Householder matrix **H** and another matrix **A**.

5-5.5. If **H** is the Householder matrix corresponding to the **w** in (38), what is $(\mathbf{Hx})_k$?

5-5.6. If **H** is the Householder matrix that annihilates components 2 through n of **x**, what is the first column of **H**? [*Hint:* Multiply the relation

$$\mathbf{Hx} = (*, 0, 0, \cdots, 0)^T$$

by **H**.] What is the dot product between **x** and each column of **H**?

5-5.7. Let $\mathbf{H}_k$ denote the Householder matrix corresponding to (38) that annihilates components $k + 1$ through n of some given vector **x** and let $\mathbf{e}_j$ denote a vector with every component zero except for component j. Show that $\mathbf{H}_k\mathbf{e}_j = \mathbf{e}_j$ for each $j < k$.

5-5.8. Consider the following vector:

$$\mathbf{x} = \begin{bmatrix} 12 \\ 5 \\ 12 \end{bmatrix}.$$

(a) Construct the Householder matrix which annihilates the third component of **x** but leaves the first component intact.

(b) Construct the Householder matrix which annihilates the first component of **x** but leaves the third component intact.

(c) What is the general formula for the vector **w** corresponding to the Householder matrix which annihilates the first $k - 1$ components of a given vector **x** while leaving the last $n - k$ components intact.

5-5.9. Show that every 2×2 Householder matrix **H** has the form

$$\mathbf{H} = \begin{bmatrix} a & b \\ b & -a \end{bmatrix}, \tag{46}$$

where a and b are any scalars with the property that $a^2 + b^2 = 1$.

5-5.10. Given an arbitrary vector **x** with two components, what is a value for a and b in (46) such that $a^2 + b^2 = 1$ and the second component of the vector **Hx** is zero?

5-5.11. Given a vector **x** with two components, there are two unit vectors **w** with the property that the second component of $(\mathbf{I} - 2\mathbf{w}\mathbf{w}^T)\mathbf{x}$ is zero:

$$\mathbf{w} = \frac{1}{\sqrt{2s(s \pm x_1)}} \begin{bmatrix} x_1 \pm s \\ x_2 \end{bmatrix} \quad \text{and} \quad s = \sqrt{x_1^2 + x_2^2}.$$

In general, it is better to use the plus sign if x_1 is positive while it is better to use the minus sign if x_1 is negative. For what type of vectors **x** is the computed $\mathbf{H} = \mathbf{I} - 2\mathbf{w}\mathbf{w}^T$ inaccurate when this sign rule is violated?

5-6. GIVENS ROTATIONS

The **QR** factorization of a matrix can also be computed using **Givens rotations**. A Givens rotation is any matrix of the form

$$
\mathbf{G}_{ij} = \begin{bmatrix}
1 \\
& \ddots \\
&& 1 \\
&&& c & 0 & \cdots & 0 & -s \\
&&& 0 & 1 &&& 0 \\
&&& \vdots && \ddots && \vdots \\
&&& 0 &&& 1 & 0 \\
&&& s & 0 & \cdots & 0 & c \\
&&&&&&&& 1 \\
&&&&&&&&& \ddots \\
&&&&&&&&&& 1
\end{bmatrix}, \tag{47}
$$

where $c^2 + s^2 = 1$. The i and j subscript in $\mathbf{G}_{ij}$ correspond to the row numbers associated with the c's: The first c is in row j and the second c is in row i. Observe that the intersection of row i and row j with column i and column j for $\mathbf{G}_{ij}$ is the 2×2 matrix

$$
\begin{bmatrix} c & -s \\ s & c \end{bmatrix}.
$$

This matrix is orthogonal since multiplication by its transpose yields $\mathbf{I}$:

$$
\begin{bmatrix} c & -s \\ s & c \end{bmatrix}^T \begin{bmatrix} c & -s \\ s & c \end{bmatrix} = \begin{bmatrix} c & s \\ -s & c \end{bmatrix} \begin{bmatrix} c & -s \\ s & c \end{bmatrix} = \begin{bmatrix} c^2+s^2 & 0 \\ 0 & c^2+s^2 \end{bmatrix} = \begin{bmatrix} 1 & 0 \\ 0 & 1 \end{bmatrix}.
$$

Similarly, matrix (47) is orthogonal since $\mathbf{G}_{ij}^T\mathbf{G}_{ij} = \mathbf{I}$. Matrix (47) is called a rotation for the following reason: If two vectors $\mathbf{x}$ and $\mathbf{y}$ satisfy the relation

$$
\begin{bmatrix} y_1 \\ y_2 \end{bmatrix} = \begin{bmatrix} \cos\theta & -\sin\theta \\ \sin\theta & \cos\theta \end{bmatrix} \begin{bmatrix} x_1 \\ x_2 \end{bmatrix}, \tag{48}
$$

then $\mathbf{y}$ is $\mathbf{x}$ rotated by the angle θ (see Figure 5-11).

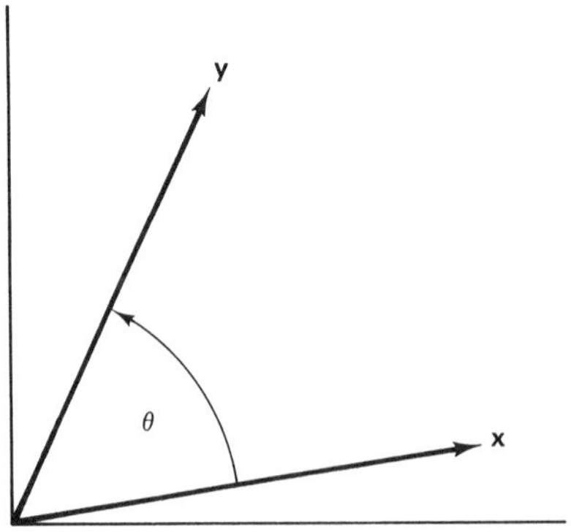

Figure 5-11 The Givens rotation (48).

Consequently, multiplying a vector by the Givens rotation (47) is equivalent to rotating two components of the vector through the angle $\theta = \arctan s/c$ while leaving the other components intact.

Given a vector **x** with two components and defining

$$c = \frac{x_1}{\sqrt{x_1^2 + x_2^2}} \quad \text{and} \quad s = -\frac{x_2}{\sqrt{x_1^2 + x_2^2}}, \tag{49}$$

observe that $c^2 + s^2 = 1$ and

$$\begin{bmatrix} c & -s \\ s & c \end{bmatrix} \begin{bmatrix} x_1 \\ x_2 \end{bmatrix} = \begin{bmatrix} \sqrt{x_1^2 + x_2^2} \\ 0 \end{bmatrix}. \tag{50}$$

Utilizing relation (50), we can construct a Givens rotation that annihilates a particular matrix element. To **QR** factor **A**, we multiply **A** by a sequence of Givens rotations, annihilating elements beneath the diagonal to obtain the upper triangular matrix **R**. Finally, **Q** is the product of rotations.

We illustrate this algorithm with the matrix

$$\mathbf{A} = \begin{bmatrix} 90 & -153 & 114 \\ 120 & -79 & -223 \\ 200 & -40 & 395 \end{bmatrix}.$$

The factorization starts with column 1 of **A**:

$$\begin{bmatrix} 90 \\ 120 \\ 200 \end{bmatrix}.$$

Referring to (49) and identifying x_1 with 90 and x_2 with 120, the Givens rotation $\mathbf{G}_{21}$ that annihilates the second component of column 1 is

$$\mathbf{G}_{21} = \begin{bmatrix} \frac{90}{150} & \frac{120}{150} & 0 \\ -\frac{120}{150} & \frac{90}{150} & 0 \\ 0 & 0 & 1 \end{bmatrix} = \begin{bmatrix} \frac{3}{5} & \frac{4}{5} & 0 \\ -\frac{4}{5} & \frac{3}{5} & 0 \\ 0 & 0 & 1 \end{bmatrix}.$$

Premultiplying **A** by $\mathbf{G}_{21}$ yields a matrix that we denote $\mathbf{A}_{21}$:

$$\mathbf{A}_{21} = \mathbf{G}_{21}\mathbf{A} = \begin{bmatrix} .6 & .8 & 0 \\ -.8 & .6 & 0 \\ 0 & 0 & 1 \end{bmatrix} \begin{bmatrix} 90 & -153 & 114 \\ 120 & -79 & -223 \\ 200 & -40 & 395 \end{bmatrix} = \begin{bmatrix} 150 & -155 & -110 \\ 0 & 75 & -225 \\ 200 & -40 & 395 \end{bmatrix}. \tag{51}$$

Since each coefficient on the right side of (51) is a multiple of 5, we extract the factor 5 to obtain

$$\mathbf{A}_{21} = 5 \begin{bmatrix} 30 & -31 & -22 \\ 0 & 15 & -45 \\ 40 & -8 & 79 \end{bmatrix}.$$

Referring to (49) and identifying x_1 with 30 and x_2 with 40, the Givens rotation $\mathbf{G}_{31}$ that annihilates the third component in column 1 of $\mathbf{A}_{21}$ is

$$\mathbf{G}_{31} = \begin{bmatrix} .6 & 0 & .8 \\ 0 & 1 & 0 \\ -.8 & 0 & .6 \end{bmatrix}.$$

Multiplying $\mathbf{A}_{21}$ by $\mathbf{G}_{31}$ gives

$$\mathbf{A}_{31} = \mathbf{G}_{31}\mathbf{A}_{21} = 5\begin{bmatrix} .6 & 0 & .8 \\ 0 & 1 & 0 \\ -.8 & 0 & .6 \end{bmatrix}\begin{bmatrix} 30 & -31 & -22 \\ 0 & 15 & -45 \\ 40 & -8 & 79 \end{bmatrix} = 25\begin{bmatrix} 10 & -5 & 10 \\ 0 & 3 & -9 \\ 0 & 4 & 13 \end{bmatrix}.$$

At this point, the subdiagonal elements in the first column of $\mathbf{A}_{31}$ are zero. Proceeding to the second column of $\mathbf{A}_{31}$, the Givens rotation $\mathbf{G}_{32}$ that annihilates the third element in the second column is

$$\mathbf{G}_{32} = \begin{bmatrix} 1 & 0 & 0 \\ 0 & .6 & .8 \\ 0 & -.8 & .6 \end{bmatrix}.$$

Premultiplying $\mathbf{A}_{31}$ by $\mathbf{G}_{32}$ yields $\mathbf{R}$:

$$\mathbf{R} = \mathbf{G}_{32}\mathbf{A}_{31} = 25\begin{bmatrix} 1 & 0 & 0 \\ 0 & .6 & .8 \\ 0 & -.8 & .6 \end{bmatrix}\begin{bmatrix} 10 & -5 & 10 \\ 0 & 3 & -9 \\ 0 & 4 & 13 \end{bmatrix} = 125\begin{bmatrix} 2 & -1 & 2 \\ 0 & 1 & 1 \\ 0 & 0 & 3 \end{bmatrix}.$$

To summarize,

$$\mathbf{R} = \mathbf{G}_{32}\mathbf{A}_{31} = \mathbf{G}_{32}\mathbf{G}_{31}\mathbf{A}_{21} = \mathbf{G}_{32}\mathbf{G}_{31}\mathbf{G}_{21}\mathbf{A}. \tag{52}$$

Since each Givens rotation $\mathbf{G}_{ij}$ is an orthogonal matrix, $\mathbf{G}_{ij}{}^T\mathbf{G}_{ij} = \mathbf{I}$. Multiplying the identity (52) by $\mathbf{G}_{21}{}^T\mathbf{G}_{31}{}^T\mathbf{G}_{32}{}^T$ gives us the relation

$$\mathbf{G}_{21}{}^T\mathbf{G}_{31}{}^T\mathbf{G}_{32}{}^T\mathbf{R} = \mathbf{G}_{21}{}^T\mathbf{G}_{31}{}^T\mathbf{G}_{32}{}^T\mathbf{G}_{32}\mathbf{G}_{31}\mathbf{G}_{21}\mathbf{A} = \mathbf{A}.$$

It follows that $\mathbf{A} = \mathbf{QR}$, where $\mathbf{Q} = \mathbf{G}_{21}{}^T\mathbf{G}_{31}{}^T\mathbf{G}_{32}{}^T$. In the example above, the $\mathbf{QR}$ factorization can be expressed

$$\mathbf{A} = \begin{bmatrix} 90 & -153 & 114 \\ 120 & -79 & -223 \\ 200 & -40 & 395 \end{bmatrix} = \left(\tfrac{1}{125}\begin{bmatrix} 45 & -108 & 44 \\ 60 & -19 & -108 \\ 100 & 60 & 45 \end{bmatrix}\right)\left(125\begin{bmatrix} 2 & -1 & 2 \\ 0 & 1 & 1 \\ 0 & 0 & 3 \end{bmatrix}\right),$$

where the first factor on the right side is $\mathbf{Q} = \mathbf{G}_{21}{}^T\mathbf{G}_{31}{}^T\mathbf{G}_{32}{}^T$, and the second factor is $\mathbf{R}$.

For a general $n \times n$ matrix, we annihilate elements 2 through n in the first column using the sequence of Givens rotations $\mathbf{G}_{21}, \mathbf{G}_{31}, \cdots, \mathbf{G}_{n1}$. Then elements 3 through n in the second column are annihilated using the sequence of Givens rotations $\mathbf{G}_{32}, \mathbf{G}_{42}, \cdots, \mathbf{G}_{n2}$. We continue in this way. The last step annihilates element n of column $n - 1$ using $\mathbf{G}_{n,n-1}$. Finally, $\mathbf{Q}$ is the product of the transpose of each rotation.

Factoring a matrix using Givens rotations is a little slower than factoring a matrix using Householder reflections since the Givens scheme requires about twice as many multiplications. We now implement the Givens scheme in a way that reduces the number of multiplications by the factor $\frac{1}{2}$. First observe that premultiplying $\mathbf{A}$ by the rotation (47) just changes the coefficients in row i and row j. In particular, if $\mathbf{A}$ has n columns, then the updated coefficients in row i and row j are given by

$$\begin{aligned} a_{jk}^{\text{new}} &\leftarrow c\,a_{ik} - s\,a_{jk} \\ a_{ik}^{\text{new}} &\leftarrow s\,a_{ik} + c\,a_{jk} \end{aligned} \tag{53}$$

for $k = 1$ to n. For each k, (53) involves four multiplications and since k ranges from 1 to n, the computation of $\mathbf{G}_{ij}\mathbf{A}$ requires $4n$ multiplications altogether. The basic observation that lets us reduce the number of multiplications is the following: If either c or s is replaced by "1," the number of multiplications in the update (53) is changed from 4 to 2. In the standard Givens scheme, we multiply **A** by a sequence of Givens rotations that reduce **A** to upper triangular form. Loosely speaking, in the fast Givens scheme, we factor a scalar from the ith row and the jth row of $\mathbf{G}_{ij}$ to obtain a simpler matrix with two more 1's. Instead of multiplying by $\mathbf{G}_{ij}$, we multiply by the simpler matrix while accumulating the row scale factors in a separate array. After achieving upper triangular form, each row is multiplied by its scale factor to obtain the final **R**. More precisely, the fast Givens scheme stores the partly processed matrix as a product **DP** between a diagonal matrix **D** (containing the row scale factors) and another matrix **P**. In each step, we update **D** and **P**. After the last update, **P** is upper triangular and **R** is the product **DP**.

To develop one step in the fast Givens scheme, let us suppose that the factorization has progressed to the point where the subdiagonal element in row i and column j of **DP** is to be annihilated. Since premultiplying **P** by a diagonal matrix multiplies each row of **P** by the corresponding diagonal element, the element in row i and column j of **DP** is $d_i p_{ij}$, where d_i is the ith diagonal element of **D**. Referring to (49), the c and the s in the Givens rotation $\mathbf{G}_{ij}$ that annihilates the (i, j) element of **DP** are

$$c = \frac{d_j p_{jj}}{\sqrt{(d_j p_{jj})^2 + (d_i p_{ij})^2}} \quad \text{and} \quad s = \frac{-d_i p_{ij}}{\sqrt{(d_j p_{jj})^2 + (d_i p_{ij})^2}} \tag{54}$$

We now compute the product $\mathbf{G}_{ij}\mathbf{DP}$ in an unusual way. Instead of computing $\mathbf{G}_{ij}(\mathbf{DP})$, we look for a diagonal matrix $\mathbf{D}^{\text{new}}$ and another matrix $\mathbf{P}^{\text{new}}$ with the property that

$$\mathbf{D}^{\text{new}}\,\mathbf{P}^{\text{new}} = \mathbf{G}_{ij}\mathbf{DP}. \tag{55}$$

Working toward this objective, observe that

$$\mathbf{G}_{ij}\mathbf{DP} = \mathbf{D}\mathbf{D}^{-1}\mathbf{G}_{ij}\mathbf{DP} = \mathbf{D}(\mathbf{D}^{-1}\mathbf{G}_{ij}\mathbf{D})\mathbf{P}. \tag{56}$$

You can verify that each element in $\mathbf{D}^{-1}\mathbf{G}_{ij}\mathbf{D}$ is equal to the corresponding element in $\mathbf{G}_{ij}$ except for the element in row i and column j and the element in row j and column i. In particular, the 2×2 submatrix of $\mathbf{D}^{-1}\mathbf{G}_{ij}\mathbf{D}$ formed by the intersection of row i and row j with column i and column j is

$$\begin{bmatrix} c & s_1 \\ s_2 & c \end{bmatrix} \tag{57}$$

where $s_1 = -sd_i/d_j$ and $s_2 = sd_j/d_i$. If $|c| \geqq 1/\sqrt{2}$, we factor c from each row of this matrix to obtain

$$\begin{bmatrix} 1 & \dfrac{s_1}{c} \\ \dfrac{s_2}{c} & 1 \end{bmatrix}, \tag{58}$$

and if $|s| > 1/\sqrt{2}$, we factor s_1 from the first row and s_2 from the second row in (57) to get

$$\begin{bmatrix} \frac{c}{s_1} & 1 \\ 1 & \frac{c}{s_2} \end{bmatrix}. \tag{59}$$

In either case, two of the four coefficients are 1, so a matrix product involving either (58) or (59) requires half as many multiplications as a matrix product involving (57).

Factoring a scalar from a row of a matrix is equivalent to multiplying by a diagonal matrix. Let **E** denote the diagonal matrix with diagonal elements e_1, $e_2, \cdots, e_n$, where every diagonal element is equal to 1 except for element i and element j, which are defined as follows: If $|c| \geqq 1/\sqrt{2}$, then $e_i = c = e_j$ and if $|s| > 1/\sqrt{2}$, then $e_j = s_1$ and $e_i = s_2$. Since $\mathbf{EE}^{-1} = \mathbf{I}$, (55) and (56) can be written

$$\mathbf{D}^{\text{new}}\mathbf{P}^{\text{new}} = \mathbf{DEE}^{-1}(\mathbf{D}^{-1}\mathbf{G}_{ij}\mathbf{D})\mathbf{P} = (\mathbf{DE})(\mathbf{E}^{-1}\mathbf{D}^{-1}\mathbf{G}_{ij}\mathbf{D})\mathbf{P}.$$

Since the product between two diagonal matrices is a diagonal matrix, the product **DE** becomes the new **D**. For the matrix $\mathbf{E}^{-1}\mathbf{D}^{-1}\mathbf{G}_{ij}\mathbf{D}$, the 2×2 submatrix formed by the intersection of row i and row j with column i and column j is either (58) or (59) and the product $(\mathbf{E}^{-1}\mathbf{D}^{-1}\mathbf{G}_{ij}\mathbf{D})\mathbf{P}$ becomes the new **P**:

$$\mathbf{D}^{\text{new}} = \mathbf{DE} \quad \text{and} \quad \mathbf{P}^{\text{new}} = \mathbf{E}^{-1}\mathbf{D}^{-1}\mathbf{G}_{ij}\mathbf{DP}.$$

Referring to (58), (59), and the definition of **E**, the elements in row i and row j of the new **D** and **P** matrices are computed by the following rules:

for $|c| \geqq 1/\sqrt{2}$

$r_1 \leftarrow (sd_i)/(cd_j)$, $r_2 \leftarrow (sd_j)/(cd_i)$
$d_j \leftarrow cd_j$, $d_i \leftarrow cd_i$
$k = j$ to n
  $t \leftarrow p_{jk} - r_1 p_{ik}$
  $p_{ik} \leftarrow p_{ik} + r_2 p_{jk}$
  $p_{jk} \leftarrow t$
next k

for $|s| > 1/\sqrt{2}$

$r_1 \leftarrow (cd_j)/(sd_i)$, $r_2 \leftarrow (cd_i)/(sd_j)$
$d_i \leftrightarrow d_j$, $d_j \leftarrow -sd_j$, $d_i \leftarrow sd_i$
$k = j$ to n
  $t \leftarrow p_{ik} - r_1 p_{jk}$
  $p_{ik} \leftarrow p_{jk} + r_2 p_{ik}$
  $p_{jk} \leftarrow t$
next k

(60)

The update formula (60) coupled with the definition (54) of c and s fully describe the rotation that annihilates the subdiagonal element in row i and column j. To summarize, the partly processed matrix is stored as the product **DP**. Instead of evaluating the product $\mathbf{G}_{ij}\mathbf{DP}$ directly, we compute a diagonal matrix $\mathbf{D}^{\text{new}}$ and a related matrix $\mathbf{P}^{\text{new}}$ with the property that $\mathbf{D}^{\text{new}}\mathbf{P}^{\text{new}} = \mathbf{G}_{ij}\mathbf{DP}$. Of course, the starting **D** and **P** are given by $\mathbf{D} = \mathbf{I}$ and $\mathbf{P} = \mathbf{A}$. The following algorithm generates the **QR** factorization of a $m \times n$ matrix **A** using the fast Givens scheme. Substituting for c and s in the update formula (60) using definition (54), it can be shown that parameters r_1 and r_2 depend only on d_i^2 and d_j^2. For this reason, our algorithm stores and updates the square of the diagonal elements. Also, note that in going from (60) to (61), we changed the signs of r_1 and r_2 to eliminate some minus signs.

$$
\begin{array}{l}
d_i \leftarrow 1 \text{ for } i = 1 \text{ to } m \\
j = 1 \text{ to } \min\{n, m-1\} \\
\quad i = j+1 \text{ to } m \\
\quad\quad \text{if } a_{ij} = 0 \text{ then go to next } i \\
\quad\quad C \leftarrow d_j a_{jj}^2,\ S \leftarrow d_i a_{ij}^2 \\
\quad\quad \text{if } S \leqq C \text{ then} \\
\quad\quad\quad r_2 \leftarrow a_{ij}/a_{jj},\ r_1 \leftarrow d_i r_2/d_j,\ C \leftarrow C/(C+S) \\
\quad\quad\quad d_j \leftarrow C d_j,\ d_i \leftarrow C d_i \\
\quad\quad\quad t \leftarrow a_{jk} + r_1 a_{ik},\ a_{ik} \leftarrow a_{ik} - r_2 a_{jk}, \text{ and } a_{jk} \leftarrow t \text{ for } k = j \text{ to } n \\
\quad\quad \text{if } S > C \text{ then} \\
\quad\quad\quad r_2 \leftarrow a_{jj}/a_{ij},\ r_1 \leftarrow d_j r_2/d_i,\ S \leftarrow S/(C+S) \\
\quad\quad\quad d_i \leftrightarrow d_j,\ d_j \leftarrow S d_j,\ d_i \leftarrow S d_i \\
\quad\quad\quad t \leftarrow r_1 a_{jk} + a_{ik},\ a_{ik} \leftarrow a_{jk} - r_2 a_{ik}, \text{ and } a_{jk} \leftarrow t \text{ for } k = j \text{ to } n \\
\quad \text{next } i \\
\text{next } j \\
a_{ij} \leftarrow d_i^{1/2} a_{ij} \text{ for } j = i \text{ to } n, \text{ for } i = 1 \text{ to } m
\end{array}
\tag{61}
$$

Each step in the fast Givens method (61) multiplies two components of **d** by a factor with magnitude c^2 or s^2, whichever is larger. Since

$$\tfrac{1}{2} \leqq \text{maximum}\,\{c^2, s^2\} \leqq 1,$$

the components of **d** typically approach zero while the elements of **A** approach $\pm\infty$ as the algorithm progresses. To avoid overflow or underflow, both **A** and **d** must be normalized occasionally. In particular, the following normalization step is performed near the start of the i loop in (61) whenever there is a danger of overflow in row k:

$$a_{kl} \leftarrow d_k^{1/2} a_{kl} \text{ for } l = j \text{ to } n \text{ and } d_k \leftarrow 1.$$

Exercises

5-6.1. Construct the Givens rotation **G** which annihilates the second component of the vector

$$\mathbf{x} = \begin{bmatrix} 5 \\ 12 \end{bmatrix}.$$

For this particular **G** and **x**, use a sketch to show how $\mathbf{y} = \mathbf{Gx}$ is generated from **x**.

5-6.2. Using a Givens rotation, compute the **QR** factorization of the matrix

$$\begin{bmatrix} 5 & -13 \\ 12 & 26 \end{bmatrix}.$$

5-6.3. Consider a vector **x** with two components. What are the c and the s corresponding to the Givens rotation that annihilates the first component of **x**? Explain how to annihilate the first k components of an arbitrary vector using a sequence of Givens rotations.

5-6.4. Many different Givens rotations can be used to annihilate one component of a vector. Given a vector with three components, describe two different rotations that annihilate the third component of the vector. Given a vector with n components, describe $n - 1$ different rotations that annihilate the last component of the vector.

5-6.5. As your observations in Exercise 5-6.4 suggest, many different sequences of Givens rotations can be used to reduce a matrix to upper triangular form. Describe another sequence of rotations that is different from the sequence employed in the text.

5-6.6. If $\mathbf{A} = \mathbf{QR}$, show that $\mathbf{A}^T\mathbf{A} = \mathbf{R}^T\mathbf{R}$.

5-6.7. If $\mathbf{Q}_1$ and $\mathbf{Q}_2$ are orthogonal matrices, then show that $(\mathbf{Q}_1\mathbf{Q}_2)^{-1} = \mathbf{Q}_2{}^T\mathbf{Q}_1{}^T$.

5-7. LEAST SQUARES AND THE QR FACTORIZATION

The **QR** factorization can be used to solve a system $\mathbf{Ax} = \mathbf{b}$ of n equations in n unknowns. The associative law for matrix multiplication implies that the factored system $(\mathbf{QR})\mathbf{x} = \mathbf{b}$ can be written $\mathbf{Q}(\mathbf{Rx}) = \mathbf{b}$. Defining $\mathbf{y} = \mathbf{Rx}$, it follows that $\mathbf{Qy} = \mathbf{b}$. Hence, $\mathbf{Ax} = \mathbf{b}$ is solved in two steps:

1. Solve $\mathbf{Qy} = \mathbf{b}$ for the unknown $\mathbf{y}$.
2. Solve $\mathbf{Rx} = \mathbf{y}$ for the unknown $\mathbf{x}$.

The second system is solved by back substitution. The first system is easy to solve since $\mathbf{Q}$ is orthogonal. Multiplying $\mathbf{Qy} = \mathbf{b}$ by $\mathbf{Q}^T$ yields $\mathbf{Q}^T\mathbf{Qy} = \mathbf{Q}^T\mathbf{b}$. Since $\mathbf{Q}$ is orthogonal, $\mathbf{Q}^T\mathbf{Q} = \mathbf{I}$ and $\mathbf{y} = \mathbf{Q}^T\mathbf{b}$. Although the **QR** factorization is more stable than the **LU** factorization, **QR** factoring a matrix takes twice as much time as **LU** factoring a matrix. Consequently, solving a linear system using a **QR** factorization is slower than solving a linear system using a **LU** factorization.

The **QR** factorization can also be used to solve the least squares problem. Let us consider an overdetermined system $\mathbf{Ax} = \mathbf{b}$, where $\mathbf{A}$ is $m \times n$ and m is greater than n. We saw in equation (44) that the upper triangular factor in the **QR** factorization of a 3×2 matrix has a row of zeros. In general, when $m > n$, the last $m - n$ rows of the upper triangular factor are completely zero and the **QR** factorization of $\mathbf{A}$ has the form

$$\mathbf{A} = \mathbf{Q}\begin{bmatrix}\mathbf{R}\\ \mathbf{0}\end{bmatrix}, \tag{62}$$

where $\mathbf{Q}$ is a $m \times m$ orthogonal matrix, $\mathbf{R}$ is a $n \times n$ upper triangular matrix, and $\mathbf{0}$ is a $(m - n) \times n$ matrix of zeros. Replacing $\mathbf{A}$ by its factorization (62), the coefficient matrix $\mathbf{A}^T\mathbf{A}$ for the normal equation $\mathbf{A}^T\mathbf{Ax} = \mathbf{A}^T\mathbf{b}$ can be expressed:

$$\mathbf{A}^T\mathbf{A} = \left(\mathbf{Q}\begin{bmatrix}\mathbf{R}\\ \mathbf{0}\end{bmatrix}\right)^T\mathbf{Q}\begin{bmatrix}\mathbf{R}\\ \mathbf{0}\end{bmatrix} = [\mathbf{R}^T \quad \mathbf{0}^T]\,\mathbf{Q}^T\mathbf{Q}\begin{bmatrix}\mathbf{R}\\ \mathbf{0}\end{bmatrix}. \tag{63}$$

Since $\mathbf{Q}$ is orthogonal, $\mathbf{Q}^T\mathbf{Q} = \mathbf{I}$ and (63) implies that

$$\mathbf{A}^T\mathbf{A} = [\mathbf{R}^T \quad \mathbf{0}^T]\mathbf{Q}^T\mathbf{Q}\begin{bmatrix}\mathbf{R}\\ \mathbf{0}\end{bmatrix} = [\mathbf{R}^T \quad \mathbf{0}^T]\begin{bmatrix}\mathbf{R}\\ \mathbf{0}\end{bmatrix} = \mathbf{R}^T\mathbf{R}. \tag{64}$$

Note that $\mathbf{R}^T\mathbf{R}$ is the Cholesky factorization of $\mathbf{A}^T\mathbf{A}$, so, in essence, the **QR** factorization is one way to compute the Cholesky decomposition of $\mathbf{A}^T\mathbf{A}$. After replacing $\mathbf{A}$ by its factorization (62), the right side $\mathbf{A}^T\mathbf{b}$ of the normal equation is

$$\mathbf{A}^T\mathbf{b} = \left(\mathbf{Q}\begin{bmatrix}\mathbf{R}\\ \mathbf{0}\end{bmatrix}\right)^T\mathbf{b} = [\mathbf{R}^T \quad \mathbf{0}^T]\,(\mathbf{Q}^T\mathbf{b}). \tag{65}$$

When evaluating the right side of (65), the last $m - n$ components of $\mathbf{Q}^T\mathbf{b}$ are multiplied by zero. Hence (65) simplifies to

$$\mathbf{A}^T\mathbf{b} = \mathbf{R}^T(\mathbf{Q}^T\mathbf{b})_{\mathrm{R}}, \tag{66}$$

where $(\mathbf{Q}^T\mathbf{b})_{\mathrm{R}}$ denotes the first n components of the vector $\mathbf{Q}^T\mathbf{b}$. Combining (64) and (66), the normal equation $\mathbf{A}^T\mathbf{A}\mathbf{x} = \mathbf{A}^T\mathbf{b}$ reduces to

$$\mathbf{R}^T\mathbf{R}\mathbf{x} = \mathbf{R}^T(\mathbf{Q}^T\mathbf{b})_{\mathrm{R}}. \tag{67}$$

Finally, multiplying (67) by $(\mathbf{R}^T)^{-1}$, we have

$$\mathbf{R}\mathbf{x} = (\mathbf{Q}^T\mathbf{b})_{\mathrm{R}}. \tag{68}$$

The solution to this upper triangular system is the least squares solution to the overdetermined linear system $\mathbf{A}\mathbf{x} = \mathbf{b}$. If the coefficient matrix $\mathbf{A}$ is **QR** factored using algorithm (45) and if each diagonal element of $\mathbf{R}$ is nonzero, the following algorithm computes $(\mathbf{Q}^T\mathbf{b})_{\mathrm{R}}$ and solves the upper triangular system (68):

```
x_i ← b_i for i = 1 to m
j = 1 to n
    t ← 0
    t ← t + x_i a_ij for i = j to m
    x_i ← x_i − t a_ij for i = j to m
next j
j = n down to 1
    x_j ← x_j/d_j
    x_i ← x_i − a_ij x_j for i = 1 to j−1
next j
```
(69)

Subroutine OVER implements algorithm (69). For illustration, the following program generates the least squares solution to the overdetermined linear system (19):

```
REAL A(20),B(3),X(3)
DATA A/3.,4.,0.,0.,5.,4./
DATA B/6.,12.,4./
CALL QR(A,3,3,2)
CALL OVER(X,A,B)
WRITE(6,*) X(1),X(2)
END
```

Note that the X argument of OVER can be identified with the B argument, although the original contents of B will be destroyed. When the X and the B arguments of OVER are distinct, X must have at least m elements, where m is the number of rows in the coefficient matrix.

Now consider an underdetermined system $\mathbf{A}\mathbf{x} = \mathbf{b}$ of m equations in n unknowns. For an underdetermined system, we **QR** factor $\mathbf{A}^T$ instead of $\mathbf{A}$. Since $n > m$ and $\mathbf{A}^T$ is $n \times m$, the **QR** factorization of $\mathbf{A}^T$ has the form

$$\mathbf{A}^T = \mathbf{Q}\begin{bmatrix}\mathbf{R}\\ \mathbf{0}\end{bmatrix}, \tag{70}$$

where $\mathbf{Q}$ is a $n \times n$ orthogonal matrix, $\mathbf{R}$ is a $m \times m$ upper triangular matrix, and $\mathbf{0}$ is a $(n - m) \times m$ matrix of zeros. The transpose of (70) is

$$\mathbf{A} = [\mathbf{R}^T \quad \mathbf{0}^T]\mathbf{Q}^T.$$

Making this substitution for $\mathbf{A}$, the equation $\mathbf{Ax} = \mathbf{b}$ becomes

$$[\mathbf{R}^T \quad \mathbf{0}^T](\mathbf{Q}^T\mathbf{x}) = \mathbf{b}. \tag{71}$$

Let us change to the variable $\mathbf{y} = \mathbf{Q}^T\mathbf{x}$. In terms of $\mathbf{y}$, equation (71) is

$$[\mathbf{R}^T \quad \mathbf{0}^T]\mathbf{y} = \mathbf{b}. \tag{72}$$

Since $\mathbf{y}^T\mathbf{y} = (\mathbf{Q}^T\mathbf{x})^T(\mathbf{Q}^T\mathbf{x}) = \mathbf{x}^T\mathbf{Q}\mathbf{Q}^T\mathbf{x} = \mathbf{x}^T\mathbf{x}$, we see that $\mathbf{x}$ and $\mathbf{y}$ have the same Euclidean length. Hence, if $\mathbf{y}$ is the shortest vector that satisfies (72), the corresponding $\mathbf{x}$ is the shortest vector that satisfies (71). To find the corresponding $\mathbf{x}$, we multiply the relation $\mathbf{y} = \mathbf{Q}^T\mathbf{x}$ by $\mathbf{Q}$ to obtain $\mathbf{Qy} = \mathbf{QQ}^T\mathbf{x} = \mathbf{x}$. Therefore, $\mathbf{x} = \mathbf{Qy}$. To solve (72), let us partition the components of $\mathbf{y}$ into two parts:

$$\mathbf{y} = \begin{bmatrix} \mathbf{y}_R \\ \mathbf{y}_Z \end{bmatrix},$$

where $\mathbf{y}_R$ denotes the components of $\mathbf{y}$ corresponding to the columns of $\mathbf{R}^T$ while $\mathbf{y}_Z$ denotes the components of $\mathbf{y}$ corresponding to the columns of $\mathbf{0}^T$ in (72). With this notation, (72) reduces to

$$\mathbf{R}^T\mathbf{y}_R = \mathbf{b}. \tag{73}$$

Clearly, the $\mathbf{y}_Z$ components of the shortest vector $\mathbf{y}$ that satisfies (73) are zero. Moreover, if the diagonal elements of $\mathbf{R}$ are nonzero, (73) determines $\mathbf{y}_R$ uniquely.

In summary, the vector $\mathbf{x}$ with smallest Euclidean length that satisfies the underdetermined system $\mathbf{Ax} = \mathbf{b}$ is given by

$$\mathbf{x} = \mathbf{Q}\begin{bmatrix} \mathbf{y}_R \\ \mathbf{0} \end{bmatrix}, \tag{74}$$

where $\mathbf{y}_R$ is the solution to (73). Suppose that $\mathbf{A}$ is replaced by its transpose and the resulting matrix is $\mathbf{QR}$ factored using algorithm (45). Assuming that the diagonal elements of $\mathbf{R}$ are nonzero, the following algorithm solves (73) and forms the product (74) to obtain the least squares solution to the underdetermined linear system $\mathbf{Ax} = \mathbf{b}$ of m equations in n unknowns.

```
x_i ← b_i for i = 1 to m
j = 1 to m
    x_j ← x_j − a_ij x_i for i = 1 to j−1
    x_j ← x_j/d_j
next j
x_j ← 0 for j = m+1 to n                          (75)
j = m down to 1
    t ← 0
    t ← t + a_ij x_i for i = j to n
    x_i ← x_i − t a_ij for i = j to n
next j
```

Subroutine UNDER implements algorithm (75). The following program solves the underdetermined system (27):

```
REAL A(20),B(2),X(3)
DATA A/3.,4.,0.,0.,5.,4./
DATA B/70.,81./
CALL QR(A,3,3,2)
CALL UNDER(X,A,B)
WRITE(6,*) X(1),X(2),X(3)
END
```

Exercises

5-7.1. Consider the following matrix

$$\mathbf{A} = \begin{bmatrix} .6 & .8 & 0 \\ .8 & -.6 & 0 \\ 0 & 0 & 1 \end{bmatrix} \begin{bmatrix} 1 & 1 \\ 0 & 2 \\ 0 & 0 \end{bmatrix}.$$

(a) Determine the least squares solution to the overdetermined linear system $\mathbf{Ax} = \mathbf{b}$, where

$$\mathbf{b} = \begin{bmatrix} 10 \\ 20 \\ 10 \end{bmatrix}.$$

(b) Determine the least squares (minimum norm) solution to the underdetermined linear system $\mathbf{A}^T\mathbf{x} = \mathbf{b}$, where

$$\mathbf{b} = \begin{bmatrix} 20 \\ 40 \end{bmatrix}.$$

5-7.2. Use subroutines QR, OVER, and UNDER to solve Exercises 5-2.3 and 5-3.1.

5-8. ORTHONORMAL BASES

In this section, we study methods to compute an orthonormal basis for the space spanned by a collection of vectors. First, let us introduce some terminology. The **space spanned** by a collection of vectors $\mathbf{a}_1, \mathbf{a}_2, \cdots, \mathbf{a}_n$ is the space $\mathbf{S}$ consisting of all linear combinations of the vectors. That is, $\mathbf{S}$ is the collection of vectors of the form

$$x_1\mathbf{a}_1 + x_2\mathbf{a}_2 + \cdots + x_n\mathbf{a}_n,$$

where $x_1, \cdots, x_n$ are arbitrary scalars. A collection of vectors $\mathbf{v}_1, \mathbf{v}_2, \cdots, \mathbf{v}_k$ is **linearly independent** if no vector in the collection can be expressed as a linear combination of the other vectors. The vectors $\mathbf{v}_1, \cdots, \mathbf{v}_k$ are a **basis** for $\mathbf{S}$ if they are linearly independent and the space spanned by $\mathbf{v}_1, \cdots, \mathbf{v}_k$ is equal to $\mathbf{S}$. An **orthonormal basis** $\mathbf{q}_1, \cdots, \mathbf{q}_k$ for $\mathbf{S}$ is a basis which satisfies the orthonormality conditions

$$\mathbf{q}_i^T\mathbf{q}_j = 0 \quad \text{for every } i \neq j \qquad \text{and} \qquad \mathbf{q}_i^T\mathbf{q}_i = 1 \quad \text{for every } i. \tag{76}$$

In Exercise 5-8.1 you will show that an orthonormal collection of vectors is always linearly independent. Hence, to show that a collection of vectors $\mathbf{q}_1, \cdots, \mathbf{q}_k$ satisfying (76) is a basis for $\mathbf{S}$, we just need to verify that their span is $\mathbf{S}$.

When the vectors $\mathbf{a}_1, \cdots, \mathbf{a}_n$ are linearly independent, the **QR** factorization can be used to obtain an orthonormal basis for $\mathbf{S}$. In particular, let $\mathbf{A}$ denote the matrix with columns $\mathbf{a}_1, \cdots, \mathbf{a}_n$. Since the columns of $\mathbf{A}$ are linearly independent, $\mathbf{A}$ has at least as many rows as columns and (as in Section 5-7) the $\mathbf{QR}$ factorization of $\mathbf{A}$ has the form

$$\mathbf{A} = \mathbf{Q}\begin{bmatrix} \mathbf{R} \\ \mathbf{0} \end{bmatrix} = \mathbf{Q}_{\mathrm{R}}\mathbf{R},$$

where $\mathbf{R}$ is a $n \times n$ upper triangular matrix and $\mathbf{Q}_{\mathrm{R}}$ is the submatrix of $\mathbf{Q}$ formed from its first n columns. We now show that the columns of $\mathbf{Q}_{\mathrm{R}}$, denoted $\mathbf{q}_1, \cdots, \mathbf{q}_n$, are an orthonormal basis for $\mathbf{S}$.

Since $\mathbf{Q}$ is an orthogonal matrix, $\mathbf{Q}^T\mathbf{Q} = \mathbf{I}$, or equivalently, column i of $\mathbf{Q}$ dot column j of $\mathbf{Q}$ is zero for $i \neq j$, while column i dot itself is 1. Thus the orthonormality condition (76) is satisfied. The equality $\mathbf{A} = \mathbf{Q}_{\mathrm{R}}\mathbf{R}$ implies that column j of $\mathbf{A}$ is equal to $\mathbf{Q}_{\mathrm{R}}$ times column j of $\mathbf{R}$: $\mathbf{a}_j = \mathbf{Q}_{\mathrm{R}}\mathbf{r}_j$ for each j. Since $\mathbf{a}_j$ can be expressed as a linear combination of the columns of $\mathbf{Q}_{\mathrm{R}}$, $\mathbf{S}$ is contained in the space spanned by the $\mathbf{q}_i$. On the other hand, $\mathbf{R}$ is invertible since the columns of $\mathbf{A}$ are linearly independent (see Exercise 5-8.8) and $\mathbf{Q}_{\mathrm{R}} = \mathbf{A}\mathbf{R}^{-1}$. As demonstrated above, this equality implies that the columns of $\mathbf{Q}_{\mathrm{R}}$ are linear combinations of the columns of $\mathbf{A}$ and the space spanned by the $\mathbf{q}_i$ is contained in $\mathbf{S}$. Since the space spanned by the $\mathbf{q}_i$ is both contained in and contains $\mathbf{S}$, the space spanned by $\mathbf{q}_1, \cdots, \mathbf{q}_n$ is precisely $\mathbf{S}$. In summary, when the columns of $\mathbf{A}$ are linearly independent, the first n columns of $\mathbf{Q}$ are an orthonormal basis for the columns of $\mathbf{A}$.

If some of $\mathbf{A}$'s columns are linear combinations of other columns, the standard $\mathbf{QR}$ factorization scheme must be slightly modified to obtain an orthonormal basis. The fundamental problem when the columns are dependent is that $\mathbf{R}$ is not invertible since some of its diagonal elements vanish. Thus the identity $\mathbf{Q}_{\mathrm{R}} = \mathbf{A}\mathbf{R}^{-1}$ is no longer valid, and the columns of $\mathbf{Q}_{\mathrm{R}}$ are not all contained in $\mathbf{S}$. An invertible $\mathbf{R}$ (with nonzero diagonal elements) can be generated if a pivot is performed before each step in the factorization process. Before the first step, where elements 2 through m in the first column are annihilated, we interchange the first column with the succeeding column which has the largest Euclidean length. Similarly, before step k in the factorization process, where elements $k + 1$ through m in column k are annihilated, we interchange the kth column of the partly processed matrix with a succeeding column so that the first column of the $(m - k + 1) \times (n - k + 1)$ submatrix in the lower right corner has the largest Euclidean length among all columns in the submatrix.

Letting $\mathbf{H}_k$ denote the Householder matrix corresponding to step k and letting $\mathbf{P}_k$ denote the matrix representation of the column interchange that precedes step k, the modified factorization scheme can be expressed $\mathbf{A}_1 = \mathbf{H}_1\mathbf{A}\mathbf{P}_1$, $\mathbf{A}_2 = \mathbf{H}_2\mathbf{A}_1\mathbf{P}_2, \cdots,$

$$\mathbf{R} = \mathbf{A}_n = \mathbf{H}_n\mathbf{A}_{n-1} = \cdots = \mathbf{H}_n \cdots \mathbf{H}_2\mathbf{H}_1\mathbf{A}\mathbf{P}_1\mathbf{P}_2 \cdots \mathbf{P}_{n-1}. \tag{77}$$

Multiplying (77) by $\mathbf{Q} = \mathbf{H}_1\mathbf{H}_2 \cdots \mathbf{H}_n$ and letting $\mathbf{\Pi}$ denote the product $\mathbf{P}_1\mathbf{P}_2 \cdots \mathbf{P}_{n-1}$, it follows that $\mathbf{A\Pi} = \mathbf{QR}$. Pictorially, this identity can be expressed

$$\mathbf{A\Pi} = [\mathbf{Q}_1 \quad \mathbf{Q}_2]\begin{bmatrix} \mathbf{R}_1 & \mathbf{R}_2 \\ \mathbf{0} & \mathbf{0} \end{bmatrix} = [\mathbf{Q}_1\mathbf{R}_1 \quad \mathbf{Q}_1\mathbf{R}_2] \tag{78}$$

where $\mathbf{R}_1$ is a square upper triangular matrix. As with the standard $\mathbf{QR}$ factorization, the columns of $\mathbf{Q}_1$ satisfy the orthonormality conditions (76) and $\mathbf{S}$ is contained in the space spanned by the columns of $\mathbf{Q}_1$. Unlike the $\mathbf{R}$ in the standard factorization, $\mathbf{R}_1$ is invertible. Letting r denote the number of rows of $\mathbf{R}_1$ and letting $(\mathbf{A\Pi})_1$ denote the first r columns of $\mathbf{A\Pi}$, (78) implies that $\mathbf{Q}_1\mathbf{R}_1 = (\mathbf{A\Pi})_1$ and $\mathbf{Q}_1 = (\mathbf{A\Pi})_1\mathbf{R}_1^{-1}$. Since the columns of $\mathbf{Q}_1$ are also linear combinations of the columns of $\mathbf{A}$, the space spanned by the columns of $\mathbf{Q}_1$ is $\mathbf{S}$. In summary, for the **QR** factorization scheme with column pivoting depicted in (78), $\mathbf{R}_1$ is invertible and the columns of $\mathbf{Q}_1$ are an orthonormal basis for the space spanned by the columns of $\mathbf{A}$.

Even though the bottom half of the $\mathbf{R}$ matrix in (78) is theoretically zero, rounding errors can lead to nonzero coefficients. When computing an orthonormal basis using a **QR** factorization with column pivoting, we often select a cutoff parameter τ (such as the machine epsilon times the 1-norm of $\mathbf{A}$). Whenever $|r_{kk}| \leqq \tau$, we stop the factorization and we set the succeeding rows of $\mathbf{R}$ to zero.

Recall that if $\mathbf{A}$ is **QR** factored using algorithm (45), then instead of storing the $\mathbf{Q}$ matrix, we essentially store the Householder matrices used to factor $\mathbf{A}$. To recover the columns of $\mathbf{Q}$, observe that the jth column of $\mathbf{Q}$ is equal to $\mathbf{Q}\mathbf{e}_j$ where $\mathbf{e}_j$ is the vector with every component equal to zero except for the jth component which is 1. Since $\mathbf{Q} = \mathbf{H}_1\mathbf{H}_2 \cdots \mathbf{H}_n$, the jth column of $\mathbf{Q}$ is $\mathbf{H}_1\mathbf{H}_2 \cdots \mathbf{H}_n\mathbf{e}_j$. In Exercise 5-5.7 you show that $\mathbf{H}_k\mathbf{e}_j = \mathbf{e}_j$ for each $k > j$. Consequently, $\mathbf{H}_1\mathbf{H}_2 \cdots \mathbf{H}_n\mathbf{e}_j = \mathbf{H}_1\mathbf{H}_2 \cdots \mathbf{H}_j\mathbf{e}_j$ and the jth column of $\mathbf{Q}$ is given by the product

$$\mathbf{H}_1\mathbf{H}_2 \cdots \mathbf{H}_j\mathbf{e}_j .$$

If the matrix $\mathbf{A}$ is **QR** factored using algorithm (45), the following algorithm stores in $\mathbf{Q}$ an orthonormal basis for the columns of $\mathbf{A}$:

```
j = 1 to k
    q_ij = 0 for i = 1 to m
    q_jj = 1
    l = j down to 1
        t ← 0
        t ← t + a_il q_ij for i = l to m
        q_ij ← q_ij − t a_il for i = l to m
    next l
next j
```
(79)

The subroutine which implements algorithm (79) is called BASIS and its arguments are the following:

B:	array to store basis (can identify with array A)
LB:	leading (row) dimension of array B
N:	number of vectors in the basis
A:	factorization computed by subroutine QR of the matrix of vectors
C:	cutoff

The following program computes an orthonormal basis for the space spanned by the columns of the coefficient matrix in (19).

```
      REAL A(3,5),B(3,2)
      DATA A/.6,.8,0.,0.,.7808688,.6246950/
      CALL QR(A,3,3,2)
      CALL BASIS(B,3,N,A,1.E-6)
      DO 10 J = 1,N
         WRITE(6,*) 'COLUMN',J
         WRITE(6,*) (B(I,J), I=1,3)
10    CONTINUE
      END
```

Observe that the input vectors for the program above are normalized. You will show in Exercise 5-8.2 that some basis vectors can be dropped due to the cutoff process if the starting vectors are not normalized.

The **Gram-Schmidt** process is another way to compute an orthonormal basis for the space spanned by a collection of vectors $\mathbf{a}_1, \mathbf{a}_2, \cdots, \mathbf{a}_n$. The Gram-Schmidt process starts by normalizing $\mathbf{a}_1$: $\mathbf{q}_1 = \mathbf{a}_1/\|\mathbf{a}_1\|_2$. We then express (see Figure 5-12) $\mathbf{a}_2$ as the sum of a vector $\mathbf{v}_2$ in the direction $\mathbf{q}_1$ and a vector $\mathbf{p}_2$ perpendicular to $\mathbf{a}_1$: $\mathbf{a}_2 = \mathbf{v}_2 + \mathbf{p}_2$. Since $\mathbf{v}_2$, the projection of $\mathbf{a}_2$ onto $\mathbf{q}_1$, is $\mathbf{a}_2$ dot $\mathbf{q}_1$ times $\mathbf{q}_1$, we have

$$\mathbf{p}_2 = \mathbf{a}_2 - \mathbf{v}_2 = \mathbf{a}_2 - (\mathbf{q}_1^T\mathbf{a}_2)\mathbf{q}_1.$$

Then $\mathbf{p}_2$ is normalized to obtain $\mathbf{q}_2$:

$$\mathbf{q}_2 = \frac{\mathbf{p}_2}{\|\mathbf{p}_2\|_2} = \frac{\mathbf{a}_2 - (\mathbf{q}_1^T\,\mathbf{a}_2)\mathbf{q}_1}{\|\mathbf{a}_2 - (\mathbf{q}_1^T\,\mathbf{a}_2)\mathbf{q}_1\|_2}.$$

Note that both $\mathbf{q}_1$ and $\mathbf{q}_2$ are unit vectors, $\mathbf{q}_1$ is orthogonal to $\mathbf{q}_2$, and the span of $\mathbf{a}_1$ and $\mathbf{a}_2$ is equal to the span of $\mathbf{q}_1$ and $\mathbf{q}_2$ (see Figure 5-12).

In general, the basis vector $\mathbf{q}_j$ is expressed in terms of $\mathbf{a}_j$ and the previously computed basis vectors $\mathbf{q}_1$ through $\mathbf{q}_{j-1}$. We write $\mathbf{a}_j = \mathbf{v}_j + \mathbf{p}_j$, where $\mathbf{v}_j$ is the projection of $\mathbf{a}_j$ onto the space spanned by $\mathbf{q}_1$ through $\mathbf{q}_{j-1}$ and $\mathbf{p}_j$ is perpendicular

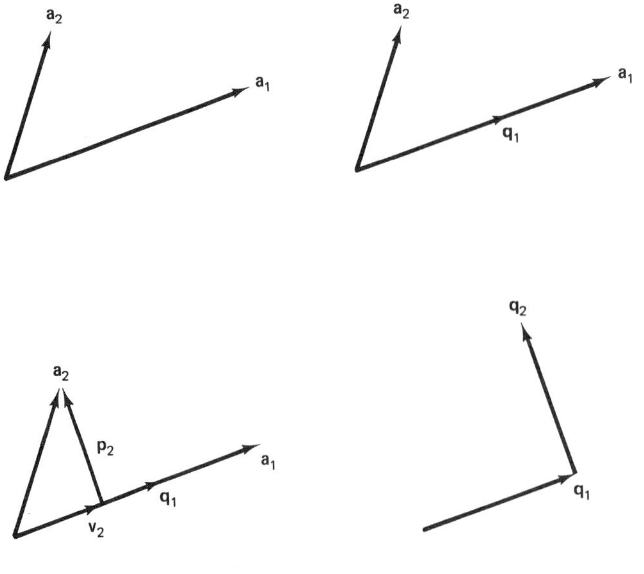

Figure 5-12 The Gram-Schmidt process.

to the projection. The projection of $\mathbf{a}_j$ onto $\mathbf{q}_i$ is $(\mathbf{q}_i^T\mathbf{a}_j)\mathbf{q}_i$, while the projection of $\mathbf{a}_j$ onto the space spanned by the previously computed basis vectors is the sum of the projections on each basis vector:

$$\mathbf{v}_j = \sum_{i=1}^{j-1} (\mathbf{q}_i^T\mathbf{a}_j)\mathbf{q}_i.$$

Therefore, $\mathbf{p}_j$ can be expressed

$$\mathbf{p}_j = \mathbf{a}_j - \mathbf{v}_j = \mathbf{a}_j - \sum_{i=1}^{j-1} (\mathbf{q}_i^T\mathbf{a}_j)\mathbf{q}_i. \tag{80}$$

Finally, $\mathbf{q}_j$ is $\mathbf{p}_j$ normalized: $\mathbf{q}_j = \mathbf{p}_j/\|\mathbf{p}_j\|_2$. Letting r_{jj} denote $\|\mathbf{p}_j\|_2$ and defining $r_{ij} = \mathbf{q}_i^T\mathbf{a}_j$, the Gram-Schmidt process can be stated in the following way:

$$\boxed{\begin{array}{l}
j = 1 \text{ to } n \\
\quad \mathbf{q}_j \leftarrow \mathbf{a}_j \\
\quad i = 1 \text{ to } j - 1 \\
\quad\quad r_{ij} \leftarrow \mathbf{q}_i^T\mathbf{a}_j \\
\quad\quad \mathbf{q}_j \leftarrow \mathbf{q}_j - r_{ij}\mathbf{q}_i \\
\quad \text{next } i \\
\quad r_{jj} \leftarrow \|\mathbf{q}_j\|_2 \\
\quad \mathbf{q}_j \leftarrow \mathbf{q}_j/r_{jj} \\
\text{next } j
\end{array}} \tag{81}$$

The Gram-Schmidt process can be viewed as a partial **QR** factorization. The identity (80) coupled with the definition $\mathbf{q}_j = \mathbf{p}_j/\|\mathbf{p}_j\|_2 = \mathbf{p}_j/r_{jj}$ implies that

$$\mathbf{p}_j = r_{jj}\,\mathbf{q}_j = \mathbf{a}_j - \sum_{i=1}^{j-1} r_{ij}\,\mathbf{q}_j.$$

Solving for $\mathbf{a}_j$, we have

$$\mathbf{a}_j = \sum_{i=1}^{j} \mathbf{q}_i\, r_{ij}.$$

Using matrix notation, this relation is expressed $\mathbf{A} = \mathbf{QR}$, where $\mathbf{Q}$ is the matrix whose columns are the $\mathbf{q}_i$ and $\mathbf{R}$ is the upper triangular matrix formed from the r_{ij}.

Although there are some similarities between the Gram-Schmidt factorization and Householder's factorization, there are also some important differences. An attractive feature of the Gram-Schmidt process is its speed—it can be twice as fast as Householder's algorithm since the columns of **Q** are computed directly. (In Householder's method, **Q** is the product of Householder matrices.) An attractive feature of Householder's method is that we generate a square orthogonal **Q** matrix while the Gram-Schmidt process generates a rectangular matrix with the same dimensions as **A**. As we will soon see, Householder's **Q** contains both a basis for the space spanned by the columns of **A** and a basis for the space perpendicular to the columns of **A**. A disadvantage with the Gram-Schmidt process is that it tends to be numerically unstable since nearly equal numbers can be subtracted. That is, if $\mathbf{a}_j$ is nearly contained in the space spanned by

$\mathbf{q}_1, \cdots, \mathbf{q}_{j-1}$, then the difference $\mathbf{a}_j - \mathbf{v}_j$ in (80) is the difference between nearly equal vectors. From the discussion in Section 1-4, we know that the relative error in this subtraction can be large.

In the modified Gram-Schmidt process, we reorganize the computation of $\mathbf{p}_j$ in (80) to improve the numerical accuracy. The modified formula of $\mathbf{p}_j$ is the following:

```
p_j ← a_j
i = 1 to j - 1
    p_j ← p_j - (q_i^T p_j) q_i
next i
```

(82)

Since the coefficient $\mathbf{q}_i^T\mathbf{p}_j$ of $\mathbf{q}_i$ in (82) is equivalent to the coefficient $\mathbf{q}_i^T\mathbf{a}_j$ of $\mathbf{q}_i$ in (80) (see Exercise 5-8.3), it follows that the formula (82) for $\mathbf{p}_j$ is equivalent to (80). It turns out that the formula (82) yields better accuracy than (80) since the successive $\mathbf{p}_j$'s generated in (82) decrease in size and the dot product $\mathbf{q}_i^T\mathbf{p}_j$ can be evaluated more accurately than the dot product $\mathbf{q}_i^T\mathbf{a}_j$. In detail, the modified Gram-Schmidt process is the following:

```
j = 1 to n
    q_j ← a_j
    i = 1 to j - 1
        r_ij ← q_i^T q_j
        q_j ← q_j - r_ij q_i
    next i
    r_jj ← ||q_j||_2
    q_j ← q_j / r_jj
next j
```

(83)

Although the modified Gram-Schmidt algorithm is more numerically stable than the original Gram-Schmidt process, Householder's method still tends to generate a basis which is more nearly orthonormal than the basis generated by the modified Gram-Schmidt process.

As noted above, Householder's **QR** factorization generates both a basis for the columns of **A** and a basis for the space perpendicular to the columns of **A**. Referring to the decomposition (78), the columns of $\mathbf{Q}_1$ are a basis for the columns of **A**. Since $\mathbf{Q} = [\mathbf{Q}_1|\mathbf{Q}_2]$ is an orthogonal matrix, each column of $\mathbf{Q}_2$ is orthogonal to each column of $\mathbf{Q}_1$. Since the columns of $\mathbf{Q}_2$ are orthogonal to the columns of $\mathbf{Q}_1$ and since the columns of $\mathbf{Q}_1$ are a basis for the columns of **A**, it follows that the columns of $\mathbf{Q}_2$ are a basis for the space of vectors perpendicular to the columns of **A**. Subroutine NULL uses Householder's method to generate an orthonormal basis for the space perpendicular to a collection of vectors.

Exercises

5-8.1. Show that an orthonormal collection of vectors $\mathbf{q}_1, \cdots, \mathbf{q}_n$ is always linearly independent.

5-8.2. Let us compute an orthonormal basis for a collection of vectors $\mathbf{a}_1, \cdots, \mathbf{a}_n$ using Householder's method and a cutoff τ which is much smaller than 1. Assume that every component of $\mathbf{a}_i$ is zero except for component i. The first component of $\mathbf{a}_1$ is 1 while the ith component of $\mathbf{a}_i$ is $\tau/2$ for $i > 1$. What is the computed orthonormal basis? What is the correct orthonormal basis corresponding to $\tau = 0$?

5-8.3. Show that the coefficient $\mathbf{q}_i^T\mathbf{p}_j$ of $\mathbf{q}_i$ in (82) is equivalent to the coefficient $\mathbf{q}_i^T\mathbf{a}_j$ of $\mathbf{q}_i$ in (80).

5-8.4. Show that the $\mathbf{p}_j$ given in (80) is perpendicular to $\mathbf{q}_1, \cdots, \mathbf{q}_{j-1}$. (*Hint:* Two vectors are perpendicular if their dot product is zero.)

5-8.5. Assuming that each vector in (81) has m components, determine the asymptotic parameter corresponding to a computer program that implements the Gram-Schmidt process. Describe the two cycle times that must be evaluated to determine the asymptotic running time of a computer program which implements algorithm (81).

5-8.6. Determine the asymptotic parameter corresponding to a computer program tha implements algorithm (79). Describe the two cycle times that must be evaluated to determine the asymptotic running time of a computer program which implements algorithm (79). [Note that the asymptotic parameter for Householder's scheme to generate an orthonormal basis is double the asymptotic parameter for (79) since we must use (45) to compute the Householder vectors before executing (79).]

5-8.7. Use subroutines QR and BASIS to compute an orthonormal basis for the space spanned by the following two vectors:

$$\begin{bmatrix} 3 \\ 4 \\ 0 \end{bmatrix} \quad \text{and} \quad \begin{bmatrix} 0 \\ 5 \\ 4 \end{bmatrix}.$$

Use subroutines QR and NULL to compute an orthonormal basis for the space perpendicular to these two vectors.

5-8.8. Suppose that the columns of $\mathbf{A}$ are linearly independent and $\mathbf{A} = \mathbf{QR}$ where $\mathbf{R}$ is a square upper triangular matrix. Show that every diagonal element of $\mathbf{R}$ is nonzero.

REVIEW PROBLEMS

5-1. Consider an underdetermined system $\mathbf{Ax} = \mathbf{b}$ of m equations in $m + 2$ unknowns, where every coefficient is zero except for a_{ii}, $a_{i,i+1}$, and $a_{i,i+2}$, $i = 1, 2, \cdots, m$. What is the structure of the coefficient matrix $\mathbf{AA}^T$ in the normal equation? When computing the least squares solution to $\mathbf{Ax} = \mathbf{b}$, which subroutines from Chapter 2 should be employed? If $\mathbf{A}^T$ is $\mathbf{QR}$ factored using Householder reflections, what is the structure of $\mathbf{R}$ and what is the structure of the vectors $\mathbf{w}$ associated with each Householder matrix?

5-2. **(a)** What is the asymptotic parameter associated with algorithm (45) which **QR** factors a matrix using Householder reflections? Describe the two cycle times that must be evaluated to determine the asymptotic running time of a computer program that implements algorithm (45).

(b) What is the asymptotic parameter associated with algorithm (61), which **QR** factors a matrix using Givens rotations? Describe the cycle time that must be evaluated to determine the asymptotic running time of a computer program that implements algorithm (61).

(c) What are the asymptotic parameters and the cycle times associated with algorithm (69), which generates the least squares solution to an overdetermined linear system?

(d) Consider an overdetermined system of m equations in n unknowns. Show that when m is much larger than n, the algorithm developed in Section 5-2 to compute the least squares solution is about twice as fast as algorithm (45) combined with algorithm (69).

5-3. Suppose that algorithm (61) is used to reduce $\mathbf{A}$ to upper triangular form with Givens rotations. Modify this algorithm so that the orthogonal factor of $\mathbf{A}$ is also computed. (*Hint:* Initialize $\mathbf{Q} = \mathbf{I}$ and apply the Givens rotations to $\mathbf{Q}$ and $\mathbf{A}$ simultaneously.)

5-4. Given a $m \times n$ matrix $\mathbf{A}$ with $m > n$, let $\mathbf{QR}$ denote the factorization generated by the Gram-Schmidt process. Show that the least squares solution to $\mathbf{Ax} = \mathbf{b}$ satisfies the equation $\mathbf{Rx} = \mathbf{Q}^T\mathbf{b}$. (*Hint:* Although the $\mathbf{Q}$ generated by the Gram-Schmidt process is not an orthogonal matrix, the columns of $\mathbf{Q}$ are orthonormal.)

5-5. Consider the matrix $\mathbf{A}$ given by

$$\mathbf{A} = \begin{bmatrix} 2 & 1 \\ 4 & 5 \\ 6 & 15 \end{bmatrix}.$$

Let us use Gaussian elimination to annihilate "subdiagonal" elements of $\mathbf{A}$. Subtracting 2 times the first row from the second row and subtracting 3 times the first row from the third row yields

$$\begin{bmatrix} 2 & 1 \\ 0 & 3 \\ 0 & 12 \end{bmatrix}.$$

Then subtracting 4 times the second row from the third row gives

$$\begin{bmatrix} 2 & 1 \\ 0 & 3 \\ 0 & 0 \end{bmatrix}. \tag{84}$$

We extract a 2×2 upper triangular matrix $\mathbf{U}$ from (84):

$$\mathbf{U} = \begin{bmatrix} 2 & 1 \\ 0 & 3 \end{bmatrix}.$$

Using the multipliers, we form a 3×2 "lower triangular" matrix $\mathbf{L}$:

$$\mathbf{L} = \begin{bmatrix} 1 & 0 \\ 2 & 1 \\ 3 & 4 \end{bmatrix}. \tag{85}$$

Verify that $\mathbf{A} = \mathbf{LU}$. Now consider an overdetermined system $\mathbf{Ax} = \mathbf{b}$ of m equations in n unknowns. If $\mathbf{A} = \mathbf{LU}$, where $\mathbf{U}$ is a $n \times n$ nonsingular matrix, and $\mathbf{L}$ is a $m \times n$ matrix like (85), show that the least squares solution to $\mathbf{Ax} = \mathbf{b}$ is

$$\mathbf{x} = \mathbf{U}^{-1}(\mathbf{L}^T\mathbf{L})^{-1}\mathbf{L}^T\mathbf{b}. \tag{86}$$

When $\mathbf{A}$ has more columns than rows, its $\mathbf{L}$ factor is $m \times m$ and its $\mathbf{U}$ factor is $m \times n$. Derive a formula similar to (86) for the least squares solution to an underdetermined system $\mathbf{Ax} = \mathbf{b}$. Show that both (86) and the related formula for an underdetermined system are special cases of the general rule

$$\mathbf{x} = \mathbf{U}^T(\mathbf{UU}^T)^{-1}(\mathbf{L}^T\mathbf{L})^{-1}\mathbf{L}^T\mathbf{b}.$$

5-6. Consider an overdetermined system $\mathbf{Ax} = \mathbf{b}$ and the factorization algorithm (45). If the parameter s is zero for some value of l, algorithm (45) skips a column of $\mathbf{A}$. Using an illustrative $\mathbf{R}$, explain why there exists an infinite number of least squares solutions when algorithm (45) skips a column of $\mathbf{A}$. When there exists an infinite number of least squares solutions, explain how to compute that least squares solution with the smallest Euclidean length.

5-7. Explain the relationship between the modified Gram-Schmidt algorithm (83) and the following implementation of the modified Gram-Schmidt process:

$$
\boxed{
\begin{array}{l}
i = 1 \text{ to } n \\
\quad r_{ii} \leftarrow \|\mathbf{a}_i\|_2 \\
\quad \mathbf{a}_i \leftarrow r_{ii}^{-1}\mathbf{a}_i \\
\quad j = i+1 \text{ to } n \\
\quad\quad r_{ij} \leftarrow \mathbf{a}_i^T\mathbf{a}_j \\
\quad\quad \mathbf{a}_j \leftarrow \mathbf{a}_j - r_{ij}\,\mathbf{a}_i \\
\quad \text{next } j \\
\text{next } i
\end{array}
}
$$

5-8. Write a program to implement either the Gram-Schmidt process or the modified Gram-Schmidt process. Using your program, compute an orthonormal basis $\mathbf{q}_1$ and $\mathbf{q}_2$ for the space spanned by the vectors

$$\mathbf{a}_1 = \begin{bmatrix} 1 + c \\ 1 \end{bmatrix} \quad \text{and} \quad \mathbf{a}_2 = \begin{bmatrix} 1 \\ 1 \end{bmatrix},$$

where c is any small number which is larger than the machine epsilon. Test the orthogonality of the computed orthonormal basis by evaluating the dot product $\mathbf{q}_1{}^T\mathbf{q}_2$ which is zero in theory.

5-9. Consider the following algorithm to generate an orthonormal basis for a pair of vectors $\mathbf{a}_1$ and $\mathbf{a}_2$: Let $\mathbf{H}$ denote the Householder matrix that annihilates components 2 through n of $\mathbf{a}_1$, let $\mathbf{z}$ be the vector whose first component is zero and whose remaining components agree with the corresponding components of $\mathbf{Ha}_2$, set $\mathbf{q}_1 = \mathbf{a}_1/\|\mathbf{a}_1\|_2$, and set $\mathbf{q}_2 = \mathbf{Hz}/\|\mathbf{Hz}\|_2$.

(a) Verify that $\mathbf{q}_1$ is perpendicular to $\mathbf{q}_2$.

(b) Write a computer program that implements this algorithm and evaluate the dot product $\mathbf{q}_1{}^T\mathbf{q}_2$ corresponding to the vectors $\mathbf{a}_1$ and $\mathbf{a}_2$ in Problem 5-8.

5-10. The algorithms developed in this chapter must be modified for a complex matrix. The complex analog of an orthogonal matrix is a **unitary matrix**. If the symbol * denotes the transpose of the complex conjugate, then a matrix $\mathbf{Q}$ is unitary if $\mathbf{Q}^*\mathbf{Q} = \mathbf{I}$. The complex analog of a Householder matrix is $\mathbf{H} = \mathbf{I} - 2\mathbf{ww}^*$, where $\mathbf{w}$ is any vector such that $\mathbf{w}^*\mathbf{w} = 1$.

(a) Show that the complex Householder matrix is unitary and $\mathbf{H}^* = \mathbf{H}$.

(b) The magnitude of a complex number c, denoted $|c|$, is the square root of c times c conjugate. Given a (complex) vector $\mathbf{x}$, let us define

$$\mathbf{w} = \frac{1}{\sqrt{2r(r + |x_k|)}} \begin{bmatrix} 0 \\ \vdots \\ 0 \\ x_k + \sigma r \\ x_{k+1} \\ \vdots \\ x_n \end{bmatrix}, \qquad \text{where } r = \sqrt{|x_k|^2 + |x_{k+1}|^2 + \cdots + |x_n|^2}$$

and $\sigma = x_k/|x_k|$ if $x_k \neq 0$ while $\sigma = 1$ otherwise. Show that $(\mathbf{Hx})_i = x_i$ for $i < k$ and $(\mathbf{Hx})_i = 0$ for $i > k$. What is $(\mathbf{Hx})_k$?

5-11. Let c and s denote any complex numbers with the property that $c^*c + s^*s = 1$.

Show that the following matrix **G** is unitary:

$$\mathbf{G} = \begin{bmatrix} c^* & s^* \\ -s & c \end{bmatrix}.$$

Given a vector $\mathbf{x} = (x_1, x_2)^T$ with two components, for what choice of c and s is the second component of the vector **Gx** annihilated? For what choice of c and s is the first component of the vector **Gx** annihilated?

REFERENCES

The book [L2] by Lawson and Hanson provides a comprehensive treatment of least squares problems. More recent results are also discussed in Chapter 6 of [G15]. Various applications of least squares to statistics are presented by Grayhill in [G16]. The **QR** factorization of a matrix using elementary reflectors is first developed by Householder in [H17]. Practical details of Householder's method are given by Golub in [G13] and by Businger and Golub in [B23]. The **QR** factorization of a matrix using plane rotations is first discussed by Givens in [G11]. The fast Givens scheme is developed by Gentleman in [G3]. Also, see Hammarling [H8]. Of course, least squares problems and orthogonal techniques were studied long before Householder and Givens. Goldstine's history of numerical analysis [G12] notes that in 1805 Legendre published (in [L3]) the solution to an overdetermined linear system although the solution had been used much earlier by Gauss. Also, in 1845 Jacobi introduced (in [J1]) rotations for least squares problems. The formula for the pseudoinverse in Problem 5-5 appears in the article [P9] by Peters and Wilkinson as well as in the book [S15] by Strang. With regard to Problem 5-9, another efficient, stable technique to compute an orthonormal basis for a pair of vectors is Kahan's reorthogonalization procedure reported by Parlett on page 107 of [P4]. Reorthogonalization is also discussed in [A2].

6

Eigenproblems

6-1. APPLICATIONS

Given a square matrix **A**, this chapter studies techniques for computing those nonzero vectors **x** and those scalars λ with the property that

$$\mathbf{A}\mathbf{x} = \lambda\mathbf{x}. \tag{1}$$

The pair (λ, **x**) is called an **eigenpair** consisting of the **eigenvalue** λ and the **eigenvector x**. The set of all eigenvalues for a matrix is its **spectrum**. Observe that if **x** is an eigenvector of **A**, so is any multiple of **x**. For example, multiplying equation (1) by 2 gives us the relation $2\mathbf{A}\mathbf{x} = 2\lambda\mathbf{x}$, which can be written

$$\mathbf{A}(2\mathbf{x}) = \lambda(2\mathbf{x}).$$

Hence (λ, 2**x**) is an eigenpair if (λ, **x**) is an eigenpair. To eliminate this multiplicity, the eigenvector can be **normalized** so that its length is 1. If **x** is normalized to have unit Euclidean length, the eigenproblem can be stated: Find those vectors **x** and those scalars λ with the property that

$$\begin{aligned} \mathbf{A}\mathbf{x} &= \lambda\mathbf{x}, \\ \mathbf{x}^T\mathbf{x} &= 1. \end{aligned} \tag{2}$$

If **A** is $n \times n$, then equation (2) comprises $n + 1$ equations in $n + 1$ unknowns; the unknowns are the n components $x_1, x_2, \cdots, x_n$ of **x** and the scalar λ. Equation (2) is nonlinear since it involves the product $\lambda\mathbf{x}$ between the unknown scalar and the unknown vector as well as the dot product $\mathbf{x}^T\mathbf{x}$. Although algorithms for solving a nonlinear system of equations can be applied to (2), more efficient techniques have been developed that exploit the structure of the eigenproblem. Before studying these techniques, we discuss some applications of eigenpairs.

The solution to a linear differential equation with constant coefficients can be expressed in terms of the eigenpairs of the coefficient matrix. Let us consider the following system of two first-order differential equations:

$$\begin{aligned} \frac{du(t)}{dt} &= 2u(t) + 6v(t), \\ \frac{dv(t)}{dt} &= -2u(t) - 5v(t). \end{aligned} \tag{3}$$

We search for a solution of the form

$$u(t) = e^{\lambda t}x_1 \quad \text{and} \quad v(t) = e^{\lambda t}x_2, \tag{4}$$

where x_1, x_2, and λ are constants independent of t. Substituting (4) into equation (3) yields

$$\begin{aligned} \lambda e^{\lambda t}x_1 &= 2e^{\lambda t}x_1 + 6e^{\lambda t}x_2, \\ \lambda e^{\lambda t}x_2 &= -2e^{\lambda t}x_1 - 5e^{\lambda t}x_2. \end{aligned}$$

Canceling the common factor $e^{\lambda t}$ gives us the relations

$$\begin{aligned} 2x_1 + 6x_2 &= \lambda x_1, \\ -2x_1 - 5x_2 &= \lambda x_2. \end{aligned} \tag{5}$$

Using matrix-vector notation, (5) is written

$$\begin{bmatrix} 2 & 6 \\ -2 & -5 \end{bmatrix} \begin{bmatrix} x_1 \\ x_2 \end{bmatrix} = \lambda \begin{bmatrix} x_1 \\ x_2 \end{bmatrix}.$$

Therefore, if $u(t) = e^{\lambda t}x_1$ and $v(t) = e^{\lambda t}x_2$ is a solution to the differential equation (3), then $(\lambda, \mathbf{x})$ is an eigenpair for the coefficient matrix

$$\mathbf{A} = \begin{bmatrix} 2 & 6 \\ -2 & -5 \end{bmatrix}.$$

Conversely, if $(\lambda, \mathbf{x})$ is an eigenpair for the coefficient matrix, then $u(t) = e^{\lambda t}x_1$ and $v(t) = e^{\lambda t}x_2$ is a solution to the differential equation.

In the next section, we show that the matrix **A** has two eigenpairs:

$$\lambda = -1, \mathbf{x} = \begin{bmatrix} 2 \\ -1 \end{bmatrix} \quad \text{and} \quad \lambda = -2, \mathbf{x} = \begin{bmatrix} -3 \\ 2 \end{bmatrix}.$$

A solution to the differential equation (3) corresponding to the eigenvalue $\lambda = -1$ is

$$u(t) = 2e^{-t} \quad \text{and} \quad v(t) = -e^{-t},$$

while a solution corresponding to the eigenvalue $\lambda = -2$ is

$$u(t) = -3e^{-2t} \quad \text{and} \quad v(t) = 2e^{-2t}.$$

Using vector notation, these solutions can be expressed

$$\begin{bmatrix} u(t) \\ v(t) \end{bmatrix} = e^{-t} \begin{bmatrix} 2 \\ -1 \end{bmatrix} \quad \text{and} \quad \begin{bmatrix} u(t) \\ v(t) \end{bmatrix} = e^{-2t} \begin{bmatrix} -3 \\ 2 \end{bmatrix}.$$

The general solution to a linear system of two first-order differential equations is a linear combination of any two (linearly independent) solutions. Thus the general solution to (3) is

$$\begin{bmatrix} u(t) \\ v(t) \end{bmatrix} = c_1 e^{-t} \begin{bmatrix} 2 \\ -1 \end{bmatrix} + c_2 e^{-2t} \begin{bmatrix} -3 \\ 2 \end{bmatrix}, \tag{6}$$

where c_1 and c_2 are arbitrary constants. If the value of u and v are specified at some time like $t = 0$, these two constants can be computed. For example, if the initial condition

$$u(0) = 7 \quad \text{and} \quad v(0) = -4 \tag{7}$$

is prescribed, then substituting $t = 0$ in (6) gives us the relation

$$\begin{bmatrix} 7 \\ -4 \end{bmatrix} = c_1 \begin{bmatrix} 2 \\ -1 \end{bmatrix} + c_2 \begin{bmatrix} -3 \\ 2 \end{bmatrix},$$

which can be written

$$\begin{bmatrix} 2 & -3 \\ -1 & 2 \end{bmatrix} \begin{bmatrix} c_1 \\ c_2 \end{bmatrix} = \begin{bmatrix} 7 \\ -4 \end{bmatrix}.$$

Since the solution to this linear system is $c_1 = 2$ and $c_2 = -1$, it follows from (6) that the solution to the differential equation (3) that satisfies the initial condition (7) is

$$\begin{bmatrix} u(t) \\ v(t) \end{bmatrix} = 2e^{-t} \begin{bmatrix} 2 \\ -1 \end{bmatrix} - e^{-2t} \begin{bmatrix} -3 \\ 2 \end{bmatrix}.$$

In some applications, the exact solution to the differential equation is not as important as the qualitative behavior of the solution. Since e^{-t} and e^{-2t} approach zero as t increases, it follows that for any choice of c_1 or c_2, the solution (6) to the differential equation (3) approaches zero as t increases. We say that $u = 0$ and $v = 0$ is a **stable equilibrium point** for the differential equation. In general, zero is a stable equilibrium point for a linear differential equation if each eigenvalue of the coefficient matrix has negative real part (eigenvalues can be complex numbers). Thus the qualitative behavior of the solution to a linear differential equation depends on the signs of the eigenvalues.

Eigenvalues are also related to the properties of structures. When a force or load is applied along a simply supported beam†, the beam buckles (see Figure 6-1) when the load reaches a critical value λ_1. As the load increases further, a second critical point is reached, where the beam can buckle into a new configuration (see Figure 6-2, on page 240). Physically, to generate this new configuration, a thin clamp must be placed at the middle of the beam. Each load where the beam buckles into a new state is an eigenvalue of a boundary-value problem. An eigenproblem for a buckling beam can have the following form: Find all

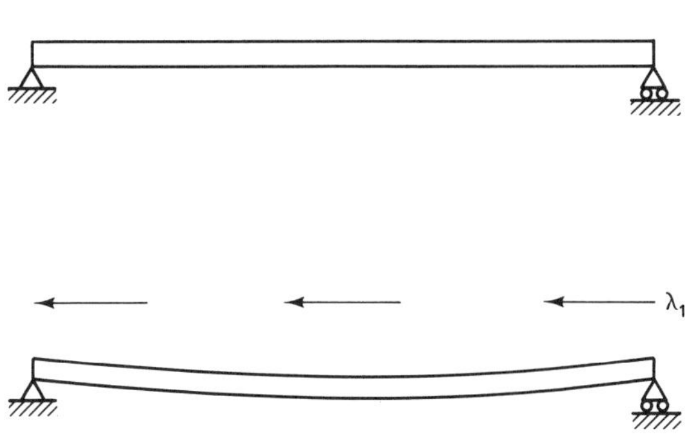

Figure 6-1 Beam buckling.

† A simply supported beam is one whose ends sit on an object which cannot move in the vertical direction.

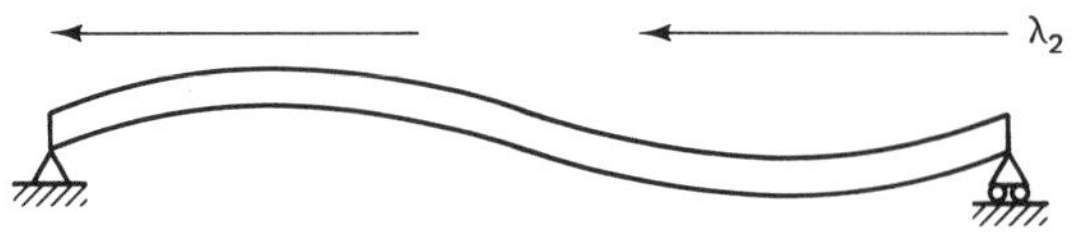

Figure 6-2 The second eigenstate.

functions $x(t)$ that are not identically zero and all scalars λ which satisfy the equation

$$-\frac{d^2x(t)}{dt^2} = \lambda x(t), \qquad 0 \leqq t \leqq 1, \tag{8}$$

subject to the boundary condition $x(0) = x(1) = 0$. In this model, t measures distance along the beam and $x(t)$ is the bending moment of the beam (see Figure 6-3). Theoretically, there are an infinite number of buckling states, but in practice, the beam may break soon after the load exceeds the smallest eigenvalue λ_1. With a buckling beam, we are often interested in the value of λ_1.

Equation (8) can be transformed into a matrix eigenproblem using the finite-difference technique developed in Section 1-5. The interval [0,1] is partitioned into N subintervals, each subinterval of width $\Delta t = 1/N$. Letting $t_i = i\Delta t$ denote the grid points and letting x_i denote $x(t_i)$, the second derivative in equation (8) has the approximation

$$\frac{d^2x(t_i)}{dt^2} \approx \frac{x_{i+1} - 2x_i + x_{i-1}}{(\Delta t)^2}.$$

This substitution in equation (8) leads to the relation $\mathbf{Ax} = \lambda\mathbf{x}$, where $\mathbf{A}$ is a $(N - 1) \times (N - 1)$ tridiagonal matrix:

$$\mathbf{A} = N^2\begin{bmatrix} 2 & -1 & & & & \\ -1 & 2 & \cdot & & & \\ & \cdot & \cdot & \cdot & & \\ & & \cdot & \cdot & \cdot & \\ & & & \cdot & \cdot & -1 \\ & & & & -1 & 2 \end{bmatrix}. \tag{9}$$

As N increases, the smallest eigenvalue of matrix $\mathbf{A}$ approaches the smallest eigenvalue of the boundary-value problem (8). Thus when N is large, the smallest eigenvalue of $\mathbf{A}$ approximates the structure's breaking point.

There are many other applications of eigenvalues. In 1940 the Tacoma Narrows Bridge collapsed due to some neglected eigenvalues. When tuning a guitar, we solve an eigenproblem. The length of pipes in a church organ are

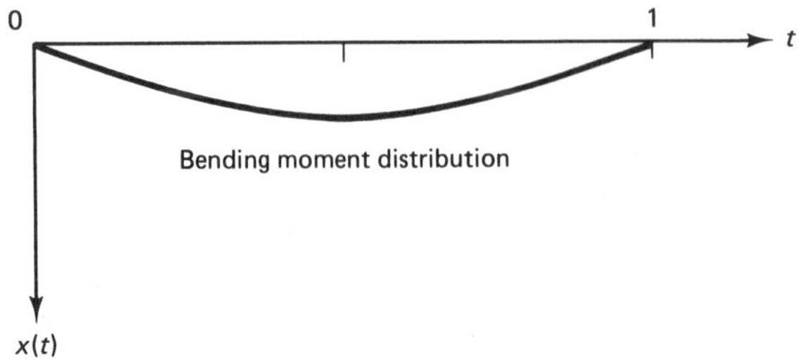

Figure 6-3 Coordinate system for the beam.

the solution to an eigenproblem. The energy states of an atom are the eigenvalues of a boundary-value problem.

Exercises

6-1.1. In the next section we observe that eigenvalues can be complex numbers. Show that the eigenvalues of a real symmetric matrix are real. [*Hint*: If the symbol * denotes the transpose of the complex conjugate and if $(\lambda, \mathbf{x})$ is an eigenpair for $\mathbf{A}$, then $\mathbf{Ax} = \lambda\mathbf{x}$ and $\mathbf{x}^*\mathbf{Ax} = \lambda\mathbf{x}^*\mathbf{x}$. Now take the conjugate transpose of this relation and remember that any complex number which equals its conjugate is a real number.]

6-1.2. A matrix $\mathbf{A}$ is **skew-symmetric** if $\mathbf{A}^T = -\mathbf{A}$. Show that every eigenvalue of a real skew-symmetric matrix is pure imaginary. (*Hint*: Again consider the relation $\mathbf{x}^*\mathbf{Ax} = \lambda\mathbf{x}^*\mathbf{x}$ and its conjugate transpose.)

6-1.3. More precisely, equation (1) defines a right eigenvector of a matrix $\mathbf{A}$ corresponding to the eigenvalue λ. The vector $\mathbf{y}$ is a **left eigenvector** corresponding to the eigenvalue μ if $\mathbf{y}^T\mathbf{A} = \mu\mathbf{y}^T$. Show that a left eigenvector and a right eigenvector corresponding to different eigenvalues are orthogonal. (*Hint*: Premultiply the relation $\mathbf{Ax} = \lambda\mathbf{x}$ by $\mathbf{y}^T$.)

6-1.4. Show that for either a real symmetric matrix or a real skew-symmetric matrix, the eigenvectors corresponding to distinct eigenvalues are orthogonal. In other words, if $\mathbf{x}$ and $\mathbf{y}$ are eigenvectors corresponding to distinct eigenvalues, show that $\mathbf{x}^*\mathbf{y} = 0$. (*Hint*: See Exercise 6-1.3.)

6-1.5. Show that each eigenvalue of a real orthogonal matrix has unit magnitude. (*Hint*: Multiply each side of the relation $\mathbf{Qx} = \lambda\mathbf{x}$ by its conjugate transpose.)

6-1.6. Show that for a real orthogonal matrix, the eigenvectors corresponding to distinct eigenvalues are orthogonal. In general, matrices with a complete set of orthonormal eigenvectors are called **normal matrices.**

6-2. DIAGONALIZATION

Since $\mathbf{x} = \mathbf{Ix}$, where $\mathbf{I}$ is the identity matrix, the eigenequation $\mathbf{Ax} = \lambda\mathbf{x}$ can be written

$$(\mathbf{A} - \lambda\mathbf{I})\mathbf{x} = \mathbf{0}. \tag{10}$$

For fixed λ, equation (10) is a linear system with coefficient matrix $\mathbf{A} - \lambda\mathbf{I}$. This system is also called a **homogeneous** linear system since the right side is $\mathbf{0}$. For fixed λ, let us **LU** factor the coefficient matrix $\mathbf{A} - \lambda\mathbf{I}$ to obtain the equivalent system $(\mathbf{LU})\mathbf{x} = \mathbf{0}$. Defining $\mathbf{y} = \mathbf{Ux}$, the factored system is decomposed into two equations: $\mathbf{Ly} = \mathbf{0}$ and $\mathbf{Ux} = \mathbf{y}$. The solution to $\mathbf{Ly} = \mathbf{0}$ is $\mathbf{y} = \mathbf{0}$ and if every diagonal element of $\mathbf{U}$ is nonzero, then the solution to $\mathbf{Ux} = \mathbf{0}$ is also $\mathbf{x} = \mathbf{0}$. But if one of $\mathbf{U}$'s diagonal elements is zero, then (10) has an infinite number of solutions. To summarize, (10) has a nonzero solution $\mathbf{x}$ if and only if some diagonal element of $\mathbf{U}$ is zero. In Chapter 2 we saw that the determinant of a matrix is the product of the diagonal elements in the upper triangular factor. Consequently, a diagonal element of $\mathbf{U}$ is zero if and only if the determinant of $\mathbf{A} - \lambda\mathbf{I}$ is zero. The equation

$$\det(\mathbf{A} - \lambda\mathbf{I}) = 0$$

is called the **characteristic equation** and its solutions are the eigenvalues of $\mathbf{A}$.

For an upper triangular matrix such as

$$\mathbf{A} = \begin{bmatrix} 1 & 1 \\ 0 & 2 \end{bmatrix}, \tag{11}$$

the determinant of $\mathbf{A} - \lambda\mathbf{I}$ is

$$\det(\mathbf{A} - \lambda\mathbf{I}) = \det\left(\begin{bmatrix} 1 & 1 \\ 0 & 2 \end{bmatrix} - \lambda\begin{bmatrix} 1 & 0 \\ 0 & 1 \end{bmatrix}\right) = \det\begin{bmatrix} 1-\lambda & 1 \\ 0 & 2-\lambda \end{bmatrix}.$$

Observe that $\mathbf{A} - \lambda\mathbf{I}$ is upper triangular when $\mathbf{A}$ is upper triangular. Since the determinant of an upper triangular matrix is the product of the diagonal elements, we have

$$\det\begin{bmatrix} 1-\lambda & 1 \\ 0 & 2-\lambda \end{bmatrix} = (1-\lambda)(2-\lambda).$$

The roots of the characteristic equation $(1 - \lambda)(2 - \lambda) = 0$ are $\lambda = 1$ and $\lambda = 2$, the diagonal elements of $\mathbf{A}$. In general, the eigenvalues of a triangular matrix are the diagonal elements.

The eigenvectors $\mathbf{x}$ corresponding to the eigenvalue λ are the solutions to $(\mathbf{A} - \lambda\mathbf{I})\mathbf{x} = \mathbf{0}$. For matrix (11), the eigenvector equation is

$$\underset{(\mathbf{A}-\lambda\mathbf{I})}{\begin{bmatrix} 1-\lambda & 1 \\ 0 & 2-\lambda \end{bmatrix}}\underset{\mathbf{x}}{\begin{bmatrix} x_1 \\ x_2 \end{bmatrix}} = \underset{\mathbf{0}}{\begin{bmatrix} 0 \\ 0 \end{bmatrix}}. \tag{12}$$

To compute an eigenvector corresponding to the eigenvalue $\lambda = 1$, we substitute $\lambda = 1$ in (12) to obtain the equations

$$0x_1 + x_2 = 0,$$
$$x_2 = 0.$$

The second equation states that $x_2 = 0$ and the first equation reduces to $0x_1 = 0$. Since x_1 can be anything and x_2 is zero, the eigenvectors corresponding to the eigenvalue $\lambda = 1$ have the form

$$\mathbf{x} = s\begin{bmatrix} 1 \\ 0 \end{bmatrix},$$

where s is an arbitrary scalar. To compute an eigenvector corresponding to the eigenvalue $\lambda = 2$, we substitute $\lambda = 2$ in (12) to obtain

$$-x_1 + x_2 = 0,$$
$$0x_2 = 0.$$

The second equation is satisfied for any choice of x_2 and the first equation implies that $x_1 = x_2$. The eigenvectors corresponding to the eigenvalue $\lambda = 2$ have the form

$$\mathbf{x} = s\begin{bmatrix} 1 \\ 1 \end{bmatrix}.$$

In general, if $\mathbf{A}$ is an upper triangular matrix, the eigenvalues are the diagonal elements $a_{11}, a_{22}, \cdots, a_{nn}$. The eigenvectors are computed by back substitution and the eigenvector $\mathbf{x}$ corresponding to the eigenvalue a_{ii} has the structure

$$\mathbf{x} = s\begin{bmatrix} * \\ \cdot \\ \cdot \\ * \\ 1 \\ 0 \\ \cdot \\ \cdot \\ 0 \end{bmatrix},$$

where the ith component on the right side is 1 and the last $n - i$ components are zero.

For a second example, let us consider the matrix

$$\mathbf{A} = \begin{bmatrix} 2 & 6 \\ -2 & -5 \end{bmatrix} \tag{13}$$

appearing in Section 6-1. The characteristic equation is

$$\det(\mathbf{A} - \lambda\mathbf{I}) = \det\begin{bmatrix} 2-\lambda & 6 \\ -2 & -5-\lambda \end{bmatrix} = 0.$$

Utilizing the formula

$$\det\begin{bmatrix} a & c \\ b & d \end{bmatrix} = ad - bc$$

for the determinant of a 2×2 matrix, we have

$$\det\begin{bmatrix} 2-\lambda & 6 \\ -2 & -5-\lambda \end{bmatrix} = (2-\lambda)(-5-\lambda) + 12 = \lambda^2 + 3\lambda + 2.$$

Since $\lambda^2 + 3\lambda + 2$ factors into $(\lambda + 1)(\lambda + 2)$, the characteristic equation is

$$(\lambda + 1)(\lambda + 2) = 0.$$

The roots of the characteristic equation are $\lambda = -1$ and $\lambda = -2$.

Again, the eigenvectors $\mathbf{x}$ corresponding to the eigenvalue λ are the solutions to $(\mathbf{A} - \lambda\mathbf{I})\mathbf{x} = \mathbf{0}$. For matrix (13), the eigenvector equation can be written:

$$\begin{bmatrix} 2-\lambda & 6 \\ -2 & -5-\lambda \end{bmatrix}\begin{bmatrix} x_1 \\ x_2 \end{bmatrix} = \begin{bmatrix} 0 \\ 0 \end{bmatrix}. \tag{14}$$

Let us consider the eigenvalue $\lambda = -1$. Inserting $\lambda = -1$ into (14) yields

$$3x_1 + 6x_2 = 0,$$
$$-2x_1 - 4x_2 = 0.$$

These equations are solved by Gaussian elimination. Subtracting $-\frac{2}{3}$ times the first equation from the second equation yields the upper triangular system

$$3x_1 + 6x_2 = 0,$$
$$0x_1 + 0x_2 = 0.$$

Using matrix-vector notation, this system is written

$$\begin{bmatrix} 3 & 6 \\ 0 & 0 \end{bmatrix}\begin{bmatrix} x_1 \\ x_2 \end{bmatrix} = \begin{bmatrix} 0 \\ 0 \end{bmatrix}. \tag{15}$$

Notice that the coefficients of $\mathbf{x}$ in the second row of the coefficient matrix are zero. The coefficient of x_1 is zero since this coefficient is beneath the diagonal in the upper triangular system. The coefficient of x_2 is zero since $\lambda = -1$ is an eigenvalue and a diagonal element of the upper triangular factor associated with $\mathbf{A} - \lambda\mathbf{I}$ must be zero. Since the second equation in (15) is satisfied trivially, it can be dropped. Therefore, $\mathbf{x}$ is an eigenvector corresponding to the eigenvalue $\lambda = -1$ if and only if

$$3x_1 + 6x_2 = 0.$$

Solving for x_1, we have

$$x_1 = -2x_2.$$

In other words, the first component of $\mathbf{x}$ is -2 times the second component and a typical eigenvector corresponding to the eigenvalue $\lambda = -1$ has the form

$$\mathbf{x} = s\begin{bmatrix} -2 \\ 1 \end{bmatrix},$$

where s is an arbitrary scalar. Taking $s = -\frac{1}{2}$ gives us the normalized eigenvector

$$\mathbf{x} = \begin{bmatrix} 1 \\ -\frac{1}{2} \end{bmatrix},$$

whose max-norm is 1. Verify that the normalized eigenvector corresponding to the eigenvalue $\lambda = -2$ is

$$\mathbf{x} = \begin{bmatrix} 1 \\ -\frac{2}{3} \end{bmatrix}.$$

Using these two eigenvectors, let us form a matrix $\mathbf{X}$ whose columns are the eigenvectors:

$$\mathbf{X} = \begin{bmatrix} 1 & 1 \\ -\frac{1}{2} & -\frac{2}{3} \end{bmatrix}.$$

The product $\mathbf{AX}$ is

$$\mathbf{AX} = \begin{bmatrix} 2 & 6 \\ -2 & -5 \end{bmatrix}\begin{bmatrix} 1 & 1 \\ -\frac{1}{2} & -\frac{2}{3} \end{bmatrix} = \begin{bmatrix} -1 & -2 \\ \frac{1}{2} & \frac{4}{3} \end{bmatrix}. \tag{16}$$

By the definition of a matrix product, the first column of $\mathbf{AX}$ is equal to $\mathbf{A}$ times the first column of $\mathbf{X}$. Since the first column of $\mathbf{X}$ is the eigenvector corresponding to the eigenvalue -1, the first column of $\mathbf{AX}$ is equal to the eigenvalue -1 times the corresponding eigenvector. Similarly, the second column of the product $\mathbf{AX}$ is equal to $\mathbf{A}$ times the second column of $\mathbf{X}$. Since the second column of $\mathbf{X}$ is the eigenvector corresponding to the eigenvalue -2, the second column of $\mathbf{AX}$ is equal to the eigenvalue -2 times the corresponding eigenvector. Let $\mathbf{\Lambda}$ denote the diagonal matrix whose diagonal elements are the eigenvalues

$$\mathbf{\Lambda} = \begin{bmatrix} -1 & 0 \\ 0 & -2 \end{bmatrix}.$$

The matrix $\mathbf{X\Lambda}$ is same as $\mathbf{X}$ except that each column of $\mathbf{X}$ is multiplied by the corresponding eigenvalue:

$$\mathbf{X\Lambda} = \begin{bmatrix} 1 & 1 \\ -\frac{1}{2} & -\frac{2}{3} \end{bmatrix}\begin{bmatrix} -1 & 0 \\ 0 & -2 \end{bmatrix} = \begin{bmatrix} -1 & -2 \\ \frac{1}{2} & \frac{4}{3} \end{bmatrix}. \tag{17}$$

Comparing (16) and (17), we see that $\mathbf{AX} = \mathbf{X\Lambda}$. Postmultiplying the identity $\mathbf{AX} = \mathbf{X\Lambda}$ by $\mathbf{X}^{-1}$ gives us the decomposition $\mathbf{A} = \mathbf{X\Lambda X}^{-1}$:

$$\underset{\mathbf{A}}{\begin{bmatrix} 2 & 6 \\ -2 & -5 \end{bmatrix}} = \underset{\mathbf{X}}{\begin{bmatrix} 1 & 1 \\ -\frac{1}{2} & -\frac{2}{3} \end{bmatrix}} \underset{\mathbf{\Lambda}}{\begin{bmatrix} -1 & 0 \\ 0 & -2 \end{bmatrix}} \underset{\mathbf{X}^{-1}}{\begin{bmatrix} 4 & 6 \\ -3 & -6 \end{bmatrix}}.$$

This representation of $\mathbf{A}$ as the product between the matrix $\mathbf{X}$ of eigenvectors, the diagonal matrix $\mathbf{\Lambda}$ of eigenvalues, and the inverse of $\mathbf{X}$ is called the **diagonalization** of $\mathbf{A}$.

Almost every matrix can be diagonalized using its eigenvectors and its eigenvalues. A matrix that cannot be diagonalized is called **defective**. The following matrix is defective:

$$\mathbf{A} = \begin{bmatrix} 1 & 1 \\ 0 & 1 \end{bmatrix}. \tag{18}$$

The characteristic polynomial is

$$\det(\mathbf{A} - \lambda\mathbf{I}) = \begin{bmatrix} 1-\lambda & 1 \\ 0 & 1-\lambda \end{bmatrix} = (1-\lambda)^2.$$

The only zero of the characteristic polynomial is $\lambda = 1$ with multiplicity 2. To compute the eigenvector(s), we form the equation $(\mathbf{A} - \lambda\mathbf{I})\mathbf{x} = \mathbf{0}$ corresponding to the eigenvalue $\lambda = 1$:

$$(\mathbf{A} - \lambda\mathbf{I})\mathbf{x} = \left(\begin{bmatrix} 1 & 1 \\ 0 & 1 \end{bmatrix} - (1)\begin{bmatrix} 1 & 0 \\ 0 & 1 \end{bmatrix}\right)\begin{bmatrix} x_1 \\ x_2 \end{bmatrix} = \begin{bmatrix} 0 & 1 \\ 0 & 0 \end{bmatrix}\begin{bmatrix} x_1 \\ x_2 \end{bmatrix} = \begin{bmatrix} 0 \\ 0 \end{bmatrix}.$$

This equation reduces to $0x_1 + x_2 = 0$. Since x_2 is zero and x_1 can be anything, the only eigenvector is a multiple of

$$\begin{bmatrix} 1 \\ 0 \end{bmatrix}.$$

The 2×2 matrix (18) cannot be diagonalized since it has one rather than two linearly independent eigenvectors.

The defective aspect of matrix (18) is not just the eigenvalue multiplicity. An example of a nondefective matrix with multiple eigenvalues is the identity matrix

$$\mathbf{I} = \begin{bmatrix} 1 & 0 \\ 0 & 1 \end{bmatrix}.$$

The characteristic equation is $(1 - \lambda)^2 = 0$ and $\lambda = 1$ is an eigenvalue of multiplicity 2. Since the eigenvector equation $(\mathbf{A} - \lambda\mathbf{I})\mathbf{x} = \mathbf{0}$ reduces to $\mathbf{0x} = \mathbf{0}$ at $\lambda = 1$, we conclude that any $\mathbf{x}$ is an eigenvector of $\mathbf{I}$. In particular, two linearly independent eigenvectors are

$$\begin{bmatrix} 1 \\ 0 \end{bmatrix} \text{ and } \begin{bmatrix} 0 \\ 1 \end{bmatrix}.$$

The eigenvector matrix $\mathbf{X}$ corresponding to these two eigenvectors is $\mathbf{X} = \mathbf{I}$ and the eigenvalue matrix $\mathbf{\Lambda}$ corresponding to the multiple eigenvalue $\lambda = 1$ is $\mathbf{\Lambda} = \mathbf{I}$. Thus the identity matrix has the diagonalization $\mathbf{I} = \mathbf{X\Lambda X}^{-1} = \mathbf{III}$. Since the identity matrix can be diagonalized, we conclude that both the eigenvalue multiplicity and the nonzero upper right corner of (18) combine to make the matrix defective—changing the corner 1 to zero yields the identity matrix which is not defective.

Finally, let us compute the eigenvalues of the matrix

$$\mathbf{A} = \begin{bmatrix} 4 & -1 \\ 2 & 2 \end{bmatrix}. \tag{19}$$

In this case, the characteristic polynomial det $(\mathbf{A} - \lambda\mathbf{I})$ is

$$\det(\mathbf{A} - \lambda\mathbf{I}) = \det \begin{bmatrix} 4-\lambda & -1 \\ 2 & 2-\lambda \end{bmatrix} = (4-\lambda)(2-\lambda) + 2 = \lambda^2 - 6\lambda + 10,$$

and the characteristic equation is

$$\lambda^2 - 6\lambda + 10 = 0.$$

The roots of this equation are given by the quadratic formula:

$$\lambda = \frac{6 \pm \sqrt{36 - 40}}{2} = 3 \pm \mathbf{i},$$

where $\mathbf{i}$ denotes $\sqrt{-1}$. Since eigenvalues are the zeros of a polynomial and since a polynomial can have complex zeros, it follows that eigenvalues can be complex numbers.

As always, the eigenvectors corresponding to a given eigenvalue are the solutions to $(\mathbf{A} - \lambda\mathbf{I})\mathbf{x} = \mathbf{0}$. For the matrix (19), the eigenvector equation is

$$\begin{bmatrix} 4-\lambda & -1 \\ 2 & 2-\lambda \end{bmatrix} \begin{bmatrix} x_1 \\ x_2 \end{bmatrix} = \begin{bmatrix} 0 \\ 0 \end{bmatrix}. \tag{20}$$

Inserting the eigenvalue $\lambda = 3 + \mathbf{i}$ into equation (20) yields

$$\begin{bmatrix} 1-\mathbf{i} & -1 \\ 2 & -1-\mathbf{i} \end{bmatrix} \begin{bmatrix} x_1 \\ x_2 \end{bmatrix} = \begin{bmatrix} 0 \\ 0 \end{bmatrix}.$$

Subtracting $2/(1 - \mathbf{i})$ times the first equation from the second equation yields

$$\begin{bmatrix} 1-\mathbf{i} & -1 \\ 0 & 0 \end{bmatrix} \begin{bmatrix} x_1 \\ x_2 \end{bmatrix} = \begin{bmatrix} 0 \\ 0 \end{bmatrix}.$$

The second equation can be dropped since it is satisfied trivially. Solving the first equation for x_2, we have

$$x_2 = (1 - \mathbf{i})x_1.$$

Since the second component of $\mathbf{x}$ is $1 - \mathbf{i}$ times the first component of $\mathbf{x}$, the eigenvectors corresponding to the eigenvalue $\lambda = 3 + \mathbf{i}$ have the form

$$\mathbf{x} = s\begin{bmatrix} 1 \\ 1-\mathbf{i} \end{bmatrix}, \tag{21}$$

where s is an arbitrary scalar. The magnitude of the complex number $z = a + b\mathbf{i}$ is

$$|z| = \sqrt{a^2 + b^2}.$$

Since the magnitude of the complex number $1 - \mathbf{i}$ appearing in (21) is $\sqrt{2}$, the second component of (21) is larger in magnitude than the first component and the normalized eigenvector is obtained by setting $s = 1/(1 - \mathbf{i})$ in (21), giving us the vector

$$\begin{bmatrix} .5(1 + \mathbf{i}) \\ 1 \end{bmatrix}. \tag{22}$$

Although the eigenvector corresponding to the conjugate eigenvalue $\lambda = 3 - \mathbf{i}$ can be computed in the same way that we computed the eigenvector (21), there is a shortcut. If $(\lambda, \mathbf{x})$ is an eigenpair for a real matrix $\mathbf{A}$, the complex conjugate of the relation $\mathbf{Ax} = \lambda\mathbf{x}$ can be expressed $\mathbf{Ay} = \mu\mathbf{y}$, where μ is the conjugate of λ and $\mathbf{y}$ is the conjugate of $\mathbf{x}$. Thus the conjugate pair $(\mu, \mathbf{y})$ is also an eigenpair of $\mathbf{A}$. Since the coefficients of $\mathbf{A}$ in (19) are real, the eigenvector corresponding to the conjugate eigenvalue $\lambda = 3 - \mathbf{i}$ is the conjugate of (22):

$$\begin{bmatrix} .5(1 - \mathbf{i}) \\ 1 \end{bmatrix}.$$

Often the eigenvector matrix $\mathbf{X}$ in the diagonalization of $\mathbf{A}$ is constructed from normalized eigenvectors. But theoretically, any linearly independent eigenvectors can be employed. For example, taking $s = 1$ in (21) gives us the eigenvector

$$\mathbf{x} = \begin{bmatrix} 1 \\ 1 - \mathbf{i} \end{bmatrix},$$

corresponding to the eigenvalue $\lambda = 3 + \mathbf{i}$. An unnormalized eigenvector corresponding to the conjugate eigenvalue $\lambda = 3 - \mathbf{i}$ is

$$\mathbf{x} = \begin{bmatrix} 1 \\ 1 + \mathbf{i} \end{bmatrix}.$$

Using these two eigenvectors, we form the eigenvector matrix

$$\mathbf{X} = \begin{bmatrix} 1 & 1 \\ 1 - \mathbf{i} & 1 + \mathbf{i} \end{bmatrix}.$$

The diagonalization of $\mathbf{A}$ is

$$\underset{\mathbf{A}}{\begin{bmatrix} 4 & -1 \\ 2 & 2 \end{bmatrix}} = \underset{\mathbf{X}}{\begin{bmatrix} 1 & 1 \\ 1 - \mathbf{i} & 1 + \mathbf{i} \end{bmatrix}} \underset{\mathbf{\Lambda}}{\begin{bmatrix} 3 + \mathbf{i} & 0 \\ 0 & 3 - \mathbf{i} \end{bmatrix}} \underset{\mathbf{X}^{-1}}{\begin{bmatrix} \frac{1 - \mathbf{i}}{2} & \frac{\mathbf{i}}{2} \\ \frac{1 + \mathbf{i}}{2} & -\frac{\mathbf{i}}{2} \end{bmatrix}}.$$

Usually, it is better to normalize the eigenvectors in $\mathbf{X}$ since the condition number for the normalized $\mathbf{X}$ can be smaller than the condition number for the unnormalized $\mathbf{X}$.

In each 2×2 example above, $\det(\mathbf{A} - \lambda\mathbf{I})$ reduced to a second degree polynomial. Recall that a determinant is a sum of products; for a $n \times n$ matrix, each term in the sum is the product of n different elements from the matrix. The term corresponding to the highest power of λ in $\det(\mathbf{A} - \lambda\mathbf{I})$ is the product of the diagonal elements $a_{ii} - \lambda$ for $i = 1$ to n. Since $\prod_{i=1}^{n} (a_{ii} - \lambda) = (-1)^n\lambda^n$ plus lower order terms, the characteristic polynomial for a $n \times n$ matrix is a nth degree polynomial.

The eigenproblem for a 2×2 matrix is pretty easy—we evaluate the characteristic polynomial and we compute its zeros using the quadratic formula. But for larger matrices, this strategy hits a snag. Even though the eigenproblem is well conditioned (small changes in the matrix coefficients produce small changes in the eigenvalues), the characteristic equation can be ill conditioned (small changes in the polynomial coefficients produce large changes in the roots). Wilkinson [W5] gives a vivid illustration: Consider the 20×20 diagonal matrix

$$\mathbf{A} = \begin{bmatrix} 1 & & & & \\ & 2 & & & \\ & & \cdot & & \\ & & & \cdot & \\ & & & & \cdot \\ & & & & & 20 \end{bmatrix}. \tag{23}$$

Since the eigenvalues of a triangular matrix are the diagonal elements, the eigenvalues of the matrix (23) are $\lambda = 1, \lambda = 2, \cdots, \lambda = 20$. Now suppose that ε is added to some element of $\mathbf{A}$. Since the resulting matrix is either upper triangular or lower triangular, the eigenvalues are the diagonal elements of the perturbed matrix. If ε is added to an off-diagonal element, the diagonal elements as well as the eigenvalues are not affected. And if ε is added to a diagonal element of $\mathbf{A}$, the corresponding eigenvalue shifts by ε. The eigenproblem corresponding to matrix (23) is well conditioned since a small change in a coefficient produces a small change in the eigenvalues.

Now consider the characteristic polynomial corresponding to (23):

$$\det(\mathbf{A} - \lambda\mathbf{I}) = (1 - \lambda)(2 - \lambda) \cdots (20 - \lambda)$$
$$= \lambda^{20} - 210\lambda^{19} + \cdots + 2432902008176640000.$$

In Section 4-7 we saw that a small change in the coefficient of λ^{19} produces a large change in the zeros of this polynomial. The roots of the characteristic polynomial are extremely sensitive to changes in the polynomial coefficients even though the eigenvalues of the corresponding matrix are not very sensitive to changes in the matrix coefficients. When computing eigenvalues, the characteristic polynomial should never be evaluated explicitly; due to rounding errors, the computed coefficients of the characteristic polynomial deviate slightly from their correct values and small errors in the coefficients can produce large errors in the eigenvalues.

Corresponding to each matrix, we have its characteristic polynomial. This relation between a matrix and a polynomial can also be reversed since every polynomial is a characteristic polynomial for the corresponding "companion matrix." For example, $\lambda^2 - 3\lambda + 2$ is the characteristic polynomial for the matrix

$$\begin{bmatrix} 0 & 1 \\ -2 & 3 \end{bmatrix}$$

since

$$\det(\mathbf{A} - \lambda\mathbf{I}) = \det\begin{bmatrix} -\lambda & 1 \\ -2 & 3 - \lambda \end{bmatrix} = -\lambda(3 - \lambda) + 2 = \lambda^2 - 3\lambda + 2.$$

Similarly, $\lambda^4 + 5\lambda^3 + 4\lambda^2 - 8\lambda + 6$ is the characteristic polynomial for the matrix

$$\begin{bmatrix} 0 & 1 & 0 & 0 \\ 0 & 0 & 1 & 0 \\ 0 & 0 & 0 & 1 \\ -6 & 8 & -4 & -5 \end{bmatrix}.$$

A matrix that is completely zero except for the superdiagonal elements, which are 1, and the elements in the last row, which are arbitrary, is called a **companion matrix**. In general, the polynomial

$$\lambda^n + a_1\lambda^{n-1} + a_2\lambda^{n-2} + \cdots + a_n$$

is the characteristic polynomial for the $n \times n$ companion matrix whose last row is

$$[-a_n \quad -a_{n-1} \quad -a_{n-2} \quad \cdots \quad -a_1].$$

This correspondence between a companion matrix and the coefficients of a polynomial explains the complexity of the eigenproblem. If there is a simple formula for the eigenvalues of a matrix, there is a simple formula for the zeros of a polynomial. Since it has been proved that the zeros of polynomials of degree greater than 4 cannot be expressed as "simple functions" of the coefficients, there is no simple formula for the eigenvalues of matrices larger than 4×4. Therefore, algorithms for the eigenproblem must be iterative.

Exercises

6-2.1 Compute the diagonalization of each of the following matrices.

$$\text{(a)} \quad \begin{bmatrix} 3 & 4 \\ 4 & -3 \end{bmatrix} \qquad \text{(b)} \quad \begin{bmatrix} 2 & -3 \\ 3 & 2 \end{bmatrix}$$

6-2.2 We saw in Chapter 5 that a typical 2×2 Householder reflection **H** and a Givens rotation **G** have the form

$$\mathbf{H} = \begin{bmatrix} a & b \\ b & -a \end{bmatrix} \quad \text{and} \quad \mathbf{G} = \begin{bmatrix} c & -s \\ s & c \end{bmatrix},$$

where $a^2 + b^2 = 1 = c^2 + s^2$. What are the eigenvalues of **H** and of **G**? How does the angle θ associated with the Givens rotation relate to the eigenvalues of **G**? What are the eigenvectors of **G**?

6-2.3 If $\mathbf{X\Lambda X}^{-1}$ is the diagonalization of **A**, show that the rows of $\mathbf{X}^{-1}$ are the left eigenvectors of **A** (left eigenvectors are defined in Exercise 6-1.3).

6-2.4 What is the relationship between the eigenpairs of **A** and the eigenpairs of $\mathbf{I} + \mathbf{A}$? What is the relationship between the eigenpairs of **A** and the eigenpairs of $\mathbf{I} + 2\mathbf{A}$?

6-2.5 If **w** is a unit vector in the Euclidean norm, what are the eigenpairs of the matrix $\mathbf{ww}^T$? (*Hint:* Consider the vector $\mathbf{x} = \mathbf{w}$ as well as the set of vectors perpendicular to **w**.)

6-2.6 If **w** is a unit vector in the Euclidean norm, what are the eigenpairs of the Householder matrix $\mathbf{H} = \mathbf{I} - 2\mathbf{ww}^T$? (*Hint:* Combine your observations from Exercises 6-2.4 and 6-2.5.)

6-2.7 Let **W** denote any matrix whose columns are orthonormal. What are the eigenpairs of $\mathbf{WW}^T$?

6-2.8 Let **A** denote any matrix with linearly independent columns. What are the eigenpairs of the projection matrix $\mathbf{P} = \mathbf{I} - \mathbf{A}(\mathbf{A}^T\mathbf{A})^{-1}\mathbf{A}^T$? (*Hint:* Consider the product **PA**.)

6-2.9 Given a rectangular matrix $\mathbf{W}$, let $\mathbf{X\Lambda X}^{-1}$ denote the diagonalization of $\mathbf{W}^T\mathbf{W}$. Show that each diagonal element of $\mathbf{\Lambda}$ is an eigenvalue of $\mathbf{WW}^T$, while the corresponding eigenvectors are the columns of $\mathbf{WX}$. Are there any additional eigenpairs associated with $\mathbf{WW}^T$? (*Hint:* Consider the null space of $\mathbf{W}^T$.)

6-2.10 Consider a symmetric matrix with distinct eigenvalues. If the eigenvector matrix $\mathbf{X}$ is normalized so that each column $\mathbf{x}_i$ is a unit vector in the Euclidean norm, show that $\mathbf{X}$ is an orthogonal matrix. (*Hint:* See Exercise 6-1.4.)

6-2.11 Consider the system of differential equations

$$\frac{du\,(t)}{dt} = 3u(t) + 4v(t),$$

$$\frac{dv(t)}{dt} = 4u(t) - 3v(t).$$

Find the solution to this system that satisfies the boundary conditions $u(0) = -3$ and $v(1) = -2(e^{-5} + e^{5})$. Is $u = 0$ and $v = 0$ a stable equilibrium point?

6-3. GERSCHGORIN'S THEOREM

Although computing the eigenvalues of a matrix may not be easy, Gerschgorin found a simple way to estimate eigenvalues. Conceptually, complex numbers are points in the plane. The complex number 2 + 3**i**, which has real part 2 and imaginary part 3, is located 2 units to the right of the origin and 3 units above the origin (see Figure 6-4). The vertical axis is called the imaginary axis and the horizontal axis is called the real axis. Given a square matrix $\mathbf{A}$, let r_i denote the sum across row i of the magnitude of each element, excluding the diagonal element. That is, r_i is defined by

$$r_i = \sum_{\substack{j=1\\ j\neq i}}^{n} |a_{ij}|.$$

And let R_i denote the collection of points in the complex plane whose distance to a_{ii} is at most r_i. In other words, R_i is the collection of complex numbers z with the property that

$$|z - a_{ii}| \leq r_i = \sum_{\substack{j=1\\ j\neq i}}^{n} |a_{ij}|.$$

Gerschgorin observed that the eigenvalues of $\mathbf{A}$ are contained in the union of the disks R_i. Moreover, if a collection of k disks does not touch the remaining disks, exactly k eigenvalues are contained in the collection.

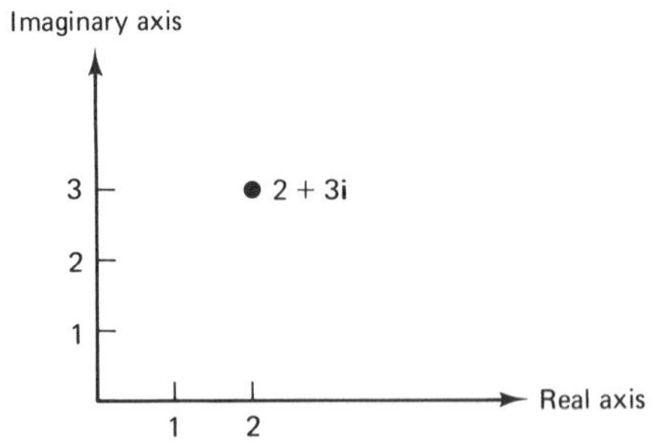

Figure 6-4 The plane of complex numbers.

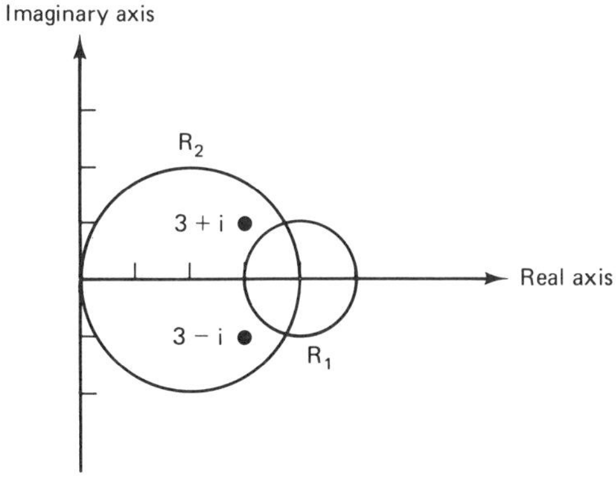

Figure 6-5 Gerschgorin disks for the matrix (24).

We illustrate Gerschgorin's theorem with the matrix

$$\mathbf{A} = \begin{bmatrix} 4 & -1 \\ 2 & 2 \end{bmatrix}, \tag{24}$$

whose eigenvalues $3 \pm \mathbf{i}$ were computed in Section 6-2. In this case, R_1 is the disk with center $4 + 0\mathbf{i}$ and radius 1 while R_2 is the disk with center $2 + 0\mathbf{i}$ and radius 2. Figure 6-5 shows that the eigenvalues $3 \pm \mathbf{i}$ are contained in the union of the disks R_1 and R_2. For a second example, consider the matrix

$$\begin{bmatrix} 1 & .2 \\ .1 & 2 \end{bmatrix}. \tag{25}$$

In Figure 6-6 we see that the disk R_1 with center $1 + 0\mathbf{i}$ and radius .2 is disjoint from the disk R_2 with center $2 + 0\mathbf{i}$ and radius .1. Therefore, R_1 contains an eigenvalue λ_1 and R_2 contains an eigenvalue λ_2. Also recall that the eigenvalues of a matrix with real coefficients occur in conjugate pairs. Since each disk in Figure 6-6 contains exactly one eigenvalue and since complex eigenvalues must occur in conjugate pairs, each eigenvalue for the matrix (25) is real. Since the eigenvalues lie both in the disks and on the real axis, it follows that

$$.8 \leqq \lambda_1 \leqq 1.2 \qquad \text{and} \qquad 1.9 \leqq \lambda_2 \leqq 2.1. \tag{26}$$

Although Gerschgorin's theorem was stated for row sums, column sums work equally well. Given a matrix $\mathbf{A}$, let C_j be the set of complex numbers z with the property that

$$|z - a_{jj}| \leqq \sum_{\substack{i=1 \\ i \neq j}}^{n} |a_{ij}|.$$

In other words, C_j is the collection of complex numbers whose distance to a_{jj} is less than or equal to the sum of the magnitudes of the elements in column j

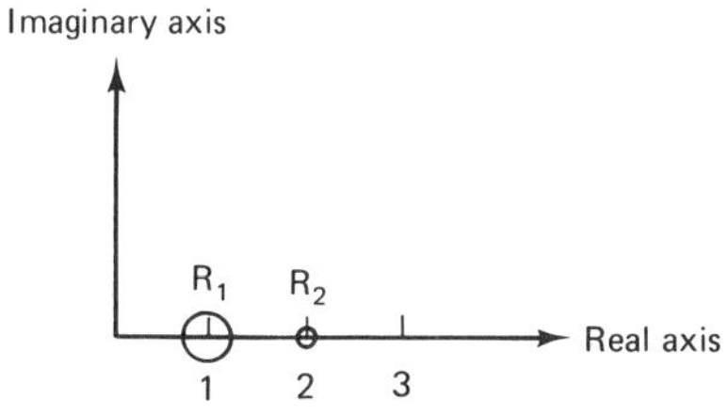

Figure 6-6 Gerschgorin disks for the matrix (25).

excluding the diagonal element. By Gerschgorin's theorem, the eigenvalues of **A** lie in the union of the C_j and if a collection of k disks does not touch the remaining disks, then exactly k eigenvalues are contained in the collection. Forming column disks for the matrix (25), we conclude that

$$.9 \leqq \lambda_1 \leqq 1.1 \quad \text{and} \quad 1.8 \leqq \lambda_2 \leqq 2.2,$$

and combining these relations with (26) yields the sharper inequalities

$$.9 \leqq \lambda_1 \leqq 1.1 \quad \text{and} \quad 1.9 \leqq \lambda_2 \leqq 2.1.$$

To verify Gerschgorin's theorem, let $(\lambda, \mathbf{x})$ be any eigenpair of **A** and let x_m be a component of **x** with largest magnitude. The mth equation in the system $\mathbf{Ax} = \lambda\mathbf{x}$ is

$$\sum_{j=1}^{n} a_{mj}\, x_j = \lambda x_m.$$

Isolating from the sum the term corresponding to $j = m$ and rearranging the equation, we have

$$x_m(\lambda - a_{mm}) = \sum_{\substack{j=1 \\ j \neq m}}^{n} a_{mj}\, x_j.$$

Dividing by x_m yields

$$\lambda - a_{mm} = \sum_{\substack{j=1 \\ j \neq m}}^{n} a_{mj} \frac{x_j}{x_m}.$$

Finally, we take the absolute value and apply the triangle inequality to obtain

$$|\lambda - a_{mm}| \leqq \sum_{\substack{j=1 \\ j \neq m}}^{n} |a_{mj}| \frac{|x_j|}{|x_m|}. \tag{27}$$

Since the magnitude of x_j is less than or equal to the magnitude of x_m, $|x_j|/|x_m| \leqq 1$ for each j, and by (27), we have

$$|\lambda - a_{mm}| \leqq \sum_{\substack{j=1 \\ j \neq m}}^{n} |a_{mj}|.$$

In other words, λ is contained in the disk R_m. But if λ is contained in one of the disks, it is contained in the union of the disks.

Now let us consider the Gerschgorin column disks. Since the determinant of a matrix is equal to the determinant of its transpose, we have

$$\det(\mathbf{A} - \lambda\mathbf{I}) = \det(\mathbf{A} - \lambda\mathbf{I})^T = \det(\mathbf{A}^T - \lambda\mathbf{I}).$$

This implies that each root of the equation $\det(\mathbf{A} - \lambda\mathbf{I}) = 0$ is also a root of the equation $\det(\mathbf{A}^T - \lambda\mathbf{I}) = 0$. Hence the eigenvalues of **A** are the same as the eigenvalues of $\mathbf{A}^T$. Since the row disks for $\mathbf{A}^T$ are the column disks for **A**, we conclude that λ is contained in the union of the column disks.

The last part of Gerschgorin's theorem states that if a collection of k disks does not touch the remaining disks, then exactly k eigenvalues are contained in the collection. To establish this result, we form a matrix $\mathbf{A}(\varepsilon)$ which is identical to **A** except that each of **A**'s off diagonal elements is multiplied by ε. Since $\mathbf{A}(0)$

is a diagonal matrix, the eigenvalues of $\mathbf{A}(0)$ are the diagonal elements a_{11}, a_{22}, $\cdots$, a_{nn}. When $\varepsilon = 0$, the Gerschgorin disks are simply the points a_{11}, a_{22}, $\cdots$, a_{nn} and each disk contains exactly one eigenvalue. As ε increases, the disks grow continuously and the eigenvalues move in a continuous fashion.† In particular, as ε varies continuously, an eigenvalue cannot jump from one collection of disks to another collection. Since the eigenvalues are contained in the union of the disks for each ε and since each disk contains an eigenvalue when ε is zero, we conclude that for any ε, a collection of k disks which is disjoint from the remaining disks contains exactly k eigenvalues. Since $\mathbf{A}(\varepsilon)$ is equal to the original matrix $\mathbf{A}$ when $\varepsilon = 1$, it follows that for the original matrix $\mathbf{A}$, a collection of k disks which does not touch the remaining disks contains exactly k eigenvalues.

Exercise

6-3.1 Consider the differential equation $\mathbf{x}'(t) = \mathbf{A}\mathbf{x}(t)$. For each of the following choices of $\mathbf{A}$, use Gerschgorin's theorem to decide whether $\mathbf{x} = \mathbf{0}$ is a stable equilibrium point of the differential equation:

$$\text{(a)}\quad \mathbf{A} = \begin{bmatrix} -3 & 2 & 4 \\ 1 & -4 & -1 \\ 1 & 1 & -6 \end{bmatrix} \qquad \text{(b)}\quad \mathbf{A} = \begin{bmatrix} -3 & 1 & 1 \\ 1 & -4 & 2 \\ 0 & 1 & 2 \end{bmatrix}$$

6-4. THE POWER METHOD

Given a matrix $\mathbf{A}$, the **power method** is defined by the iteration

$$\mathbf{x}^{\text{new}} = \mathbf{A}\mathbf{x}^{\text{old}}.$$

Except for special starting points, the iterations converge to an eigenvector corresponding to the eigenvalue of $\mathbf{A}$ with largest magnitude. The eigenvalue of largest magnitude is often called the **dominant eigenvalue**. Let us apply the power method to the matrix

$$\mathbf{A} = \begin{bmatrix} 2 & 6 \\ -2 & -5 \end{bmatrix} \tag{28}$$

encountered in Sections 6-1 and 6-2. Making the starting guess

$$\mathbf{x} = \begin{bmatrix} 1 \\ 0 \end{bmatrix},$$

we obtain the iterations in Table 6-1 (see page 254).

Since the normalized eigenpairs for the matrix (28) are

$$\lambda = -1, \quad \mathbf{x} = \begin{bmatrix} 1 \\ -\frac{1}{2} \end{bmatrix} \qquad \text{and} \qquad \lambda = -2, \quad \mathbf{x} = \begin{bmatrix} 1 \\ -\frac{2}{3} \end{bmatrix},$$

the dominant eigenvalue is $\lambda = -2$. In what sense are the iterations of Table 6-1 (on page 254) converging to the eigenvector corresponding to $\lambda = -2$? In Table 6-2 (on page 254) we normalize the power iterations by dividing with the absolute largest component each step.

† For a proof that the eigenvalues of a matrix depend continuously on the coefficients, see page 191 of [F8].

TABLE 6-1
POWER ITERATIONS
FOR THE MATRIX (28)

x_1	x_2
1	0
2	−2
−8	6
20	−14
−44	30
92	−62
−188	126

Clearly, the normalized iterations converge to the normalized eigenvector. Since unnormalized power iterations can generate overflows or underflows on a computer, the current iteration is often multipled by a scale factor s whenever the components of **x** are large or small in magnitude. Arithmetic errors associated with multiplication by s can be eliminated if s is a power of 2. For example, if some component of **x** is greater than 2^{70}, then multiply **x** by $s = 2^{-70}$. If every component of **x** is less than 2^{-70} in magnitude, then multiply **x** by $s = 2^{70}$.

To understand why the power method works, we must expand the iterations in terms of the eigenvectors. Suppose that **A** is a $n \times n$ matrix. Since the degree of **A**'s characteristic polynomial is n, **A** has n eigenvalues $\lambda_1, \lambda_2, \cdots, \lambda_n$ counting multiplicities. Let us assume that the eigenvalues are indexed according to their magnitude and λ_1 is larger in magnitude than the other eigenvalues:

$$|\lambda_1| > |\lambda_2| \geqq \cdots \geqq |\lambda_n|.$$

Provided **A** is not defective, any vector $\mathbf{x}_0$ can be expressed as a sum of eigenvectors:

$$\mathbf{x}_0 = \mathbf{e}_1 + \mathbf{e}_2 + \cdots + \mathbf{e}_n,$$

where either $\mathbf{e}_i$ is zero or $\mathbf{e}_i$ is an eigenvector corresponding to the eigenvalue λ_i. In the power method, the starting **x**, denoted $\mathbf{x}_0$, is multiplied by **A** to obtain the new **x**, denoted $\mathbf{x}_1$:

$$\mathbf{x}_1 = \mathbf{A}\mathbf{x}_0 = \mathbf{A}(\mathbf{e}_1 + \mathbf{e}_2 + \cdots + \mathbf{e}_n) = \mathbf{A}\mathbf{e}_1 + \mathbf{A}\mathbf{e}_2 + \cdots + \mathbf{A}\mathbf{e}_n$$
$$= \lambda_1\mathbf{e}_1 + \lambda_2\mathbf{e}_2 + \cdots + \lambda_n\mathbf{e}_n.$$

TABLE 6-2 NORMALIZED
POWER ITERATIONS
FOR MATRIX (28)

x_1	x_2
1.000	0.000
1.000	−1.000
1.000	−.750
1.000	−.700
1.000	−.682
1.000	−.674
1.000	−.670

In the next iteration, we multiply $\mathbf{x}_1$ by $\mathbf{A}$ to obtain $\mathbf{x}_2$:

$$\mathbf{x}_2 = \mathbf{A}\mathbf{x}_1 = \mathbf{A}(\lambda_1\mathbf{e}_1 + \lambda_2\mathbf{e}_2 + \cdots + \lambda_n\mathbf{e}_n) = \lambda_1\mathbf{A}\mathbf{e}_1 + \lambda_2\mathbf{A}\mathbf{e}_2 + \cdots + \lambda_n\mathbf{A}\mathbf{e}_n$$
$$= \lambda_1^2\mathbf{e}_1 + \lambda_2^2\mathbf{e}_2 + \cdots + \lambda_n^2\mathbf{e}_n.$$

After k iterations, we obtain the vector

$$\mathbf{x}_k = \lambda_1^k\mathbf{e}_1 + \lambda_2^k\mathbf{e}_2 + \cdots + \lambda_n^k\mathbf{e}_n.$$

Multiplying a vector by a scalar just changes the vector's length, not its direction. Multiplying $\mathbf{x}_k$ by λ_1^{-k} yields

$$\lambda_1^{-k}\mathbf{x}_k = \mathbf{e}_1 + \left(\frac{\lambda_2}{\lambda_1}\right)^k \mathbf{e}_2 + \cdots + \left(\frac{\lambda_n}{\lambda_1}\right)^k \mathbf{e}_n. \tag{29}$$

Since λ_1 is larger than the other eigenvalues in magnitude, the ratios λ_i/λ_1 in (29) are less than 1 in magnitude and the coefficients of $\mathbf{e}_2$ through $\mathbf{e}_n$ approach zero as k increase. Since the direction of $\mathbf{x}_k$ approaches the direction of $\mathbf{e}_1$ as k increases, the power method converges to an eigenvector associated with the dominant eigenvalue λ_1. Of course, the power iterations can be computed without knowing the eigenpairs of $\mathbf{A}$. The eigenpairs were introduced to help us understand why the algorithm works.

Since $|\lambda_2|$ is greater than or equal to $|\lambda_3|$ through $|\lambda_n|$, it follows that the coefficient of $\mathbf{e}_2$ in (29) is at least as large in magnitude as the other coefficients in (29). Hence the convergence speed of the power method is governed by the ratio $|\lambda_2|/|\lambda_1|$. The convergence is slow if the ratio is near 1 and the convergence is fast if the ratio is near zero. Let us consider the matrix

$$\mathbf{A} = \begin{bmatrix} 1 & 0 \\ 0 & r \end{bmatrix},$$

where $|r|$ is less than 1. Starting from

$$\mathbf{x}_0 = \begin{bmatrix} 1 \\ 1 \end{bmatrix},$$

the power iterations are

$$\mathbf{x}_1 = \begin{bmatrix} 1 \\ r \end{bmatrix}, \quad \mathbf{x}_2 = \begin{bmatrix} 1 \\ r^2 \end{bmatrix}, \quad \ldots, \quad \mathbf{x}_k = \begin{bmatrix} 1 \\ r^k \end{bmatrix}.$$

Since the normalized eigenpairs are

$$\lambda_1 = 1, \quad \mathbf{e}_1 = \begin{bmatrix} 1 \\ 0 \end{bmatrix} \quad \text{and} \quad \lambda_2 = r, \quad \mathbf{e}_2 = \begin{bmatrix} 0 \\ 1 \end{bmatrix},$$

the eigenvalue ratio is $|\lambda_2|/|\lambda_1| = |r|$. The $\mathbf{x}_k$ converge quickly to $\mathbf{e}_1$ when r is near zero since the second component of $\mathbf{x}_k$ approaches zero quickly. But if r is near 1, many iterations are needed before $\mathbf{x}_k$ is near $\mathbf{e}_1$. In particular, if $r = .99$, the second component of $\mathbf{x}_k$ is less than 10^{-6} after 1375 iterations. As r approaches 1, the power method converges more slowly.

The power method also works for a defective matrix, but the convergence is very slow. Let us consider the defective matrix

$$\mathbf{A} = \begin{bmatrix} 1 & 1 \\ 0 & 1 \end{bmatrix}$$

and the starting guess

$$\mathbf{x}_0 = \begin{bmatrix} 0 \\ 1 \end{bmatrix}.$$

The power iterations $\mathbf{x}^{\text{new}} = \mathbf{A}\mathbf{x}^{\text{old}}$ are

$$\mathbf{x}_1 = \begin{bmatrix} 1 \\ 1 \end{bmatrix}, \quad \mathbf{x}_2 = \begin{bmatrix} 2 \\ 1 \end{bmatrix}, \cdots, \quad \mathbf{x}_k = \begin{bmatrix} k \\ 1 \end{bmatrix}.$$

To normalize x_k, we multiply by k^{-1}:

$$k^{-1}\mathbf{x}_k = \begin{bmatrix} 1 \\ k^{-1} \end{bmatrix}.$$

The power iterations are converging to the eigenvector

$$\begin{bmatrix} 1 \\ 0 \end{bmatrix}$$

corresponding to the eigenvalue $\lambda = 1$ and the error in the normalized eigenvector $\mathbf{x}_k$ is $1/k$. Therefore, 1 million iterations are needed to reduce the error to 10^{-6}.

The power method generates an approximate eigenvector. To estimate the corresponding eigenvalue, we return to the relation $\mathbf{A}\mathbf{e}_1 = \lambda_1\mathbf{e}_1$ between the dominant eigenvalue λ_1 and the corresponding eigenvector $\mathbf{e}_1$. Since the iterations $\mathbf{x}_1, \mathbf{x}_2, \cdots$ generated by the power method are converging to $\mathbf{e}_1$,

$$\mathbf{A}\mathbf{x}_k \approx \lambda_1\mathbf{x}_k. \tag{30}$$

In each iteration of the power method, $\mathbf{x}_k$ is multiplied by $\mathbf{A}$ to obtain $\mathbf{x}_{k+1}$. Substituting $\mathbf{x}_{k+1}$ for $\mathbf{A}\mathbf{x}_k$ in (30), we have

$$\lambda_1\mathbf{x}_k \approx \mathbf{x}_{k+1}.$$

Therefore, the least squares solution μ to the overdetermined linear system

$$\mu\mathbf{x}_k = \mathbf{x}_{k+1}$$

is an estimate for λ_1. In Exercise 5-2.2, you show that

$$\mu = \frac{\mathbf{x}_k^T\mathbf{x}_{k+1}}{\mathbf{x}_k^T\mathbf{x}_k}.$$

This eigenvalue estimate is called the **Raleigh quotient**. Table 6-3 evaluates the Raleigh quotients for the power iterations of Table 6-1.

TABLE 6-3 RALEIGH QUOTIENTS CORRESPONDING TO TABLE 6-1

x_1	x_2	μ
1	0	
2	−2	2.00
−8	6	−3.50
20	−14	−2.44
−44	30	−2.18
92	−62	−2.08
−188	126	−2.04

For example, the first μ in Table 6-3 is given by

$$\mu = \frac{[1 \quad 0]\begin{bmatrix} 2 \\ -2 \end{bmatrix}}{[1 \quad 0]\begin{bmatrix} 1 \\ 0 \end{bmatrix}} = \frac{2}{1} = 2.$$

Clearly, the Raleigh quotients are converging to $\lambda = -2$, the dominant eigenvalue.

The power method behaves differently for the matrix

$$\mathbf{A} = \begin{bmatrix} 4 & -1 \\ 2 & 2 \end{bmatrix} \tag{31}$$

studied in Section 6-2. With the starting guess

$$\mathbf{x} = \begin{bmatrix} 1 \\ 10 \end{bmatrix},$$

the iterations in Table 6-4 wander aimlessly. Recall that matrix (31) has complex eigenvectors. Since the coefficients of **A** are real, the power iteration $\mathbf{x}^{\text{new}} = \mathbf{A}\mathbf{x}^{\text{old}}$ generates real vectors when the starting guess is real and these real vectors can never converge to the complex eigenvectors of **A**. In our theoretical justification of the power method, we assumed that λ_1 is larger in magnitude than the other eigenvalues. This assumption is crucial since it implies that $|\lambda_2|/|\lambda_1|$ is less than 1 and the coefficients in (29) approach zero as k increases. Since λ_1 is the conjugate of λ_2 and $|\lambda_1| = |\lambda_2|$ for the matrix (31), it appears that the power method cannot work.

Actually, the method works fine if we interpret the output correctly. Recall that k power iterations generate the vector

$$\mathbf{x}_k = \lambda_1^k\mathbf{e}_1 + \lambda_2^k\mathbf{e}_2 + \cdots + \lambda_n^k\mathbf{e}_n.$$

If λ_1 is larger in magnitude than the other eigenvalues, $\mathbf{x}_k$ points in the direction of $\mathbf{e}_1$ as k increases. But if $|\lambda_1| = |\lambda_2|$ while the other eigenvalues are smaller in magnitude, then $\mathbf{x}_k$ wanders in the space spanned by the vectors $\mathbf{e}_1$ and $\mathbf{e}_2$. Even though the $\mathbf{x}_k$ never converge, they approach the space spanned by $\mathbf{e}_1$ and $\mathbf{e}_2$.

A procedure similar to that employed for a single real eigenvalue can be

TABLE 6-4 POWER ITERATIONS FOR THE MATRIX (31)

unnormalized		normalized	
x_1	x_2	x_1	x_2
1	10	.10	1.00
−6	22	−.27	1.00
−46	32	1.00	−.70
−216	−28	1.00	.13
−836	−488	1.00	.58
−2856	−2648	1.00	.93
−8776	−11008	.80	1.00

used to compute a complex conjugate pair of eigenvalues $\lambda_1 = a + b\mathbf{i}$ and $\lambda_2 = a - b\mathbf{i}$. It can be shown that λ_1 and λ_2 are the roots of the quadratic equation

$$\lambda^2 + c_1\lambda + c_2 = 0, \tag{32}$$

where c_1 and c_2 have the property that for any three consecutive iterations $\mathbf{x}_k$, $\mathbf{x}_{k+1}$, and $\mathbf{x}_{k+2}$ in the power method, we have

$$\mathbf{x}_{k+2} + c_1\mathbf{x}_{k+1} + c_2\mathbf{x}_k \approx \mathbf{0}. \tag{33}$$

To estimate c_1 and c_2, we use the least squares solution d_1 and d_2 to the overdetermined linear system

$$d_1\mathbf{x}_{k+1} + d_2\mathbf{x}_k = -\mathbf{x}_{k+2}.$$

Referring to Section 5-2, d_1 and d_2 are the solution to the normal equation

$$\begin{bmatrix} \mathbf{x}_{k+1}\cdot\mathbf{x}_{k+1} & \mathbf{x}_{k+1}\cdot\mathbf{x}_k \\ \mathbf{x}_{k+1}\cdot\mathbf{x}_k & \mathbf{x}_k\cdot\mathbf{x}_k \end{bmatrix}\begin{bmatrix} d_1 \\ d_2 \end{bmatrix} = -\begin{bmatrix} \mathbf{x}_{k+2}\cdot\mathbf{x}_{k+1} \\ \mathbf{x}_{k+2}\cdot\mathbf{x}_k \end{bmatrix}, \tag{34}$$

where $\cdot$ denotes the dot product. Corresponding to the first three iterations in Table 6-4, the elements of the coefficient matrix in (34) are the following:

$$\mathbf{x}_1\cdot\mathbf{x}_1 = [-6 \quad 22]\begin{bmatrix} -6 \\ 22 \end{bmatrix} = 520,$$

$$\mathbf{x}_1\cdot\mathbf{x}_0 = [-6 \quad 22]\begin{bmatrix} 1 \\ 10 \end{bmatrix} = 214,$$

$$\mathbf{x}_0\cdot\mathbf{x}_0 = [1 \quad 10]\begin{bmatrix} 1 \\ 10 \end{bmatrix} = 101,$$

$$\mathbf{x}_2\cdot\mathbf{x}_1 = [-46 \quad 32]\begin{bmatrix} -6 \\ 22 \end{bmatrix} = 980,$$

$$\mathbf{x}_2\cdot\mathbf{x}_0 = [-46 \quad 32]\begin{bmatrix} 1 \\ 10 \end{bmatrix} = 274.$$

With these substitutions, the linear system (34) becomes

$$\begin{bmatrix} 520 & 214 \\ 214 & 101 \end{bmatrix}\begin{bmatrix} d_1 \\ d_2 \end{bmatrix} = \begin{bmatrix} -980 \\ -274 \end{bmatrix}.$$

The solution is $d_1 = -6$ and $d_2 = 10$, and by (32), the eigenvalues of matrix (31) are approximately the roots of the equation

$$\lambda^2 - 6\lambda + 10 = 0. \tag{35}$$

Since this quadratic equation is precisely the characteristic equation computed in Section 6-2, the roots of (35) are exactly equal to the eigenvalues of matrix (31). A 2×2 matrix is special and the coefficients of the approximating quadratic actually equal the coefficients of the true quadratic. In general, the coefficients d_1 and d_2 of the approximating quadratic merely approach the coefficients c_1 and c_2 of the true quadratic (32) as k increases.

To summarize, solving (34) gives us the coefficients of a quadratic and the zeros of this quadratic approximate the two dominant eigenvalues. When these dominant eigenvalues are a complex conjugate pair $\lambda_1 = a + b\mathbf{i}$ and $\lambda_2 = a - b\mathbf{i}$, it can be shown that the corresponding eigenvectors are approximately equal to

$$b\mathbf{x}^{\text{old}} \pm (a\mathbf{x}^{\text{old}} - \mathbf{x}^{\text{new}})\mathbf{i}$$

provided the power iterations $\mathbf{x}^{\text{new}} = \mathbf{A}\mathbf{x}^{\text{old}}$ are real. Unfortunately, this eigenvector estimate is inaccurate when b is small. As b goes to zero, $\lambda_1 = a + b\mathbf{i}$ and $\lambda_2 = a - b\mathbf{i}$ approach a and the two eigenvalues coalesce into a single eigenvalue of multiplicity 2. As multiple zeros gave problems (see Section 4-2), multiple eigenvalues are also troublesome.

We now modify the power method to obtain better approximations to the eigenvectors when b is near zero. The iterations start with the standard power method. After each iteration, the eigenvalues are estimated by the roots of (32), where the coefficients c_1 and c_2 are the solutions d_1 and d_2 to (34). When the eigenvalue approximations begin to converge, the standard iteration is stopped, say at iteration k, and a matrix $\mathbf{X}_k$ with two columns is formed. The first column of $\mathbf{X}_k$ is $\mathbf{x}_k$ and the second column is $\mathbf{x}_{k-1}$. The power method now continues, but in a modified format. Each iteration has two steps:

$$\mathbf{Q}_k\mathbf{R}_k = \mathbf{X}_k \qquad \text{and} \qquad \mathbf{X}_{k+1} = \mathbf{A}\mathbf{Q}_k. \tag{36}$$

In the first step, $\mathbf{X}_k$ is factored (using one of the algorithms from Chapter 5) into the product $\mathbf{Q}_k\mathbf{R}_k$ between a matrix $\mathbf{Q}_k$ with two orthonormal columns and an upper triangular matrix $\mathbf{R}_k$. Since the power iterations wander in the space spanned by the eigenvectors corresponding to dominant eigenvalues and since the columns of $\mathbf{Q}_k$ are an orthonormal basis for the space spanned by the columns of $\mathbf{X}_k$, the first step of (36) approximates an orthonormal basis for the space spanned by the eigenvectors corresponding to dominant eigenvalues. In the second step, $\mathbf{Q}_k$ is multiplied by $\mathbf{A}$ to obtain $\mathbf{X}_{k+1}$. As the iterations progress, the space spanned by the columns of $\mathbf{X}_k$ and $\mathbf{Q}_k$ approach the space spanned by the eigenvectors $\mathbf{e}_1$ and $\mathbf{e}_2$ corresponding to the dominant eigenvalues.

Estimates for the dominant eigenpairs are obtained in the following way: Since the columns of $\mathbf{Q}_k$ are orthonormal, the projection of a vector $\mathbf{x}$ into the space spanned by the columns of $\mathbf{Q}_k$ is given by $\mathbf{Q}_k\mathbf{Q}_k{}^T\mathbf{x}$. In the same fashion, the projection of the columns of $\mathbf{X}_{k+1}$ into the space spanned by the columns of $\mathbf{Q}_k$ is $\mathbf{Q}_k\mathbf{Q}_k{}^T\mathbf{X}_{k+1}$. Since the space spanned by the columns of $\mathbf{X}_{k+1}$ is nearly equal to the space spanned by the columns of $\mathbf{Q}_k$, it follows that

$$\mathbf{Q}_k\mathbf{Q}_k{}^T\mathbf{X}_{k+1} \approx \mathbf{X}_{k+1} = \mathbf{A}\mathbf{Q}_k.$$

Replacing $\mathbf{Q}_k{}^T\mathbf{X}_{k+1}$ by its diagonalization $\mathbf{P}\mathbf{\Lambda}\mathbf{P}^{-1}$, we have

$$\mathbf{Q}_k\mathbf{P}\mathbf{\Lambda}\mathbf{P}^{-1} \approx \mathbf{A}\mathbf{Q}_k.$$

Finally, postmultiplying by $\mathbf{P}$ yields

$$(\mathbf{Q}_k\mathbf{P})\mathbf{\Lambda} \approx \mathbf{A}(\mathbf{Q}_k\mathbf{P}).$$

The diagonal elements of $\mathbf{\Lambda}$ approximate the dominant eigenvalues and the columns of $\mathbf{Q}_k\mathbf{P}$ approximate the corresponding eigenvectors. To summarize, given the starting matrix $\mathbf{X}_k$, each iteration of the power method factors $\mathbf{X}_k$ into the product $\mathbf{Q}_k\mathbf{R}_k$ and multiples $\mathbf{Q}_k$ by $\mathbf{A}$ to obtain $\mathbf{X}_{k+1}$. The dominant eigenvalues of $\mathbf{A}$ are approximated by the eigenvalues of the 2×2 matrix $\mathbf{Q}_k{}^T\mathbf{X}_{k+1}$. Moreover, if $\mathbf{P}\mathbf{\Lambda}\mathbf{P}^{-1}$ is the diagonalization of $\mathbf{Q}_k{}^T\mathbf{X}_{k+1}$, the columns of $\mathbf{Q}_k\mathbf{P}$ approximate the eigenvectors of $\mathbf{A}$ corresponding to the dominant eigenvalues.

Although we have focused on the iteration (36) in the case where $\mathbf{X}_k$ has two columns, the power method also applies when $\mathbf{X}_k$ has more than two columns. In particular, if $\mathbf{X}_k$ has m columns, then (typically) the m eigenvalues of $\mathbf{A}$ with largest magnitude are approximated by the eigenvalues of the $m \times m$ matrix $\mathbf{Q}_k{}^T\mathbf{X}_{k+1}$. Moreover, if $\mathbf{P\Lambda P}^{-1}$ is the diagonalization of $\mathbf{Q}_k{}^T\mathbf{X}_{k+1}$, the columns of $\mathbf{Q}_k\mathbf{P}$ approximate the corresponding eigenvectors of $\mathbf{A}$.

The power method approximates the eigenvalue(s) of largest magnitude. Suppose that we want to compute the smallest eigenvalue or the eigenvalue nearest 2. To handle these cases and others, the power method is altered in the following way: The iteration $\mathbf{x}^{\text{new}} = \mathbf{A}\mathbf{x}^{\text{old}}$ is replaced by $\mathbf{x}^{\text{new}} = \mathbf{B}\mathbf{x}^{\text{old}}$, where $\mathbf{A}$ and $\mathbf{B}$ have the same eigenvectors, but different eigenvalues. The power iteration $\mathbf{x}^{\text{new}} = \mathbf{B}\mathbf{x}^{\text{old}}$ converges to the eigenvector associated with $\mathbf{B}$'s dominant eigenvalue. But since $\mathbf{A}$ and $\mathbf{B}$ have the same eigenvectors, we have also computed an eigenvector of $\mathbf{A}$. Letting σ denote a scalar, there are three common choices for $\mathbf{B}$: $\mathbf{B} = \mathbf{A} - \sigma\mathbf{I}$ which is called the **shifted power method**, $\mathbf{B} = \mathbf{A}^{-1}$ which is called the **inverse power method**, and $\mathbf{B} = (\mathbf{A} - \sigma\mathbf{I})^{-1}$ which is called the **inverse shifted power method**.

To show that a shift does not change the eigenvectors, let us suppose that $(\lambda, \mathbf{x})$ is an eigenpair for $\mathbf{A}$. Subtracting the identity $\sigma\mathbf{I}\mathbf{x} = \sigma\mathbf{x}$ from the relation $\mathbf{A}\mathbf{x} = \lambda\mathbf{x}$, we have

$$(\mathbf{A} - \sigma\mathbf{I})\mathbf{x} = (\lambda - \sigma)\mathbf{x}.$$

This shows that $\mathbf{x}$ is an eigenvector of the shifted matrix $\mathbf{A} - \sigma\mathbf{I}$ corresponding to the eigenvalue $\lambda - \sigma$. The eigenvectors of the shifted matrix are identical to the eigenvectors of the original matrix while the eigenvalues are shifted by σ. To show that inverting does not change the eigenvalues, we again start with the relation $\mathbf{A}\mathbf{x} = \lambda\mathbf{x}$. Multiplying by $\mathbf{A}^{-1}$ and dividing by λ yields

$$\mathbf{A}^{-1}\mathbf{x} = \lambda^{-1}\mathbf{x}.$$

This shows that $\mathbf{x}$ is an eigenvector of the inverted matrix $\mathbf{A}^{-1}$ corresponding to the eigenvalue $1/\lambda$. Thus the eigenvectors of the inverted matrix are identical to the eigenvectors of the original matrix while the eigenvalues are inverted. Table 6-5 summarizes the relationship between the eigenvalues of $\mathbf{A}$ and the eigenvalues of $\mathbf{B}$. In this table, λ_{A} and λ_{B} are eigenvalues of $\mathbf{A}$ and $\mathbf{B}$, respectively. Column 2 gives the eigenvalue of $\mathbf{B}$ corresponding to the eigenvalue λ_{A} of $\mathbf{A}$ while column 3 gives the eigenvalue of $\mathbf{A}$ corresponding to the eigenvalue λ_{B} of $\mathbf{B}$. For example, if λ_{A} is an eigenvalue of $\mathbf{A}$, then $1/\lambda_{\text{A}}$ is an eigenvalue of

TABLE 6-5 RELATIONSHIP BETWEEN EIGENVALUES OF **A** AND **B**

B	Eigenvalue of **B**	Eigenvalue of **A**
$\mathbf{A}^{-1}$	$\dfrac{1}{\lambda_{\text{A}}}$	$\dfrac{1}{\lambda_{\text{B}}}$
$\mathbf{A} - \sigma\mathbf{I}$	$\lambda_{\text{A}} - \sigma$	$\lambda_{\text{B}} + \sigma$
$(\mathbf{A} - \sigma\mathbf{I})^{-1}$	$\dfrac{1}{\lambda_{\text{A}} - \sigma}$	$\sigma + \dfrac{1}{\lambda_{\text{B}}}$

$\mathbf{B} = \mathbf{A}^{-1}$. Conversely, if λ_B is an eigenvalue of $\mathbf{B} = \mathbf{A}^{-1}$, the $1/\lambda_B$ is an eigenvalue of $\mathbf{A}$.

Now suppose that we want to compute the eigenvector corresponding to $\mathbf{A}$'s eigenvalue of smallest magnitude. Letting $(\lambda_1, \mathbf{e}_1)$ through $(\lambda_n, \mathbf{e}_n)$ denote the eigenpairs of $\mathbf{A}$, the corresponding eigenpairs of $\mathbf{B} = \mathbf{A}^{-1}$ are $(1/\lambda_1, \mathbf{e}_1)$ through $(1/\lambda_n, \mathbf{e}_n)$. If λ_n is $\mathbf{A}$'s eigenvalue of smallest magnitude, then $1/\lambda_n$ is $\mathbf{B}$'s eigenvalue of largest magnitude and the power iteration $\mathbf{x}^{\text{new}} = \mathbf{A}^{-1}\mathbf{x}^{\text{old}}$ converges to the eigenvector $\mathbf{e}_n$ corresponding to the eigenvalue $1/\lambda_n$ of $\mathbf{B} = \mathbf{A}^{-1}$. When implementing the inverse power method, we multiply by $\mathbf{A}$ to express the iteration $\mathbf{x}^{\text{new}} = \mathbf{A}^{-1}\mathbf{x}^{\text{old}}$ in the form $\mathbf{A}\mathbf{x}^{\text{new}} = \mathbf{x}^{\text{old}}$. Replacing $\mathbf{A}$ by its $\mathbf{LU}$ factorization yields

$$(\mathbf{LU})\mathbf{x}^{\text{new}} = \mathbf{x}^{\text{old}}. \tag{37}$$

In each iteration of the inverse power method, the new $\mathbf{x}$ is obtained from the old $\mathbf{x}$ by forward and back solving the factored system (37).

In a similar manner, the inverse shifted power iteration

$$\mathbf{x}^{\text{new}} = (\mathbf{A} - \sigma\mathbf{I})^{-1}\mathbf{x}^{\text{old}}$$

converges to the eigenvector of $\mathbf{A}$ whose corresponding eigenvalue is closest to σ. To implement the inverse shifted power method, we multiply by $\mathbf{A} - \sigma\mathbf{I}$ to express the iteration in the form

$$(\mathbf{A} - \sigma\mathbf{I})\mathbf{x}^{\text{new}} = \mathbf{x}^{\text{old}}.$$

If the shifted matrix $\mathbf{A} - \sigma\mathbf{I}$ is $\mathbf{LU}$ factored, then the inverse shifted power iteration assumes the form (37).

On the surface, the inverse shifted power method is an unstable algorithm. To compute the eigenpair $(\lambda, \mathbf{e})$ of $\mathbf{A}$, we select a shift σ near λ. Although the convergence of the inverse shifted power method accelerates as σ approaches λ, the numerical errors also increase as σ approaches λ. In Table 6-5 we see that $1/(\lambda - \sigma)$ is an eigenvalue of $\mathbf{B} = (\mathbf{A} - \sigma\mathbf{I})^{-1}$. By Exercise 3-3.5, the norm of $(\mathbf{A} - \sigma\mathbf{I})^{-1}$ is at least $1/|\lambda - \sigma|$. Hence, the condition number for $\mathbf{A} - \sigma\mathbf{I}$ is at least $\|\mathbf{A} - \sigma\mathbf{I}\|/|\lambda - \sigma|$. Since the conditioning of the shifted matrix worsens as σ approaches λ, we expect that the new $\mathbf{x}$ computed by the inverse shifted power method is relatively inaccurate when σ is close to λ. Fortunately, the error in the new $\mathbf{x}$ points predominantly in the direction of the desired eigenvector $\mathbf{e}$. In other words,

$$\text{computed } \mathbf{x}^{\text{new}} \approx \text{correct } \mathbf{x}^{\text{new}} + s\mathbf{e},$$

where s is some scalar. In theory, the $\mathbf{x}$'s approach $\mathbf{e}$ as the iterations progress. Computationally, the error in the correct $\mathbf{x}^{\text{new}}$ is nearly a multiple of $\mathbf{e}$ and the ill conditioning associated with the inverse shifted power method does not impede the convergence to $\mathbf{e}$.

The inverse shifted power method can be combined with the Rayleigh quotient estimate for an eigenvalue to obtain an algorithm called the **Rayleigh quotient method**, which is locally cubically convergent. When the coefficient matrix $\mathbf{A}$ is symmetric, this iteration is expressed:

$$\mathbf{w} = \frac{\mathbf{x}^{\text{old}}}{\|\mathbf{x}^{\text{old}}\|_2}, \qquad \sigma = \mathbf{w}^T\mathbf{A}\mathbf{w}, \qquad (\mathbf{A} - \sigma\mathbf{I})\mathbf{x}^{\text{new}} = \mathbf{w}.$$

This iteration can also be applied to an unsymmetric matrix, but the convergence rate is quadratic rather than cubic. To obtain cubic convergence in the unsymmetric case, we must simultaneously approximate both the right eigenvector $\mathbf{x}$ and the left eigenvector $\mathbf{y}$ as in the iteration

$$\mathbf{w} = \frac{\mathbf{x}^{\text{old}}}{\|\mathbf{x}^{\text{old}}\|_2}, \qquad \mathbf{z} = \frac{\mathbf{y}^{\text{old}}}{\|\mathbf{y}^{\text{old}}\|_2}, \qquad \sigma = \frac{\mathbf{z}^T\mathbf{A}\mathbf{w}}{\mathbf{z}^T\mathbf{w}},$$

$$(\mathbf{A} - \sigma\mathbf{I})\mathbf{x}^{\text{new}} = \mathbf{w}, \qquad (\mathbf{A} - \sigma\mathbf{I})^T\mathbf{y}^{\text{new}} = \mathbf{z}.$$

Although the Rayleigh quotient iteration converges quickly, we often do not know which eigenvalue it is converging to. In practice, a slow algorithm that picks out a specific eigenvalue can be preferable to a fast algorithm that converges to some arbitrary eigenvalue.

Subroutine POWER uses the power method to compute the dominant eigenvalue(s) and the corresponding eigenvector(s) for a matrix with real coefficients. The subroutine arguments are

EX,X:	first eigenpair (complex variable and complex array)
EY,Y:	second eigenpair (complex variable and complex array)
N:	matrix dimension
DIF:	input for subroutine WHATIS
SIZE:	input for subroutine WHATIS
NDIGIT:	desired number of correct digits
LIMIT:	maximum number of iterations
MULT:	name of subroutine to multiply matrix by vector

The subroutine to multiply a matrix by a vector has two arguments. The second argument is a real array containing a vector $\mathbf{v}$ and the first argument is a real array to store the product $\mathbf{p} = \mathbf{B}\mathbf{v}$ between the vector $\mathbf{v}$ and the matrix $\mathbf{B}$ to which the power iteration is applied. Before invoking subroutine POWER, a real approximation to an eigenvector corresponding to the dominant eigenvalue(s) of $\mathbf{B}$ must be stored in the complex array X. If no good starting guess is known, the elements of X can be assigned random values. Subroutine POWER computes an eigenvector corresponding to the dominant eigenvalue. The approximation to the dominant eigenvalue is stored in the variable EX while the corresponding eigenvector is stored in the array X. If EY is nonzero, a second eigenpair is stored in the variable EY and in the array Y.

For illustration, we present a program to compute the smallest eigenvalue of the beam buckling matrix (9) generated by 10 mesh intervals. Since the eigenvalues of a symmetric matrix are real (see Exercise 6-1.1), the eigenvalues of the beam buckling matrix are real. Also by Gerschgorin's theorem, the eigenvalues of (9) are positive. The following program uses the inverse power method to compute the smallest eigenvalue of the beam buckling matrix $\mathbf{A}$. Since the absolute largest eigenvalue of $\mathbf{B} = \mathbf{A}^{-1}$ is the reciprocal of the absolute smallest eigenvalue of $\mathbf{A}$, the reciprocal of the eigenvalue computed by subroutine POWER is the desired eigenvalue of $\mathbf{A}$.

```
      EXTERNAL PROD
      COMMON D(12),U(9)
      COMPLEX EX,X(9),EY,Y(9)
      N = 10
      M = N - 1
      T = N*N
      DO 10 I = 1,M
        D(I) = T + T
        U(I) = -T
        X(I) = 1.
10    CONTINUE
      CALL TFACT(U,D,U,M)
      CALL POWER(EX,X,EY,Y,M,DIF,SIZE,5,100,PROD)
      CALL WHATIS(DIF,SIZE)
      E = 1./EX
      WRITE(6,*) 'EIGENVALUE:',E
      WRITE(6,*) 'EIGENVECTOR:'
      DO 20 I = 1,M
20      WRITE(6,*) I,REAL(X(I))
      END

      SUBROUTINE PROD(P,V)
      COMMON D(12),U(9)
      REAL P(1),V(1)
      CALL TSOLVE(P,U,D,U,V)
      RETURN
      END
```

Exercises

6-4.1. Suppose it is known that every eigenvalue of a matrix **A** is real. To compute the largest eigenvalue of **A** using the shifted power method, what shift can be employed? To compute the smallest eigenvalue of **A** using the shifted power method, what shift can be employed? Use Gerschgorin's theorem to estimate these shifts for the matrices in Exercise 6-3.1.

6-4.2. Suppose that a matrix has a dominant pair of real eigenvalues which are the negatives of each other. That is, $\lambda_1 = -\lambda_2$ and $|\lambda_1| > |\lambda_i|$ for $i > 2$. As discussed above, λ_1 and λ_2 are the roots of (32) where c_1 and c_2 can be approximated by the solution to (34). Explain how an approximation to the corrresponding eigenvectors can be obtained from two successive power iterations $\mathbf{x}_{k+1}$ and $\mathbf{x}_k$. (*Hint:* Consider $\mathbf{x}_{k+1} \pm \lambda_1\mathbf{x}_k$.)

6-4.3. Use the power method to compute the largest eigenvalue λ_1 of the beam buckling matrix (9) when $N = 10$. Then employ the shifted power method with the shift $\sigma = \lambda_1$ to compute the smallest eigenvalue of the beam buckling matrix. For comparison, compute the smallest eigenvalue of the beam buckling matrix using the inverse power method. Which scheme for computing the smallest eigenvalue is faster, the shifted power method with shift $\sigma = \lambda_1$ or the inverse power method?

6-5. DEFLATION

Suppose that we want to compute the two smallest eigenvalues of the beam buckling matrix. The inverse power method gives us the smallest eigenvalue.

Now our goal is to deflate the computed eigenvalue from **A** by constructing a smaller matrix **F** which has the remaining eigenvalues of **A**. The deflated matrix **F** has one less column and one less row than **A**, and the eigenvalues of **F** are the same as the eigenvalues of **A** except that the previously computed eigenvalue is missing from **F**. The inverse power method applied to **F** yields the smallest eigenvalue of **F**, which is the second smallest eigenvalue of **A**.

The deflated matrix **F** is constructed using Householder matrices. Let $(\lambda_1, \mathbf{x})$ denote the computed eigenpair of the $n \times n$ matrix **A** and let $\mathbf{H} = \mathbf{I} - 2\mathbf{w}\mathbf{w}^T$ denote the Householder matrix which annihilates components 2 through n of **x**. If **w** is computed using formula (38) of Chapter 5, then **Hx** has the following structure:

$$\mathbf{Hx} = \begin{bmatrix} \sigma\sqrt{x_1^2 + \cdots + x_n^2} \\ 0 \\ \vdots \\ 0 \end{bmatrix} = \sigma \, \|\mathbf{x}\|_2 \begin{bmatrix} 1 \\ 0 \\ \vdots \\ 0 \end{bmatrix}, \tag{38}$$

Where $\sigma = -\operatorname{sign}(x_1)$. Letting $\mathbf{u}_1$ denote the vector with every component equal to zero except for the first component, which is 1, (38) can be written

$$\mathbf{Hx} = \sigma \, \|\mathbf{x}\|_2 \mathbf{u}_1 .$$

Multiplying the identity $\mathbf{Ax} = \lambda_1 \mathbf{x}$ by **H** and utilizing (38), we have

$$\mathbf{HAx} = \lambda_1 \mathbf{Hx} = \sigma\lambda_1 \|\mathbf{x}\|_2 \mathbf{u}_1 . \tag{39}$$

Since $\mathbf{HH} = \mathbf{I}$, the left side of (39) can be expressed

$$\mathbf{HAx} = \mathbf{HA(HH)x} = \mathbf{(HAH)Hx} = \mathbf{(HAH)}(\sigma\|\mathbf{x}\|_2 \mathbf{u}_1). \tag{40}$$

Combining (39) and (40) yields

$$\mathbf{(HAH)u}_1 = \lambda_1 \mathbf{u}_1 . \tag{41}$$

Since the first component of $\mathbf{u}_1$ is 1 and the remaining components are zero, the product $\mathbf{(HAH)u}_1$ is the first column of **HAH**. By (41) the first column of **HAH** is $\lambda_1 \mathbf{u}_1$, so **HAH** has the structure sketched below:

$$\mathbf{HAH} = \begin{bmatrix} \lambda_1 & r_1 & r_2 & \cdots & r_{n-1} \\ 0 & & & & \\ 0 & & & & \\ \vdots & & \mathbf{F} & & \\ 0 & & & & \end{bmatrix}. \tag{42}$$

The submatrix **F** above is $(n - 1) \times (n - 1)$ and the elements r_1 through r_{n-1} in the first row of **HAH** are not necessarily zero.

We claim that **F** has the same eigenvalues as **A** except that λ_1 is missing. To establish this result, let us first show that the eigenvalues of **A** and **HAH** are identical. If $(\lambda, \mathbf{x})$ is an eigenpair for **A**, then $\mathbf{Ax} = \lambda\mathbf{x}$. Multiplying by **H** yields $\mathbf{HAx} = \lambda\mathbf{Hx}$, and replacing **HAx** by **HAHHx**, we obtain the relation $\mathbf{HAHHx} = \lambda\mathbf{Hx}$, or equivalently,

$$\mathbf{HAH(Hx)} = \lambda\mathbf{(Hx)}.$$

This shows that $(\lambda, \mathbf{Hx})$ is an eigenpair for **HAH** and the eigenvalues of **A** are the same as the eigenvalues of **HAH**. Since the eigenvalues of **HAH** are the zeros of the characteristic polynomial, let us examine the characteristic polynomial for **HAH**:

$$\det(\mathbf{HAH} - \lambda\mathbf{I}) = \det\begin{bmatrix} \lambda_1 - \lambda & r_1 & \cdots & r_{n-1} \\ 0 & & & \\ \cdot & & \mathbf{F} - \lambda\mathbf{I} & \\ \cdot & & & \\ \cdot & & & \\ 0 & & & \end{bmatrix}. \tag{43}$$

The determinant (43) can be expressed

$$\det(\mathbf{HAH} - \lambda\mathbf{I}) = (\lambda_1 - \lambda)\det(\mathbf{F} - \lambda\mathbf{I}).$$

One zero of the characteristic polynomial for **HAH** is $\lambda = \lambda_1$ while the remaining zeros are the roots of the equation $\det(\mathbf{F} - \lambda\mathbf{I}) = 0$. Since this equation is the characteristic equation for **F**, the roots are the eigenvalues of **F**. In summary, the eigenvalues of **A** are the same as the eigenvalues of **HAH** and the eigenvalues of **HAH** are λ_1 combined with the eigenvalues of **F**. Therefore, the eigenvalues of **F** are the same as the eigenvalues of **A** except that λ_1 is missing.

For a simple illustration, we consider the matrix

$$\mathbf{A} = \begin{bmatrix} 23 & 36 \\ 36 & 2 \end{bmatrix}$$

and the eigenpair

$$\lambda = 50 \qquad \text{and} \qquad \mathbf{x} = \begin{bmatrix} 4 \\ 3 \end{bmatrix}.$$

The Householder matrix which annihilates the second component of **x** is

$$\mathbf{H} = -\tfrac{1}{5}\begin{bmatrix} 4 & 3 \\ 3 & -4 \end{bmatrix}.$$

Computing **HAH**, we have

$$\mathbf{HAH} = \begin{bmatrix} 50 & 0 \\ 0 & -25 \end{bmatrix}.$$

In this case **F** is the 1×1 matrix -25 whose eigenvalue is -25.

It is not necessary to construct **F** when performing the power iteration $\mathbf{y}^{\text{new}} = \mathbf{F}\mathbf{y}^{\text{old}}$. If **A** is $n \times n$, then **F** is $(n - 1) \times (n - 1)$ and the power iteration for **F** involves the product between **F** and a vector **y** with $n - 1$ components. Let us form a n component vector **z** by putting a zero above **y**:

$$\mathbf{z} = \begin{bmatrix} 0 \\ \mathbf{y} \end{bmatrix}.$$

The product (**HAH**)**z** is pictured below:

$$(\mathbf{HAH})\mathbf{z} = (\mathbf{HAH})\begin{bmatrix} 0 \\ \mathbf{y} \end{bmatrix} = \begin{bmatrix} \lambda_1 & r_1 & r_2 & \cdots & r_{n-1} \\ 0 & & & & \\ 0 & & & & \\ \cdot & & \mathbf{F} & & \\ \cdot & & & & \\ \cdot & & & & \\ 0 & & & & \end{bmatrix}\begin{bmatrix} 0 \\ \mathbf{y} \end{bmatrix} = \begin{bmatrix} \mathbf{r}^T\mathbf{y} \\ \mathbf{F}\mathbf{y} \end{bmatrix}.$$

Thus the product $\mathbf{Fy}$ can be computed in the following way:

1. Form the vector $\mathbf{z}$.
2. Compute the product $(\mathbf{HAH})\mathbf{z}$. (44)
3. Discard the first component of $(\mathbf{HAH})\mathbf{z}$ to obtain $\mathbf{Fy}$.

This procedure computes $\mathbf{Fy}$ without constructing $\mathbf{F}$. Furthermore, the product $(\mathbf{HAH})\mathbf{z}$ in step 2 of (44) can be computed without constructing $\mathbf{H}$. Using the associative law for matrix multiplication, $(\mathbf{HAH})\mathbf{z}$ is expressed $\mathbf{H}(\mathbf{A}(\mathbf{Hz}))$. To compute $(\mathbf{HAH})\mathbf{z}$, multiply $\mathbf{H}$ by $\mathbf{z}$ to obtain the vector $\mathbf{Hz}$, multiply $\mathbf{A}$ by $\mathbf{Hz}$ to obtain the vector $\mathbf{AHz}$, and multiply $\mathbf{H}$ by $\mathbf{AHz}$ to obtain the final result $(\mathbf{HAH})\mathbf{z}$. To multiply the Householder matrix $\mathbf{H} = \mathbf{I} - 2\mathbf{w}\mathbf{w}^T$ by a vector $\mathbf{z}$, we use formula (34) of Chapter 5:

$$\mathbf{Hz} = (\mathbf{I} - 2\mathbf{w}\mathbf{w}^T)\mathbf{z} = \mathbf{z} - 2(\mathbf{w}^T\mathbf{z})\mathbf{w}.$$

In other words, the product $\mathbf{Hz}$ is equal to $\mathbf{z}$ minus the scalar $2\mathbf{w}^T\mathbf{z}$ times $\mathbf{w}$.

Suppose that the power iterations $\mathbf{y}^{\text{new}} = \mathbf{F}\mathbf{y}^{\text{old}}$ converge to an eigenpair $(\lambda_2, \mathbf{y})$ for $\mathbf{F}$. To determine the corresponding eigenpair for $\mathbf{A}$, we first augment the vector $\mathbf{y}$, which has $n - 1$ components, to form a n component vector $\mathbf{x}$:

$$\mathbf{x} = \begin{bmatrix} x_1 \\ \mathbf{y} \end{bmatrix}.$$

If $\mathbf{x}$ is an eigenvector for the matrix $\mathbf{HAH}$ corresponding to the eigenvalue λ_2, then x_1 must be chosen so that $(\mathbf{HAH})\mathbf{x} = \lambda_2\mathbf{x}$. Referring to (42), the equation $(\mathbf{HAH})\mathbf{x} = \lambda_2\mathbf{x}$ can be written

$$\begin{bmatrix} \lambda_1 x_1 + \mathbf{r}^T\mathbf{y} \\ \mathbf{Fy} \end{bmatrix} = \lambda_2 \begin{bmatrix} x_1 \\ \mathbf{y} \end{bmatrix}. \tag{45}$$

Since $\mathbf{Fy} = \lambda_2\mathbf{y}$, the bottom half of equation (45) is satisfied trivially. Solving for x_1 in the top half of equation (45), gives us

$$x_1 = \frac{\mathbf{r}^T\mathbf{y}}{\lambda_2 - \lambda_1}. \tag{46}$$

In summary, we start with an eigenpair $(\lambda_2, \mathbf{y})$ for $\mathbf{F}$ and we augment $\mathbf{y}$ to form $\mathbf{x}$. If x_1 satisfies (46), then $(\lambda_2, \mathbf{x})$ is an eigenpair for $\mathbf{HAH}$. Premultiplying the eigenvector relation $(\mathbf{HAH})\mathbf{x} = \lambda_2\mathbf{x}$ by $\mathbf{H}$ gives $\mathbf{A}(\mathbf{Hx}) = \lambda_2(\mathbf{Hx})$. Therefore, $\mathbf{Hx}$ is an eigenvector of $\mathbf{A}$ corresponding to the eigenvalue λ_2.

The deflation process can be applied repeatedly to compute several eigenpairs of $\mathbf{A}$. Starting with an eigenpair $(\lambda_1, \mathbf{x}_1)$, we form a matrix $\mathbf{H}_1$ that annihilates components 2 through n of $\mathbf{x}_1$. The lower right corner of the matrix $\mathbf{H}_1\mathbf{A}\mathbf{H}_1$ contains the deflated matrix $\mathbf{F}_1$. After computing an eigenpair for $\mathbf{H}_1\mathbf{A}\mathbf{H}_1$ we deflate λ_2 using a Householder matrix $\mathbf{H}_2$ that annihilates components 3 through n of the eigenvector. The deflated matrix $\mathbf{F}_2$ sits in the lower right corner of $\mathbf{H}_2\mathbf{H}_1\mathbf{A}\mathbf{H}_1\mathbf{H}_2$. The kth step of the deflation process involves the matrix $\mathbf{H}_k$ that annihilates components $k + 1$ through n of an eigenvector for

$$\mathbf{H}_{k-1} \cdots \mathbf{H}_2\mathbf{H}_1\mathbf{A}\mathbf{H}_1\mathbf{H}_2 \cdots \mathbf{H}_{k-1},$$

and the deflated matrix $\mathbf{F}_k$ sits in the lower right corner of

$$\mathbf{H}_k \cdots \mathbf{H}_2\mathbf{H}_1\mathbf{A}\mathbf{H}_1\mathbf{H}_2 \cdots \mathbf{H}_k.$$

Although all the eigenpairs can be generated by this process, the **QR** method developed in the next section computes every eigenpair of a matrix more efficiently. The power method is best suited for finding a few eigenpairs. On the other hand, if this deflation process is carried out to completion, we obtain a decomposition with theoretical significance. Observe that deflation step k essentially annihilates the coefficients in column k beneath the diagonal. Consequently, for a $n \times n$ matrix, $n - 1$ deflation steps yield an upper triangular matrix $\mathbf{U}$:

$$\mathbf{U} = \mathbf{H}_{n-1} \cdots \mathbf{H}_2\mathbf{H}_1\mathbf{A}\mathbf{H}_1\mathbf{H}_2 \cdots \mathbf{H}_{n-1}.$$

Letting $\mathbf{P}$ denote the product $\mathbf{H}_1\mathbf{H}_2 \cdots \mathbf{H}_{n-1}$, it follows that $\mathbf{U} = \mathbf{P}^{-1}\mathbf{A}\mathbf{P}$. Solving for $\mathbf{A}$, we have $\mathbf{A} = \mathbf{P}\mathbf{U}\mathbf{P}^{-1}$, where $\mathbf{P}$ is an orthogonal matrix when every eigenvector is real while $\mathbf{P}$ is unitary when some eigenvector is complex. The decomposition $\mathbf{A} = \mathbf{P}\mathbf{U}\mathbf{P}^{-1}$ is called the **Schur decomposition**. In contrast to the diagonalization, which does not exist when the matrix is defective, the Schur upper triangular decomposition always (surely) exists.

Finally, we mention subroutine MPOWER, which uses the power method and deflation to compute several eigenpairs of a matrix with real coefficients. The subroutine arguments are

EX,X:	first eigenpair (complex variable and complex array)
EY,Y:	second eigenpair (complex variable and complex array)
N:	matrix dimension
NE:	number of eigenpairs to compute (input) and number of eigenpairs left to compute (output)
DIF:	input for subroutine WHATIS
SIZE:	input for subroutine WHATIS
NDIGIT:	desired number of correct digits
LIMIT:	maximum number of iterations per eigenpair
MULT:	name of subroutine to multiply matrix by vector
W:	work array [length at least NE(N + 2) − 2]

The subroutine to multiply a matrix by a vector has two arguments. The second argument is a real array containing a vector $\mathbf{v}$ and the first argument is a real array to store the product $\mathbf{p} = \mathbf{B}\mathbf{v}$ between the vector $\mathbf{v}$ and the matrix $\mathbf{B}$ to which the power iteration is applied. Before invoking MPOWER, a real approximation to an eigenvector corresponding to the current dominant eigenvalue(s) of $\mathbf{B}$ must be stored in the complex array X. If no good starting guess is known, the elements of X can be assigned random values. Each time that subroutine MPOWER is invoked, a new eigenpair is computed. The variable EX contains the eigenvalue, while the array X contains the eigenvector. If EY is nonzero, a second eigenpair is stored in EY and Y. Below we compute the first three eigenpairs of the beam buckling matrix (9) corresponding to a 10-interval mesh.

```
      EXTERNAL PROD
      COMMON D(12),U(9)
      COMPLEX EX,X(9),EY,Y(9)
      REAL W(40),T
      N = 10
      M = N - 1
      T = N*N
      DO 10 I = 1,M
        D(I) = T + T
10      U(I) = -T
      CALL TFACT(U,D,U,M)
      NE = 3
20    T = (4-NE)*ARCOS(-1.)/N
      DO 30 I = 1,M
30       X(I) = SIN(I*T)
      CALL MPOWER(EX,X,EY,Y,M,NE,DIF,SIZE,5,100,PROD,W)
      CALL WHATIS (DIF,SIZE)
      E = 1./EX
      WRITE(6,*) 'EIGENVALUE:',E
      WRITE(6,*) 'EIGENVECTOR:'
      DO 40 I = 1,M
40      WRITE(6,*) I,REAL(X(I))
      IF ( CABS(EY) .EQ. 0.) GOTO 60
      E = 1./EY
      WRITE(6,*) 'EIGENVALUE:',E
      WRITE(6,*) 'EIGENVECTOR:'
      DO 50 I = 1,M
50      WRITE(6,*) I,REAL(Y(I))
60    IF (NE .GT. 0 ) GOTO 20
      END

      SUBROUTINE PROD(P,V)
      COMMON D(12),U(9)
      REAL P(1),V(1)
      CALL TSOLVE(P,U,D,U,V)
      RETURN
      END
```

Since we are using the inverse power method and since the eigenvalues for the beam buckling matrix are real, the reciprocal of EX (E in the program above) is the desired eigenvalue. Observe that the starting guess used above for the jth eigenvector is given by $x_i = \sin(ij\pi/N)$, which resembles the sine functions depicted in Figures 6-1 and 6-2. For the beam buckling problem, a starting guess such as $x_i = 1$ for every i is not good in general, since a vector with each component the same is orthogonal to the even numbered eigenvectors.

Exercise

6-5.1. Explain why every real symmetric matrix can be diagonalized by an orthogonal eigenvector matrix.

6-6. THE QR METHOD

The **QR** method for computing all the eigenvalues of a matrix originates from the work of Rutishauser [R10], Francis [F6], and Kublanovskaya [K7] in the late 1950s and in the early 1960s. Each step of the **QR** algorithm factors a matrix and interchanges the factors:

$$\mathbf{A}^{\text{old}} = \mathbf{QR} \quad \text{and} \quad \mathbf{A}^{\text{new}} = \mathbf{RQ}. \tag{47}$$

These iterations typically converge to a "quasi-triangular" matrix whose diagonal blocks contain the eigenvalues. For example, starting with the matrix

$$\begin{bmatrix} 5 & -8 \\ -4 & 1 \end{bmatrix},$$

the three-digit **QR** factorization is

$$\underset{\mathbf{A}}{\begin{bmatrix} 5 & -8 \\ -4 & 1 \end{bmatrix}} \approx \underset{\mathbf{Q}}{\begin{bmatrix} -.780 & .624 \\ .624 & .780 \end{bmatrix}} \underset{\mathbf{R}}{\begin{bmatrix} -6.40 & 6.86 \\ 0 & -4.21 \end{bmatrix}}.$$

The new **A** is the product **RQ**:

$$\mathbf{A}^{\text{new}} = \underset{\mathbf{R}}{\begin{bmatrix} -6.40 & 6.86 \\ 0 & -4.21 \end{bmatrix}} \underset{\mathbf{Q}}{\begin{bmatrix} -.780 & .624 \\ .624 & .780 \end{bmatrix}} \approx \begin{bmatrix} 9.29 & 1.37 \\ -2.63 & -3.29 \end{bmatrix}. \tag{48}$$

The second iteration just repeats these steps. The **QR** factorization of **A** in (48) is

$$\underset{\mathbf{A}^{\text{old}}}{\begin{bmatrix} 9.29 & 1.37 \\ -2.63 & -3.29 \end{bmatrix}} = \underset{\mathbf{Q}}{\begin{bmatrix} -.962 & .273 \\ .273 & .962 \end{bmatrix}} \underset{\mathbf{R}}{\begin{bmatrix} -9.65 & -2.22 \\ 0 & -2.79 \end{bmatrix}},$$

and the new **A** is **RQ**:

$$\mathbf{A}^{\text{new}} = \underset{\mathbf{R}}{\begin{bmatrix} -9.65 & -2.22 \\ 0 & -2.79 \end{bmatrix}} \underset{\mathbf{Q}}{\begin{bmatrix} -.962 & .273 \\ .273 & .962 \end{bmatrix}} \approx \begin{bmatrix} 8.69 & -4.76 \\ -.76 & -2.69 \end{bmatrix}.$$

Continuing for two more iterations, we obtain

$$\mathbf{A}^{\text{new}} = \begin{bmatrix} -8.72 & 4.51 \\ 0 & -3.09 \end{bmatrix} \begin{bmatrix} -.996 & .087 \\ .087 & .996 \end{bmatrix} \approx \begin{bmatrix} 9.08 & 3.73 \\ -.27 & -3.08 \end{bmatrix}$$

and

$$\mathbf{A}^{\text{new}} = \begin{bmatrix} -9.08 & -3.82 \\ 0 & -2.97 \end{bmatrix} \begin{bmatrix} -.9995 & .0298 \\ .0298 & .9995 \end{bmatrix} \approx \begin{bmatrix} 8.97 & -4.09 \\ -.09 & -2.97 \end{bmatrix}.$$

Observe that the iterations are converging to the upper triangular matrix

$$\begin{bmatrix} 9 & -4 \\ 0 & -3 \end{bmatrix}.$$

The eigenvalues of the starting **A** are the diagonal elements 9 and -3 of the upper triangular limit.

As the previous example illustrates, the **QR** iterations are real when the starting matrix is real. Consequently, if the starting matrix is real but some of the eigenvalues are complex, the **QR** algorithm cannot converge to an upper triangular matrix with the complex eigenvalues on the diagonal. When the starting matrix is real but some of the eigenvalues are complex, the **QR** algorithm typically converges to a quasi-triangular matrix whose diagonal blocks contain the eigenvalues. The following matrix is quasi-triangular:

$$\begin{bmatrix} 5 & 4 & 6 & 7 & 2 & 2 \\ 0 & 2 & -3 & 2 & 9 & 6 \\ 0 & 2 & 1 & 3 & 7 & 6 \\ 0 & 0 & 0 & 4 & -3 & 4 \\ 0 & 0 & 0 & 1 & 2 & 3 \\ 0 & 0 & 0 & 0 & 0 & 8 \end{bmatrix}.$$

The eigenvalues of this matrix are 5 and 8 along with the two pairs of complex conjugate eigenvalues associated with the 2×2 diagonal blocks:

$$\begin{bmatrix} 2 & -3 \\ 2 & 1 \end{bmatrix} \quad \text{and} \quad \begin{bmatrix} 4 & -3 \\ 1 & 2 \end{bmatrix}.$$

In general, a matrix is **quasi-upper triangular** if every element beneath the subdiagonal is zero and at least every other subdiagonal element is zero.

There does not seem to be a simple explanation for the convergence of the **QR** method. A rigorous convergence proof for the algorithm is given by Wilkinson [W6]. Basically the scheme is a power method (see Watkins [W3] and Exercise 6-6.1). To simplify the **QR** iteration (47), let us combine the two steps into one step. Solving for **R** using the first equation in (47), we have

$$\mathbf{R} = \mathbf{Q}^{-1}\mathbf{A}^{\text{old}},$$

and substituting this into the second equation in (47) gives

$$\mathbf{A}^{\text{new}} = \mathbf{Q}^{-1}\mathbf{A}^{\text{old}}\mathbf{Q}. \tag{49}$$

If two matrices **B** and **C** satisfy the relation $\mathbf{C} = \mathbf{Q}^{-1}\mathbf{B}\mathbf{Q}$ for some invertible matrix **Q**, we say that **B** and **C** are **similar** and the product $\mathbf{Q}^{-1}\mathbf{B}\mathbf{Q}$ is called a **similarity transformation**. It follows from (49) that the new **A** is similar to the old **A**.

There is a simple relationship between the eigenpairs of **B** and the eigenpairs of a similar matrix $\mathbf{C} = \mathbf{Q}^{-1}\mathbf{B}\mathbf{Q}$. An eigenpair $(\lambda, \mathbf{x})$ for **C** satisfies the equation $\mathbf{C}\mathbf{x} = \lambda\mathbf{x}$, or equivalently, $(\mathbf{Q}^{-1}\mathbf{B}\mathbf{Q})\mathbf{x} = \lambda\mathbf{x}$. Multiplying by **Q** yields $\mathbf{B}\mathbf{Q}\mathbf{x} = \lambda\mathbf{Q}\mathbf{x}$, which can be written

$$\mathbf{B}(\mathbf{Q}\mathbf{x}) = \lambda(\mathbf{Q}\mathbf{x}).$$

This shows that if $(\lambda, \mathbf{x})$ is an eigenpair for **C**, then $(\lambda, \mathbf{Q}\mathbf{x})$ is an eigenpair for **B**. Thus the eigenvalues of two similar matrices are identical. Since each **QR** iteration is a similarity transformation, the new **A** has the same eigenvalues as the old **A** and if **x** is an eigenvector for $\mathbf{A}^{\text{new}}$, then $\mathbf{Q}\mathbf{x}$ is an eigenvector for $\mathbf{A}^{\text{old}}$. If $\mathbf{A}_m$ is the mth matrix generated by the **QR** algorithm and if $\mathbf{Q}_1, \mathbf{Q}_2, \cdots, \mathbf{Q}_m$ denote the orthogonal matrices corresponding to the first m iterations, then $\mathbf{A}_m = (\mathbf{Q}_m^{-1} \cdots \mathbf{Q}_2^{-1}\mathbf{Q}_1^{-1})\mathbf{A}\,(\mathbf{Q}_1\mathbf{Q}_2 \cdots \mathbf{Q}_m)$. Letting **P** denote the product $\mathbf{Q}_1\mathbf{Q}_2 \cdots \mathbf{Q}_m$, it follows that $\mathbf{A}_m = \mathbf{P}^{-1}\mathbf{A}\mathbf{P}$. Hence, if $(\lambda, \mathbf{x})$ is an eigenpair for $\mathbf{A}_m$, then $(\lambda, \mathbf{P}\mathbf{x})$ is an eigenpair for the starting **A**. The product **P** of the $\mathbf{Q}_i$ can be constructed in the following way: Initialize **P** to the identity **I**, then at step i, update **P** by the rule

$$\mathbf{P}^{\text{new}} = \mathbf{P}^{\text{old}}\mathbf{Q}_i.$$

After m steps, **P** is the product $\mathbf{Q}_1\mathbf{Q}_2 \cdots \mathbf{Q}_m$.

For an arbitrary matrix, the **QR** method is terribly expensive since the time to **QR** factor a general $n \times n$ matrix is proportional to n^3. However, we show in the next section that any matrix is similar to one with zeros beneath the subdiagonal. A matrix with zeros beneath the subdiagonal is called **upper Hessenberg** and a 5×5 illustration appears below:

$$\begin{bmatrix} * & * & * & * & * \\ * & * & * & * & * \\ 0 & * & * & * & * \\ 0 & 0 & * & * & * \\ 0 & 0 & 0 & * & * \end{bmatrix}.$$

The time to **QR** factor an upper Hessenberg matrix is proportional to n^2. A sequence of $n - 1$ Givens rotations can be used to annihilate the subdiagonal coefficients: A Givens rotation $\mathbf{G}_{21}$ is used to annihilate a_{21} and another Givens rotation $\mathbf{G}_{32}$ is used to annihilate a_{32}. Continuing down the diagonal, we annihilate subdiagonal elements until a rotation $\mathbf{G}_{n,n-1}$ annihilates the final subdiagonal coefficient $a_{n,n-1}$. The resulting upper triangular matrix is **R**. The time to multiply a matrix by a Givens rotation is proportional to n and there are $n - 1$ rotations altogether. Thus the time to **QR** factor an upper Hessenberg matrix is proportional to n^2. Similarly, the time to compute the product **RQ** in (47) is proportional to n^2.

Reducing a matrix to upper Hessenberg form certainly speeds up each iteration of the **QR** method. Nonetheless, algorithm (47) is still slow when a large number of iterations are required. The convergence of the **QR** algorithm is often accelerated using a shift technique which we describe shortly. The total number of iterations required by the shifted **QR** algorithm is often less than $2n$. Since the time per iteration is proportional to n^2, the time to compute every eigenvalue of a general matrix is proportional to n^3.

An iteration of the shifted **QR** algorithm is given by

$$\mathbf{A}^{\text{new}} = \mathbf{Q}^{-1}\mathbf{A}^{\text{old}}\mathbf{Q}, \tag{50}$$

where **Q** comes from the **QR** factorization of the shifted matrix $\mathbf{A} - \sigma\mathbf{I}$. The shifted matrix is obtained by subtracting the shift σ from each diagonal element of **A**. A very effective shift strategy, developed by Francis [F6], involves the eigenvalues of 2×2 diagonal submatrices. The first few iterations of the **QR** algorithm focus on the lower right corner of **A**:

$$\begin{bmatrix} a_{n-1,n-1} & a_{n-1,n} \\ a_{n,n-1} & a_{n,n} \end{bmatrix}.$$

In each iteration we shift using the eigenvalue of this 2×2 matrix which is closest to a_{nn}. Typically, a few shifted **QR** iterations drive $a_{n,n-1}$ to zero, so that the new **A** has the following structure:

$$\mathbf{A}^{\text{new}} = \begin{bmatrix} * & * & * & * & * \\ * & * & * & * & * \\ 0 & * & * & * & * \\ 0 & 0 & * & * & * \\ 0 & 0 & 0 & \varepsilon & a_{nn} \end{bmatrix},$$

where ε is a relatively small number. A good approximation to an eigenvalue is a_{nn}. After dropping the last row and column of $\mathbf{A}^{\text{new}}$, the shifted **QR** method is applied to the remaining $(n-1) \times (n-1)$ submatrix. Again, we shift, using the eigenvalue of the trailing 2×2 submatrix which is closest to $a_{n-1,n-1}$ and we drive $a_{n-1,n-2}$ to zero. Continuing in this way, the subdiagonal elements are driven toward zero, leaving us with a matrix that is essentially upper triangular. The diagonal elements of this upper triangular matrix are the eigenvalues.

Subroutine VALS computes the eigenvalues of a real matrix by reducing the matrix to upper Hessenberg form and applying the shifted **QR** algorithm. The arguments of VALS are

- E: complex array of eigenvalues
- A: real array containing matrix [length at least N(N + 7) − 2]
- LA: leading (row) dimension of array A
- N: dimension of matrix stored in A
- V: work array which can be identified with A [length at least N(N + 2) + 1]

The analogous subroutine for a matrix with complex coefficients is called CVALS. Since each **QR** iteration overwrites the coefficients of the old **A** with the coefficients of the new **A**, the original contents of array A are destroyed by the subroutine. When the V argument above is not identified with the A argument, information needed to compute eigenvectors will be stored in V. Subroutine VECT (or CVECT when the coefficient matrix is complex) can be used to compute the eigenvector corresponding to a given eigenvalue. The arguments of VECT are

- E: estimate of eigenvalue (complex variable)
- X: eigenvector (complex array)
- V: output from subroutine VALS
- W: work array [length at least N(N + 9) − 6 if AIMAG(E) ≠ 0 and length at least .5N(N + 9) − 3 if AIMAG(E) = 0]

Subroutine VECT uses the inverse power method to estimate the eigenvector corresponding to the given eigenvalue. When the Rayleigh quotient yields a better approximation to the eigenvalue than the input value of E, subroutine VECT will replace the value of E by the improved estimate for the eigenvalue. For illustration, the eigenpairs of the matrix

$$\begin{bmatrix} 32 & 19 & 1 \\ -2 & 23 & -4 \\ 4 & 17 & 26 \end{bmatrix} \tag{51}$$

are computed by the following program:

```
      REAL A(30),E(3),V(20),W(30)
      COMPLEX E(3),X(3)
      DATA A/32.,-2.,4.,19.,23.,17.,1.,-4.,26./
      CALL VALS(E,A,3,3,V)
      DO 10 J = 1,3
        CALL VECT(E(J),X,V,W)
        WRITE(6,*) 'EIGENVALUE NUMBER',J,'IS',E(J)
        WRITE(6,*) 'EIGENVECTOR:'
        DO 10 I = 1,3
10        WRITE(6,*) I,X(I)
      END
```

When some of the eigenvalues are close to each other, it may be better to compute the diagonalization directly rather than use the inverse power method to estimate the eigenvectors. This direct algorithm is called DIAG and its arguments are

E: complex array of eigenvalues
V: complex array of eigenvectors
LV: leading (row) dimension of array V
A: real array containing matrix [length at least N(N + 7) − 2]
LA: leading (row) dimension of array A
N: dimension of matrix stored in A

The following program computes the diagonalization of the matrix (51).

```
      REAL A(3,10),E(3)
      COMPLEX V(3,3)
      DATA A/32.,-2.,4.,19.,23.,17.,1.,-4.,26./
      CALL DIAG(E,V,3,A,3,3)
      DO 10 J = 1,3
        WRITE(6,*) 'EIGENVALUE NUMBER',J,'IS',E(J)
        WRITE)6,*) 'EIGENVECTOR:'
        DO 10 I= 1,3
10        WRITE(6,*) I,V(I,J)
      END
```

Exercises

6-6.1. Consider the power method (36) in the special case where $\mathbf{X}_k$ is a square matrix. Defining $\mathbf{A}_k = \mathbf{Q}_{k-1}^T\mathbf{X}_k$ and $\mathbf{P}_k = \mathbf{Q}_{k-1}^T\mathbf{Q}_k$, establish the following three identities: (a) $\mathbf{A}_{k+1} = \mathbf{P}_k^T \mathbf{A}_k\mathbf{P}_k$, (b) $\mathbf{A}_k = \mathbf{P}_k\mathbf{R}_k$, and (c) $\mathbf{A}_{k+1} = \mathbf{R}_k\mathbf{P}_k$. This shows that the **QR** method is essentially the power method when $\mathbf{X}_k$ is a square matrix.

6-6.2. The first few iterations of the unshifted **QR** method are

$$\mathbf{A}_1 = \mathbf{Q}_1\mathbf{R}_1, \qquad \mathbf{A}_2 = \mathbf{R}_1\mathbf{Q}_1, \qquad \mathbf{A}_2 = \mathbf{Q}_2\mathbf{R}_2, \qquad \mathbf{A}_3 = \mathbf{R}_2\mathbf{Q}_2,$$

where $\mathbf{A}_1 = \mathbf{A}$ is the starting matrix. Verify the following identities:

$$\mathbf{Q}_1\mathbf{Q}_2\mathbf{R}_2\mathbf{R}_1 = \mathbf{A}^2 \tag{52}$$

and

$$\mathbf{A}_3 = \mathbf{Q}_2{}^T\mathbf{Q}_1{}^T\mathbf{A}\mathbf{Q}_1\mathbf{Q}_2.$$

Since the product of orthogonal matrices is orthogonal and the product of upper triangular matrices is upper triangular, (52) implies that $\mathbf{P} = \mathbf{Q}_1\mathbf{Q}_2$ is the orthogonal factor in a $\mathbf{QR}$ factorization of $\mathbf{A}^2$. Now consider the first two iterations of the shifted $\mathbf{QR}$ algorithm:

$$\mathbf{A}_1 - \sigma_1\mathbf{I} = \mathbf{Q}_1\mathbf{R}_1, \qquad \mathbf{A}_2 = \mathbf{Q}_1^T\mathbf{A}_1\mathbf{Q}_1, \qquad \mathbf{A}_2 - \sigma_2\mathbf{I} = \mathbf{Q}_2\mathbf{R}_2.$$

Show that

$$\mathbf{Q}_1\mathbf{Q}_2\mathbf{R}_2\mathbf{R}_1 = (\mathbf{A} - \sigma_2\mathbf{I})(\mathbf{A} - \sigma_1\mathbf{I}). \tag{53}$$

This identity is the basis for a real two-step version of the $\mathbf{QR}$ method which applies whenever the coefficient matrix is real. It follows from (53) that for two steps of the shifted $\mathbf{QR}$ method, the product $\mathbf{Q}_1\mathbf{Q}_2$ is the orthogonal factor in the $\mathbf{QR}$ factorization of $(\mathbf{A} - \sigma_2\mathbf{I})(\mathbf{A} - \sigma_1\mathbf{I}) = \mathbf{A}^2 - (\sigma_1 + \sigma_2)\mathbf{A} + \sigma_1\sigma_2\mathbf{I}$. If σ_2 is the conjugate of σ_1, both $\sigma_1 + \sigma_2$ and $\sigma_1\sigma_2$ are real, which implies that the product $(\mathbf{A} - \sigma_2\mathbf{I})(\mathbf{A} - \sigma_1\mathbf{I})$ is real when $\mathbf{A}$ is real. Therefore, the orthogonal factor $\mathbf{Q}_1\mathbf{Q}_2$ of $(\mathbf{A} - \sigma_2\mathbf{I})(\mathbf{A} - \sigma_1\mathbf{I})$ is real and $\mathbf{A}_3 = (\mathbf{Q}_1\mathbf{Q}_2)^T\mathbf{A}(\mathbf{Q}_1\mathbf{Q}_2)$ is real. In other words, a double $\mathbf{QR}$ iteration with complex conjugate shifts applied to a real matrix generates real iterations.

6-7. REDUCTION TO HESSENBERG FORM

In Section 6-6 we observed that it is much more efficient to apply the $\mathbf{QR}$ algorithm to a Hessenberg matrix than to a general matrix. It is now demonstrated that any square matrix $\mathbf{A}$ is similar to an upper Hessenberg matrix. That is, there exists an invertible matrix $\mathbf{P}$ such that $\mathbf{P}^{-1}\mathbf{AP}$ is upper Hessenberg. After computing an eigenpair $(\lambda, \mathbf{x})$ for $\mathbf{P}^{-1}\mathbf{AP}$, we multiply the eigenvector by $\mathbf{P}$ to obtain the corresponding eigenpair $(\lambda, \mathbf{Px})$ of $\mathbf{A}$.

The reduction of a matrix to upper Hessenberg form is illustrated with

$$\mathbf{A} = \begin{bmatrix} 450 & 75 & -525 & 150 \\ 75 & 253 & 380 & -79 \\ 150 & 5 & 325 & -215 \\ 150 & -604 & 160 & 322 \end{bmatrix}. \tag{54}$$

We start with the first column of $\mathbf{A}$:

$$\begin{bmatrix} 450 \\ 75 \\ 150 \\ 150 \end{bmatrix}.$$

Let $\mathbf{H}_1$ denote the Householder matrix which annihilates the third and fourth components of the first column while leaving the first component intact. Referring to equation (38) of Chapter 5, this Householder matrix is

$$\mathbf{H}_1 = \begin{bmatrix} 1 & 0 & 0 & 0 \\ 0 & -\frac{1}{3} & -\frac{2}{3} & -\frac{2}{3} \\ 0 & -\frac{2}{3} & \frac{2}{3} & -\frac{1}{3} \\ 0 & -\frac{2}{3} & -\frac{1}{3} & \frac{2}{3} \end{bmatrix}.$$

Premultiplying $\mathbf{A}$ by $\mathbf{H}_1$ yields

$$\mathbf{H}_1\mathbf{A} = \begin{bmatrix} 450 & 75 & -525 & 150 \\ -225 & 315 & -450 & -45 \\ 0 & 36 & -90 & -198 \\ 0 & -573 & -255 & 339 \end{bmatrix}.$$

Then postmultiplying $\mathbf{H}_1\mathbf{A}$ by $\mathbf{H}_1$ gives us a matrix that we denote $\mathbf{A}_1$:

$$\mathbf{A}_1 = \mathbf{H}_1\mathbf{A}\mathbf{H}_1 = \begin{bmatrix} 450 & 225 & -450 & 225 \\ -225 & 225 & -495 & -90 \\ 0 & 180 & -18 & -126 \\ 0 & 135 & 99 & 693 \end{bmatrix}.$$

Notice that the first column of $\mathbf{H}_1\mathbf{A}$ is equal to the first column of $\mathbf{H}_1\mathbf{A}\mathbf{H}_1$. The reason for this is that the first column of $\mathbf{H}_1$ contains a 1 followed by zeros and postmultiplying $\mathbf{H}_1\mathbf{A}$ by $\mathbf{H}_1$ leaves the first column intact. At this point, the first column of $\mathbf{A}_1$ has the proper structure for an upper Hessenberg matrix.

Next we focus on the second column of $\mathbf{A}_1$:

$$\begin{bmatrix} 225 \\ 225 \\ 180 \\ 135 \end{bmatrix}.$$

The Householder matrix that annihilates the fourth component of the second column while leaving the first two components intact is

$$\mathbf{H}_2 = \begin{bmatrix} 1 & 0 & 0 & 0 \\ 0 & 1 & 0 & 0 \\ 0 & 0 & -.8 & -.6 \\ 0 & 0 & -.6 & .8 \end{bmatrix}.$$

Premultiplying $\mathbf{A}_1$ by $\mathbf{H}_2$, we have

$$\mathbf{H}_2\mathbf{A}_1 = \begin{bmatrix} 450 & 225 & -450 & 225 \\ -225 & 225 & -495 & -90 \\ 0 & -225 & -45 & -315 \\ 0 & 0 & 90 & 630 \end{bmatrix}.$$

By the structure of $\mathbf{H}_2$, the first column of $\mathbf{H}_2\mathbf{A}_1$ is equal to the first column of $\mathbf{A}_1$ and the work done to the first column during the first step (annihilate the third and fourth components of the first column) is not disturbed by the second step. The matrix $\mathbf{H}_2\mathbf{A}_1$ is now upper Hessenberg. To complete the similarity transformation, $\mathbf{H}_2\mathbf{A}_1$ is postmultiplied by $\mathbf{H}_2$ to obtain the final matrix $\mathbf{A}_2$:

$$\mathbf{A}_2 = \mathbf{H}_2\mathbf{A}_1\mathbf{H}_2 = \begin{bmatrix} 450 & 225 & 225 & 450 \\ -225 & 225 & 450 & 225 \\ 0 & -225 & 225 & -225 \\ 0 & 0 & -450 & 450 \end{bmatrix}.$$

You can see that the Hessenberg structure of $\mathbf{H}_2\mathbf{A}_1$ is preserved when we postmultiply by $\mathbf{H}_2$; in fact, postmultiplying $\mathbf{H}_2\mathbf{A}_1$ by $\mathbf{H}_2$ does not alter the first two columns of $\mathbf{H}_2\mathbf{A}_1$.

To summarize, $\mathbf{A}_2 = \mathbf{H}_2\mathbf{A}_1\mathbf{H}_2$ is upper Hessenberg, where $\mathbf{A}_1$ is the product $\mathbf{H}_1\mathbf{A}\mathbf{H}_1$. Substituting $\mathbf{H}_1\mathbf{A}\mathbf{H}_1$ for $\mathbf{A}_1$, we have

$$\mathbf{A}_2 = \mathbf{H}_2\mathbf{H}_1\mathbf{A}\mathbf{H}_1\mathbf{H}_2. \tag{55}$$

Let $\mathbf{P}$ denote the product $\mathbf{H}_1\mathbf{H}_2$. The inverse of $\mathbf{P}$ is

$$\mathbf{P}^{-1} = (\mathbf{H}_1\mathbf{H}_2)^{-1} = \mathbf{H}_2^{-1}\mathbf{H}_1^{-1}.$$

Since a Householder matrix is its own inverse, $\mathbf{H}_1^{-1} = \mathbf{H}_1$, $\mathbf{H}_2^{-1} = \mathbf{H}_2$, and $\mathbf{P}^{-1} =$

$\mathbf{H}_2\mathbf{H}_1$. Hence, by (55) $\mathbf{A}_2$ is equal to $\mathbf{P}^{-1}\mathbf{AP}$ and the upper Hessenberg matrix $\mathbf{A}_2$ is similar to the starting matrix $\mathbf{A}$.

For a general $n \times n$ matrix $\mathbf{A}$, the reduction to upper Hessenberg form proceeds in the same way. In the first step, we construct the Householder matrix $\mathbf{H}_1$ that annihilates elements 3 through n in the first column of $\mathbf{A}$ and we set $\mathbf{A}_1 = \mathbf{H}_1\mathbf{A}\mathbf{H}_1$. In the kth step, we construct the Householder matrix $\mathbf{H}_k$ that annihilates elements $k + 2$ through n in the kth column of $\mathbf{A}_{k-1}$ and we set $\mathbf{A}_k = \mathbf{H}_k\mathbf{A}_{k-1}\mathbf{H}_k$. Step $n - 2$ generates the upper Hessenberg matrix $\mathbf{A}_{n-2}$. The relation $\mathbf{A}_k = \mathbf{H}_k\mathbf{A}_{k-1}\mathbf{H}_k$ implies that

$$\mathbf{A}_{n-2} = \mathbf{H}_{n-2} \cdots \mathbf{H}_2\mathbf{H}_1\mathbf{A}\mathbf{H}_1\mathbf{H}_2 \cdots \mathbf{H}_{n-2}.$$

Letting $\mathbf{P}$ be the product of the Householder matrices:

$$\mathbf{P} = \mathbf{H}_{n-2} \cdots \mathbf{H}_2\mathbf{H}_1,$$

$\mathbf{A}_{n-2}$ is equal to $\mathbf{P}^{-1}\mathbf{AP}$ and $\mathbf{A}$ is similar to an upper Hessenberg matrix.

Each step in the reduction of $\mathbf{A}$ to an upper Hessenberg matrix is given by

$$\mathbf{A}^{\text{new}} = \mathbf{H}\mathbf{A}^{\text{old}}\mathbf{H}, \tag{56}$$

where $\mathbf{H} = \mathbf{I} - 2\mathbf{w}\mathbf{w}^T$ is a Householder matrix. Let us now express (56) in terms of $\mathbf{w}$. First, the factor 2 in the definition of $\mathbf{H}$ is absorbed into $\mathbf{w}$ by writing $\mathbf{H} = \mathbf{I} - \mathbf{v}\mathbf{v}^T$, where $\mathbf{v} = \sqrt{2}\mathbf{w}$. Substituting $\mathbf{H} = \mathbf{I} - \mathbf{v}\mathbf{v}^T$ in (56), we have

$$\begin{aligned}
\mathbf{A}^{\text{new}} &= (\mathbf{I} - \mathbf{v}\mathbf{v}^T)\mathbf{A}^{\text{old}}(\mathbf{I} - \mathbf{v}\mathbf{v}^T) \\
&= \mathbf{A}^{\text{old}} - \mathbf{v}\mathbf{v}^T\mathbf{A}^{\text{old}} - \mathbf{A}^{\text{old}}\mathbf{v}\mathbf{v}^T + \mathbf{v}(\mathbf{v}^T\mathbf{A}^{\text{old}}\mathbf{v})\mathbf{v}^T \\
&= \mathbf{A}^{\text{old}} - \mathbf{v}\mathbf{v}^T\mathbf{A}^{\text{old}} - \mathbf{A}^{\text{old}}\mathbf{v}\mathbf{v}^T + \alpha\mathbf{v}\mathbf{v}^T,
\end{aligned}$$

where $\alpha = \mathbf{v}^T\mathbf{A}\mathbf{v}$. Defining the vectors $\mathbf{a}$ and $\mathbf{b}$ by

$$\mathbf{a}^T = \frac{\alpha}{2}\mathbf{v}^T - \mathbf{v}^T\mathbf{A}^{\text{old}} \quad \text{and} \quad \mathbf{b} = \frac{\alpha}{2}\mathbf{v} - \mathbf{A}^{\text{old}}\mathbf{v},$$

(56) is written

$$\mathbf{A}^{\text{new}} = \mathbf{A}^{\text{old}} + \mathbf{v}\mathbf{a}^T + \mathbf{b}\mathbf{v}^T. \tag{57}$$

Since column j of the matrix $\mathbf{v}\mathbf{a}^T$ is a_j times $\mathbf{v}$, the element in row i and column j for the matrix $\mathbf{v}\mathbf{a}^T$ is $v_i a_j$. Similarly, the element in row i and column j for the matrix $\mathbf{b}\mathbf{v}^T$ is $b_i v_j$. In component form, (57) is expressed:

$$a_{ij}^{\text{new}} = a_{ij}^{\text{old}} + v_i a_j + b_i v_j. \tag{58}$$

To summarize, each step in the reduction to Hessenberg form involves multiplying $\mathbf{v}^T$ by $\mathbf{A}^{\text{old}}$ in the computation of $\mathbf{a}$, multiplying $\mathbf{A}^{\text{old}}$ by $\mathbf{v}$ in the computation of $\mathbf{b}$, and the update (58). The following algorithm reduces a matrix to upper Hessenberg form using the update (58). Since step k annihilates $n - k - 1$ coefficients using a Householder matrix whose corresponding vector $\mathbf{w}$ has $n - k$ nonzero components, the nonzero components of $\mathbf{w}$ do not fit in the space occupied by the annihilated coefficients. For this reason, the subdiagonal of the upper Hessenberg matrix is stored in a separate array $\mathbf{l}$.

$$
\begin{array}{l}
k = 1 \text{ to } n-2 \\
\quad s \leftarrow (a_{k+1,k}^2 + a_{k+2,k}^2 + \cdots + a_{nk}^2)^{1/2} \\
\quad \text{if } s = 0 \text{ then } l_k \leftarrow 0 \text{ and go to next } k \\
\quad t \leftarrow a_{k+1,k},\ r \leftarrow 1/(s(s+|t|))^{1/2} \\
\quad \text{if } t < 0 \text{ then } s \leftarrow -s \\
\quad l_k \leftarrow -s,\ a_{k+1,k} \leftarrow r(t+s),\ s \leftarrow 0 \\
\quad a_{ik} = r a_{ik} \text{ for } i = k+2 \text{ to } n \\
\quad b_i \leftarrow 0 \text{ for } i = 1 \text{ to } n \\
\quad b_i \leftarrow b_i + a_{ij} a_{jk} \text{ for } i = 1 \text{ to } n, \text{ for } j = k+1 \text{ to } n \\
\quad s \leftarrow s + b_i a_{ik} \text{ for } i = k+1 \text{ to } n \\
\quad s \leftarrow s/2 \\
\quad b_i \leftarrow s a_{ik} - b_i \text{ for } i = k+1 \text{ to } n \\
\quad j = k+1 \text{ to } n \\
\qquad a_{ij} \leftarrow a_{ij} - b_i a_{jk} \text{ for } i = 1 \text{ to } k \\
\qquad t \leftarrow s a_{jk} \\
\qquad t \leftarrow t - a_{ij} a_{ik} \text{ for } i = k+1 \text{ to } n \\
\qquad a_{ij} \leftarrow a_{ij} + t a_{ik} + b_i a_{jk} \text{ for } i = k+1 \text{ to } n \\
\quad \text{next } j \\
\text{next } k
\end{array}
\tag{59}
$$

The subroutine that implements algorithm (59) is called HESS and the subroutine arguments are

A: coefficient matrix [length at least 2 + N(N + 1)]
LA: leading (row) dimension of array A
N: dimension of matrix stored in A
W: work array with at least N elements

Note that HESS stores the **l** of algorithm (59) within the A array and the elements of the coefficient matrix are rearranged to make room for **l**. Subroutines VALS and DIAG, the routines of Section 6-6 that implement the **QR** method, use (59) to initially reduce a general matrix to upper Hessenberg form. The similarity transformation $\mathbf{P} = \mathbf{H}_{n-2} \cdots \mathbf{H}_2\mathbf{H}_1$ involved in the reduction to Hessenberg form is computed by subroutine SIM whose input is the array **A** output by subroutine HESS.

There is another strategy for reducing a matrix to a similar Hessenberg matrix. This scheme is based on elimination and employs partial pivoting. At step k in the elimination scheme, row $k + 1$ is interchanged with the row beneath it, say row m, which has the absolute largest coefficient in column k. To annihilate the coefficients in column k beneath the subdiagonal coefficient $a_{k+1,k}$, we subtract from row i, $l_{ik} = a_{ik}/a_{k+1,k}$ times row $k + 1$ for i between $k + 2$ and n. To complete the similarity transformation, column $k + 1$ is interchanged with column m and l_{ik} times column i is added to column $k + 1$ for i between $k + 2$ and n. In summary, the following algorithm transforms **A** to a similar upper Hessenberg matrix using partial pivoting and elimination:

$$
\begin{array}{l}
k = 1 \text{ to } n-2 \\
\quad m = \arg\max \{|a_{ik}| : i = k+1 \text{ to } n\} \\
\quad \text{if } a_{mk} = 0 \text{ then go to next } k \\
\quad a_{mk} \leftrightarrow a_{k+1,k} \\
\quad a_{ik} \leftarrow a_{ik}/a_{k+1,k} \text{ for } i = k+2 \text{ to } n \\
\quad j = k+1 \text{ to } n \\
\quad\quad a_{mj} \leftrightarrow a_{k+1,j} \\
\quad\quad a_{ij} \leftarrow a_{ij} - a_{ik}a_{k+1,j} \text{ for } i = k+2 \text{ to } n \\
\quad \text{next } j \\
\quad a_{im} \leftrightarrow a_{i,k+1} \text{ for } i = 1 \text{ to } n \\
\quad a_{i,k+1} \leftarrow a_{i,k+1} + a_{jk}a_{ij} \text{ for } i = 1 \text{ to } n, \text{ for } j = k+2 \text{ to } n \\
\text{next } k
\end{array}
\tag{60}
$$

Although this scheme for reducing a matrix to a similar Hessenberg matrix is about twice as fast as Householder's method, Householder's method is more stable than (60) in the following sense: The coefficients in the upper Hessenberg matrix generated by elimination can be much larger than the coefficients of the starting matrix while for Householder's method, the Euclidean norm of the starting matrix is equal to the Euclidean norm of the final Hessenberg matrix (see Exercise 6-7.1). Another disadvantage in the elimination scheme (60) is that symmetry is not preserved by the reduction process.

Exercises

6-7.1. Show that Householder's scheme for reducing a matrix to a similar upper Hessenberg matrix has the property that the Euclidean norm of the starting coefficient matrix is equal to the Euclidean norm of the final upper Hessenberg matrix.

6-7.2. Suppose that **A** is an upper Hessenberg matrix with each subdiagonal coefficient nonzero and suppose that **A** can be diagonalized. Show that no two eigenvalues (appearing on the diagonal in the diagonalization) are the same. (*Hint:* How many linearly independent columns are contained in $\mathbf{A} - \lambda\mathbf{I}$?)

6-7.3. Evaluate the asymptotic parameters corresponding to the deepest loops in (59) and (60).

6-8. SYMMETRIC REDUCTION AND SYMMETRIC QR

This section, along with Sections 6-9 and 6-10, presents special eigenvalue routines for a symmetric matrix. The technique developed in Section 6-7 to reduce a matrix to Hessenberg form simplifies when the starting matrix **A** is symmetric since each intermediate matrix $\mathbf{A}_k$ is also symmetric. To demonstrate this symmetry, let us consider the first matrix $\mathbf{A}_1 = \mathbf{H}_1\mathbf{A}\mathbf{H}_1$ generated during the reduction to Hessenberg form. Since the transpose of the product of matrices is the product of the transposes in the reverse order, the transpose of $\mathbf{A}_1$ is

$$\mathbf{A}_1^T = \mathbf{H}_1^T\mathbf{A}^T\mathbf{H}_1^T.$$

Since a Householder matrix is symmetric, $\mathbf{H}_1^T = \mathbf{H}_1$. If **A** is also symmetric, then $\mathbf{A}^T = \mathbf{A}$ and

$$\mathbf{A}_1^T = \mathbf{H}_1\mathbf{A}\mathbf{H}_1 = \mathbf{A}_1.$$

Since $\mathbf{A}_1^T = \mathbf{A}_1$, the first step in the reduction process preserves symmetry. In a similar fashion, each of the intermediate matrices $\mathbf{A}_k$ is symmetric. Since $\mathbf{A}_{n-2}$ is upper Hessenberg, coefficients beneath the subdiagonal are zero. By symmetry, the corresponding coefficients above the superdiagonal are zero. A matrix that is zero beneath the subdiagonal and zero above the superdiagonal is tridiagonal. Therefore, a symmetric matrix can be reduced to a symmetric tridiagonal using a sequence of Householder matrices.

Now let us examine the update formula (58) when the coefficient matrix is symmetric. Referring to the definition of **a** and **b**, we see that **a** = **b** for a symmetric matrix. Replacing **a** by **b** in (58), one step in the reduction to tridiagonal form is

$$a_{ij}^{\text{new}} = a_{ij}^{\text{old}} + v_i b_j + b_i v_j, \tag{61}$$

where

$$\mathbf{b} = \frac{\alpha}{2}\mathbf{v} - \mathbf{A}^{\text{old}}\mathbf{v}.$$

The following algorithm uses (61) to reduce a symmetric matrix **A** to tridiagonal form. The diagonal and superdiagonal of the tridiagonal matrix are stored in **d** and **u**, respectively, while the vector **w** characterizing each Householder matrix is stored above the diagonal.

$k = 1$ to $n-2$
  $s \leftarrow (a_{k,k+1}^2 + a_{k,k+2}^2 + \cdots + a_{kn}^2)^{1/2}$
  $d_k \leftarrow a_{kk}$
  if $s = 0$ then $u_k \leftarrow 0$ and go to next k
  $t \leftarrow a_{k,k+1}$, $r \leftarrow 1/(s(s+|t|))^{1/2}$
  if $t < 0$ then $s \leftarrow -s$
  $u_k \leftarrow -s$, $a_{k,k+1} \leftarrow r(t+s)$, $s \leftarrow 0$
  $a_{ki} \leftarrow r a_{ki}$ for $i = k+2$ to n
  $b_i \leftarrow 0$ for $i = k+1$ to n
  $i = k+1$ to n (62)
    $t \leftarrow a_{ii} a_{ki}$
    $t \leftarrow t + a_{ij} a_{kj}$ and $b_j \leftarrow b_j + a_{ij} a_{ki}$ for $j = i+1$ to n
    $b_i \leftarrow b_i + t$ and $s \leftarrow s + b_i a_{ki}$
  next i
  $s \leftarrow s/2$
  $b_i \leftarrow s a_{ki} - b_i$ for $i = k+1$ to n
  $a_{ij} \leftarrow a_{ij} + a_{ki} b_j + b_i a_{kj}$ for $j = i$ to n, for $i = k+1$ to n
next k
$d_{n-1} \leftarrow a_{n-1,n-1}$, $d_n \leftarrow a_{nn}$, $u_{n-1} \leftarrow a_{n-1,n}$

Subroutine SHESS, which uses algorithm (62) to reduce a symmetric matrix to a tridiagonal matrix, has the following arguments:

D: diagonal of tridiagonal matrix
U: superdiagonal of tridiagonal matrix
A: coefficient matrix in compressed format
N: matrix dimension

The input matrix must be stored using the symmetric mode discussed in Section 2-4.3.

Another type of matrix whose structure can be preserved during the reduction process is a symmetric band matrix. As with a symmetric matrix, the reduced upper Hessenberg matrix is symmetric and tridiagonal. In addition, each intermediate matrix generated during the reduction process is a band matrix. The subroutine which reduces a symmetric band matrix to tridiagonal form is called HHESS and the subroutine arguments are

D: diagonal of tridiagonal matrix
U: superdiagonal of tridiagonal matrix
A: array containing diagonal and subdiagonal bands of coefficient matrix
LA: leading (row) dimension of array A
N: matrix dimension
H: half band width
W: work array which can be identified with A although the original coefficients are destroyed [length at least (H + 1)(N + 6 − H/2) − 12]

Table 6-6 lists the asymptotic parameters and effective cycle times associated with subroutines HESS, SHESS, and HHESS. These cycle times correspond to an IBM 3090 computer and the Fortran IV-G compiler. As usual, the asymptotic running time for a program is the product between the asymptotic parameter and the effective cycle time. To explain how the effective cycle times in Table 6-6 are derived, let us consider algorithm (59). The asymptotic time to reduce a general matrix to Hessenberg form is

$$\tfrac{1}{2} n^3 C_1 + \tfrac{1}{6} n^3 C_2 + \tfrac{1}{3} n^3 C_3 + \tfrac{1}{3} n^3 C_4, \tag{63}$$

where C_1 through C_4 are the respective cycle times for the following statements:

$$\begin{aligned} b_i &\leftarrow b_i - a_{ij} a_{jk}, \\ a_{ij} &\leftarrow a_{ij} - b_i a_{jk}, \\ t &\leftarrow t - a_{ij} a_{ik}, \\ a_{ij} &\leftarrow a_{ij} + t a_{ik} + b_i a_{jk}. \end{aligned}$$

If each of these cycle times were the same, then (63) would reduce to $\frac{4}{3}n^3C$, where C is the common cycle time and $\frac{4}{3}n^3$ is the number of cycles. Although

TABLE 6-6 HOUSEHOLDER'S REDUCTION TO HESSENBERG FORM

Matrix Type	Asymptotic Parameter	Cycle Time (microseconds)
General	$\frac{4}{3}n^3$	.73
Symmetric	$\frac{1}{2}n^3$	.61
Symmetric h-band	$2hn^2$	1.14

these cycle times are not the same, the expression $\frac{4}{3}n^3C$ is certainly more convenient than (63). With this motivation, let us factor $\frac{4}{3}n^3$ from (63) to obtain

$$\tfrac{4}{3}n^3(\tfrac{3}{8}C_1 + \tfrac{1}{8}C_2 + \tfrac{1}{4}C_3 + \tfrac{1}{4}C_4).$$

Letting C denote the effective cycle time defined by

$$C = \tfrac{3}{8}C_1 + \tfrac{1}{8}C_2 + \tfrac{1}{4}C_3 + \tfrac{1}{4}C_4,$$

the asymptotic time to reduce a general matrix to Hessenberg form is $\frac{4}{3}n^3C$.

Now let us examine how the **QR** algorithm changes when the coefficient matrix is symmetric. Before starting the **QR** iteration, the original matrix should be reduced to tridiagonal form since the **QR** iteration simplifies greatly when the coefficient matrix is symmetric and tridiagonal. The algorithms of Reinsch [R3] and Pal, Walker, and Kahan (see page 169 of [P4]) are the culmination of several attempts to simplify the **QR** algorithm while preserving numerical stability. Letting **d** denote a vector that stores the diagonal of a $n \times n$ symmetric tridiagonal matrix and letting **p** denote a vector that stores the square of each subdiagonal element, one iteration of the Pal-Walker-Kahan implicit **QR** algorithm with shift σ appears below.

$c \leftarrow 1,\ s \leftarrow 0,\ g \leftarrow d_1 - \sigma,\ q \leftarrow g^2$
$i = 1$ to $n-1$
  $t \leftarrow p_i,\ r \leftarrow q + t,\ p_{i-1} \leftarrow sr,\ b \leftarrow c,\ c \leftarrow q/r,\ s \leftarrow t/r$
  $f \leftarrow g,\ z \leftarrow d_{i+1},\ g \leftarrow c(z - \sigma) - sf,\ d_i \leftarrow f + (z - g)$
  if $c \neq 0$ then $q \leftarrow g^2/c$ (64)
  if $c = 0$ then $q \leftarrow bt$
next i
$p_{n-1} \leftarrow sq$
$d_n \leftarrow \sigma + g$

Again, a very effective shift utilizes an eigenvalue of the 2×2 trailing submatrix. That eigenvalue of the 2×2 trailing submatrix which is closest to d_n is often called the Wilkinson shift. After a little algebra, the Wilkinson shift is expressed:

$$\sigma = d_n - \frac{\operatorname{sign}(\delta)\, p_{n-1}}{|\delta| + \sqrt{\delta^2 + p_{n-1}}}, \tag{65}$$

where $\delta = (d_{n-1} - d_n)/2$. The **QR** method is cubically convergent with the Wilkinson shift. Typically, a couple iterations of (64) drive p_{n-1} toward zero. Changing the upper limit of the i loop in (64) from $n - 1$ to $n - 2$, a couple more iterations drive p_{n-2} toward zero. Continuing in this way, the elements of **p** are driven toward zero and the eigenvalues are left in **d**.

The implementation of the **QR** algorithm described in Section 6-6 is called an **explicit** implementation of the **QR** algorithm since the shifted matrix is factored in each iteration. The Pal-Walker-Kahan scheme, on the other hand, is an **implicit** implementation of the **QR** algorithm. In this implicit scheme, we subtract the shift σ from the leading diagonal element a_{11} and determine the Givens rotation $\mathbf{G}_1$ that annihilates the (2,1) element of the shifted matrix. Premultiplying the tridiagonal matrix $\mathbf{A}$ by $\mathbf{G}_1$ and postmultiplying by $\mathbf{G}_1^T$ yields a matrix with the following structure:

$$\mathbf{A}^{\text{new}} = \mathbf{G}_1\mathbf{A}\mathbf{G}_1^T = \begin{bmatrix} * & * & * & & & \\ * & * & * & & & \\ * & * & * & * & & \\ & & * & * & * & \\ & & & * & * & \cdot \\ & & & & \cdot & \cdot \end{bmatrix}. \tag{66}$$

In the next step, we construct a Givens rotation $\mathbf{G}_2$ that annihilates the a_{31} element in (66). Premultiplying by $\mathbf{G}_2$ and postmultiplying by $\mathbf{G}_2^T$ give us

$$\mathbf{A}^{\text{new}} = \mathbf{G}_2\mathbf{A}^{\text{old}}\mathbf{G}_2^T = \begin{bmatrix} * & * & & & & \\ * & * & * & * & & \\ & * & * & * & & \\ & * & * & * & * & \\ & & & * & * & \cdot \\ & & & & \cdot & \cdot \end{bmatrix}.$$

Continuing in this way, we multiply by a sequence of Givens rotations, chasing the "bulge" down the diagonal, eventually obtaining a tridiagonal matrix which (in most cases) is identical to the tridiagonal matrix obtained by the explicit algorithm.

The one troublesome case for the implicit scheme is the case where some subdiagonal element is zero at the start. If the implicit **QR** scheme is used when a subdiagonal element vanishes, then convergence can be slow since the vanishing element washes out the impact of the shift. Notice though that if a subdiagonal element vanishes, then the coefficient matrix **A** has the following structure:

$$\mathbf{A} = \begin{bmatrix} \mathbf{B} & \mathbf{0} \\ \mathbf{0} & \mathbf{C} \end{bmatrix},$$

where both **B** and **C** are tridiagonal. The characteristic equation for a matrix with this structure is

$$\det(\mathbf{B} - \lambda\mathbf{I})\det(\mathbf{C} - \lambda\mathbf{I}) = 0,$$

and the eigenvalues are the root of the two equations

$$\det(\mathbf{B} - \lambda\mathbf{I}) = 0 \qquad \text{and} \qquad \det(\mathbf{C} - \lambda\mathbf{I}) = 0.$$

Therefore, when a subdiagonal element vanishes, the eigenproblem uncouples into two smaller problems, and the eigenvalues of **A** are the same as the eigenvalues of **B** combined with the eigenvalues of **C**. To summarize, for an implicit **QR** algorithm such as (64), it is important to check for small subdiagonal elements. If p_j is relatively small, the lower limit for the i loop in (64) is changed from 1 to $j + 1$.

Subroutine TVALS implements algorithm (64). The subroutine arguments are

E: eigenvalues
L: subdiagonal (can be identified with U)
D: diagonal
U: superdiagonal
N: matrix dimension
W: work array (length at least N)

As you show in Exercise 6-8.2, algorithm (64) can be applied to any tridiagonal matrix provided the cross-diagonal products $p_k = a_{k,k-1}a_{k-1,k}$ are nonnegative. For this reason, both the subdiagonal and the superdiagonal are inputs to subroutine TVALS. Also note that the original contents of arrays L, D, and U are unaltered by the subroutine so that they can be used for input to other subroutines. The eigenvalues are arranged in E in order from smallest to largest. In the following program, we reduce the matrix

$$\begin{bmatrix} 4 & 3 & 2 & 1 \\ 3 & 4 & 3 & 2 \\ 2 & 3 & 4 & 3 \\ 1 & 2 & 3 & 4 \end{bmatrix}$$

to tridiagonal form using SHESS, we compute the eigenvalues using TVALS, and we compute the eigenvectors using subroutine SVECT.

```
      REAL A(10),D(4),U(4),X(4),E(4),W(16)
      DATA A/4.,3.,2.,1.,4.,3.,2.,4.,3.,4./
      CALL SHESS(D,U,A,4)
      CALL TVALS(E,U,D,U,4,W)
      DO 20 J = 1,4
         CALL SVECT(E(J),X,A,D,U,W)
         WRITE(6,*) 'EIGENVALUE:',E(J)
         WRITE(6,*) 'EIGENVECTOR:'
         DO 10 I = 1,4
10          WRITE(6,*) X(I)
20    CONTINUE
```

In general, the work argument of SVECT must have at least 4N elements. If the starting matrix is a symmetric band matrix that is tridiagonalized using HHESS, subroutine HVECT computes the eigenvector corresponding to a given eigenvalue.

Exercises

6-8.1. Construct a similarity transformation **P** that is a diagonal matrix and with the property that $\mathbf{P}^{-1}\mathbf{AP}$ is symmetric for the matrix

$$\mathbf{A} = \begin{bmatrix} 1 & 4 & 0 \\ 1 & 2 & 9 \\ 0 & 1 & 3 \end{bmatrix}.$$

6-8.2. Given a tridiagonal matrix **A** with positive cross-diagonal products, write a program that generates the diagonal of a diagonal matrix **P** for which $\mathbf{P}^{-1}\mathbf{AP}$ is symmetric. How does the diagonal of **A** relate to the diagonal of $\mathbf{P}^{-1}\mathbf{AP}$? How do the cross-diagonal products $a_{i,i-1}a_{i-1,i}$ for **A** relate to the corresponding cross-diagonal products for $\mathbf{P}^{-1}\mathbf{AP}$?

6-8.3. Consider the following 2 × 2 matrix:

$$\begin{bmatrix} a & b \\ b & c \end{bmatrix}. \tag{67}$$

Analogous to formula (65), show that the eigenvalue σ of (67) closest to c is given by

$$\sigma = c - \frac{\text{sign}(\delta)b^2}{|\delta| + \sqrt{\delta^2 + b^2}},$$

where $\delta = (a - c)/2$.

6-8.4. Evaluate the asymptotic parameters corresponding to the deepest loop in (64).

6-9. BISECTION METHOD

The **QR** method generates all the eigenvalues of a matrix. Let us now study methods to compute a few eigenvalues of a symmetric matrix. Although the power method can be used to compute a few eigenvalues, some special techniques are often more efficient for a symmetric matrix. The **bisection scheme** is based on the following property: If $\mathbf{A} = \mathbf{LU}$ is the factorization without pivots for a symmetric matrix, the number of positive, zero, and negative eigenvalues of **A** (the **inertia** of **A**) is equal to the number of positive, zero, and negative diagonal elements of **U**. This property also applies to the shifted matrix $\mathbf{A}_\sigma = \mathbf{A} - \sigma\mathbf{I}$ since subtracting σ from each diagonal element of **A** does not disturb symmetry. Since the eigenvalues of $\mathbf{A}_\sigma$ are the eigenvalues of **A** minus σ, the number of positive, zero, and negative eigenvalues of $\mathbf{A}_\sigma$ is equal to the number of **A**'s eigenvalues that are greater than σ, equal to σ, and less than σ, respectively.

The bisection method for computing the kth largest eigenvalue of a symmetric matrix **A** can be implemented in the following way: Using Gerschgorin's theorem, we find an interval $[a, b]$ that contains the kth largest eigenvalue. Now let σ be the midpoint of the interval $[a, b]$:

$$\sigma = \frac{a + b}{2}$$

Factoring $\mathbf{A}_\sigma$, we determine the number of **A**'s eigenvalues which are greater than σ. If fewer than k eigenvalues are greater than σ, the kth largest eigenvalue lies in the interval $[a, \sigma]$. If k or more eigenvalues are greater than σ, the kth largest eigenvalue lies in the interval $[\sigma, b]$. In either case, we find a smaller interval that contains the kth largest eigenvalue and the process repeats. Before starting the bisection process, it may help to reduce **A** to tridiagonal form since tridiagonal matrices can be factored quickly.

To illustrate the bisection scheme, let us compute the second largest eigenvalue of the matrix

$$\mathbf{A} = \begin{bmatrix} 2 & 1 & 0 \\ 1 & 3 & 1 \\ 0 & 1 & 6 \end{bmatrix}. \qquad (68)$$

In Figure 6-7 we project the three Gerschgorin disks associated with **A** onto the real axis. Since the projected disks extend from 1 to 7, the eigenvalues of **A** lie on the interval [1, 7]. Setting $\sigma = (1 + 7)/2 = 4$, we factor $\mathbf{A}_\sigma$ to obtain

$$\mathbf{A}_4 = \begin{bmatrix} -2 & 1 & 0 \\ 1 & -1 & 1 \\ 0 & 1 & 2 \end{bmatrix} = \begin{bmatrix} 1 & 0 & 0 \\ -\frac{1}{2} & 1 & 0 \\ 0 & -2 & 1 \end{bmatrix} \begin{bmatrix} -2 & 1 & 0 \\ 0 & -\frac{1}{2} & 1 \\ 0 & 0 & 4 \end{bmatrix}.$$

Since the diagonal elements -2, $-\frac{1}{2}$, and 4 of the upper triangular factor for $\mathbf{A}_4$ exhibit the sign pattern $[-, -, +]$, **A** has one eigenvalue larger than 4 and two eigenvalues smaller than 4. In particular, the second largest eigenvalue is less than 4. Since the second largest eigenvalue lies on the interval [1, 7] but

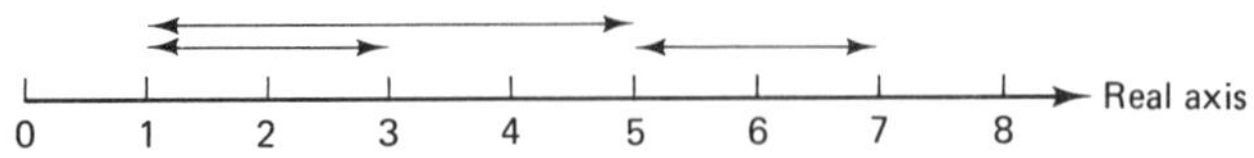

Figure 6-7 Projected Gerschgorin disks for the matrix (68).

is less than 4, we conclude that the second largest eigenvalue lies on the interval [1, 4]. In the next iteration, σ is $(1 + 4)/2 = 2.5$ and the sign pattern for the upper triangular factor of $\mathbf{A}_\sigma$ is $[-,+,+]$. Since **A** has two eigenvalues larger than 2.5, the second largest eigenvalue is contained in the interval [2.5,4]. Continuing this process, we obtain the bracketing intervals shown in Table 6-7. The second largest eigenvalue is near 3.358.

Since pivoting is not allowed in the bisection method, the factorization process breaks down when a zero is encountered on the diagonal. One way to cope with a zero pivot is to replace the zero diagonal element by a small positive number and continue the factorization. However, if the symmetric matrix is also tridiagonal, there is another solution based on a simple recurrence. Given a tridiagonal matrix **A**, let f_k denote the determinant of the kth **leading submatrix**. That is, f_k is the determinant of the submatrix formed by the elements common to the first k rows and the first k columns of **A**. For the 3×3 tridiagonal matrix

$$\mathbf{A} = \begin{bmatrix} a_{11} & a_{12} & 0 \\ a_{21} & a_{22} & a_{23} \\ 0 & a_{32} & a_{33} \end{bmatrix},$$

there are three leading submatrices and the corresponding determinants are

$$f_1 = \det [a_{11}], \qquad f_2 = \det \begin{bmatrix} a_{11} & a_{12} \\ a_{21} & a_{22} \end{bmatrix}, \qquad f_3 = \det \begin{bmatrix} a_{11} & a_{12} & 0 \\ a_{21} & a_{22} & a_{23} \\ 0 & a_{32} & a_{33} \end{bmatrix}.$$

It can be shown that these determinants satisfy a three-term recurrence:

$$f_k = a_{kk} f_{k-1} - p_k f_{k-2}, \tag{69}$$

where $p_k = a_{k,k-1} a_{k-1,k}$ is the product between elements on opposite sides of the diagonal and the initial values for the f_k are $f_0 = 1$ and $f_{-1} = 0$. Applying the recurrence (69) to the matrix

$$\mathbf{A} = \begin{bmatrix} 3 & 1 & 0 \\ 1 & 5 & 2 \\ 0 & 2 & 4 \end{bmatrix}, \tag{70}$$

TABLE 6-7 BISECTION METHOD FOR MATRIX (68)

Sign Pattern	a	b	σ
− − +	1.000	7.000	4.000
− + +	1.000	4.000	2.500
− + +	2.500	4.000	3.250
− − +	3.250	4.000	3.625
− + −	3.250	3.625	3.438
− + +	3.250	3.438	3.344
− + −	3.344	3.438	3.391
− + −	3.344	3.391	3.367
− + +	3.344	3.367	3.356
− + −	3.356	3.367	3.361
− + −	3.356	3.361	3.358
− + +	3.356	3.358	3.357
− + +	3.357	3.358	3.358

we have

$$f_1 = a_{11}f_0 = 3 \times 1 = 3,$$
$$f_2 = a_{22}f_1 - a_{21}a_{12}f_0 = 5 \times 3 - 1 \times 1 \times 1 = 14,$$
$$f_3 = a_{33}f_2 - a_{32}a_{23}f_1 = 4 \times 14 - 2 \times 2 \times 3 = 44.$$

Since $f_3 = 44$, the determinant of the matrix (70) is 44.

The recurrence (69) also provides a formula for the characteristic polynomial det $(\mathbf{A} - \lambda\mathbf{I})$. If $\mathbf{A}$ is a tridiagonal matrix, $\mathbf{A} - \lambda\mathbf{I}$ is the tridiagonal matrix obtained by subtracting λ from each diagonal element of $\mathbf{A}$. Letting $f_k(\lambda)$ denote the determinant of the kth leading submatrix for $\mathbf{A} - \lambda\mathbf{I}$, $f_k(\lambda)$ satisfies recurrence (69) except that a_{kk} is replaced by $a_{kk} - \lambda$:

$$f_k(\lambda) = (a_{kk} - \lambda)f_{k-1}(\lambda) - p_k f_{k-2}(\lambda), \tag{71}$$

where $f_0(\lambda) = 1$ and $f_{-1}(\lambda) = 0$. For example, applying (71) to the matrix (70) with $\lambda = 2$ yields

$$f_1(2) = (a_{11} - 2)f_0(2) = 1 \times 1 = 1,$$
$$f_2(2) = (a_{22} - 2)f_1(2) - a_{21}a_{12}f_0(2) = 3 \times 1 - 1 \times 1 \times 1 = 2,$$
$$f_3(2) = (a_{33} - 2)f_2(2) - a_{32}a_{23}f_1(2) = 2 \times 2 - 2 \times 2 \times 1 = 0.$$

Since $f_3(2) = 0$, the determinant of $\mathbf{A} - 2\mathbf{I}$ is zero and $\lambda = 2$ is an eigenvalue of the matrix (70).

When each product p_k is greater than or equal to zero, the recurrence (71) not only evaluates the characteristic polynomial, but also indicates the number of eigenvalues smaller than λ. It can be shown that for a $n \times n$ tridiagonal matrix with each p_k nonnegative, the number of eigenvalues smaller than λ is equal to the number of changes in sign for the sequence $f_0(\lambda), f_1(\lambda), \cdots, f_n(\lambda)$. This property is called the **Sturm sequence property**. For a symmetric matrix, the p_k are always nonnegative: $p_k = a_{k,k-1}a_{k-1,k} = (a_{k,k-1})^2$. Hence this relationship between the number of eigenvalues smaller than λ and the number of sign changes always applies to a symmetric matrix. For illustration, let us consider the matrix (70) with $\lambda = 2$. Since $f_0(2) = 1$, $f_1(2) = 1$, $f_2(2) = 2$, and $f_3(2) = 0$, there are no changes in sign and every eigenvalue is greater than or equal to 2. On the other hand, for $\lambda = 6$, we have $f_0(6) = 1$, $f_1(6) = -3$, $f_2(6) = 2$, and $f_3(6) = 8$. The sign pattern $[+,-,+,+]$ changes from + to − to +. Since there are two sign changes, the matrix (70) has two eigenvalues smaller than 6.

This relationship between sign changes in the sequence $f_0(\sigma), f_1(\sigma), \cdots, f_n(\sigma)$ and the number of eigenvalues smaller than σ can be incorporated in a bisection procedure to compute a particular eigenvalue of a symmetric tridiagonal matrix. For example, to compute the kth smallest eigenvalue, we first use Gerschgorin's theorem to obtain an interval $[a, b]$ which contains the kth smallest eigenvalue. Letting σ be $(a + b)/2$, the midpoint of the interval $[a, b]$, we count the number m of sign changes in the sequence $f_0(\sigma), f_1(\sigma), \cdots, f_n(\sigma)$. If m is less than k, the kth smallest eigenvalue lies in the interval $[\sigma, b]$. Conversely, if m is greater than or equal to k, the kth smallest eigenvalue lies in the interval $[a, \sigma]$. In either case, we obtain a smaller interval that contains the kth smallest eigenvalue and the bisection process repeats.

Comparing the bisection scheme based on recurrence (71) to the bisection scheme based upon the factorization of the shifted matrix $\mathbf{A}_\sigma$, we note that the recurrence (71) involves no divisions while the factorization of a matrix requires division by a diagonal element during elimination. Although we are never concerned with zero pivots when evaluating the recurrence (71), the $f_k(\lambda)$ tend to approach zero or infinity as k increases. To prevent underflow or overflow, we must multiply $f_k(\lambda)$ by a scale factor whenever $|f_k(\lambda)|$ or its reciprocal approaches the largest possible floating-point number.

When implementing the bisection method using recurrence (71), convergence can be accelerated. Recall that the eigenvalues of a matrix are the roots of the characteristic equation det $(\mathbf{A} - \lambda\mathbf{I}) = 0$. Since det $(\mathbf{A} - \lambda\mathbf{I}) = f_n(\lambda)$, the eigenvalues are also the roots of the equation $f_n(\lambda) = 0$. Therefore, after finding an interval that contains just one eigenvalue, the "Newton type" procedure employed by subroutine ROOT can be used to solve $f_n(\lambda) = 0$. Subroutine TVAL (not to be confused with TVALS) computes the kth largest or the kth smallest eigenvalue of a symmetric tridiagonal matrix using recurrence (71) coupled with a routine similar to ROOT. The subroutine arguments are

E: eigenvalue (output)
K: index of desired eigenvalue
L: subdiagonal
D: diagonal
U: superdiagonal
N: matrix dimension
W: work array (length at least N)

Setting K = 2 generates the second smallest eigenvalue while setting K = −2 generates the second largest eigenvalue. The following program computes the smallest eigenvalue of the matrix (70) while the corresponding eigenvector is evaluated using TVECT.

```
      REAL D(3),U(3),X(3),W(12)
      DATA D/3.,5.,4./U/1.,2./
      CALL TVAL(SMALL,1,U,D,U,3,W)
      CALL TVECT(SMALL,X,U,D,U,3,W)
      WRITE(6,*)'SMALLEST EIGENVALUE:',SMALL
      WRITE(6,*) 'EIGENVECTOR:'
      DO 10 I = 1,3
10       WRITE(6,*) I,X(I)
      END
```

In general, the work argument for TVECT must have at least 4N elements.

Another subroutine called SLICE uses the same algorithm employed by TVAL to compute the eigenvalues of a symmetric tridiagonal matrix which lie on a given interval. The subroutine arguments are

E: array of eigenvalues (output)
K: number of eigenvalues on specified interval (output)
A: lower limit of interval
B: upper limit of interval

L: subdiagonal (can be identified with U)
D: diagonal
U: superdiagonal
N: matrix dimension
W: work array (length at least N + 9K)

Exercises

6-9.1. Formula (71) shows that for a tridiagonal matrix, the characteristic polynomial is determined by the diagonal elements and the cross-diagonal products p_k. Construct a symmetric matrix with the same eigenvalues as those for the matrix **A** of Exercise 6-8.1.

6-9.2. What is the relationship between the eigenvalues of the following two matrices:

$$\mathbf{A} = \begin{bmatrix} 2 & 10 & 0 \\ 2 & 2 & 1 \\ 0 & 1 & 2 \end{bmatrix} \quad \text{and} \quad \mathbf{B} = \begin{bmatrix} 2 & 4 & 0 \\ 5 & 2 & -1 \\ 0 & -1 & 2 \end{bmatrix}.$$

6-9.3. Does the following matrix have any negative eigenvalues?

$$\begin{bmatrix} 1 & 1 & 0 \\ 1 & 1 & 1 \\ 0 & 1 & 2 \end{bmatrix}$$

6-9.4. Suppose that **A** is a real symmetric tridiagonal matrix with $a_{i,i-1} \neq 0$ for every i. Explain why each eigenvalue of **A** is distinct. (*Hint*: See Exercises 6-7.2 and 6-5.1.)

6-10. LANCZOS METHOD

Lanczos method [L1] reduces a symmetric matrix to a tridiagonal matrix using an orthogonal similarity transformation. That is, given a symmetric matrix **A**, an orthogonal matrix **Q** is constructed with the property that

$$\mathbf{Q}^{-1}\mathbf{A}\mathbf{Q} = \mathbf{T}, \tag{72}$$

where **T** is a symmetric tridiagonal matrix. In Householder's reduction scheme, we premultiply and postmultiply **A** by Householder matrices, reducing **A** to a tridiagonal matrix. In Lanczos scheme, we start with (72) and we derive a recurrence for the columns of **Q** and the elements of **T**. Multiplying (72) by **Q** gives us the relation **AQ** = **QT**. Therefore, each column of the matrix **AQ** is equal to the corresponding column of the matrix **QT**. If $\mathbf{q}_j$ denotes the jth column of **Q**, the jth column of **AQ** is $\mathbf{A}\mathbf{q}_j$. Letting **d** and **u** denote the diagonal and the superdiagonal of **T**, we have

$$\mathbf{T} = \begin{bmatrix} d_1 & u_1 & & & & \\ u_1 & d_2 & \cdot & & & \cdot \\ & \cdot & \cdot & \cdot & & \\ & & \cdot & \cdot & \cdot & \\ & & & \cdot & \cdot & u_{n-1} \\ & & & & u_{n-1} & d_n \end{bmatrix}.$$

The jth column of $\mathbf{T}$ is

$$\begin{bmatrix} 0 \\ \cdot \\ \cdot \\ \cdot \\ 0 \\ u_{j-1} \\ d_j \\ u_j \\ 0 \\ \cdot \\ \cdot \\ \cdot \\ 0 \end{bmatrix}.$$

Since the jth column of $\mathbf{QT}$ is equal to $\mathbf{Q}$ times the jth column of $\mathbf{T}$, the jth column of $\mathbf{QT}$ can be expressed

$$u_{j-1}\mathbf{q}_{j-1} + d_j\mathbf{q}_j + u_j\mathbf{q}_{j+1}.$$

Equating the jth column of $\mathbf{AQ}$ to the jth column of $\mathbf{QT}$, we have:

$$\mathbf{A}\mathbf{q}_j = u_{j-1}\mathbf{q}_{j-1} + d_j\mathbf{q}_j + u_j\mathbf{q}_{j+1}. \tag{73}$$

With the convention that $\mathbf{q}_0$ is zero, this relation also holds for $j = 1$. Since $\mathbf{Q}$ is an orthogonal matrix,

$$\mathbf{q}_j^T\mathbf{q}_{j-1} = \mathbf{q}_j^T\mathbf{q}_{j+1} = 0 \qquad \text{and} \qquad \mathbf{q}_j^T\mathbf{q}_j = 1.$$

Premultiplying (73) by $\mathbf{q}_j^T$, we conclude that

$$d_j = \mathbf{q}_j^T\mathbf{A}\mathbf{q}_j.$$

Defining the vector $\mathbf{r}_j$ by

$$\mathbf{r}_j = \mathbf{A}\mathbf{q}_j - d_j\mathbf{q}_j - u_{j-1}\mathbf{q}_{j-1},$$

(73) implies that $u_j\mathbf{q}_{j+1} = \mathbf{r}_j$. Taking the Euclidean norm, we have

$$|u_j|\,\|\mathbf{q}_{j+1}\|_2 = \|\mathbf{r}_j\|_2.$$

Since $\mathbf{q}_{j+1}$ is a unit vector in the Euclidean norm, u_j is plus or minus the Euclidean length of $\mathbf{r}_j$. Finally, solving the identity $u_j\mathbf{q}_{j+1} = \mathbf{r}_j$ for $\mathbf{q}_{j+1}$ gives $\mathbf{q}_{j+1} = \mathbf{r}_j/u_j$. These formulas for d_j, u_j, and $\mathbf{q}_{j+1}$ constitute Lanczos method for reducing a symmetric matrix to a tridiagonal matrix. In summary, starting from an arbitrary nonzero vector $\mathbf{q}_1$, Lanczos method is the following:

$$\boxed{\begin{array}{l} \mathbf{q}_0 = \mathbf{0},\ \mathbf{r} = \mathbf{q}_1,\ u_0 = \|\mathbf{r}\|_2 \\ j = 1 \text{ to } n \\ \quad \mathbf{q}_j = \mathbf{r}/u_{j-1},\ d_j = \mathbf{q}_j^T\mathbf{A}\mathbf{q}_j \\ \quad \mathbf{r} = (\mathbf{A} - d_j\mathbf{I})\mathbf{q}_j - u_{j-1}\mathbf{q}_{j-1} \\ \quad u_j = \|\mathbf{r}\|_2 \\ \text{next } j \end{array}} \tag{74}$$

Observe that when $\mathbf{q}_j$ is computed, we divide each component of $\mathbf{r}$ by the superdiagonal element u_{j-1}. Obviously, this division is impossible when $u_{j-1} = 0$.

If a superdiagonal element vanishes, then to continue the tridiagonalization process, we replace $\mathbf{q}_j$ by any unit vector orthogonal to $\mathbf{q}_1$ through $\mathbf{q}_{j-1}$ and we continue to execute (74).

Recall that Householder's method for reduction to tridiagonal form involves the update

$$\mathbf{A}^{\text{new}} = \mathbf{H}\mathbf{A}^{\text{old}}\mathbf{H},$$

which changes the coefficient matrix in each iteration. If $\mathbf{A}$ is sparse, nonzero elements are often generated where there had been zeros. In contrast, Lanczos method does not alter the coefficient matrix. Each iteration of Lanczos method multiplies $\mathbf{A}$ by a vector $\mathbf{q}_j$ and $\mathbf{A}$ does not change. Theoretically, Lanczos method is faster than Householder's method when $\mathbf{A}$ is a large sparse matrix since a sparse matrix can be multiplied by a vector quicker than the update $\mathbf{A}^{\text{new}} = \mathbf{H}\mathbf{A}^{\text{old}}\mathbf{H}$ can be applied to a general matrix. Unfortunately, Lanczos method suffers from an instability. Typically, a small but nonzero superdiagonal element u_{j-1} is encountered at some point and thereafter, the $\mathbf{q}_j$ lose their orthogonality. In theory, Lanczos method reduces $\mathbf{A}$ to the tridiagonal matrix $\mathbf{T}$ and the eigenvalues of $\mathbf{T}$ can be computed using the **QR** algorithm. But due to numerical instability, Lanczos method is better suited for approximating the most positive and the most negative eigenvalues of $\mathbf{A}$. The most positive and the most negative eigenvalues of a matrix are called the **extreme eigenvalues**.

The instability in Lanczos method is related to the convergence of the power method. Referring to (73), $\mathbf{q}_2$ is given by

$$\mathbf{q}_2 = u_1^{-1}(\mathbf{A}\mathbf{q}_1 - d_1\mathbf{q}_1).$$

Thus $\mathbf{q}_2$ lies in the space spanned by the vectors $\mathbf{q}_1$ and $\mathbf{A}\mathbf{q}_1$. In the next iteration of Lanczos method, $\mathbf{q}_2$ is multipled by $\mathbf{A}$ to obtain $\mathbf{q}_3$. Consequently, $\mathbf{q}_3$ lies in the space spanned by the vectors $\mathbf{q}_1$, $\mathbf{A}\mathbf{q}_1$, and $\mathbf{A}^2\mathbf{q}_1$. In general, $\mathbf{q}_j$ lies in the space spanned by the vectors $\mathbf{q}_1, \mathbf{A}\mathbf{q}_1, \cdots, \mathbf{A}^{j-1}\mathbf{q}_1$. This space, denoted $\mathbf{K}^j(\mathbf{q}_1)$, is called the jth **Krylov space** associated with $\mathbf{q}_1$. Lanczos method essentially generates an orthonormal basis for the Krylov spaces. From our analysis of the power method, we know that as j increases, the vector $\mathbf{A}^j\mathbf{q}_1$ usually approaches the eigenvector corresponding to $\mathbf{A}$'s dominant eigenvalue. Consequently, for large j, the vector $\mathbf{A}^j\mathbf{q}_1$ in the $(j+1)$st Krylov space is nearly contained in the jth Krylov space.

Returning to equation (73), let $\mathbf{v}$ denote the sum $d_j\mathbf{q}_j + u_{j-1}\mathbf{q}_{j-1}$. Since $\mathbf{q}_j$ and $\mathbf{q}_{j-1}$ are contained in the jth Krylov space and since $\mathbf{q}_{j+1}$ is perpendicular to the jth Krylov space, equation (73) expresses $\mathbf{A}\mathbf{q}_j$ as the sum of a vector $\mathbf{v}$ in the jth Krylov space and a vector $u_j\mathbf{q}_{j+1}$ perpendicular to the jth Krylov space. Thus $\mathbf{v}$ is the projection of $\mathbf{A}\mathbf{q}_j$ into the jth Krylov space. If $\mathbf{A}\mathbf{q}_j$ is nearly contained in the jth Krylov space, then $\mathbf{v}$ is nearly equal to $\mathbf{A}\mathbf{q}_j$. Since $\mathbf{q}_{j+1} = (\mathbf{A}\mathbf{q}_j - \mathbf{v})/u_j$, nearly equal vectors are subtracted when $\mathbf{q}_{j+1}$ is computed and the error in the computed $\mathbf{q}_{j+1}$ is relatively large. Moreover, taking the 2-norm of the identity $\mathbf{q}_{j+1} = (\mathbf{A}\mathbf{q}_j - \mathbf{v})/u_j$ yields

$$|u_j| = \|\mathbf{A}\mathbf{q}_j - \mathbf{v}\|_2/\|\mathbf{q}_{j+1}\|_2 = \|\mathbf{A}\mathbf{q}_j - \mathbf{v}\|_2.$$

Hence, $|u_j|$ is small whenever $\mathbf{v}$ is nearly equal to $\mathbf{A}\mathbf{q}_j$. This shows that a small superdiagonal element u_j and a relatively large error in the computed $\mathbf{q}_{j+1}$ go hand in hand.

We now explain how Lanczos method can be used to compute a few of the extreme eigenvalues of a symmetrix matrix $\mathbf{A}$. Let $\mathbf{T}_j$ denote the intersection of the first j rows and the first j columns of $\mathbf{T}$:

$$\mathbf{T} = \begin{bmatrix} d_1 & u_1 & & & \\ u_1 & d_2 & \cdot & & \\ & \cdot & \cdot & \cdot & \\ & & \cdot & \cdot & \cdot \\ & & & \cdot & \cdot & u_{j-1} \\ & & & & u_{j-1} & d_j \end{bmatrix}.$$

As j increases, the extreme eigenvalues of $\mathbf{T}_j$ will approach the extreme eigenvalues of $\mathbf{A}$ much faster than the power method with starting guess $\mathbf{q}_1$ will approach the dominant eigenvalue of $\mathbf{A}$. Thus one way to approximate, say, the dominant eigenvalue of $\mathbf{A}$ is to compute the dominant eigenvalue α_j of $\mathbf{T}_j$ for $j = 1, j = 2$, and so on. We stop the computation when the sequence $\alpha_1, \alpha_2, \alpha_3, \cdots$ of approximations to the dominant eigenvalue converges to within some specified tolerance.

To carry out this procedure, we need an efficient method to compute α_j given α_{j-1}. One strategy is to apply a single iteration of the **QR** method to $\mathbf{T}_j$ using the shift $\sigma = \alpha_{j-1}$. Another strategy is to apply a single iteration of Newton's method to the characteristic equation det $(\mathbf{T}_j - \lambda\mathbf{I}) = 0$. Letting f_j denote det $(\mathbf{T}_j - \lambda\mathbf{I})$ and letting g_j denote the derivative of det $(\mathbf{T}_j - \lambda\mathbf{I})$, recall that Newton's method is given by the iteration

$$\alpha^{\text{new}} = \alpha^{\text{old}} - \frac{f_j(\alpha^{\text{old}})}{g_j(\alpha^{\text{old}})}. \tag{75}$$

In Section 6-9 we saw that $f_j(\lambda)$ can be evaluated using the recurrence

$$f_k(\lambda) = (d_k - \lambda)f_{k-1}(\lambda) - u_{k-1}^2 f_{k-2}(\lambda) \tag{76}$$

with the starting conditions $f_0(\lambda) = 1$ and $f_{-1}(\lambda) = 0$. Moreover, differentiating (76) with respect to λ, we obtain a recurrence for $g_j(\lambda)$:

$$g_k(\lambda) = (d_k - \lambda)g_{k-1}(\lambda) - u_{k-1}^2\, g_{k-2}(\lambda) - f_{k-1}(\lambda), \tag{77}$$

where $g_0(\lambda) = g_{-1}(\lambda) = 0$. In summary, to approximate α_j we can apply a single Newton iteration (75) with the starting guess $\alpha^{\text{old}} = \alpha_{j-1}$ and with f_j and g_j evaluated using (76) and (77), respectively.

After computing an approximation to an eigenvalue, there are two different ways to compute the corresponding eigenvector. One method is to apply the inverse shifted power method, where the shift is the approximate eigenvalue. Another method is to compute the eigenvector $\mathbf{x}$ of $\mathbf{T}_j$ corresponding to the eigenvalue α_j. It can be shown that an approximation to the corresponding eigenvector of $\mathbf{A}$ is

$$x_1\mathbf{q}_1 + x_2\mathbf{q}_2 + \cdots + x_j\mathbf{q}_j$$

Where $\mathbf{q}_1, \mathbf{q}_2, \cdots, \mathbf{q}_j$ denote the vectors generated by the Lanczos process.

To illustrate this Lanczos process for computing an extreme eigenvalue, let us consider the problem of computing the smallest eigenvalue and the corresponding eigenvector for the 100×100 symmetric pentadiagonal matrix $\mathbf{A}$ with every diagonal element equal to 6.01, with every superdiagonal element equal to -4, and with every element on the second superdiagonal equal to 1:

$$\mathbf{A} = \begin{bmatrix} 6.01 & -4 & 1 & & & & \\ -4 & 6.01 & -4 & \cdot & & & \\ 1 & -4 & 6.01 & \cdot & \cdot & & \\ & \cdot & \cdot & \cdot & \cdot & \cdot & \\ & & \cdot & \cdot & \cdot & \cdot & 1 \\ & & & \cdot & \cdot & \cdot & -4 \\ & & & & 1 & -4 & 6.01 \end{bmatrix}. \tag{78}$$

Since this matrix comes from the discretization of a boundary-value problem with positive eigenvalues, it is known in advance that the eigenvalues are positive. We apply the Lanczos process to the matrix $\mathbf{A}^{-1}$ whose dominant eigenvalue is the reciprocal of the smallest eigenvalue of $\mathbf{A}$. In the first phase of the program that follows, the dominant eigenvalue α_8 of $\mathbf{T}_8$ is computed using TVAL (the bisection subroutine). In the second phase, the Newton process outlined above is used to estimate α_9, α_{10}, $\cdots$. After evaluating α_{12}, the relative difference $|\alpha_j - \alpha_{j-1}|/|\alpha_j|$ is at most 10^{-4}. Using the shift $\sigma = \alpha_{12}$, the third phase applies four iterations of the inverse shifted power method to obtain an approximation to desired eigenpair. For comparison, the inverse power method computes an approximation to the smallest eigenvalue and the corresponding eigenvector with comparable accuracy in about 2000 iterations. The computer program listed below contains two new subroutines. Subroutine LANCZ performs a single iteration of Lanczos method. The arguments of LANCZ are

- Q: array to store jth column of $\mathbf{Q}$
- D: diagonal of $\mathbf{T}$
- U: superdiagonal of $\mathbf{T}$
- P: superdiagonal squared
- J: index of diagonal element in $\mathbf{T}$ to be computed
- N: dimension of $\mathbf{A}$
- MULT: name of subroutine to multiply matrix by vector
- W: work array with at least 2N elements

Subroutine NEWTON uses (76) and (77) to perform a single Newton iteration. The subroutine arguments are

- E: eigenvalue estimate (input) and new eigenvalue estimate (output)
- S: the step (old estimate minus new estimate)
- K: number of eigenvalues less than or equal to the old E
- D: diagonal
- P: superdiagonal squared
- N: matrix dimension

Finally, we list the program that computes the smallest eigenvalue and the corresponding eigenvector for the pentadiagonal matrix (78).

```
      DOUBLE PRECISION A(8,101),D(100),P(100),Q(100)
      DOUBLE PRECISION U(100),W(200),E,S,T,DIF,SIZE
```

```
      COMPLEX*16 EX,X(100),EY,Y(100)
      COMMON A
      EXTERNAL PRODH,PRODB
      N = 100
      M = 8
      DO 10 J = 1,100
         Q(J) = 1.
         A(1,J) = 6.01D0
         A(2,J) = -4.
10       A(3,J) = 1.
      CALL HFACT(A,8,N,2)

       *** PHASE ONE ***

      DO 20 J = 1,M
20       CALL LANCZ(Q,D,U,P,J,N,PRODH,W)
      CALL TVAL(E,-1,U,D,U,M,Q)

       *** PHASE TWO ***

      J = M
30       J = J + 1
         CALL LANCZ(Q,D,U,P,J,N,PRODH,W)
         CALL NEWTON(E,S,K,D,P,J)
         IF (K .LT. J-1 ) CALL TVAL(E,-1,U,D,U,J,Q)
         DIF = S
         SIZE = E
         CALL STOPIT(DIF,SIZE,4,100)
         CALL WHATIS(DIF,SIZE)
         IF ( DIF .GT. 0. ) GOTO 30

        *** PHASE THREE ***

      S = 1./E
      DO 40 J = 1,100
         X(J) = 1.
         A(1,J) = 1.
         A(2,J) = -4.
         A(3,J) = 6.01D0 - S
         A(4,J) = -4.
40       A(5,J) = 1.
      CALL BFACT(A,8,100,2,2)
      CALL POWER(EX,X,EY,Y,100,DIF,SIZE,4,10,PRODB)
      CALL WHATIS(DIF,SIZE)
      E = S + 1./EX
      WRITE(6,*) 'EIGENVALUE:',E
      WRITE(6,*) 'EIGENVECTOR:'
      DO 50 I = 1,N
50       WRITE(6,*) X(I)
      STOP
      END
```

```
      SUBROUTINE PRODH(Y,X)
      DOUBLE PRECISION X(1),Y(1),A(8,101)
      COMMON A
      CALL HSOLVE(Y,A,X)
      RETURN
      END

      SUBROUTINE PRODB(Y,X)
      DOUBLE PRECISION X(1),Y(1),A(8,101)
      COMMON A
      CALL BSOLVE(Y,A,X)
      RETURN
      END
```

The program above computes the smallest eigenvalue of the matrix (78). To compute the second-smallest eigenvalue, we could deflate the known eigenvalue using the deflation process of Section 6-4 and then apply Lanczos method to the deflated matrix. However, there are more efficient ways to compute several eigenpairs. Some excellent references for these Lanczos methods include the books [C9] by Cullum and Willoughby and the paper [P6] by Parlett and Scott. Theoretically, not only do the extreme eigenvalues of $\mathbf{T}_j$ approach the extreme eigenvalues of $\mathbf{A}$, but so do the near extreme eigenvalues. Some work which illuminates this convergence property includes [K2] by Kaniel, [P1] by Paige, and [S1] by Saad. Unfortunately, rounding errors interfere with these beautiful convergence results. It turns out that the matrices $\mathbf{T}_j$ have some eigenvalues which are good approximations to eigenvalues of $\mathbf{A}$, while other eigenvalues are labeled "spurious" by Cullum and Willoughby. Cullum and Willoughby develop a method for distinguishing the spurious eigenvalues from the good ones. Parlett and Scott, on the other hand, deal with the numerical instability of the Lanczos process using selective reorthogonalization.

6-11. THE SINGULAR VALUE DECOMPOSITION AND THE PSEUDOINVERSE

In Chapter 5 we focused on both overdetermined linear systems where the columns of the coefficient matrix are linearly independent, and underdetermined linear systems where the rows of the coefficient matrix are linearly independent. Now let us examine matrices where neither the rows nor the columns are independent. When the coefficient matrix has this property, we say that the system is **singular**. For the system

$$\underset{\mathbf{A}}{\begin{bmatrix} 1 & 2 \\ -2 & -4 \\ 2 & 4 \end{bmatrix}} \underset{\mathbf{x}}{\begin{bmatrix} x_1 \\ x_2 \end{bmatrix}} = \underset{\mathbf{b}}{\begin{bmatrix} 4 \\ -5 \\ 5 \end{bmatrix}}, \tag{79}$$

each row of $\mathbf{A}$ is a multiple of the first row since -2 times row 1 equals row 2 and 2 times row 1 equals row 3. And the second column is a multiple of the first column since 2 times column 1 equals column 2. Using Householder's scheme to compute the least squares solution to the overdetermined system (79), we first form the Householder matrix **H** that annihilates the second and third elements in the first column of **A**:

$$\mathbf{H} = \frac{1}{3}\begin{bmatrix} -1 & 2 & -2 \\ 2 & 2 & 1 \\ -2 & 1 & 2 \end{bmatrix}.$$

Premultiplying equation (79) by **H**, we obtain

$$\begin{bmatrix} -3 & -6 \\ 0 & 0 \\ 0 & 0 \end{bmatrix}\begin{bmatrix} x_1 \\ x_2 \end{bmatrix} = \begin{bmatrix} -8 \\ 1 \\ -1 \end{bmatrix}. \tag{80}$$

Since the last two rows of the coefficient matrix in (80) are zero, we conclude that any **x** which satisfies the equation

$$-3x_1 - 6x_2 = -8 \tag{81}$$

has the property that the Euclidean norm of the residual associated with (79) is as small as possible. Since (81) has an infinite number of solutions, the theory of Chapter 5 tells us that the solution to (81) with smallest Euclidean norm is

$$\begin{bmatrix} x_1 \\ x_2 \end{bmatrix} = \frac{-8}{3^2 + 6^2}\begin{bmatrix} -3 \\ -6 \end{bmatrix} = \begin{bmatrix} \frac{8}{15} \\ \frac{16}{15} \end{bmatrix}.$$

In general, when the linear system $\mathbf{Ax} = \mathbf{b}$ has linearly dependent rows and linearly dependent columns, there are an infinite number of **x**'s which make the Euclidean norm of the residual as small as possible. Of those **x**'s which minimize the residual, that **x** with the smallest Euclidean norm is the least squares solution to $\mathbf{Ax} = \mathbf{b}$. The least squares solution to a linear system can be represented using the singular value decomposition of a matrix. The **singular value decomposition** of a $m \times n$ matrix has the form

$$\mathbf{A} = \mathbf{QSP}^T,$$

where **Q** is a $m \times m$ orthogonal matrix, **S** is a $m \times n$ "diagonal matrix," and **P** is a $n \times n$ orthogonal matrix. If m is larger than n, then **S** has the structure illustrated by the following 4×2 matrix:

$$\begin{bmatrix} s_1 & 0 \\ 0 & s_2 \\ 0 & 0 \\ 0 & 0 \end{bmatrix},$$

and if n is larger than m, then **S** has the structure illustrated by the following 2×4 matrix:

$$\begin{bmatrix} s_1 & 0 & 0 & 0 \\ 0 & s_2 & 0 & 0 \end{bmatrix}.$$

The "diagonal elements" of **S**, also called the **singular values** of **A**, are denoted s_1 $s_2, \cdots, s_l$ where l is either m or n, whichever is smaller. The columns $\mathbf{q}_1$, $\mathbf{q}_2$, $\cdots$, $\mathbf{q}_m$ of **Q** and the columns $\mathbf{p}_1$, $\mathbf{p}_2$, $\cdots$, $\mathbf{p}_n$ of **P** are called the **singular vectors**. By an appropriate choice of **Q** and **P**, the singular values can be arranged so that $s_1 \geqq s_2 \geqq \cdots \geqq s_l \geqq 0$. Postmultiplying the relation $\mathbf{A} = \mathbf{QSP}^T$ by **P** and equating columns, we obtain the following equation relating singular values and singular vectors:

$$\mathbf{Ap}_i = s_i\mathbf{q}_i \qquad \text{for } i = 1 \text{ to } l.$$

Similarly, premultiplying the relation $\mathbf{A} = \mathbf{QSP}^T$ by $\mathbf{Q}^T$ and equating rows gives

$$\mathbf{q}_i^{\ T}\mathbf{A} = s_i\mathbf{p}_i^{\ T} \qquad \text{for } i = 1 \text{ to } l.$$

The singular value decomposition $\mathbf{QSP}^T$ provides much information about the structure of a matrix. If $\mathbf{A}$ is $m \times n$ and $\mathbf{A}$ has p positive singular values, the rank (the number of linearly independent columns) of $\mathbf{A}$ is p and the Euclidean norm of $\mathbf{A}$ is the largest singular value s_1 (see Exercise 6-11.1). Moreover, the first p columns of $\mathbf{Q}$ are an orthonormal basis for the space spanned by the columns of $\mathbf{A}$. This space is called either the **range space** or the **column space** of $\mathbf{A}$. The first p columns of $\mathbf{P}$ are an orthonormal basis for the space spanned by the rows of $\mathbf{A}$. This space is called either the **row space** of $\mathbf{A}$ or the range space of $\mathbf{A}^T$. The last $n - p$ columns of $\mathbf{P}$ are an orthonormal basis for the space of vectors orthogonal to the rows of $\mathbf{A}$. This space is called the **null space** of $\mathbf{A}$. Finally, the last $m - p$ columns of $\mathbf{Q}$ are an orthonormal basis for the null space of $\mathbf{A}^T$. As explained later, the singular value decomposition of a matrix can be computed using the **QR** algorithm. For the moment, let us study the connection between the singular value decomposition and the least squares problem.

Given a linear system $\mathbf{Ax} = \mathbf{b}$, we saw in Chapter 5 that $\mathbf{x}$ minimizes the Euclidean norm of the residual $\mathbf{r} = \mathbf{b} - \mathbf{Ax}$ if and only if the normal equation

$$\mathbf{A}^T\mathbf{Ax} = \mathbf{A}^T\mathbf{b}$$

is satisfied. Replacing $\mathbf{A}$ by its singular value decomposition, the normal equation becomes

$$(\mathbf{QSP}^T)^T\mathbf{QSP}^T\mathbf{x} = (\mathbf{QSP}^T)^T\mathbf{b}. \tag{82}$$

Since $(\mathbf{QSP}^T)^T = \mathbf{PS}^T\mathbf{Q}^T$ and $\mathbf{Q}^T\mathbf{Q} = \mathbf{I}$, (82) reduces to

$$\mathbf{PS}^T\mathbf{SP}^T\mathbf{x} = \mathbf{PS}^T\mathbf{Q}^T\mathbf{b}. \tag{83}$$

Premultiplying (83) by the inverse of $\mathbf{P}$, we conclude that those $\mathbf{x}$ which minimize the Euclidean norm of the residual are solutions to

$$\mathbf{S}^T\mathbf{SP}^T\mathbf{x} = \mathbf{S}^T\mathbf{Q}^T\mathbf{b}. \tag{84}$$

Making the orthogonal change of variables $\mathbf{y} = \mathbf{P}^T\mathbf{x}$, (84) takes the form

$$\mathbf{S}^T\mathbf{Sy} = \mathbf{S}^T\mathbf{Q}^T\mathbf{b}. \tag{85}$$

Since the diagonal elements of $\mathbf{S}$ are nonnegative and decreasing, $\mathbf{S}$ can be partitioned in the form

$$\mathbf{S} = \begin{bmatrix} \mathbf{S}_+ & \mathbf{0} \\ \mathbf{0} & \mathbf{0} \end{bmatrix},$$

where $\mathbf{S}_+$ is a square diagonal matrix whose diagonal contains the positive singular values. The product $\mathbf{S}^T\mathbf{S}$ has the following structure:

$$\mathbf{S}^T\,\mathbf{S} = \begin{bmatrix} \mathbf{S}_+^2 & \mathbf{0} \\ \mathbf{0} & \mathbf{0} \end{bmatrix}.$$

The vector $\mathbf{y}$ in (85) has one component corresponding to each column of $\mathbf{S}$. Partition the components of $\mathbf{y}$ into $\mathbf{y}_+$ and $\mathbf{y}_0$, where $\mathbf{y}_+$ denotes those components corresponding to columns of $\mathbf{S}$ with positive singular values and $\mathbf{y}_0$ denotes those components corresponding to the zero columns of **S**. With this notation, (85) can be expressed

$$\mathbf{S}_+^2\mathbf{y}_+ + \mathbf{0}\mathbf{y}_0 = \mathbf{S}_+(\mathbf{Q}^T\mathbf{b})_+ \qquad (86)$$
$$\mathbf{0} = \mathbf{0},$$

where $(\mathbf{Q}^T\mathbf{b})_+$ denotes those components of $\mathbf{Q}^T\mathbf{b}$ corresponding to columns of $\mathbf{S}^T$ that contain positive singular values. Since the coefficient of $\mathbf{y}_0$ in (86) is zero, that **y** with smallest Euclidean norm which satisfies (86) must have $\mathbf{y}_0 = \mathbf{0}$. Solving (86) for $\mathbf{y}_+$ yields

$$\mathbf{y}_+ = \mathbf{S}_+^{-1}(\mathbf{Q}^T\mathbf{b})_+ .$$

If **A** has p positive singular values, it follows that the components of **y** are given by

$$y_i = \begin{cases} \dfrac{1}{s_i}(\mathbf{Q}^T\,\mathbf{b})_i & \text{for } i = 1, \cdots, p, \\ 0 & \text{for } i > p. \end{cases} \qquad (87)$$

Knowing **y**, we can solve the relation $\mathbf{y} = \mathbf{P}^T\mathbf{x}$ for **x**. In particular, since **P** is orthogonal, we multiply $\mathbf{y} = \mathbf{P}^T\mathbf{x}$ by **P** to obtain $\mathbf{Py} = \mathbf{PP}^T\mathbf{x} = \mathbf{x}$ and $\mathbf{x} = \mathbf{Py}$.

Let us now express the least squares solution to a linear system using matrix notation:

$$\mathbf{x} = \mathbf{Py} = \mathbf{P}\begin{bmatrix} \mathbf{y}_+ \\ \mathbf{y}_0 \end{bmatrix} = \mathbf{P}\begin{bmatrix} \mathbf{S}_+^{-1}(\mathbf{Q}^T\mathbf{b})_+ \\ \mathbf{0} \end{bmatrix}. \qquad (88)$$

If **A** is $m \times n$, then **S** is $m \times n$. Formula (88) is simplified if we define a $n \times m$ matrix $\mathbf{S}^{-1}$ by the following rule:

$$\mathbf{S}^{-1} = \begin{bmatrix} \mathbf{S}_+^{-1} & \mathbf{0} \\ \mathbf{0} & \mathbf{0} \end{bmatrix}.$$

This definition implies that

$$\mathbf{S}^{-1}\mathbf{Q}^T\mathbf{b} = \begin{bmatrix} \mathbf{S}_+^{-1}(\mathbf{Q}^T\mathbf{b})_+ \\ \mathbf{0} \end{bmatrix}. \qquad (89)$$

Combining (88) and (89), we see that the least squares solution to $\mathbf{Ax} = \mathbf{b}$ is

$$\mathbf{x} = \mathbf{PS}^{-1}\mathbf{Q}^T\mathbf{b}. \qquad (90)$$

This **x** makes the residual $\mathbf{r} = \mathbf{b} - \mathbf{Ax}$ as small as possible and among all **x**'s which minimize the residual, the **x** given by (90) has the smallest Euclidean norm. The matrix $\mathbf{PS}^{-1}\mathbf{Q}^T$, called the **pseudoinverse** of **A**, will be denoted $\mathbf{A}^{-1}$. Formally, the pseudoinverse can be derived in the following way: Since **A** is the product $\mathbf{QSP}^T$ and the inverse of a product is the product of the inverses arranged in the reverse order, we have $\mathbf{A}^{-1} = (\mathbf{P}^T)^{-1}\mathbf{S}^{-1}\mathbf{Q}^{-1}$. Since both **Q** and **P** are orthogonal, $(\mathbf{P}^T)^{-1}$ can be replaced by **P** and $\mathbf{Q}^{-1}$ can be replaced by $\mathbf{Q}^T$ to obtain $\mathbf{A}^{-1} = \mathbf{PS}^{-1}\mathbf{Q}^T$. Unfortunately, this slick derivation of the pseudoinverse overlooks an important restriction in the rule for inverting a matrix product. The inverse of a product is the product of the inverses when the matrices are square and nonsingular. Since **S** may not be square, we cannot use the standard rule for inverting a product. The formula $\mathbf{PS}^{-1}\mathbf{Q}^T$ for the pseudoinverse of a matrix resembles the product of the inverses only because of the *definition* for $\mathbf{S}^{-1}$.

Let us illustrate the pseudoinverse using the linear system

$$\underset{\mathbf{A}}{\begin{bmatrix} 4 & 3 \\ 8 & 6 \\ 8 & 6 \end{bmatrix}} \underset{\mathbf{x}}{\begin{bmatrix} x_1 \\ x_2 \end{bmatrix}} = \underset{\mathbf{b}}{\begin{bmatrix} 5 \\ 13 \\ 15 \end{bmatrix}}. \tag{91}$$

The singular value decomposition of the coefficient matrix is

$$\underset{\mathbf{A}}{\begin{bmatrix} 4 & 3 \\ 8 & 6 \\ 8 & 6 \end{bmatrix}} = \underset{\mathbf{Q}}{\begin{bmatrix} \frac{1}{3} & \frac{2}{3} & \frac{2}{3} \\ \frac{2}{3} & -\frac{2}{3} & \frac{1}{3} \\ \frac{2}{3} & \frac{1}{3} & -\frac{2}{3} \end{bmatrix}} \underset{\mathbf{S}}{\begin{bmatrix} 15 & 0 \\ 0 & 0 \\ 0 & 0 \end{bmatrix}} \underset{\mathbf{P}^T}{\begin{bmatrix} \frac{4}{5} & \frac{3}{5} \\ \frac{3}{5} & -\frac{4}{5} \end{bmatrix}}.$$

The pseudoinverse of **A** is

$$\underset{\mathbf{P}}{\begin{bmatrix} \frac{4}{5} & \frac{3}{5} \\ \frac{3}{5} & -\frac{4}{5} \end{bmatrix}} \underset{\mathbf{S}^{-1}}{\begin{bmatrix} \frac{1}{15} & 0 & 0 \\ 0 & 0 & 0 \end{bmatrix}} \underset{\mathbf{Q}^T}{\begin{bmatrix} \frac{1}{3} & \frac{2}{3} & \frac{2}{3} \\ \frac{2}{3} & -\frac{2}{3} & \frac{1}{3} \\ \frac{2}{3} & \frac{1}{3} & -\frac{2}{3} \end{bmatrix}} = \underset{\mathbf{A}^{-1}}{\begin{bmatrix} \frac{4}{225} & \frac{8}{225} & \frac{8}{225} \\ \frac{1}{75} & \frac{2}{75} & \frac{2}{75} \end{bmatrix}}.$$

The least squares solution to (91) is

$$\mathbf{x} = \mathbf{A}^{-1}\mathbf{b} = \begin{bmatrix} \frac{4}{225} & \frac{8}{225} & \frac{8}{225} \\ \frac{1}{75} & \frac{2}{75} & \frac{2}{75} \end{bmatrix} \begin{bmatrix} 5 \\ 13 \\ 15 \end{bmatrix} = \begin{bmatrix} \frac{244}{225} \\ \frac{61}{75} \end{bmatrix}.$$

In this example, the last two columns of **Q** and the last column of **P** are immaterial since they are multiplied by zero when the product $\mathbf{P}^T\mathbf{S}^{-1}\mathbf{Q}$ is computed. In general, if **A** is $m \times n$ and **A** has p positive singular values, only the first p columns of **P** and the first p columns of **Q** are needed to evaluate the product $\mathbf{P}^T\mathbf{S}^{-1}\mathbf{Q}$ and to compute the least squares solution to $\mathbf{Ax} = \mathbf{b}$.

For a singular matrix, both the rows and the columns of **A** are linearly dependent, p is less than both m and n, and some of **A**'s singular values are zero. Computing the least squares solution to a linear system when some singular values vanish is not easy since the computed singular values are usually small nonzero numbers. For example, if the **S** matrix in **A**'s singular value decomposition is

$$\mathbf{S} = \begin{bmatrix} 1 & 0 \\ 0 & 0 \\ 0 & 0 \end{bmatrix},$$

the computed **S** may be

$$\mathbf{S} = \begin{bmatrix} 1 & 0 \\ 0 & \varepsilon \\ 0 & 0 \end{bmatrix},$$

where ε is a small number. The true pseudoinverse of **S** is

$$\mathbf{S}^{-1} = \begin{bmatrix} 1 & 0 & 0 \\ 0 & 0 & 0 \end{bmatrix},$$

while the computed pseudoinverse is

$$\begin{bmatrix} 1 & 0 & 0 \\ 0 & \varepsilon^{-1} & 0 \end{bmatrix}.$$

Hence the small error ε in a singular value is inverted when we compute $\mathbf{S}^{-1}$. Due to the $1/\varepsilon$ coefficient in the computed $\mathbf{S}^{-1}$, the computed least squares solution $\mathbf{x} = \mathbf{PS}^{-1}\mathbf{Q}^T\mathbf{b}$ to the linear system $\mathbf{Ax} = \mathbf{b}$ deviates substantially from the true least squares solution.

To compute the least squares solution to a nearly singular linear system, a small singular value s_i is often replaced by zero whenever s_i is smaller than some **cutoff** value. Since s_1 is the Euclidean norm of $\mathbf{A}$, a typical cutoff criterion is to replace s_i with zero whenever s_i is less than or equal to s_1 times the machine epsilon. Golub and Van Loan [G15] point out that the cutoff should be consistent with the data accuracy. That is, if the coefficient matrix is only accurate to two decimal places, a natural cutoff is $10^{-2}s_1$. The small singular values are cut off since the error in the computed singular value is larger than the actual singular value and the actual singular value may be zero even though the computed singular value is nonzero. By replacing a relatively small singular value with zero, the numerical error that results from taking the reciprocal of a small number is eliminated.

Another strategy for handling a nearly singular linear system is to compute a **regularized solution**. The least squares solution to a linear system has the property that the Euclidean norm of the residual is as small as possible and among those $\mathbf{x}$'s which minimize the residual, the least squares solution is that $\mathbf{x}$ with the smallest Euclidean norm. As an alternative to this least squares approach, we propose to minimize the expression

$$\|\mathbf{b} - \mathbf{Ax}\|_2^2 + r\|\mathbf{x}\|_2^2 \tag{92}$$

over $\mathbf{x}$ where r is a small constant that reflects the importance of a small residual relative to the norm of $\mathbf{x}$. If it is important that the computed $\mathbf{x}$ has a small residual, then take r small. If it is important that the computed $\mathbf{x}$ have a small norm, then make r larger. Letting $\mathbf{x}_r$ denote the solution to (92), we will soon see that the standard least squares solution to $\mathbf{Ax} = \mathbf{b}$ is the limit of $\mathbf{x}_r$ as r tends to zero. The $\|\mathbf{x}\|_2$ term in (92) is a stabilizing term that prevents the minimizer $\mathbf{x}_r$ from being too large. In particular, inserting $\mathbf{x} = \mathbf{0}$ in (92), we see that the minimum is bounded by $\|\mathbf{b}\|_2^2$. Consequently, at the minimum $\mathbf{x} = \mathbf{x}_r$, we have

$$\|\mathbf{b} - \mathbf{Ax}_r\|_2^2 + r\|\mathbf{x}_r\|_2^2 \leqq \|\mathbf{b}\|_2^2. \tag{93}$$

Dropping the first term in (93) and taking square roots gives us the relation $\|\mathbf{x}_r\|_2 \leqq \|\mathbf{b}\|_2/\sqrt{r}$. In other words, the Euclidean norm of the minimizer associated with (92) is bounded by the Euclidean norm of the right side $\mathbf{b}$ divided by the square root of r. In contrast, the Euclidean norm of the least squares solution is bounded by one over s_p times the Euclidean norm of $\mathbf{b}$, where s_p is the smallest positive singular value. As the smallest singular value approaches zero, the least squares solution approaches infinity while $\mathbf{x}_r$ is bounded by $\|\mathbf{b}\|_2/\sqrt{r}$, independent of the singular values. We often select a value for r that is consistent with the accuracy in the coefficients. For example, if the coefficients are known precisely, then one possibility is to choose $r = b^{-n}\|\mathbf{A}\|_2^2$ where b is the base and n is the mantissa corresponding to floating point numbers—the 2-norm of $\mathbf{A}$ is squared to be consistent with the square of $\mathbf{A}$ contained in the first term of (93). On the other hand, if the coefficients are only accurate to two significant digits, we might set $r = 10^{-2}\|\mathbf{A}\|_2^2$. For an enlightening application of the regularized solution, see Exercise 6-11.7.

The formula for $\mathbf{x}_r$ is similar to the formula for the least squares solution to $\mathbf{Ax} = \mathbf{b}$. If $\mathbf{QSP}^T$ is the singular value decomposition of $\mathbf{A}$ and $\mathbf{A}$ has p positive singular values, then $\mathbf{x}_r = \mathbf{Py}$ where,

$$y_i = \begin{cases} \dfrac{s_i}{s_i^2 + r}(\mathbf{Q}^T\mathbf{b})_i & \text{for } i = 1, \cdots, p, \\ 0 & \text{for } i > p. \end{cases} \tag{94}$$

Comparing (94) to (87), we see that the regularized solution $\mathbf{x}_r$ is equal to the least squares solution when r is zero. But when r is positive, the coefficient $s_i/(s_i^2 + r)$ in (94) approaches zero as s_i approaches zero while the least squares $\mathbf{y}$ in (87) approaches infinity.

In this section we have expressed the pseudoinverse of matrix and the least squares solution to a linear system in terms of the singular value decomposition. In contrast, Chapter 5 develops even simpler ways to compute the least squares solution to a linear system. For example, if the columns of $\mathbf{A}$ are linearly independent, the least squares solution to $\mathbf{Ax} = \mathbf{b}$ is $\mathbf{x} = (\mathbf{A}^T\mathbf{A})^{-1}\mathbf{A}^T\mathbf{b}$. And if the rows of $\mathbf{A}$ are linearly independent, the least squares solution is $\mathbf{x} = \mathbf{A}^T(\mathbf{AA}^T)^{-1}\mathbf{b}$. Since the general formula for the least squares solution to $\mathbf{Ax} = \mathbf{b}$ is $\mathbf{x} = \mathbf{A}^{-1}\mathbf{b}$, where $\mathbf{A}^{-1}$ is the pseudoinverse of $\mathbf{A}$, we conclude that $\mathbf{A}^{-1} = (\mathbf{A}^T\mathbf{A})^{-1}\mathbf{A}^T$ when the columns of $\mathbf{A}$ are independent and $\mathbf{A}^{-1} = \mathbf{A}^T(\mathbf{AA}^T)^{-1}$ when the rows of $\mathbf{A}$ are independent. Although these formulas for the pseudoinverse as well as others derived in Chapter 5 may be computationally simpler than the formula $\mathbf{A}^{-1} = \mathbf{PS}^{-1}\mathbf{Q}^T$, these simpler formulas are inaccurate when $\mathbf{A}$ is nearly singular. The singular value decomposition, on the other hand, alerts us to the fact that the matrix is nearly singular since a nearly singular matrix has relatively small singular values. And when the matrix is nearly singular, corrective action can be taken—the small singular values can be cut off or the regularized solution can be computed.

Theoretically, the existence of the singular value decomposition is related to two separate diagonalizations. Let $\mathbf{P}\Lambda_1\mathbf{P}^{-1}$ denote the diagonalization of $\mathbf{A}^T\mathbf{A}$. By Exercise 6-5.1, the symmetric matrix $\mathbf{A}^T\mathbf{A}$ can be diagonalized with an orthogonal eigenvector matrix so that $\mathbf{P}^{-1} = \mathbf{P}^T$. If $\mathbf{p}_i$ denotes column i of $\mathbf{P}$, you will show in Exercise 6-11.4 that the vector $\mathbf{r}_i = \mathbf{Ap}_i$ is an eigenvector of $\mathbf{AA}^T$ whenever $\mathbf{r}_i \neq \mathbf{0}$; furthermore, $\mathbf{r}_i$ is orthogonal to $\mathbf{r}_j$ whenever $i \neq j$. Letting $\mathbf{Q}\Lambda_2\mathbf{Q}^T$ denote the diagonalization of the symmetric matrix $\mathbf{AA}^T$, it can be arranged so that each $\mathbf{r}_i$ is a nonnegative multiple of column i of $\mathbf{Q}$. Since the $\mathbf{r}_i$ form the columns of $\mathbf{AP}$, it follows that there exists a diagonal matrix $\mathbf{S}$ such that $\mathbf{AP} = \mathbf{QS}$. Finally, postmultiplying by $\mathbf{P}^T$ yields $\mathbf{A} = \mathbf{QSP}^T$, the singular value decomposition.

To compute the singular value decomposition of a matrix, a method (developed by Golub and Kahn in [G14]) similar to the $\mathbf{QR}$ algorithm is employed. Recall that each iteration of the $\mathbf{QR}$ algorithm takes less computer time for a Hessenberg matrix than for a general matrix. Similarly, the singular value decomposition is easier to compute if the starting matrix is bidiagonal. An upper bidiagonal matrix is completely zero except for the diagonal elements or the superdiagonal elements. A 4×4 upper bidiagonal matrix appears below:

$$\begin{bmatrix} * & * & 0 & 0 \\ 0 & * & * & 0 \\ 0 & 0 & * & * \\ 0 & 0 & 0 & * \end{bmatrix}.$$

The computation of the singular value decomposition has two phases. In the first phase, we compute orthogonal matrices $\mathbf{Q}_0$ and $\mathbf{P}_0$ and a bidiagonal matrix $\mathbf{B}$ such that $\mathbf{A} = \mathbf{Q}_0\mathbf{B}\mathbf{P}_0^T$. In the second phase, the singular value decomposition $\tilde{\mathbf{Q}}\mathbf{S}\tilde{\mathbf{P}}^T$ of $\mathbf{B}$ is computed using the $\mathbf{QR}$ algorithm. Since $\mathbf{A} = \mathbf{Q}_0\mathbf{B}\mathbf{P}_0 = \mathbf{Q}_0\tilde{\mathbf{Q}}\mathbf{S}\tilde{\mathbf{P}}^T\mathbf{P}_0{}^T$, the $\mathbf{Q}$ and the $\mathbf{P}$ factors in the singular value decomposition $\mathbf{QSP}^T$ of $\mathbf{A}$ are

$$\mathbf{Q} = \mathbf{Q}_0\tilde{\mathbf{Q}} \qquad \text{and} \qquad \mathbf{P} = \mathbf{P}_0\tilde{\mathbf{P}}.$$

The reduction of $\mathbf{A}$ to the bidiagonal form $\mathbf{B}$ is illustrated using a 4×4 matrix. In the first step, we premultiply $\mathbf{A}$ by a Householder matrix $\mathbf{Q}_1$ that annihilates all elements in the first column except for the first element:

$$\mathbf{Q}_1\mathbf{A} = \begin{bmatrix} * & * & * & * \\ 0 & * & * & * \\ 0 & * & * & * \\ 0 & * & * & * \end{bmatrix}.$$

We postmultiply $\mathbf{Q}_1\mathbf{A}$ by a Householder matrix $\mathbf{P}_1$ that annihilates all elements in the first row except for the first two elements giving us a matrix that we denote $\mathbf{A}_1$:

$$\mathbf{A}_1 = \mathbf{Q}_1\mathbf{A}\mathbf{P}_1 = \begin{bmatrix} * & * & 0 & 0 \\ 0 & * & * & * \\ 0 & * & * & * \\ 0 & * & * & * \end{bmatrix}.$$

In the second step, $\mathbf{A}_1$ is premultiplied by the Householder matrix $\mathbf{Q}_2$ that annihilates elements 3 and 4 in column 2:

$$\mathbf{Q}_2\mathbf{A}_1 = \begin{bmatrix} * & * & 0 & 0 \\ 0 & * & * & * \\ 0 & 0 & * & * \\ 0 & 0 & * & * \end{bmatrix}.$$

Then $\mathbf{Q}_2\mathbf{A}_1$ is postmultiplied by the Householder matrix $\mathbf{P}_2$ that annihilates element 4 in row 2, giving us a matrix that we denote $\mathbf{A}_2$:

$$\mathbf{A}_2 = \mathbf{Q}_2\mathbf{A}_1\mathbf{P}_2 = \begin{bmatrix} * & * & 0 & 0 \\ 0 & * & * & 0 \\ 0 & 0 & * & * \\ 0 & 0 & * & * \end{bmatrix}.$$

Finally, we premultiply $\mathbf{A}_2$ by the Householder matrix $\mathbf{Q}_3$ that annihilates element 4 in column 3 to obtain the bidiagonal form $\mathbf{B}$:

$$\mathbf{B} = \mathbf{Q}_3\mathbf{A}_2 = \begin{bmatrix} * & * & 0 & 0 \\ 0 & * & * & 0 \\ 0 & 0 & * & * \\ 0 & 0 & 0 & * \end{bmatrix}.$$

To summarize,

$$\mathbf{B} = \mathbf{Q}_3\mathbf{A}_2 = \mathbf{Q}_3\mathbf{Q}_2\mathbf{A}_1\mathbf{P}_2 = \mathbf{Q}_3\mathbf{Q}_2\mathbf{Q}_1\mathbf{A}\mathbf{P}_1\mathbf{P}_2.$$

Solving for $\mathbf{A}$ and recalling that each Householder matrix is its own inverse, we have

$$\mathbf{A} = \mathbf{Q}_1\mathbf{Q}_2\mathbf{Q}_3\mathbf{B}\mathbf{P}_2\mathbf{P}_1.$$

Letting $\mathbf{Q}_0$ denote the product $\mathbf{Q}_1\mathbf{Q}_2\mathbf{Q}_3$ and letting $\mathbf{P}_0^T$ denote the product $\mathbf{P}_2\mathbf{P}_1$, it follows that $\mathbf{A} = \mathbf{Q}_0\mathbf{B}\mathbf{P}_0^T$.

Subroutine BIDAG uses this algorithm to reduce a matrix to bidiagonal form. The subroutine arguments are

- D: diagonal of bidiagonal matrix
- B: superdiagonal of bidiagonal matrix if M ≧ N and subdiagonal if M < N
- A: array containing coefficient matrix
- LA: leading (row) dimension of array A
- M: row dimension of matrix stored in array A
- N: column dimension of matrix stored in array A

BIDAG essentially overwrites the starting coefficients with the vectors **w** characterizing each Householder matrix.

After reducing **A** to bidiagonal form **B**, the singular value decomposition of **B** can be computed using an implementation (see [G14]) of the **QR** algorithm. Subroutine SING utilizes this method coupled with subroutine BIDAG to compute the singular value decomposition of a matrix. The arguments of SING are

- Q: array containing first factor in the singular value decomposition
- LQ: leading (row) dimension of array Q
- IQ: integer between 0 and 3 defined below
- S: singular values in decreasing order
- P: array containing second factor in the singular value decomposition
- LP: leading (row) dimension of array P
- IP: integer between 0 and 3 defined below
- A: array containing coefficient matrix
- LA: leading (row) dimension of array A
- M: row dimension of matrix stored in A
- N: column dimension of matrix stored in A
- W: work array [length at least max(M, 3L − 1), where L = MIN(M,N)]

If the integer IQ is zero, the **Q** factor in the singular value decomposition is not computed. Letting L denote M or N, whichever is smaller, set IQ = 1 to compute the first L columns of **Q**, set IQ = 2 to compute the last M − L columns of **Q**, and set IQ = 3 to compute all M columns of **Q**. Similarly, if IP is zero, the **P** factor in the singular value decomposition is not computed; we set IP equal to 1, 2, or 3 to compute the first L columns, the last N − L columns, or all N columns of **P**, respectively.

Subroutine RSOLVE uses the output of SING to compute the regularized solution to a linear system. The subroutine arguments are

- X: regularized solution
- B: right side
- Q: first factor in the singular value decomposition
- LQ: leading (row) dimension of array Q
- MQ: number of rows for matrix stored in Q
- S: singular values
- P: last factor in the singular value decomposition
- LP: leading (row) dimension of array P

MP: number of rows for matrix stored in P
R: regularization parameter

Subroutine PSEUDO computes the regularized pseudoinverse of a matrix. The arguments of PSEUDO are the same as the arguments of RSOLVE except that X is replaced by A, an array to store the pseudoinverse, and B is replaced by LA, the leading (row) dimension of array A. Note that when argument R is zero, RSOLVE generates the standard least squares solution to a linear system, while PSEUDO generates the standard pseudoinverse.

Exercises

6-11.1. Show that the Euclidean norm of a matrix is the largest singular value. (*Hint:* See Exercises 6-7.1 and 3-3.8.)

6-11.2. In terms of **A**'s singular values, what is the 2-norm of the pseudoinverse $\mathbf{A}^{-1}$?

6-11.3. Show that the 2-norm condition number of an invertible matrix is one if and only if the matrix is a multiple of an orthogonal matrix.

6-11.4. Let $\mathbf{P\Lambda P}^T$ denote the diagonalization of $\mathbf{A}^T\mathbf{A}$, where **P** is an orthogonal matrix, let $\mathbf{p}_i$ denote column i of **P**, and define $\mathbf{q}_i = \mathbf{A}\mathbf{p}_i$. Show that $\mathbf{q}_i$ is an eigenvector of the matrix $\mathbf{AA}^T$ whenever $\mathbf{q}_i \neq \mathbf{0}$ and $\mathbf{q}_i$ is orthogonal to $\mathbf{q}_j$ whenever $i \neq j$.

6-11.5. Show that for the pseudoinverse, $(\mathbf{A}^{-1})^{-1} = \mathbf{A}$.

6-11.6. Show that the pseudoinverse $\mathbf{A}^{-1}$ of a rectangular matrix satisfies the following four (Moore-Penrose) conditions: (a) $\mathbf{AA}^{-1}\mathbf{A} = \mathbf{A}$; (b) $\mathbf{A}^{-1}\mathbf{AA}^{-1} = \mathbf{A}^{-1}$; (c) $(\mathbf{AA}^{-1})^T = \mathbf{AA}^{-1}$; (d) $(\mathbf{A}^{-1}\mathbf{A})^T = \mathbf{A}^{-1}\mathbf{A}$. Since there exists a unique matrix $\mathbf{A}^{-1}$ that satisfies (a)–(d), these conditions can be considered an alternative definition for the pseudoinverse of a matrix.

6-11.7. Using RSOLVE, compute the least squares solution to the following underdetermined system:

$$\begin{aligned} x_1 + 2x_2 + 3x_3 &= 1, \\ 4x_1 + 5x_2 + 6x_3 &= 1. \end{aligned} \tag{95}$$

(Remember that the least squares solution is obtained by setting the regularization parameter to zero.) Then consider the equivalent system:

$$\begin{aligned} x_1 + 2x_2 + 3x_3 &= 1, \\ 4x_1 + 5x_2 + 6x_3 &= 1, \\ x_1 + 2x_2 + 3x_3 &= 1. \end{aligned} \tag{96}$$

First compute the least squares solution to (96) using RSOLVE. Then compute the regularized solution where the regularization parameter is b^{-n} times the 2-norm of the coefficient matrix squared. Recall that b is the base and n is the mantissa length corresponding to your computer, and by Exercise 6-11.1, the 2-norm of a matrix is its largest singular value (the first element of the S array in subroutine SING). Theoretically, the least squares solution to (96) should be the same as the least squares solution to (95). Comparing the computed least squares solution for (96) to the computed regularized solution, which solution better approximates the true least squares solution?

6-12. THE GENERALIZED EIGENPROBLEM

Given two $n \times n$ matrices **A** and **B**, the **generalized eigenproblem** can be stated: Find those scalars λ and those nonzero vectors **x** with the property that

$$\mathbf{Ax} = \lambda\mathbf{Bx}. \tag{97}$$

Observe that the standard eigenproblem is a special case of the generalized eigenproblem corresponding to $\mathbf{B} = \mathbf{I}$. Generalized eigenproblems are often generated by Ritz and Galerkin approximations to eigenproblems corresponding to a differential equation. Abstractly, an eigenproblem associated with a linear operator L is to find those scalars λ and those nonzero functions x for which $Lx = \lambda x$. For the beam buckling problem of Section 6-1, $Lx = -x''$. Given a basis $\phi_1, \cdots, \phi_n$ for the solution space, we approximate x by a linear combination of the basis functions:

$$x = \sum_{i=1}^{n} x_i \phi_i . \tag{98}$$

Given another collection of functions $\psi_1, \cdots, \psi_n$ called the test functions, we take the inner product between the equation $Lx = \lambda x$, where x is given by (98), and the test functions to obtain the approximating eigenproblem: Find $x \neq 0$ of the form (98) such that

$$<\psi_i, Lx> = \lambda <\psi_i, x> \qquad \text{for } i = 1 \text{ to } n. \tag{99}$$

Since $<\psi_i, Lx> = <\psi_i, \sum_{j=1}^{n} x_j L\phi_j> = \sum_{j=1}^{n} <\psi_i, L\phi_j> x_j$, and

$$<\psi_i, x> = \sum_{j=1}^{n} <\psi_i, \phi_j> x_j ,$$

equation (99) is equivalent to

$$\sum_{j=1}^{n} <\psi_i, L\phi_j> x_j = \lambda \sum_{j=1}^{n} <\psi_i, \phi_j> x_j \qquad \text{for } i = 1 \text{ to } n.$$

On the other hand, the ith component of (97) can be expressed

$$\sum_{j=1}^{n} a_{ij} x_j = \lambda \sum_{j=1}^{n} b_{ij} x_j .$$

Therefore, equation (99) is equivalent to equation (97) with $a_{ij} = <\psi_i, L\phi_j>$ and with $b_{ij} = <\psi_i, \phi_j>$.

For the eigenproblem $-x''(t) = \lambda x(t)$, $0 \leqq t \leqq 1$, with the boundary conditions $x(0) = x(1) = 0$, the following inner product is frequently employed:

$$<x, y> = \int_0^1 x(t) y(t) \, dt .$$

A popular choice for the basis functions and test functions are the piecewise linear polynomials depicted in Figure 6-8. When the basis functions and test functions are piecewise linear polynomials, the matrices $\mathbf{A}$ and $\mathbf{B}$ in (97) have

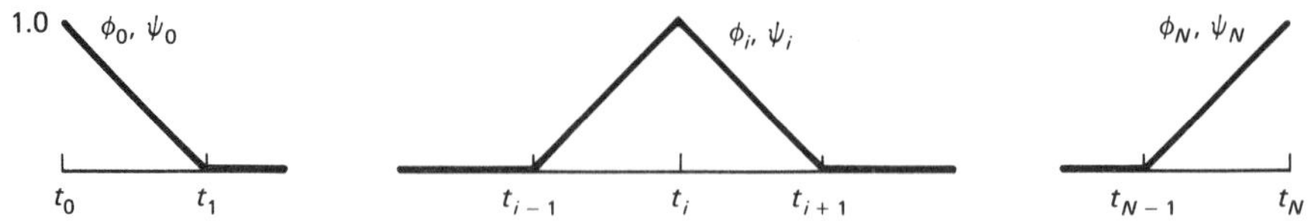

Figure 6-8 Piecewise linear polynomials.

the form

$$\mathbf{A} = N\begin{bmatrix} 2 & -1 & & & & \\ -1 & 2 & \cdot & & & \\ & \cdot & \cdot & \cdot & & \\ & & \cdot & \cdot & \cdot & \\ & & & \cdot & \cdot & -1 \\ & & & & -1 & 2 \end{bmatrix} \quad \text{and}$$

$$\mathbf{B} = \frac{1}{6N}\begin{bmatrix} 4 & 1 & & & & \\ 1 & 4 & \cdot & & & \\ & \cdot & \cdot & \cdot & & \\ & & \cdot & \cdot & \cdot & \\ & & & \cdot & \cdot & 1 \\ & & & & 1 & 4 \end{bmatrix},$$

where N is the number of mesh intervals in the discrete approximation and both **A** and **B** are $(N - 1) \times (N - 1)$. For more information concerning the connection between generalized eigenproblems and Ritz-Galerkin approximations, see the book [S17] by Strang and Fix.

Returning to (97), we rearrange terms to obtain $(\mathbf{A} - \lambda\mathbf{B})\mathbf{x} = \mathbf{0}$. Since $\mathbf{x}$ is nonzero, the determinant of the matrix $\mathbf{A} - \lambda\mathbf{B}$ must vanish. Thus λ is an eigenvalue for the generalized eigenproblem (97) if and only if the **generalized characteristic equation**

$$\det(\mathbf{A} - \lambda\mathbf{B}) = 0 \tag{100}$$

is satisfied. When **A** and **B** are upper triangular, the generalized eigenvalues can be expressed in terms of the diagonal elements. For example, if **A** and **B** are 2×2 upper triangular matrices:

$$\mathbf{A} = \begin{bmatrix} a_{11} & a_{12} \\ 0 & a_{22} \end{bmatrix} \quad \text{and} \quad \mathbf{B} = \begin{bmatrix} b_{11} & b_{12} \\ 0 & b_{22} \end{bmatrix},$$

then equation (100) takes the form

$$\det\begin{bmatrix} a_{11} - \lambda b_{11} & a_{12} - \lambda b_{12} \\ 0 & a_{22} - \lambda b_{22} \end{bmatrix} = (a_{11} - \lambda b_{11})(a_{22} - \lambda b_{22}) = 0.$$

The generalized eigenvalues are $\lambda = a_{11}/b_{11}$ and $\lambda = a_{22}/b_{22}$ provided that both b_{11} and b_{22} are nonzero. In Exercise 6-12.1 you will show that if either b_{11} or b_{22} is zero, the generalized eigenproblem has either no eigenvalues, one eigenvalue, or an infinite number of eigenvalues.

For 2×2 matrices, the generalized eigenproblem can be solved using determinants. For example, if **A** and **B** are defined by

$$\mathbf{A} = \begin{bmatrix} 2 & -3 \\ 6 & -4 \end{bmatrix} \quad \text{and} \quad \mathbf{B} = \begin{bmatrix} 1 & -1 \\ 2 & -1 \end{bmatrix},$$

equation (100) is

$$\det\begin{bmatrix} 2 - \lambda & \lambda - 3 \\ 6 - 2\lambda & \lambda - 4 \end{bmatrix} = (2 - \lambda)(\lambda - 4) - (6 - 2\lambda)(\lambda - 3) = \lambda^2 - 6\lambda + 10 = 0.$$

Applying the quadratic formula to the equation $\lambda^2 - 6\lambda + 10 = 0$, we obtain the generalized eigenvalues $\lambda = 3 \pm \mathbf{i}$. The corresponding generalized eigenvectors are the solutions to $(\mathbf{A} - \lambda\mathbf{B})\mathbf{x} = \mathbf{0}$.

Although the characteristic polynomial can be used to compute the eigenvalues of a 2×2 matrix, we discovered in Section 6-2 that this method is unstable for large matrices. Assuming that $\mathbf{B}$ is invertible, one strategy for solving the generalized eigenproblem $\mathbf{Ax} = \lambda\mathbf{Bx}$ is to premultiply by $\mathbf{B}^{-1}$, giving us the equation

$$\mathbf{B}^{-1}\mathbf{Ax} = \lambda\mathbf{x}.$$

Letting $\mathbf{C}$ denote $\mathbf{B}^{-1}\mathbf{A}$, we have $\mathbf{Cx} = \lambda\mathbf{x}$. This shows that the generalized eigenproblem $\mathbf{Ax} = \lambda\mathbf{Bx}$ is equivalent to the standard eigenproblem $\mathbf{Cx} = \lambda\mathbf{x}$ with $\mathbf{C} = \mathbf{B}^{-1}\mathbf{A}$ and the eigenpairs for $\mathbf{C}$ are the generalized eigenpairs for $\mathbf{Ax} = \lambda\mathbf{Bx}$. Thus one strategy for solving the generalized eigenproblem is to form the matrix $\mathbf{C} = \mathbf{B}^{-1}\mathbf{A}$ and compute its eigenpairs using the routines developed for the standard eigenproblem. Some deficiencies with this strategy are that the computation of $\mathbf{C} = \mathbf{B}^{-1}\mathbf{A}$ introduces new numerical errors and destroys symmetries and sparseness. On the other hand, if $\mathbf{A}$ is symmetric and $\mathbf{B}$ is symmetric and positive definite, symmetry can be preserved with the Cholesky factorization $\mathbf{LL}^T$ of $\mathbf{B}$. Replacing $\mathbf{B}$ by its Cholesky factorization, the generalized eigenproblem is written

$$\mathbf{Ax} = \lambda\mathbf{LL}^T\mathbf{x}.$$

Premultiplying by $\mathbf{L}^{-1}$ yields

$$\mathbf{L}^{-1}\mathbf{Ax} = \lambda\mathbf{L}^T\mathbf{x}.$$

Then replacing $\mathbf{Ax}$ with $\mathbf{AL}^{-T}\mathbf{L}^T\mathbf{x}$, where $\mathbf{L}^{-T}$ denotes the inverse of $\mathbf{L}^T$ gives

$$\mathbf{L}^{-1}\mathbf{AL}^{-T}\mathbf{L}^T\mathbf{x} = \lambda\mathbf{L}^T\mathbf{x}.$$

Finally, letting $\mathbf{y}$ denote $\mathbf{L}^T\mathbf{x}$, we obtain the standard eigenproblem

$$\mathbf{Cy} = \lambda\mathbf{y} \quad \text{with} \quad \mathbf{C} = \mathbf{L}^{-1}\mathbf{AL}^{-T}. \tag{101}$$

Since $\mathbf{y} = \mathbf{L}^T\mathbf{x}$, it follows that if $(\lambda,\mathbf{y})$ is an eigenpair for $\mathbf{C}$, then $(\lambda,\mathbf{L}^{-T}\mathbf{y})$ is an eigenpair for the generalized eigenproblem $\mathbf{Ax} = \lambda\mathbf{Bx}$. The transformed problem (101) preserves the symmetry of the original problem $\mathbf{Ax} = \lambda\mathbf{Bx}$ since the matrix $\mathbf{C} = \mathbf{L}^{-1}\mathbf{AL}^T$ is symmetric:

$$\mathbf{C}^T = (\mathbf{L}^{-1}\mathbf{AL}^{-T})^T = (\mathbf{L}^{-T})^T\mathbf{A}^T(\mathbf{L}^{-1})^T = \mathbf{L}^{-1}\mathbf{A}^T\mathbf{L}^{-T} = \mathbf{L}^{-1}\mathbf{AL}^{-T}.$$

(This chain of equalities utilizes the property that the inverse of the transpose is equal to the transpose of the inverse.) To summarize, if $\mathbf{A}$ is symmetric and $\mathbf{B}$ is symmetric and positive definite, the generalized eigenproblem $\mathbf{Ax} = \lambda\mathbf{Bx}$ is equivalent to finding the eigenpairs of the symmetric matrix $\mathbf{L}^{-1}\mathbf{AL}^{-T}$, where $\mathbf{LL}^T$ is the Cholesky factorization of $\mathbf{B}$. Moreover, if $(\lambda, \mathbf{y})$ is an eigenpair for $\mathbf{C} = \mathbf{L}^{-1}\mathbf{AL}^{-T}$, then λ is a generalized eigenvalue for (97) while the corresponding generalized eigenvector is the solution $\mathbf{x}$ to $\mathbf{L}^T\mathbf{x} = \mathbf{y}$.

For arbitrary $\mathbf{A}$ and $\mathbf{B}$, more accurate approximations to the generalized eigenvalues are often generated if we work with the original equation $\mathbf{Ax} = \lambda\mathbf{Bx}$ rather than the transformed equation $\mathbf{B}^{-1}\mathbf{Ax} = \lambda\mathbf{x}$. Recall that with the standard eigenproblem $\mathbf{Ax} = \lambda\mathbf{x}$, the **QR** algorithm reduces $\mathbf{A}$ to an upper triangular matrix with the eigenvalues on the diagonal. A similar algorithm called the **QZ** algorithm has been developed for the generalized eigenproblem. The **QZ** algorithm reduces $\mathbf{A}$ and $\mathbf{B}$ to upper triangular form. The generalized eigenvalues are obtained by dividing each diagonal element of the reduced $\mathbf{A}$ by the corresponding diagonal element of the reduced $\mathbf{B}$. Recall that each step of the **QR** algorithm is a similarity transformation: $\mathbf{A}^{\text{new}} = \mathbf{Q}^{-1}\mathbf{A}^{\text{old}}\mathbf{Q}$. The analogous transformation for the generalized eigenproblem involves two invertible matrices $\mathbf{P}$ and $\mathbf{Q}$. Premultiplying $\mathbf{Ax} = \lambda\mathbf{Bx}$ by $\mathbf{Q}^{-1}$ and replacing $\mathbf{x}$ by $\mathbf{PP}^{-1}\mathbf{x}$, we obtain the equivalent equation

$$\mathbf{Q}^{-1}\mathbf{APP}^{-1}\mathbf{x} = \lambda\mathbf{Q}^{-1}\mathbf{BPP}^{-1}\mathbf{x}.$$

Letting $\mathbf{y}$ denote the product $\mathbf{P}^{-1}\mathbf{x}$, this reduces to

$$(\mathbf{Q}^{-1}\mathbf{AP})\mathbf{y} = \lambda(\mathbf{Q}^{-1}\mathbf{BP})\mathbf{y}.$$

Hence the generalized eigenvalues corresponding to the pair $(\mathbf{A}, \mathbf{B})$ are the same as the generalized eigenvalues corresponding to the pair $(\mathbf{Q}^{-1}\mathbf{AP}, \mathbf{Q}^{-1}\mathbf{BP})$. Moreover, if $\mathbf{y}$ is a generalized eigenvector corresponding to the pair $(\mathbf{Q}^{-1}\mathbf{AP}, \mathbf{Q}^{-1}\mathbf{BP})$, the corresponding generalized eigenvector for $(\mathbf{A}, \mathbf{B})$ is $\mathbf{x} = \mathbf{Py}$. Each iteration in the $\mathbf{QZ}$ algorithm is given by

$$\mathbf{A}^{\text{new}} = \mathbf{Q}^{-1}\mathbf{A}^{\text{old}}\mathbf{P} \quad \text{and} \quad \mathbf{B}^{\text{new}} = \mathbf{Q}^{-1}\mathbf{B}^{\text{old}}\mathbf{P},$$

where $\mathbf{Q}$ and $\mathbf{P}$ are orthogonal matrices chosen to drive $\mathbf{A}$ and $\mathbf{B}$ to upper triangular form. Although each $\mathbf{QZ}$ iteration is equivalent to a $\mathbf{QR}$ iteration applied to the matrix $\mathbf{B}^{-1}\mathbf{A}$, the matrix $\mathbf{B}^{-1}\mathbf{A}$ is never explicitly computed. For a detailed description of the $\mathbf{QZ}$ iteration, see [M9] or [G15].

Before starting the $\mathbf{QZ}$ iteration, $\mathbf{A}$ should be reduced to upper Hessenberg form and $\mathbf{B}$ should be reduced to upper triangular form. To accomplish this reduction, we construct orthogonal matrices $\mathbf{Q}_0$ and $\mathbf{P}_0$ with the property that $\mathbf{Q}_0^T\mathbf{AP}_0$ is upper Hessenberg and $\mathbf{Q}_0^T\mathbf{BP}_0$ is upper triangular. To obtain $\mathbf{Q}_0$ and $\mathbf{P}_0$, we first factor $\mathbf{B}$ into the product $\mathbf{Q}_1\mathbf{R}_1$ between an orthogonal matrix $\mathbf{Q}_1$ and an upper triangular matrix $\mathbf{R}_1$ and then compute

$$\mathbf{A}_1 = \mathbf{Q}_1^T\mathbf{A}.$$

Then $\mathbf{A}_1$ is reduced to upper Hessenberg form without destroying the upper triangular structure of $\mathbf{R}_1$. The reduction is accomplished using Givens rotations. In each step, we premultiply by a Givens rotation $\mathbf{Q}_{ij}$ to annihilate the (i, j) coefficient in $\mathbf{A}_1$ and we postmultiply by a different Givens rotation $\mathbf{P}_{ij}$ to smooth the bulge created in $\mathbf{R}_1$. This process is illustrated using 4×4 matrices:

$$\mathbf{A}_1 = \begin{bmatrix} * & * & * & * \\ * & * & * & * \\ * & * & * & * \\ * & * & * & * \end{bmatrix} \quad \text{and} \quad \mathbf{R}_1 = \begin{bmatrix} * & * & * & * \\ 0 & * & * & * \\ 0 & 0 & * & * \\ 0 & 0 & 0 & * \end{bmatrix}.$$

Starting at the lower left corner of $\mathbf{A}_1$, we premultiply by the Givens rotation $\mathbf{G}_{43}$ [see equation (47) of Chapter 5] that annihilates the coefficient in column 1 and row 4. If $\mathbf{Q}_{41}$ denotes this rotation, we have

$$\mathbf{Q}_{41}\mathbf{A}_1 = \begin{bmatrix} * & * & * & * \\ * & * & * & * \\ * & * & * & * \\ 0 & * & * & * \end{bmatrix} \quad \text{and} \quad \mathbf{Q}_{41}\mathbf{R}_1 = \begin{bmatrix} * & * & * & * \\ 0 & * & * & * \\ 0 & 0 & * & * \\ 0 & 0 & * & * \end{bmatrix}.$$

To smooth the bulge in the upper triangular matrix, postmultiply by the Givens rotation $\mathbf{G}_{43}$ that annihilates the coefficient in row 4 and column 3. Letting $\mathbf{P}_{41}$ denote this rotation, we have

$$\mathbf{A}_2 = \mathbf{Q}_{41}\mathbf{A}_1\mathbf{P}_{41} = \begin{bmatrix} * & * & * & * \\ * & * & * & * \\ * & * & * & * \\ 0 & * & * & * \end{bmatrix} \quad \text{and} \quad \mathbf{R}_2 = \mathbf{Q}_{41}\mathbf{R}_1\mathbf{P}_{41} = \begin{bmatrix} * & * & * & * \\ 0 & * & * & * \\ 0 & 0 & * & * \\ 0 & 0 & 0 & * \end{bmatrix}.$$

The next step uses a Givens rotation $\mathbf{G}_{32}$ to annihilate the coefficient in row 3 and column 1 of $\mathbf{A}_2$ and a different rotation $\mathbf{G}_{32}$ to smooth the bulge in $\mathbf{R}_2$. Letting $\mathbf{Q}_{31}$ and $\mathbf{P}_{31}$ denote these rotations, we have

$$\mathbf{A}_3 = \mathbf{Q}_{31}\mathbf{A}_2\mathbf{P}_{31} = \begin{bmatrix} * & * & * & * \\ * & * & * & * \\ 0 & * & * & * \\ 0 & * & * & * \end{bmatrix}$$

and $\mathbf{R}_3 = \mathbf{Q}_{31}\mathbf{R}_2\mathbf{P}_{31}$. At this point, the first column of $\mathbf{A}_3$ has the correct structure for an upper Hessenberg matrix. Proceeding to the second column, we start at the bottom and premultiply by a Givens rotations $\mathbf{G}_{43}$ that annihilates the coefficient in row 4 and column 2. Then we postmultiply by a different rotation $\mathbf{G}_{43}$ that smoothes the bulge in the upper triangular matrix. If $\mathbf{Q}_{42}$ and $\mathbf{P}_{42}$ denote these rotations, we have

$$\mathbf{A}_4 = \mathbf{Q}_{42}\mathbf{A}_3\mathbf{P}_{42} = \begin{bmatrix} * & * & * & * \\ * & * & * & * \\ 0 & * & * & * \\ 0 & 0 & * & * \end{bmatrix} \tag{102}$$

and $\mathbf{R}_4 = \mathbf{Q}_{42}\mathbf{R}_3\mathbf{P}_{42}$. Observe that $\mathbf{A}_4$ is upper Hessenberg. Since $\mathbf{A}_4 = \mathbf{Q}_{42}\mathbf{A}_3\mathbf{P}_{42}$, $\mathbf{A}_3 = \mathbf{Q}_{31}\mathbf{A}_2\mathbf{P}_{31}$, $\mathbf{A}_2 = \mathbf{Q}_{41}\mathbf{A}_1\mathbf{P}_{41}$, and $\mathbf{A}_1 = \mathbf{Q}_1{}^T\mathbf{A}$, it follows that the upper Hessenberg matrix $\mathbf{A}_4$ is equal to $\mathbf{Q}^T\mathbf{A}\mathbf{P}$, where

$$\mathbf{Q}^T = \mathbf{Q}_{42}\mathbf{Q}_{31}\mathbf{Q}_{41}\mathbf{Q}_1^T \quad \text{and} \quad \mathbf{P} = \mathbf{P}_{41}\mathbf{P}_{31}\mathbf{P}_{42}.$$

Similarly, the upper triangular matrix $\mathbf{R}_4$ is equal to $\mathbf{Q}^T\mathbf{B}\mathbf{P}$. After reducing $\mathbf{A}$ to upper Hessenberg form and $\mathbf{B}$ to upper triangular form, the **QZ** algorithm uses a sequence of **QR**-type iterations to reduce $\mathbf{A}_4$ to upper triangular form while preserving the upper triangular form of $\mathbf{R}_4$.

Exercises

6-12.1. Consider the generalized eigenproblem $\mathbf{Ax} = \lambda\mathbf{Bx}$, where $\mathbf{A}$ and $\mathbf{B}$ are 2×2 triangular matrices. Verify the following:

(a) If $a_{11} = b_{11} = 0$ or $a_{22} = b_{22} = 0$, there are an infinite number of eigenvalues.

(b) If $b_{11} = 0$ and $b_{22} = 0$ but both a_{11} and a_{22} are nonzero, there are no eigenvalues.

(c) If $b_{11} = 0$ but both a_{11} and b_{22} are nonzero, there is exactly one eigenvalue. In what other situation is there exactly one eigenvalue?

In each of the cases (a), (b), and (c), what can be said about the inverse of $\mathbf{B}$?

6-12.2. Given two square matrices $\mathbf{A}$ and $\mathbf{B}$, consider the eigenproblem $(\mathbf{AB})\mathbf{x} = \lambda\mathbf{x}$. If $\mathbf{B}$ is symmetric and positive definite and $\mathbf{LL}^T$ is its Cholesky factorization, show that the eigenvalues of $\mathbf{AB}$ and the eigenvalues of $\mathbf{L}^T\mathbf{AL}$ are the same. What is the relationship between the eigenvectors of $\mathbf{AB}$ and the eigenvectors of $\mathbf{L}^T\mathbf{AL}$?

6-13. CONDITIONING AND REFINEMENT

The conditioning of the eigenproblem is related to the change in the eigenpairs relative to a change in the coefficient matrix. Suppose that $\mathbf{A}$ is a nondefective matrix with eigenvalues $\lambda_1, \lambda_2, \cdots, \lambda_n$ and with corresponding linearly independent eigenvectors $\mathbf{x}_1, \mathbf{x}_2, \cdots, \mathbf{x}_n$. If μ denotes any eigenvalue of the perturbed matrix $\mathbf{A} + \delta\mathbf{A}$, the Bauer-Fike theorem bounds the distance between μ and the spectrum of $\mathbf{A}$:

$$\text{minimum } \{|\lambda_j - \mu| : j = 1, \cdots, n\} \leqq c_p(\mathbf{X})\|\delta\mathbf{A}\|_p, \tag{103}$$

where $\mathbf{X}$ is the matrix whose columns are the eigenvectors of $\mathbf{A}$, $\|\cdot\|_p$ denotes the p-norm, and $c_p(\mathbf{X}) = \|\mathbf{X}\|_p\|\mathbf{X}^{-1}\|_p$ is the p-norm condition number of $\mathbf{X}$.

To establish (103), we form the matrix

$$\mathbf{X}^{-1}(\mathbf{A} + \delta\mathbf{A} - \mu\mathbf{I})\mathbf{X}.$$

Since $\mathbf{X}^{-1}\mathbf{A}\mathbf{X} = \mathbf{\Lambda}$, where $\mathbf{\Lambda}$ is the diagonal matrix with diagonal elements λ_1, $\lambda_2, \cdots, \lambda_n$, we have

$$\mathbf{X}^{-1}(\mathbf{A} + \delta\mathbf{A} - \mu\mathbf{I})\mathbf{X} = \mathbf{\Lambda} - \mu\mathbf{I} + \mathbf{X}^{-1}\delta\mathbf{A}\mathbf{X}. \tag{104}$$

If μ is an eigenvalue for $\mathbf{A} + \delta\mathbf{A}$, the matrix $\mathbf{A} + \delta\mathbf{A} - \mu\mathbf{I}$ as well as the product $\mathbf{X}^{-1}(\mathbf{A} + \delta\mathbf{A} - \mu\mathbf{I})\mathbf{X}$ is singular. Hence there exists a nonzero vector $\mathbf{y}$ such that $\mathbf{X}^{-1}(\mathbf{A} + \delta\mathbf{A} - \mu\mathbf{I})\mathbf{X}\mathbf{y} = \mathbf{0}$. By (104) we conclude that

$$(\mathbf{\Lambda} - \mu\mathbf{I})\mathbf{y} = -\mathbf{X}^{-1}\delta\mathbf{A}\mathbf{X}\mathbf{y}. \tag{105}$$

In Exercise 6-13.1 you show that if $\mathbf{D}$ is a diagonal matrix, then

$$\|\mathbf{D}\mathbf{y}\|_p \geqq \text{minimum } \{|d_{11}|, |d_{22}|, \cdots, |d_{nn}|\}\ \|\mathbf{y}\|_p. \tag{106}$$

Taking the p-norm of (105) and utilizing (106) yields

$$\underset{j=1,\cdots,n}{\text{minimum}} \{|\lambda_j - \mu|\}\ \|\mathbf{y}\|_p \leqq \|(\mathbf{\Lambda} - \mu\mathbf{I})\mathbf{y}\|_p = \|\mathbf{X}^{-1}\delta\mathbf{A}\mathbf{X}\mathbf{y}\|_p \leqq \|\mathbf{X}^{-1}\|_p\|\delta\mathbf{A}\|_p\|\mathbf{X}\|_p\|\mathbf{y}\|_p.$$

Dividing by $\|\mathbf{y}\|_p$, we get (103).

If $\mathbf{A}$ is symmetric and p is 2, the Bauer-Fike theorem can be simplified since the eigenvector matrix $\mathbf{X}$ corresponding to a symmetric matrix is orthogonal (see Exercise 6-5.1). Also, in Exercise 6-11.3, you saw that the 2-norm condition number of an orthogonal matrix is 1. Thus (103) implies that

$$\text{minimum } \{|\lambda_j - \mu| : j = 1, \cdots, n\} \leqq \|\delta\mathbf{A}\|_2$$

when $\mathbf{A}$ is symmetric. Since the condition number of a matrix is at least 1, it follows that a symmetric matrix has the smallest possible value for the parameter $c_2(\mathbf{X})$. In general, a matrix with a complete set of orthonormal eigenvectors is called a **normal matrix** and for any normal matrix, $c_2(\mathbf{X}) = 1$. The class of normal matrices includes those that are either symmetric, skew-symmetric, orthogonal, or unitary.

An example of a matrix with ill-conditioned eigenvalues is

$$\mathbf{A} = \begin{bmatrix} 101 & 110 \\ -90 & -98 \end{bmatrix}.$$

Since the diagonalization $\mathbf{X}\mathbf{\Lambda}\mathbf{X}^{-1}$ of $\mathbf{A}$ is

$$\begin{bmatrix} 101 & 110 \\ -90 & -98 \end{bmatrix} = \begin{bmatrix} -11 & 10 \\ 10 & -9 \end{bmatrix}\begin{bmatrix} 1 & 0 \\ 0 & 2 \end{bmatrix}\begin{bmatrix} 9 & 10 \\ 10 & 11 \end{bmatrix},$$

the 1-norm of $\mathbf{X}$ and $\mathbf{X}^{-1}$ are both 21 and the 1-norm condition number for $\mathbf{X}$ is $c_1(\mathbf{X}) = 21^2 = 441$. The Bauer-Fike theorem tells us that a change ε in a coefficient of $\mathbf{A}$ can produce a change 441ε in an eigenvalue. Changing the a_{11} coefficient of $\mathbf{A}$ from 101 to 100, the eigenvalues $\lambda_1 = 1$ and $\lambda_2 = 2$ of $\mathbf{A}$ change to $\lambda_1 \approx 1 + 10\mathbf{i}$ and $\lambda_2 \approx 1 - 10\mathbf{i}$. Thus a relatively small change in a coefficient generated a relatively large change in the eigenvalues. (Note though that the Bauer-Fike theorem overestimated the actual change in the eigenvalues.)

The Bauer-Fike theorem is in some sense a macroscopic estimate since the change in all the eigenvalues is estimated simultaneously. Let us now estimate the conditioning of a specific eigenvalue. We consider perturbations of the form $\boldsymbol{\delta}\mathbf{A} = \varepsilon\mathbf{E}$, where ε is a small parameter and $\mathbf{E}$ is a fixed matrix. For convenience, assume that the Euclidean norm of $\mathbf{E}$ is equal to $\|\mathbf{A}\|_2$. With this assumption, the relative change $\|\boldsymbol{\delta}\mathbf{A}\|_2/\|\mathbf{A}\|_2$ in the coefficient matrix is ε. If λ is a simple zero for the characteristic polynomial of $\mathbf{A}$ and $\mathbf{x}$ is the corresponding normalized eigenvector, then for small values of ε, the matrix $\mathbf{A} + \varepsilon\mathbf{E}$ has an eigenvalue $\lambda(\varepsilon)$ near λ and a corresponding normalized eigenvector $\mathbf{x}(\varepsilon)$. Differentiating the relationship

$$(\mathbf{A} + \varepsilon\mathbf{E})\mathbf{x}(\varepsilon) = \lambda(\varepsilon)\mathbf{x}(\varepsilon)$$

with respect to ε and setting $\varepsilon = 0$, we have

$$\mathbf{A}\mathbf{x}'(0) + \mathbf{E}\mathbf{x} = \lambda'(0)\mathbf{x} + \lambda(0)\mathbf{x}'(0). \tag{107}$$

Equation (107) will be simplified using left eigenvectors. Let $\mathbf{A} = \mathbf{X\Lambda X}^{-1}$ be the diagonalization of $\mathbf{A}$. Premultiplying the equation $\mathbf{A} = \mathbf{X\Lambda X}^{-1}$ by $\mathbf{X}^{-1}$ yields $\mathbf{X}^{-1}\mathbf{A} = \mathbf{\Lambda X}^{-1}$. Therefore, the ith row of $\mathbf{X}^{-1}\mathbf{A}$ is equal to the ith row of $\mathbf{\Lambda X}^{-1}$. Let $\mathbf{y}_i^T$ denote the ith row of $\mathbf{X}^{-1}$. The ith row of $\mathbf{X}^{-1}\mathbf{A}$ is the ith row of $\mathbf{X}^{-1}$ (or $\mathbf{y}_i^T$) times $\mathbf{A}$ while the ith row of $\mathbf{\Lambda X}^{-1}$ is equal to λ_i times $\mathbf{y}_i^T$. Equating the ith row of $\mathbf{X}^{-1}\mathbf{A}$ to the ith row of $\mathbf{\Lambda X}^{-1}$ yields the relation

$$\mathbf{y}_i^T\mathbf{A} = \lambda_i\mathbf{y}_i^T.$$

In other words, $\mathbf{y}_i^T$ times $\mathbf{A}$ is equal to λ_i times $\mathbf{y}_i^T$. The rows of $\mathbf{X}^{-1}$ are **left eigenvectors** of $\mathbf{A}$. Letting $\mathbf{y}$ denote the left eigenvector of $\mathbf{A}$ corresponding to the eigenvalue $\lambda(0)$, we premultiply (107) by $\mathbf{y}^T$. The $\mathbf{x}'(0)$ terms cancel to give

$$(\mathbf{y}^T\mathbf{x})\lambda'(0) = \mathbf{y}^T\mathbf{E}\mathbf{x}.$$

Finally, we solve for $\lambda'(0)$:

$$\lambda'(0) = \frac{\mathbf{y}^T\mathbf{E}\mathbf{x}}{\mathbf{y}^T\mathbf{x}}. \tag{108}$$

Since the norm of a product is less than or equal to the product of the norms and since the Euclidean norm of $\mathbf{E}$ is $\|\mathbf{A}\|_2$, it follows that

$$|\mathbf{y}^T\mathbf{E}\mathbf{x}| \leq \|\mathbf{y}\|_2\|\mathbf{x}\|_2\|\mathbf{A}\|_2. \tag{109}$$

Combining (108) and (109), we have

$$|\lambda'(0)| \leq \|\mathbf{A}\|_2 \frac{\|\mathbf{x}\|_2\|\mathbf{y}\|_2}{|\mathbf{y}^T\mathbf{x}|}.$$

Since the ratio $\mathbf{y}^T\mathbf{x}/\|\mathbf{x}\|_2\|\mathbf{y}\|_2$ is the cosine of the angle θ between $\mathbf{x}$ and $\mathbf{y}$, this bound for $|\lambda'(0)|$ can be expressed:

$$|\lambda'(0)| \leq \frac{\|\mathbf{A}\|_2}{|\cos\theta|}. \tag{110}$$

To relate (110) to the change $\lambda(\varepsilon) - \lambda(0)$ in the eigenvalue, we expand $\lambda(\varepsilon)$ in a Taylor series about $\varepsilon = 0$:

$$\lambda(\varepsilon) = \lambda(0) + \varepsilon\lambda'(0) + O(\varepsilon^2).$$

Dropping the order ε^2 term gives the approximation

$$\lambda(\varepsilon) \approx \lambda(0) + \varepsilon\lambda'(0).$$

Since the change $\lambda(\varepsilon) - \lambda(0)$ in the eigenvalue is approximately equal to $\varepsilon\lambda'(0)$, (110) implies that if $\|\mathbf{A}\|_2/|\cos\theta|$ is small, a small ε yields a small change $|\lambda(\varepsilon) - \lambda(0)|$ in the eigenvalue. Conversely, if $\|\mathbf{A}\|_2/|\cos\theta|$ is large, a small ε can generate a large change in the eigenvalue. The conditioning of an eigenvalue is related to the relative change

$$\frac{\lambda(\varepsilon) - \lambda(0)}{\lambda(0)} \approx \varepsilon\frac{\lambda'(0)}{\lambda(0)}.$$

If the ratio $|\lambda'(0)|/|\lambda(0)|$ is large, the eigenvalue is ill conditioned. If the ratio $|\lambda'(0)|/|\lambda(0)|$ is small, the eigenvalue is well conditioned. Dividing (110) by $|\lambda(0)|$ yields

$$\frac{|\lambda'(0)|}{|\lambda(0)|} \leq \frac{\|\mathbf{A}\|_2}{|\lambda(0)\cos\theta|}.$$

Therefore, the best-conditioned eigenvalues are those for which the ratio $\|\mathbf{A}\|_2/|\lambda(0)\cos\theta|$ is near 1 or less than 1. Since the left and the right eigenvectors of a symmetric matrix are equal, $\cos\theta = 1$ and the conditioning of an eigenvalue for a symmetric matrix only depends on the ratio $\|\mathbf{A}\|_2/|\lambda(0)|$.

Formula (108) for the derivative $\lambda'(0)$ coupled with the first-order Taylor expansion

$$\lambda(\varepsilon) \approx \lambda(0) + \varepsilon\lambda'(0)$$

can be used to estimate the movement of eigenvalues. Substituting for $\varepsilon\lambda'(0)$ using (108) yields:

$$\lambda(\varepsilon) \approx \lambda(0) + \varepsilon\frac{\mathbf{y}^T\mathbf{E}\mathbf{x}}{\mathbf{y}^T\mathbf{x}} = \lambda(0) + \frac{\mathbf{y}^T(\boldsymbol{\delta}\mathbf{A})\mathbf{x}}{\mathbf{y}^T\mathbf{x}}, \tag{111}$$

where $\boldsymbol{\delta}\mathbf{A} = \varepsilon\mathbf{E}$ is the perturbation in the coefficients. In other words, to first order, the change $\boldsymbol{\delta}\mathbf{A}$ in the coefficient matrix generates a change $\mathbf{y}^T(\boldsymbol{\delta}\mathbf{A})\mathbf{x}/\mathbf{y}^T\mathbf{x}$ in the eigenvalue.

For example, let us consider the coefficient matrix

$$\mathbf{A} = \begin{bmatrix} 1 & 1 \\ 0 & 2 \end{bmatrix}. \tag{112}$$

Since $\mathbf{A}$ is upper triangular, its eigenvalues are the diagonal elements $a_{11} = 1$ and $a_{22} = 2$. We will estimate the change in the eigenvalue $\lambda = 2$ when the coefficient matrix is perturbed to

$$\mathbf{A} + \boldsymbol{\delta}\mathbf{A} = \begin{bmatrix} 1 & 1 \\ 0 & 2 \end{bmatrix} + \begin{bmatrix} 0 & 0 \\ .01 & 0 \end{bmatrix} = \begin{bmatrix} 1 & 1 \\ .01 & 2 \end{bmatrix}. \tag{113}$$

Since $\mathbf{A}$ is upper triangular, its left and right eigenvectors corresponding to the eigenvalue $\lambda = 2$ can be computed by forward and back substitution:

$$\mathbf{y} = \begin{bmatrix} 0 \\ 1 \end{bmatrix} \quad \text{and} \quad \mathbf{x} = \begin{bmatrix} 1 \\ 1 \end{bmatrix}.$$

Applying (111), the change $\delta\lambda$ in the eigenvalue $\lambda = 2$ that results from the change $\boldsymbol{\delta}\mathbf{A}$ in the coefficient matrix is

$$\delta\lambda = \frac{\mathbf{y}^T\boldsymbol{\delta}\mathbf{A}\mathbf{x}}{\mathbf{y}^T\mathbf{x}} = \frac{[0 \quad 1]\begin{bmatrix} 0 & 0 \\ .01 & 0 \end{bmatrix}\begin{bmatrix} 1 \\ 1 \end{bmatrix}}{[0 \quad 1]\begin{bmatrix} 1 \\ 1 \end{bmatrix}} = .01.$$

To first order in $\boldsymbol{\delta}\mathbf{A}$, $\mu = \lambda + \delta\lambda = 2 + .01 = 2.01$ is an eigenvalue of the matrix $\mathbf{A} + \boldsymbol{\delta}\mathbf{A}$.

This procedure for estimating the eigenvalue change associated with a coefficient change can be applied to the **QR** method. The **QR** method starts with a Hessenberg matrix and drives subdiagonal elements to zero. At some point, we stop the iterations and replace the small subdiagonal elements with zero to obtain an upper triangular matrix whose diagonal elements approximate the eigenvalues of the original matrix. Relation (111) estimates the change in the eigenvalues when some subdiagonal elements are replaced with zero. For a second application of (111), suppose that an approximate diagonalization $\mathbf{X}\boldsymbol{\Lambda}\mathbf{X}^{-1}$ of $\mathbf{B}$ has been computed. We apply (111) where $\mathbf{A}$ is identified with $\mathbf{X}\boldsymbol{\Lambda}\mathbf{X}^{-1}$ and $\boldsymbol{\delta}\mathbf{A}$ is identified with $\mathbf{B} - \mathbf{X}\boldsymbol{\Lambda}\mathbf{X}^{-1}$. Letting $\mathbf{x}_i$ denote the ith column of $\mathbf{X}$ and letting $\mathbf{y}_i^T$ denote the ith row of $\mathbf{X}^{-1}$, (111) tells us that to first order, the ith eigenvalue of the matrix $\mathbf{A} + \boldsymbol{\delta}\mathbf{A} = \mathbf{B}$ is $\mathbf{y}_i^T\mathbf{B}\mathbf{x}_i$.

Using equation (107), we can also estimate the change in the eigenvectors due to a perturbation $\boldsymbol{\delta}\mathbf{A} = \varepsilon\mathbf{E}$ in the coefficient matrix. Let us assume that the eigenvalues $\lambda_1, \lambda_2, \cdots, \lambda_n$ of $\mathbf{A}$ are distinct and let $\mathbf{x}(\varepsilon)$ denote the eigenvector of $\mathbf{A} + \varepsilon\mathbf{E}$ corresponding to the jth eigenvalue of $\mathbf{A}$. Since the eigenvectors of $\mathbf{A}$ are linearly independent, we can expand $\mathbf{x}(\varepsilon)$ in terms of the eigenvectors of $\mathbf{A}$:

$$\mathbf{x}(\varepsilon) = \sum_{i=1}^{n} a_i(\varepsilon)\mathbf{x}_i. \tag{114}$$

Since a scalar multiple of an eigenvector is also an eigenvector, let us normalize $\mathbf{x}(\varepsilon)$ so that $a_j(\varepsilon) = 1$. Differentiating (114) with respect to ε, we have

$$\mathbf{x}'(0) = \sum_{\substack{i=1 \\ i\neq j}}^{n} a_i'(0)\mathbf{x}_i. \tag{115}$$

The derivative of $a_j(\varepsilon)$ is zero since $a_j(\varepsilon) = 1$ independent of ε. Letting b_i denote $a_i'(\varepsilon)$ and substituting (115) into (107), we have

$$\sum_{\substack{i=1 \\ i\neq j}}^{n} b_i(\lambda_i - \lambda_j)\mathbf{x}_i + \mathbf{E}\mathbf{x}_j = \lambda_j'(0)\mathbf{x}_j. \tag{116}$$

To solve for the unknown coefficients b_i, we utilize the orthogonality between left and right eigenvectors. Since $\mathbf{X}^{-1}\mathbf{X} = \mathbf{I}$, the dot product between row i of $\mathbf{X}^{-1}$ and column j of $\mathbf{X}$ is zero for $i \neq j$, while the dot product between row i of $\mathbf{X}^{-1}$ and column i of $\mathbf{X}$ is 1. Since the ith row of $\mathbf{X}^{-1}$ is the ith left eigenvector $\mathbf{y}_i$ and the jth column of $\mathbf{X}$ is the jth right eigenvector $\mathbf{x}_j$, the identity $\mathbf{X}^{-1}\mathbf{X} = \mathbf{I}$

is equivalent to

$$\mathbf{y}_i^T\mathbf{x}_j = \begin{cases} 0 & \text{for } i \neq j, \\ 1 & \text{for } i = j. \end{cases}$$

Multiplying (116) by $\mathbf{y}_k^T$ and solving for b_k yields

$$b_k = \frac{\mathbf{y}_k^T\mathbf{E}\mathbf{x}_j}{\lambda_j - \lambda_k}.$$

This formula for b_k coupled with equation (115) tell us that

$$\mathbf{x}'(0) = \sum_{\substack{i=1 \\ i\neq j}}^{n} \frac{\mathbf{y}_i^T\mathbf{E}\mathbf{x}_j}{\lambda_j - \lambda_i}\mathbf{x}_i. \tag{117}$$

Like the conditioning of an eigenvalue, the conditioning of an eigenvector is related to the ratio $\|\mathbf{x}'(0)\|/\|\mathbf{x}(0)\|$. Taking the norm of (117), it follows that

$$\frac{\|\mathbf{x}'(0)\|}{\|\mathbf{x}(0)\|} \leqq \sum_{\substack{i=1 \\ i\neq j}}^{n} \frac{\|\mathbf{y}_i\|_2\|\mathbf{E}\|_2\|\mathbf{x}_j\|_2\|\mathbf{x}_i\|_2}{|\lambda_j - \lambda_i|\,\|\mathbf{x}_j\|_2} = \sum_{\substack{i=1 \\ i\neq j}}^{n} \frac{\|\mathbf{y}_i\|_2\|\mathbf{x}_i\|_2\|\mathbf{A}\|_2}{|\lambda_j - \lambda_i|}$$

$$= \sum_{\substack{i=1 \\ i\neq j}}^{n} \frac{\|\mathbf{y}_i\|_2\|\mathbf{x}_i\|_2\|\mathbf{A}\|_2}{|\lambda_j - \lambda_i|\,\mathbf{x}_i^T\mathbf{y}_i} = \sum_{\substack{i=1 \\ i\neq j}}^{n} \frac{\|\mathbf{A}\|_2}{|\lambda_j - \lambda_i|\cos\theta_i},$$

where θ_i is the angle between $\mathbf{x}_i$ and $\mathbf{y}_i$, $\mathbf{x}_i^T\mathbf{y}_i = 1$ by the orthogonality property, and $\|\mathbf{E}\|_2 = \|\mathbf{A}\|_2$ by assumption. In summary, the conditioning of the jth eigenvalue depends on the ratio $\|\mathbf{A}\|_2/|\lambda_j|\,|\cos\theta_j|$, while the conditioning of the jth eigenvector depends on the ratio

$$\frac{\|\mathbf{A}\|_2}{|\lambda_i - \lambda_j|\cos\theta_i}$$

for $i \neq j$.

In the same way that equation (108) coupled with a first-order Taylor expansion led us to an estimate for the eigenvalues of a perturbed matrix, relation (117) coupled with the first-order Taylor approximation

$$\mathbf{x}(\varepsilon) \approx \mathbf{x}(0) + \varepsilon\mathbf{x}'(0)$$

give us an estimate for the eigenvector of the perturbed matrix $\mathbf{A} + \boldsymbol{\delta}\mathbf{A}$:

$$\mathbf{x}(\varepsilon) \approx \mathbf{x}(0) + \sum_{\substack{i=1 \\ i\neq j}}^{n} \frac{\mathbf{y}_i^T(\boldsymbol{\delta}\mathbf{A})\mathbf{x}_j}{\lambda_j - \lambda_i}\mathbf{x}_i. \tag{118}$$

Suppose that $\mathbf{X\Lambda X}^{-1}$ is an approximate diagonalization of $\mathbf{B}$. Applying (118) where $\mathbf{A}$ is identified with $\mathbf{X\Lambda X}^{-1}$ and $\boldsymbol{\delta}\mathbf{A}$ is identified with $\mathbf{B} - \mathbf{X\Lambda X}^{-1}$, the jth eigenvector of $\mathbf{B}$ is, to first order,

$$\mathbf{x}_j + \sum_{\substack{i=1 \\ i\neq j}}^{n} \frac{\mathbf{y}_i^T\mathbf{B}\mathbf{x}_j}{\lambda_j - \lambda_i}\mathbf{x}_i.$$

This approximation to an eigenvector coupled with the approximation $\mathbf{y}_j^T\mathbf{B}\mathbf{x}_j$ to the corresponding eigenvalue lead to a rapidly convergent algorithm to compute the eigenvalues and the eigenvectors of a matrix:

$$\begin{aligned}
\lambda_j^{\text{new}} &= (\mathbf{y}_j^{\text{old}})^T \mathbf{B}\mathbf{x}_j^{\text{old}} && \text{for } j = 1 \text{ to } n, \\
\mathbf{x}_j^{\text{new}} &= \mathbf{x}_j^{\text{old}} + \sum_{\substack{i=1 \\ i \neq j}}^{n} \frac{(\mathbf{y}_i^{\text{old}})^T \mathbf{B}\mathbf{x}_j^{\text{old}}}{\lambda_j^{\text{new}} - \lambda_i^{\text{new}}} \mathbf{x}_i^{\text{old}} && \text{for } j = 1 \text{ to } n, \\
\mathbf{Y}^{\text{new}} &= (\mathbf{X}^{\text{new}})^{-1}.
\end{aligned} \tag{119}$$

Unless good approximations to the eigenvectors of $\mathbf{B}$ are known, the $\mathbf{QR}$ algorithm computes the diagonalization of $\mathbf{B}$ more efficiently than (119).

The iteration (119) can be used to obtain a better approximation to all the eigenpairs. Similarly, the inverse power method can be used to obtain a better approximation to a single eigenpair. In either case, the accuracy in the computed eigenpairs is limited by the computing precision. We now develop a refinement technique to obtain high-accuracy approximations to a given eigenpair. The fundamental idea is to apply Newton's method to the equation $\mathbf{f}(\mathbf{x},\lambda) = \mathbf{0}$, where $\mathbf{f}$ is defined by

$$\mathbf{f}(\mathbf{x},\lambda) = \begin{bmatrix} \lambda\mathbf{x} - \mathbf{A}\mathbf{x} \\ \frac{1}{2}(\mathbf{x}^T\mathbf{x} - 1) \end{bmatrix}.$$

The top component of the equation $\mathbf{f}(\mathbf{x},\lambda) = \mathbf{0}$ is the eigenproblem $\mathbf{A}\mathbf{x} = \lambda\mathbf{x}$ and the bottom component is the normalization condition $\mathbf{x}^T\mathbf{x} = 1$. The Jacobian $\mathbf{J}$ of $\mathbf{f}$ has the following structure:

$$\mathbf{J}(\mathbf{x},\lambda) = \begin{bmatrix} \lambda\mathbf{I} - \mathbf{A} & \mathbf{x} \\ \mathbf{x}^T & 0 \end{bmatrix}.$$

Referring to Chapter 4, each iteration of Newton's method can be expressed $\mathbf{x}^{\text{new}} = \mathbf{x} + \mathbf{y}$ and $\lambda^{\text{new}} = \lambda + \mu$, where $\mathbf{y}$ and μ are the solution to

$$\begin{bmatrix} \lambda\mathbf{I} - \mathbf{A} & \mathbf{x} \\ \mathbf{x}^T & 0 \end{bmatrix} \begin{bmatrix} \mathbf{y} \\ \mu \end{bmatrix} = \begin{bmatrix} \mathbf{g} \\ h \end{bmatrix}, \qquad \mathbf{g} = \mathbf{A}\mathbf{x} - \lambda\mathbf{x}, \qquad h = \tfrac{1}{2}(1 - \mathbf{x}^T\mathbf{x}). \tag{120}$$

Using the first equation in (120) to express $\mathbf{y}$ in terms of μ and then solving for μ, we obtain

$$\mu = \frac{h + \mathbf{x}^T(\mathbf{A} - \lambda\mathbf{I})^{-1}\mathbf{g}}{\mathbf{x}^T(\mathbf{A} - \lambda\mathbf{I})^{-1}\mathbf{x}} \quad \text{and} \quad \mathbf{y} = (\mathbf{A} - \lambda\mathbf{I})^{-1}(\mu\mathbf{x} - \mathbf{g}).$$

When implementing this refinement technique, it is important to evaluate $\mathbf{g}$ and h using high precision, while $\mathbf{y}$ and μ can be computed using low precision. In summary, the following algorithm computes a high-precision approximation to an eigenpair:

1. **LU** factor $\mathbf{A} - \lambda\mathbf{I}$ in low precision.
2. Normalize $\mathbf{x}$ in high precision: $\mathbf{x} \leftarrow \mathbf{x}/\|\mathbf{x}\|_2$.
3. Compute $\mathbf{g} = \mathbf{A}\mathbf{x} - \lambda\mathbf{x}$ in high precision.
4. Using low precision, solve $\mathbf{LUq} = \mathbf{g}$ for the unknown $\mathbf{q}$ and solve $\mathbf{LUp} = \mathbf{x}$ for the unknown $\mathbf{p}$.
5. Using low precision, compute $\mu = \mathbf{x}^T\mathbf{q}/\mathbf{x}^T\mathbf{p}$ and $\mathbf{y} = \mu\mathbf{p} - \mathbf{q}$.
6. Set $\mathbf{x}^{\text{new}} = \mathbf{x} + \mathbf{y}$ and $\lambda^{\text{new}} = \lambda + \mu$ in high precision.
7. Return to step 1 unless the iterations have converged.

(121)

The variables $\mathbf{x}$ and λ in the refinement algorithm are high-precision variables while all the other variables can be low precision. In particular, $\mathbf{g}$ can be low precision, although the expression $\mathbf{Ax} - \lambda\mathbf{x}$ must be evaluated using high-precision arithmetic. In some situations, it is more efficient to return to step 2 after each iteration instead of returning to step 1. The convergence is slower if we return to step 2; however, the repeated factorization of $\mathbf{A} - \lambda\mathbf{I}$ is eliminated.

Exercise

6-13.1. Show that if $\mathbf{D}$ is a diagonal matrix, then

$$\|\mathbf{Dy}\|_p \geqq \text{minimum } \{|d_{11}|, |d_{22}|, \cdots, |d_{nn}|\} \, \|\mathbf{y}\|_p.$$

REVIEW PROBLEMS

6-1. Let $\mathbf{A}$ denote the $n \times n$ symmetric tridiagonal matrix with every diagonal element equal to zero and with every superdiagonal element equal to -1:

$$\mathbf{A} = \begin{bmatrix} 0 & -1 & & & & \\ -1 & 0 & \cdot & & & \\ & \cdot & \cdot & \cdot & & \\ & & \cdot & \cdot & \cdot & \\ & & & \cdot & \cdot & -1 \\ & & & & -1 & 0 \end{bmatrix}.$$

Using TVALS, compute the eigenvalues of $\mathbf{A}$ taking $n = 20$. What symmetry do the eigenvalues of $\mathbf{A}$ possess? If $(\lambda, \mathbf{x})$ is an eigenpair for $\mathbf{A}$, show that $(-\lambda, \mathbf{y})$ is also an eigenpair $\mathbf{A}$ where $y_i = (-1)^i \, \mathbf{x}_i$. Now let $\mathbf{B}$ denote the matrix which is the same as $\mathbf{A}$ except that every diagonal element is equal to 2:

$$\mathbf{B} = \begin{bmatrix} 2 & -1 & & & & \\ -1 & 2 & \cdot & & & \\ & \cdot & \cdot & \cdot & & \\ & & \cdot & \cdot & \cdot & \\ & & & \cdot & \cdot & -1 \\ & & & & -1 & 2 \end{bmatrix}.$$

What symmetry do the eigenvalues of $\mathbf{B}$ possess? Judging from the distribution of eigenvalues, will the power method applied to $\mathbf{B}$ or the inverse power method applied to $\mathbf{B}$ converge faster?

6-2. Let $\mathbf{A}_N$ denote the $(N - 1) \times (N - 1)$ matrix defined by

$$\mathbf{A}_N = N^2 \begin{bmatrix} 2 & -1 & & & & \\ -1 & 2 & \cdot & & & \\ & \cdot & \cdot & \cdot & & \\ & & \cdot & \cdot & \cdot & \\ & & & \cdot & \cdot & -1 \\ & & & & -1 & 2 \end{bmatrix}.$$

Loosely speaking, as N increases, the eigenvalues of $\mathbf{A}_N$ approach the eigenvalues of the boundary-value problem

$$-x''(t) = \lambda x(t), \qquad 0 \leqq t \leqq 1,$$

with the boundary conditions $x(0) = 0$ and $x(1) = 0$. The eigenpairs of this boundary-value problem are

$$x_j(t) = \sin j\pi t \qquad \text{and} \qquad \lambda_j = (j\pi)^2$$

for $j = 1, 2, \cdots$. Using subroutine TVALS, compute the eigenvalues of $\mathbf{A}_N$ when N is 20 and determine which of $\mathbf{A}_N$'s eigenvalues are good approximations to eigenvalues of the boundary-value problem. How do the eigenvalues of $\mathbf{A}_N$ relate to the eigenvalues of the matrix $\mathbf{B}$ in Problem 6-1?

6-3. Consider the generalized eigenproblem $\mathbf{Ax} = \lambda\mathbf{Bx}$, where $\mathbf{A}$ and $\mathbf{B}$ are the $(N-1)\times(N-1)$ matrices defined by

$$\mathbf{A} = N\begin{bmatrix} 2 & -1 & & & \\ -1 & 2 & \cdot & & \\ & \cdot & \cdot & \cdot & \\ & & \cdot & \cdot & \cdot \\ & & & \cdot & \cdot & -1 \\ & & & & -1 & 2 \end{bmatrix} \quad \text{and}$$

$$\mathbf{B} = \frac{1}{6N}\begin{bmatrix} 4 & 1 & & & \\ 1 & 4 & \cdot & & \\ & \cdot & \cdot & \cdot & \\ & & \cdot & \cdot & \cdot \\ & & & \cdot & \cdot & 1 \\ & & & & 1 & 4 \end{bmatrix}.$$

Compute the generalized eigenvalues corresponding to $N = 20$ using subroutine MPOWER and the inverse power method applied to $\mathbf{C} = \mathbf{B}^{-1}\mathbf{A}$. [*Hint:* Initially LU factor $\mathbf{A}$. Each iteration of the inverse power method applied to $\mathbf{B}^{-1}\mathbf{A}$ involves multiplying the matrix $(\mathbf{B}^{-1}\mathbf{A})^{-1} = \mathbf{A}^{-1}\mathbf{B}$ by a vector $\mathbf{x}$. This product is computed in two steps: Compute $\mathbf{y} = \mathbf{Bx}$ then solve $(\mathbf{LU})\mathbf{z} = \mathbf{y}$ for $\mathbf{z}$.] Is there any similarity between the eigenvalues computed in this problem and those computed in Problem 6-2?

6-4. If $\mathbf{X\Lambda X}^{-1}$ is the diagonalization of $\mathbf{A}$, what is the diagonalization of $\mathbf{A}^2$? What is the diagonalization of $\mathbf{A}^p$? How do the eigenvalues and the eigenvectors of $\mathbf{A}^p$ relate to the eigenvalues and the eigenvectors of $\mathbf{A}$?

6-5. One method to compute $\mathbf{A}^p$ is to set $\mathbf{B} = \mathbf{A}$ and then successively multiply $\mathbf{B}$ by $\mathbf{A}$ to obtain $\mathbf{A}^p$:

$$\mathbf{B} \leftarrow \mathbf{AB} \qquad \text{for } j = 1 \text{ to } p - 1.$$

If $\mathbf{A}$ is $n \times n$, how many times are two numbers multiplied together when computing $\mathbf{A}^p$ by this procedure? Now consider the following alternative method: We compute the diagonalization of $\mathbf{A}$ and we apply the formula for the diagonalization of $\mathbf{A}^p$ derived in Problem 6-4. Excluding the cost of computing the diagonalization of $\mathbf{A}$, how many multiplications are needed to compute $\mathbf{A}^p$ by this method?

6-6. If σ denotes some arbitrary scalar, what are the eigenpairs of the following 2×2 matrix?

$$\begin{bmatrix} 0 & \sigma \\ \sigma & 0 \end{bmatrix}$$

If $\mathbf{S}$ denotes a square matrix with every element equal to zero except for the diagonal elements s_{ii}, what are the eigenpairs of the following matrix?

$$\begin{bmatrix} \mathbf{0} & \mathbf{S}^T \\ \mathbf{S} & \mathbf{0} \end{bmatrix} \tag{122}$$

If $\mathbf{S}$ is a rectangular $m \times n$ matrix with $n > m$ and with every element equal to zero except for the diagonal elements s_{ii}, what are the eigenpairs of the matrix (122)?

6-7. Suppose that a matrix $\mathbf{B}$ has the following structure:

$$\mathbf{B} = \begin{bmatrix} \mathbf{0} & \mathbf{A}^T \\ \mathbf{A} & \mathbf{0} \end{bmatrix},$$

where $\mathbf{A}$ is $m \times n$ and $n > m$. Letting $\mathbf{QSP}^T$ denote the singular value decomposition of $\mathbf{A}$, express the eigenpairs of $\mathbf{B}$ in terms of $\mathbf{Q}$, $\mathbf{S}$, and $\mathbf{P}$. *Hint:* Consider the similarity transformation

$$\begin{bmatrix} \mathbf{P}^T & \mathbf{0} \\ \mathbf{0} & \mathbf{Q}^T \end{bmatrix} \begin{bmatrix} \mathbf{0} & \mathbf{A}^T \\ \mathbf{A} & \mathbf{0} \end{bmatrix} \begin{bmatrix} \mathbf{P} & \mathbf{0} \\ \mathbf{0} & \mathbf{Q} \end{bmatrix}.$$

6-8. Consider the equation

$$\mathbf{AX} + \mathbf{XB} = \mathbf{C},$$

where $\mathbf{X}$ is an unknown square matrix and $\mathbf{A}$, $\mathbf{B}$, and $\mathbf{C}$ are given matrices. If $\mathbf{PDP}^{-1}$ is the diagonalization of $\mathbf{A}$ and $\mathbf{QEQ}^{-1}$ is the diagonalization of $\mathbf{B}$, show that $\mathbf{X} = \mathbf{PYQ}^{-1}$, where

$$y_{ij} = \frac{(\mathbf{P}^{-1}\mathbf{CQ})_{ij}}{d_{ii} + e_{jj}}.$$

6-9. Write a symbolic program to compute $\mathbf{L}^{-1}\mathbf{AL}^{-T}$, where $\mathbf{L}$ is an invertible lower triangular matrix and $\mathbf{A}$ is a symmetric matrix. Since $\mathbf{A}$ is symmetric, we can write $\mathbf{A} = \mathbf{B} + \mathbf{B}^T$, where $\mathbf{B}$ is lower triangular, and the product $\mathbf{L}^{-1}\mathbf{AL}^{-T}$ can be expressed

$$\mathbf{L}^{-1}\mathbf{AL}^{-T} = \mathbf{L}^{-1}\mathbf{BL}^{-T} + \mathbf{L}^{-1}\mathbf{B}^T\mathbf{L}^{-T}.$$

Since $\mathbf{L}^{-1}\mathbf{BL}^{-T}$ is the transpose of $\mathbf{L}^{-1}\mathbf{B}^T\mathbf{L}^{-T}$, only $\mathbf{L}^{-1}\mathbf{BL}^{-T}$ needs to be computed. In your program, first compute the product $\mathbf{L}^{-1}\mathbf{B}$, which is lower triangular, and then form the product $(\mathbf{L}^{-1}\mathbf{B})\mathbf{L}^{-T}$.

6-10. The matrix $\mathbf{A}$ is **Hermitian** if it is equal to its conjugate transpose $\mathbf{A}^*$. Show that the eigenvalues of a Hermitian matrix are real.

6-11. If $\mathbf{A}$ is a Hermitian tridiagonal matrix, give an algorithm for generating the diagonal elements of a diagonal matrix $\mathbf{D}$ with the following properties: $\mathbf{DD}^* = \mathbf{I}$ and $\mathbf{D}^*\mathbf{AD}$ is a real symmetric tridiagonal matrix.

6-12. Show that every eigenvalue of a real symmetric positive definite matrix is positive. (*Hint:* See Exercise 2-4.12.)

6-13. If $\mathbf{A}$ is a symmetric positive definite matrix and the columns of $\mathbf{B}$ are linearly independent, show that the matrix $\mathbf{B}^T\mathbf{AB}$ is symmetric and positive definite.

6-14. Given a symmetric positive definite matrix $\mathbf{A}$, construct a symmetric positive definite matrix $\mathbf{B}$ with the property that $\mathbf{B}^2 = \mathbf{A}$. The matrix $\mathbf{B}$ that you construct, called the square root of $\mathbf{A}$, is often denoted $\mathbf{A}^{\frac{1}{2}}$. (*Hint:* Consider an orthogonal diagonalization of $\mathbf{A}$ and take the square root of the diagonal elements in the diagonal factor.)

6-15. Suppose that $\mathbf{A}$ and $\mathbf{B}$ are symmetric and positive definite. Show that every eigenvalue of $\mathbf{AB}$ is positive. (*Hint:* Consider a similarity transformation involving $\mathbf{B}^{\frac{1}{2}}$.)

6-16. Let $\mathbf{A}$ and $\mathbf{B}$ denote real matrices and define the complex matrix $\mathbf{C} = \mathbf{A} + \mathbf{B}i$. If $\mathbf{QSP}^*$ is the singular value decomposition of $\mathbf{C}$, both $\mathbf{Q}$ and $\mathbf{P}$ are complex unitary matrices while the singular values are real. Letting $\mathbf{Q}_R$ and $\mathbf{Q}_I$ denote the real and the imaginary parts of $\mathbf{Q}$ and letting $\mathbf{P}_R$ and $\mathbf{P}_I$ denote the real and the imaginary parts of $\mathbf{P}$, show that

$$\begin{bmatrix} \mathbf{A} & -\mathbf{B} \\ \mathbf{B} & \mathbf{A} \end{bmatrix} = \begin{bmatrix} \mathbf{Q}_R & -\mathbf{Q}_I \\ \mathbf{Q}_I & \mathbf{Q}_R \end{bmatrix} \begin{bmatrix} \mathbf{S} & \mathbf{0} \\ \mathbf{0} & \mathbf{S} \end{bmatrix} \begin{bmatrix} \mathbf{P}_R & -\mathbf{P}_I \\ \mathbf{P}_I & \mathbf{P}_R \end{bmatrix}^T. \tag{123}$$

Verify that both the first and the last factor on the right side of (123) are orthogonal matrices. Thus the singular value decomposition of the matrix on the left side of (123) can be expressed in terms of the singular value decomposition of the complex matrix $\mathbf{C}$.

REFERENCES

The classic reference for the numerical treatment of eigenvalue problems is Wilkinson's book [W6]. The book [G15] by Golub and Van Loan and the book [S14] by Stewart also provide an excellent treatment of eigenproblems while Parlett gives a lively treatment of symmetric eigenproblems in [P4]. An excellent discussion of Gerschgorin's theorem and related issues is found in the book [H16] by Horn and Johnson. The convergence of the Rayleigh quotient iteration is studied on page 636 of [W6] and in the article [P3]. The block power method (mentioned in Section 6-4) is an alternative to the deflation strategy developed in Section 6-5 (see ACM algorithm 570). The eigenvalues of a real matrix are typically computed using Francis' implicit double-shift version of the **QR** algorithm. Note though that for a $n \times n$ matrix, one iteration of the Francis scheme requires about $6n^2$ real multiplications. On the other hand, if an explicit **QR** scheme is used and if the factorizations are implemented using the fast Givens scheme, then two iterations require about $4n^2$ complex multiplications. Therefore, if hardware is available to multiply complex numbers, the complex fast Givens approach is competitive with the real Francis scheme. A procedure to reduce a symmetric band matrix to tridiagonal form is developed by Schwarz in [S8]. The Lanczos process for a nonsymmetric matrix is an area of active research. Some references include [C8], [C10], and [P7]. The diagonalization algorithm (119) may be useful in the context of the continuation method. That is, if the coefficients of a matrix depend on a parameter and the diagonalization has been computed for one value of the parameter, (119) can be used to determine the diagonalization for neighboring parameter values. Also, note that (119) is well suited for either a vector or a parallel processor. For more details concerning algorithm (119), see [H7]. With regard to the refinement algorithm (121), another method to refine eigenpairs iteratively is developed by Dongarra, Moler, and Wilkinson in [D5] and [D9].

7

Iterative Methods

7-1. JUSTIFICATION

Many important practical problems give rise to large systems of linear equations. Even though the physical problem may involve just a few parameters, the discretization process can lead to equations with many unknowns. For example, suppose that we wish to model the temperature in a three-dimensional rectangular region. Assuming that the temperature depends on position in the region, we can approximate the continuously varying temperature field by introducing a mesh and by associating a temperature with each mesh point. If each side of the rectangular region is divided into 101 pieces and if a mesh is constructed throughout the interior of the region, there are $100 \times 100 \times 100 = 1{,}000{,}000$ mesh points altogether. Even though there is just one physical parameter, the temperature, there are a million numerical parameters, the temperature at each mesh point.

For some large linear systems, Gaussian elimination is impractical. To illustrate this property, let us examine how the time to factor a general matrix depends on the matrix dimension n. Referring to Table 2-13, the asymptotic time to factor a matrix is $\frac{1}{3}Cn^3$, where C is the loop cycle time. Since C is .65 microsecond for an IBM 370 model 3090 computer (see Table 2-14), the asymptotic running time grows with the matrix dimension in the following way:

Dimension	Time
100	.2 second
1,000	3.6 minutes
10,000	2.5 days

Thus on an IBM 370 model 3090 computer, factoring a square matrix with more than 1000 rows can take a long time.

In contrast, the cycle time for the Radio Shack TRS 80 model III microcomputer is .016 second and FACT's asymptotic time is $.016n^3/3$ seconds. For matrices with dimensions ranging from 100 to 10,000, we obtain the following times:

Dimension	Time
100	89 minutes
1,000	62 days
10,000	169 years

In this case, 100 equations is a big problem. In summary, as n grows we eventually reach a point, which depends on the computer's speed, where Gaussian elimination is impractical. This chapter discusses another way to solve a linear system: Iterative methods.

An iterative method generates a sequence $\mathbf{x}_1, \mathbf{x}_2, \mathbf{x}_3, \cdots$ converging to the solution of the linear system. The sequence is terminated when the error is sufficiently small. If each iteration is faster than elimination and if the sequence converges quickly, then the iterative approach can be faster than elimination. A fundamental difference between these two approaches is that with exact arithmetic, Gaussian elimination produces the exact solution, while an iterative method almost never produces the exact solution. On the other hand, in the real world, the solution computed by elimination is polluted with rounding errors so the solution computed by elimination is inexact as well.

7-2. SPLITTING TECHNIQUES

Given the system $\mathbf{Ax} = \mathbf{b}$ and the approximate solution $\bar{\mathbf{x}}$, let us determine the correction $\mathbf{e}$ which can be added to $\bar{\mathbf{x}}$ to obtain $\mathbf{x}$. Recall that the residual corresponding to $\bar{\mathbf{x}}$ is defined by

$$\mathbf{r} = \mathbf{b} - \mathbf{A}\bar{\mathbf{x}}. \tag{1}$$

Replacing $\mathbf{b}$ by $\mathbf{Ax}$, it follows that

$$\mathbf{A}(\mathbf{x} - \bar{\mathbf{x}}) = \mathbf{r}.$$

If $\mathbf{e}$ denotes the difference $\mathbf{x} - \bar{\mathbf{x}}$, then (1) implies that $\mathbf{Ae} = \mathbf{r}$. The identity $\mathbf{e} = \mathbf{x} - \bar{\mathbf{x}}$ tells us that $\mathbf{x} = \bar{\mathbf{x}} + \mathbf{e}$. Combining these observations, $\mathbf{x}$ is equal to $\bar{\mathbf{x}}$ plus the correction $\mathbf{e}$, where $\mathbf{e}$ is the solution to $\mathbf{Ae} = \mathbf{r}$. In iterative methods to solve $\mathbf{Ax} = \mathbf{b}$, we start with an approximation to $\mathbf{x}$ and we form the corresponding residual $\mathbf{r}$; however, instead of solving $\mathbf{Ae} = \mathbf{r}$ for the correction $\mathbf{e}$, we solve $\mathbf{Se} = \mathbf{r}$, where $\mathbf{S}$ is an approximation to $\mathbf{A}$ which is easier to factor than $\mathbf{A}$ itself. The solution to $\mathbf{Se} = \mathbf{r}$ is an approximation to the true correction. Adding this approximate correction to the approximate $\mathbf{x}$ gives what we hope is a better estimate for the true solution $\mathbf{x}$. In other words, if $\mathbf{x}^{\text{old}}$ is the current approximation to $\mathbf{x}$, the new approximation $\mathbf{x}^{\text{new}}$ is evaluated in three steps:

(a) Compute the residual $\mathbf{r} = \mathbf{b} - \mathbf{A}\mathbf{x}^{\text{old}}$. (b) Solve $\mathbf{S}\mathbf{e} = \mathbf{r}$ for the unknown $\mathbf{e}$. (c) Set $\mathbf{x}^{\text{new}} = \mathbf{x}^{\text{old}} + \mathbf{e}$.	(2)

The three steps of algorithm (2) are now reduced to one step. Multiplying the relation $\mathbf{x}^{\text{new}} = \mathbf{x}^{\text{old}} + \mathbf{e}$ of step (c) by $\mathbf{S}$ yields

$$\mathbf{S}\mathbf{x}^{\text{new}} = \mathbf{S}\mathbf{x}^{\text{old}} + \mathbf{S}\mathbf{e}.$$

Since $\mathbf{S}\mathbf{e} = \mathbf{r} = \mathbf{b} - \mathbf{A}\mathbf{x}^{\text{old}}$, we have

$$\mathbf{S}\mathbf{x}^{\text{new}} = \mathbf{S}\mathbf{x}^{\text{old}} + \mathbf{b} - \mathbf{A}\mathbf{x}^{\text{old}},$$

or equivalently,

$$\mathbf{S}\mathbf{x}^{\text{new}} = (\mathbf{S}-\mathbf{A})\mathbf{x}^{\text{old}} + \mathbf{b}.$$

Letting $\mathbf{T}$ denote $\mathbf{S} - \mathbf{A}$, the iteration becomes

$$\mathbf{S}\mathbf{x}^{\text{new}} = \mathbf{T}\mathbf{x}^{\text{old}} + \mathbf{b}. \tag{3}$$

The solution $\mathbf{x}^{\text{new}}$ to equation (3) is the same as the $\mathbf{x}^{\text{new}}$ generated by algorithm (2). Since the original coefficient matrix $\mathbf{A}$ is equal to the difference $\mathbf{S} - \mathbf{T}$, the matrices $\mathbf{S}$ and $\mathbf{T}$ are often called splitting matrices. Note that if the iterations converge to the limit $\mathbf{x}$, then $\mathbf{x}^{\text{new}} = \mathbf{x}^{\text{old}} = \mathbf{x}$ and by (3), $\mathbf{S}\mathbf{x} = \mathbf{T}\mathbf{x} + \mathbf{b}$. Moving the $\mathbf{T}\mathbf{x}$ term to the left side of this equation yields $(\mathbf{S} - \mathbf{T})\mathbf{x} = \mathbf{b}$. Since $\mathbf{S} - \mathbf{T} = \mathbf{A}$, we obtain the original equation $\mathbf{A}\mathbf{x} = \mathbf{b}$. In summary, a limit of iteration (3) must satisfy the original equation $\mathbf{A}\mathbf{x} = \mathbf{b}$. Three classical iterative schemes are now discussed.

7-2.1 Jacobi Method

In Jacobi's scheme, $\mathbf{S}$ is the diagonal matrix formed from $\mathbf{A}$'s diagonal elements. For the system

$$\underset{\mathbf{A}}{\begin{bmatrix} 2 & 1 & 0 \\ 1 & 3 & 1 \\ 0 & 1 & 2 \end{bmatrix}} \underset{\mathbf{x}}{\begin{bmatrix} x_1 \\ x_2 \\ x_3 \end{bmatrix}} = \underset{\mathbf{b}}{\begin{bmatrix} 7 \\ 8 \\ 5 \end{bmatrix}}, \tag{4}$$

Jacobi's choice is

$$\mathbf{S} = \begin{bmatrix} 2 & 0 & 0 \\ 0 & 3 & 0 \\ 0 & 0 & 2 \end{bmatrix}. \tag{5}$$

Hence, the matrix $\mathbf{T} = \mathbf{S} - \mathbf{A}$ is given by

$$\mathbf{T} = \begin{bmatrix} 2 & 0 & 0 \\ 0 & 3 & 0 \\ 0 & 0 & 2 \end{bmatrix} - \begin{bmatrix} 2 & 1 & 0 \\ 1 & 3 & 1 \\ 0 & 1 & 2 \end{bmatrix} = -\begin{bmatrix} 0 & 1 & 0 \\ 1 & 0 & 1 \\ 0 & 1 & 0 \end{bmatrix}, \tag{6}$$

and the iterations (3) are

$$\begin{bmatrix} 2 & 0 & 0 \\ 0 & 3 & 0 \\ 0 & 0 & 2 \end{bmatrix} \begin{bmatrix} x_1^{\text{new}} \\ x_2^{\text{new}} \\ x_3^{\text{new}} \end{bmatrix} = \begin{bmatrix} 7 \\ 8 \\ 5 \end{bmatrix} - \begin{bmatrix} 0 & 1 & 0 \\ 1 & 0 & 1 \\ 0 & 1 & 0 \end{bmatrix} \begin{bmatrix} x_1^{\text{old}} \\ x_2^{\text{old}} \\ x_3^{\text{old}} \end{bmatrix}.$$

This matrix-vector system is equivalent to three separate equations,

$$2x_1^{\text{new}} = 7 - x_2^{\text{old}},$$
$$3x_2^{\text{new}} = 8 - x_1^{\text{old}} - x_3^{\text{old}},$$
$$2x_3^{\text{new}} = 5 - x_2^{\text{old}}.$$

Dividing these equations by 2, 3, and 2, respectively, we have

$$x_1^{\text{new}} = \frac{7 - x_2^{\text{old}}}{2},$$
$$x_2^{\text{new}} = \frac{8 - x_1^{\text{old}} - x_3^{\text{old}}}{3}, \tag{7}$$
$$x_3^{\text{new}} = \frac{5 - x_2^{\text{old}}}{2}.$$

Observe that each numerator in equation (7) is equal to a component of $\mathbf{b}$ minus the product between the components of $\mathbf{x}^{\text{old}}$ and the off-diagonal elements in the corresponding row of the coefficient matrix. For a general system $\mathbf{Ax} = \mathbf{b}$ of n equations, the Jacobi iteration can be stated:

$$x_i^{\text{new}} = \frac{b_i - \sum_{j=1}^{i-1} a_{ij}x_j^{\text{old}} - \sum_{j=i+1}^{n} a_{ij}x_j^{\text{old}}}{a_{ii}} \quad \text{for } i = 1 \text{ to } n. \tag{8}$$

The first sum in (8) is the product between coefficients to the left of the main diagonal and the corresponding components of $\mathbf{x}^{\text{old}}$, while the second sum in (8) is the product between coefficients to the right of the main diagonal and the corresponding components of $\mathbf{x}^{\text{old}}$. A row-oriented code for Jacobi iterations appears in Table 7-1. You formulate the corresponding column-oriented program in Exercise 7-2.1. The loop ending at statement 20 of Table 7-1 computes the first sum in (8) while the loop ending at statement 40 computes the second sum in (8). Statement 50 evaluates x_i^{new}. In many applications, $\mathbf{A}$ is a large sparse matrix with special structure and the sums in (8) are computed by a special

TABLE 7-1 JACOBI ITERATIONS

```
10    DO 50 I = 1,N
         T = 0.
         K = I - 1
         IF ( K .EQ. 0 ) GOTO 30
         DO 20 J = 1,K
20          T = T + A(I,J)*OLD(J)
30       K = I + 1
         IF ( K .GT. N ) GOTO 50
         DO 40 J = K,N
40          T = T + A(I,J)*OLD(J)
50       NEW(I) = (B(I) - T)/A(I,I)
      CALL UPDATE(DIF,SIZE,NEW,OLD,N)
      CALL STOPIT(DIF,SIZE,NDIGIT,LIMIT)
      IF ( DIF .GT. 0. ) GOTO 10
```

routine that takes advantage of the matrix structure. For example, if **A** is a band matrix, the zero coefficients outside the central bands can be skipped when the sums in (8) are evaluated.

Table 7-2 shows the Jacobi iteration for equation (4) starting from $\mathbf{x} = \mathbf{0}$. Observe that the iterations are converging to the solution $x_1 = 3$, $x_2 = 1$, and $x_3 = 2$. Like all iterative methods, Jacobi's scheme is not guaranteed to converge, but for many matrices that arise in practice, it will converge. The fundamental principles governing the convergence of iterative methods are developed in Section 7-3.

7-2.2 Gauss-Seidel Method

The Jacobi scheme (8) uses $\mathbf{x}^{\text{old}}$ to compute $x_1^{\text{new}}, x_2^{\text{new}}, \cdots, x_n^{\text{new}}$. Note though that when x_i^{new} is computed using (8), we already have values for $x_1^{\text{new}}, \cdots, x_{i-1}^{\text{new}}$. A simple alteration of the Jacobi scheme is to replace x_j^{old} by x_j^{new} in the first sum of (8), giving us the iteration

$$x_i^{\text{new}} = \frac{b_i - \sum_{j=1}^{i-1} a_{ij}x_j^{\text{new}} - \sum_{j=i+1}^{n} a_{ij}x_j^{\text{old}}}{a_{ii}} \quad \text{for } i = 1 \text{ to } n. \tag{9}$$

This new iteration, called Gauss-Seidel, is implemented by the code in Table 7-3. To obtain this Gauss-Seidel code, we simply replaced the array OLD in statement 20 of Table 7-1 by the array NEW.

The storage structure in Table 7-3 (on page 324) can be simplified using a single array BOTH to store the elements of NEW in locations 1 through I − 1 and the elements of OLD in locations I + 1 through N. One iteration of the simplified program appears in Table 7-4 (on page 324). To complete the code in Table 7-4, we must append statements to compute the iteration difference and to test for convergence (see Exercise 7-2-3). Table 7-5 (on page 325) gives the Gauss-Seidel iterations corresponding to equation (4) and the initial guess $\mathbf{x} = \mathbf{0}$. Although there is no general rule relating the convergence speed of Gauss-Seidel

TABLE 7-2 JACOBI ITERATIONS FOR EQUATION (4)

Iteration	x_1	x_2	x_3
1	0.00	0.00	0.00
2	3.50	2.67	2.50
3	2.17	0.67	1.17
4	3.17	1.56	2.17
5	2.72	0.89	1.72
6	3.06	1.19	2.06
7	2.91	0.96	1.91
8	3.02	1.06	2.02
9	2.97	0.99	1.97
10	3.01	1.02	2.01
11	2.99	1.00	1.99
12	3.00	1.01	2.00
13	3.00	1.00	2.00

TABLE 7-3 GAUSS-SEIDEL ITERATIONS

```
10    DO 50 I = 1,N
         T = 0.
         K = I - 1
         IF ( K .EQ. 0 ) GOTO 30
         DO 20 J = 1,K
20          T = T + A(I,J)*NEW(J)
30       K = I + 1
         IF ( K .GT. N ) GOTO 50
         DO 40 J = K,N
40          T = T + A(I,J)*OLD(J)
50       NEW(I) = (B(I) - T)/A(I,I)
      CALL UPDATE(DIF,SIZE,NEW,OLD,N)
      CALL STOPIT(DIF,SIZE,NDIGIT,LIMIT)
      IF ( DIF .GT. 0. ) GOTO 10
```

and Jacobi iterations, it turns out that for equation (4) and in many other applications, the Gauss-Seidel scheme is faster than the Jacobi scheme.

Referring to equation (3), we now derive the matrix **S**, which corresponds to the Gauss-Seidel scheme. Multiplying equation (9) by a_{ii} and moving the x_j^{new} terms to the left side, we have

$$a_{ii}x_i^{\text{new}} + \sum_{j=1}^{i-1} a_{ij}x_j^{\text{new}} = b_i - \sum_{j=i+1}^{n} a_{ij}x_j^{\text{old}} \qquad \text{for } i = 1 \text{ to } n.$$

Changing the lower limit of the first sum from $i - 1$ to i and dropping the $a_{ii}x_i^{\text{new}}$ term yields

$$\sum_{j=1}^{i} a_{ij}x_j^{\text{new}} = b_i - \sum_{j=i+1}^{n} a_{ij}x_j^{\text{old}} \qquad \text{for } i = 1 \text{ to } n. \tag{10}$$

For j between 1 and i, a_{ij} lies in row i to the left of the main diagonal. For j between $i + 1$ and n, a_{ij} lies in row i to the right of the main diagonal. Thus

TABLE 7-4 COMPACT GAUSS-SEIDEL ITERATION

```
      DO 50 I = 1,N
         T = 0.
         K = I - 1
         IF ( K .EQ. 0 ) GOTO 30
         DO 20 J = 1,K
20          T = T + A(I,J)*BOTH(J)
30       K = I + 1
         IF ( K .GT. N ) GOTO 50
         DO 40 J = K,N
40          T = T + A(I,J)*BOTH(J)
50       BOTH(I) = (B(I) - T)/A(I,I)
```

TABLE 7-5 GAUSS-SEIDEL ITERATIONS FOR EQUATION (4)

Iteration	x_1	x_2	x_3
1	0.00	0.00	0.00
2	3.50	1.50	1.75
3	2.75	1.17	1.92
4	2.92	1.06	1.97
5	2.97	1.02	1.99
6	2.99	1.01	2.00
7	3.00	1.00	2.00

when n is 3, the matrix-vector analog of equation (10) is

$$\underset{\mathbf{S}}{\begin{bmatrix} a_{11} & 0 & 0 \\ a_{21} & a_{22} & 0 \\ a_{31} & a_{32} & a_{33} \end{bmatrix}} \underset{\mathbf{x}^{\text{new}}}{\begin{bmatrix} x_1^{\text{new}} \\ x_2^{\text{new}} \\ x_3^{\text{new}} \end{bmatrix}} = \underset{\mathbf{b}}{\begin{bmatrix} b_1 \\ b_2 \\ b_3 \end{bmatrix}} - \underset{\mathbf{T}}{\begin{bmatrix} 0 & a_{12} & a_{13} \\ 0 & 0 & a_{23} \\ 0 & 0 & 0 \end{bmatrix}} \underset{\mathbf{x}^{\text{old}}}{\begin{bmatrix} x_1^{\text{old}} \\ x_2^{\text{old}} \\ x_3^{\text{old}} \end{bmatrix}}. \tag{11}$$

In other words, for the Gauss-Seidel iteration, the coefficient matrix $\mathbf{S}$ of $\mathbf{x}^{\text{new}}$ is formed from elements that are on the diagonal or beneath the diagonal of $\mathbf{A}$ while the coefficient matrix $\mathbf{T}$ of $\mathbf{x}^{\text{old}}$ is formed from elements above $\mathbf{A}$'s diagonal. Since Jacobi's $\mathbf{S}$ is formed from $\mathbf{A}$'s diagonal, a small change in the Jacobi iteration has caused a major change in the splitting matrix $\mathbf{S}$.

7-2.3 Relaxation Schemes

The Gauss-Seidel iteration (9) can be written

$$x_i^{\text{new}} = x_i^{\text{old}} + \frac{b_i - \sum_{j=1}^{i-1} a_{ij} x_j^{\text{new}} - \sum_{j=i}^{n} a_{ij} x_j^{\text{old}}}{a_{ii}} \qquad \text{for } i = 1 \text{ to } n.$$

Relaxation schemes insert a parameter ω to obtain the iteration

$$x_i^{\text{new}} = x_i^{\text{old}} + \frac{\omega\left(b_i - \sum_{j=1}^{i-1} a_{ij} x_j^{\text{new}} - \sum_{j=i}^{n} a_{ij} x_j^{\text{old}}\right)}{a_{ii}} \qquad \text{for } i = 1 \text{ to } n. \tag{12}$$

Setting $\omega = 1$, we have the Gauss-Seidel scheme. When $\omega > 1$, the scheme is called successive overrelaxation or SOR. When $\omega < 1$, the scheme is called successive underrelaxation. In Table 7-6 we modify the compact Gauss-Seidel code of Table 7-4 to perform one relaxed iteration.

In the same way that we derived the Gauss-Seidel $\mathbf{S}$, it can be shown that the $\mathbf{S}$ for the relaxed iteration corresponding to a 3×3 matrix is given by

$$\mathbf{S} = \begin{bmatrix} \sigma a_{11} & 0 & 0 \\ a_{21} & \sigma a_{22} & 0 \\ a_{31} & a_{32} & \sigma a_{33} \end{bmatrix},$$

where $\sigma = 1/\omega$ is the reciprocal of the relaxation parameter. Thus the $\mathbf{S}$ for a relaxed iteration is the Gauss-Seidel $\mathbf{S}$ with its diagonal multiplied by σ. For

TABLE 7-6 ONE RELAXED ITERATION

```
      DO 50 I = 1,N
         T = 0.
         K = I - 1
         IF ( K .EQ. 0 ) GOTO 30
         DO 20 J = 1,K
20          T = T + A(I,J)*BOTH(J)
30       K = I + 1
         IF ( K .GT. N ) GOTO 50
         DO 40 J = K,N
40          T = T + A(I,J)*BOTH(J)
50       BOTH(I) = (1-W)*BOTH(I) + W*(B(I) - T)/A(I,I)
```

some values of ω, the relaxed scheme is faster than the Gauss-Seidel scheme. In particular, applying SOR to equation (4) with $\omega = 1.10102$, we obtain the rapid convergence observed in Table 7-7. In contrast, we see in Table 7-8 that the iterations diverge when ω is 3. The convergence speed depends on the value of ω and not every ω works. Techniques for selecting ω are discussed in the next section.

Although the Jacobi scheme is slower than Gauss-Seidel or SOR in many applications, Jacobi's $\mathbf{S}$ has one attractive feature: It is symmetric. In some applications, this symmetry can be exploited to accelerate convergence. To correct for the lack of symmetry in SOR, the symmetric successive overrelaxation scheme (or SSOR) was developed. For a 3×3 matrix, the splitting matrix $\mathbf{S}$ corresponding to SSOR is

$$\mathbf{S} = \frac{1}{2\sigma - 1}\begin{bmatrix} \sigma a_{11} & 0 & 0 \\ a_{21} & \sigma a_{22} & 0 \\ a_{31} & a_{32} & \sigma a_{33} \end{bmatrix}\begin{bmatrix} a_{11}^{-1} & 0 & 0 \\ 0 & a_{22}^{-1} & 0 \\ 0 & 0 & a_{33}^{-1} \end{bmatrix}\begin{bmatrix} \sigma a_{11} & a_{12} & a_{13} \\ 0 & \sigma a_{22} & a_{23} \\ 0 & 0 & \sigma a_{33} \end{bmatrix},$$

where $\sigma = 1/\omega$ is the reciprocal of the relaxation parameter. If $\mathbf{A}$ is symmetric, $\mathbf{S}$ is symmetric. Although SSOR preserves symmetry while SOR does not preserve symmetry, SSOR is not widely used due to the computational effort required each iteration and due to convergence limitations.

TABLE 7-7 SOR ITERATIONS FOR (4) WITH $\omega = 1.10102$.

Iteration	x_1	x_2	x_3
1	.000000	.000000	.000000
2	3.853570	1.521767	1.914802
3	2.626534	1.115624	1.944955
4	2.974075	1.018036	1.995632
5	2.992690	1.002464	1.999085
6	2.999382	1.000314	1.999920
7	2.999890	1.000038	1.999987
8	2.999990	1.000005	1.999999
9	2.999999	1.000001	2.000000
10	3.000000	1.000000	2.000000

TABLE 7-8 SOR ITERATIONS FOR (4) WITH $\omega = 3$

Iteration	x_1	x_2	x_3
1	.000000	.000000	.000000
2	10.500000	−2.500000	11.250000
3	−6.750000	8.500000	−27.750000
4	11.250000	7.500000	51.750000
5	−23.250000	−35.500000	−42.750000
6	110.250000	11.500000	75.750000
7	−227.250000	136.500000	−348.750000
8	260.250000	−176.500000	969.750000
9	−245.250000	−363.500000	−1386.750000

7-2.4 Block Methods

In some applications, the unknowns and the coefficients group together in a natural way. For example, the matrix

$$\mathbf{A} = \begin{bmatrix} 4 & 1 & 0 & 0 & 1 \\ 1 & 4 & 0 & 0 & 0 \\ 0 & 0 & 4 & 1 & 0 \\ 0 & 0 & 1 & 4 & 1 \\ 1 & 0 & 0 & 1 & 4 \end{bmatrix}$$

can be **block partitioned** in the following way:

$$\mathbf{A} = \begin{bmatrix} \mathbf{A}_{11} & \mathbf{A}_{12} \\ \mathbf{A}_{21} & \mathbf{A}_{22} \end{bmatrix}, \tag{13}$$

where the four blocks forming **A** are defined by

$$\mathbf{A}_{11} = \begin{bmatrix} 4 & 1 \\ 1 & 4 \end{bmatrix}, \qquad \mathbf{A}_{12} = \begin{bmatrix} 0 & 0 & 1 \\ 0 & 0 & 0 \end{bmatrix},$$

$$\mathbf{A}_{21} = \begin{bmatrix} 0 & 0 \\ 0 & 0 \\ 1 & 0 \end{bmatrix}, \qquad \mathbf{A}_{22} = \begin{bmatrix} 4 & 1 & 0 \\ 1 & 4 & 1 \\ 0 & 1 & 4 \end{bmatrix}.$$

Of course, there are many ways to block partition a $n \times n$ matrix **A**. Given a sequence $n_1, n_2, \cdots, n_p$ of p positive integers which sum to n, **A** can be partitioned into a $p \times p$ block matrix where the element $\mathbf{A}_{ij}$ in row i and column j is the corresponding $n_i \times n_j$ submatrix of **A**. In the partitioning (13), n is 5, p is 2, n_1 is 2, and n_2 is 3. The element $\mathbf{A}_{12}$ of (13) is the $n_1 \times n_2 = 2 \times 3$ submatrix in the upper right corner of the original matrix.

Each class of matrices introduced in Chapter 2 has its block analog. For example, a $p \times p$ block tridiagonal partitioning of **A** has the form

$$\mathbf{A} = \begin{bmatrix} \mathbf{A}_{11} & \mathbf{A}_{12} & & & & \\ \mathbf{A}_{21} & \mathbf{A}_{22} & \cdot & & & \\ & \cdot & \cdot & \cdot & & \\ & & \cdot & \cdot & \cdot & \\ & & & \cdot & \cdot & \mathbf{A}_{p-1,p} \\ & & & & \mathbf{A}_{p,p-1} & \mathbf{A}_{p,p} \end{bmatrix} \tag{14}$$

where each element $\mathbf{A}_{ij}$ is a submatrix of the original $\mathbf{A}$. In the same spirit, each iterative method discussed above has its block analog where the matrix element a_{ij} employed in the description of the scheme is replaced by the block $\mathbf{A}_{ij}$. For example, if $\mathbf{A}$ is partitioned into the 2×2 block matrix (13), the corresponding block Jacobi iteration is

$$\begin{bmatrix} \mathbf{A}_{11} & \mathbf{0} \\ \mathbf{0} & \mathbf{A}_{22} \end{bmatrix} \begin{bmatrix} \mathbf{x}_1^{\text{new}} \\ \mathbf{x}_2^{\text{new}} \end{bmatrix} = \begin{bmatrix} \mathbf{b}_1 \\ \mathbf{b}_2 \end{bmatrix} - \begin{bmatrix} \mathbf{0} & \mathbf{A}_{12} \\ \mathbf{A}_{21} & \mathbf{0} \end{bmatrix} \begin{bmatrix} \mathbf{x}_1^{\text{old}} \\ \mathbf{x}_2^{\text{old}} \end{bmatrix},$$

where the coefficient matrix for $\mathbf{x}^{\text{new}}$ is formed from the diagonal blocks of the partitioned matrix. This matrix-vector system is equivalent to the two equations

$$\mathbf{A}_{11}\mathbf{x}_1^{\text{new}} = \mathbf{b}_1 - \mathbf{A}_{12}\mathbf{x}_2^{\text{old}},$$
$$\mathbf{A}_{22}\mathbf{x}_2^{\text{new}} = \mathbf{b}_2 - \mathbf{A}_{21}\mathbf{x}_1^{\text{old}}.$$

To compute $\mathbf{x}_1^{\text{new}}$ and $\mathbf{x}_2^{\text{new}}$, we must solve two systems of equations. The coefficient matrix in the first system is $\mathbf{A}_{11}$ and the coefficient matrix in the second system is $\mathbf{A}_{22}$. When the structure of the diagonal blocks is simple, the block diagonal system associated with the block Jacobi iteration is easily solved.

Exercises

7-2.1. Modify the program in Table 7-1 to perform column-oriented Jacobi iterations.

7-2.2. Adding and subtracting x_i^{old} on the right side of (8), we obtain

$$x_i^{\text{new}} = x_i^{\text{old}} + \frac{b_i - \sum_{j=1}^{n} a_{ij}x_j^{\text{old}}}{a_{ii}} \quad \text{for } i = 1 \text{ to } n. \tag{15}$$

Write a program to implement the iteration (15). If $\mathbf{A}$ is a large matrix with special structure and if a subroutine is available to compute the matrix-vector product $\mathbf{A}\mathbf{x}^{\text{old}}$, then (15) is easier to implement than (8).

7-2.3. Alter the program in Table 7-4 so that the iteration difference and iteration size are computed while the elements of BOTH are evaluated. Complete the program by testing for convergence and branching to the start of the iteration if the convergence test fails.

7-3. CONVERGENCE

In analyzing the convergence of iterations, we must refer to the individual elements in the iteration sequence. We will let $\mathbf{x}_0$ denote the starting guess, $\mathbf{x}_1$ denote the first iteration, $\mathbf{x}_2$ denote the second iteration and so on. Relation (3) connecting $\mathbf{x}^{\text{new}}$ and $\mathbf{x}^{\text{old}}$ implies that the $(k + 1)$st iterate is related to the kth iterate through the recurrence

$$\mathbf{S}\mathbf{x}_{k+1} = \mathbf{T}\mathbf{x}_k + \mathbf{b}. \tag{16}$$

To begin our analysis of iteration (16), let us consider the special case of one equation. For one equation and one unknown, $\mathbf{S}$, $\mathbf{T}$, and $\mathbf{b}$ are scalars, and we can divide (16) by $\mathbf{S}$ to obtain

$$\mathbf{x}_{k+1} = \frac{\mathbf{T}}{\mathbf{S}}\mathbf{x}_k + \frac{\mathbf{b}}{\mathbf{S}}. \tag{17}$$

If $\mathbf{S} = 10$, $\mathbf{T} = 1$, and $\mathbf{b} = 10$ and if the starting guess $\mathbf{x}_0$ is 0, then (17) generates the sequence $\mathbf{x}_1 = 1$, $\mathbf{x}_2 = 1.11$, $\mathbf{x}_3 = 1.111, \cdots$, which converges to 10/9. On the other hand, if $\mathbf{S} = 1$, $\mathbf{T} = 2$, and $\mathbf{b} = 1$, the iterations diverge: $\mathbf{x}_0 = 0$, $\mathbf{x}_1 = 1$, $\mathbf{x}_2 = 3$, $\mathbf{x}_3 = 7$, and $\mathbf{x}_k = 2^k - 1$. If the k in (17) is replaced by $k - 1$, we obtain $\mathbf{x}_k = (\mathbf{T}/\mathbf{S})\mathbf{x}_{k-1} + \mathbf{b}/\mathbf{S}$. Subtracting this relation from (17) yields

$$\mathbf{x}_{k+1} - \mathbf{x}_k = \frac{\mathbf{T}}{\mathbf{S}}(\mathbf{x}_k - \mathbf{x}_{k-1}),$$

or equivalently, $\mathbf{d}_{k+1} = \mathbf{M}\mathbf{d}_k$ where $\mathbf{d}_k = \mathbf{x}_k - \mathbf{x}_{k-1}$ is the iteration difference and $\mathbf{M} = \mathbf{T}/\mathbf{S}$. In each iteration, the iteration difference is multiplied by the factor $\mathbf{M}$. If the magnitude of $\mathbf{M}$ is less than 1, then the differences decrease in magnitude and the iterations converge. If the magnitude of $\mathbf{M}$ is larger than 1, the differences increase in magnitude and the iterations diverge.

The cobwebs developed in Chapter 1 provide a geometric picture of the iterations. To graph the iteration (17), we draw two lines, the line $\mathbf{y} = \mathbf{x}$ and the line $\mathbf{y} = \mathbf{M}\mathbf{x} + \mathbf{b}/\mathbf{S}$. Beginning at the point with x-coordinate $\mathbf{x}_0$ on the line $\mathbf{y} = \mathbf{x}$, the cobweb moves vertically to the line $\mathbf{y} = \mathbf{M}\mathbf{x} + \mathbf{b}/\mathbf{S}$, then advances horizontally to the line $\mathbf{y} = \mathbf{x}$; from there the cobweb moves vertically to the line $\mathbf{y} = \mathbf{M}\mathbf{x} + \mathbf{b}/\mathbf{S}$ and then advances horizontally to the line $\mathbf{y} = \mathbf{x}$. Four different cobwebs are depicted in Figure 7-1 (on page 330). The iterations converge when $\mathbf{M} = \pm 1/2$, the iterations diverge when $\mathbf{M} = -2$, and the iterations cycle when $\mathbf{M} = -1$.

To analyze the iteration (16) corresponding to a system of equations, we first derive an equation for the error $\mathbf{e}_k = \mathbf{x}_k - \mathbf{x}$. Recall that the coefficient matrix $\mathbf{A}$ is related to the splitting matrices $\mathbf{S}$ and $\mathbf{T}$ by $\mathbf{A} = \mathbf{S} - \mathbf{T}$. Thus the original equation $\mathbf{A}\mathbf{x} = \mathbf{b}$ is equivalent to $(\mathbf{S} - \mathbf{T})\mathbf{x} = \mathbf{b}$. Moving the $\mathbf{T}\mathbf{x}$ term to the right side, we have

$$\mathbf{S}\mathbf{x} = \mathbf{T}\mathbf{x} + \mathbf{b}. \tag{18}$$

Subtracting (18) from (16) yields

$$\mathbf{S}(\mathbf{x}_{k+1} - \mathbf{x}) = \mathbf{T}(\mathbf{x}_k - \mathbf{x}).$$

It follows that the error at step $k + 1$ and the error at step k satisfy

$$\mathbf{S}\mathbf{e}_{k+1} = \mathbf{T}\mathbf{e}_k.$$

Finally, multiplying by $\mathbf{S}^{-1}$, we have $\mathbf{e}_{k+1} = \mathbf{S}^{-1}\mathbf{T}\mathbf{e}_k$. Letting $\mathbf{M}$ denote the matrix $\mathbf{S}^{-1}\mathbf{T}$, the error satisfies the recurrence

$$\mathbf{e}_{k+1} = \mathbf{M}\mathbf{e}_k. \tag{19}$$

Inserting $k = 0$, $k = 1$, and $k = 2$ in (19), we obtain the relations $\mathbf{e}_1 = \mathbf{M}\mathbf{e}_0$, $\mathbf{e}_2 = \mathbf{M}\mathbf{e}_1$, and $\mathbf{e}_3 = \mathbf{M}\mathbf{e}_2$. Combining these relations, it follows that

$$\mathbf{e}_2 = \mathbf{M}\mathbf{e}_1 = \mathbf{M}(\mathbf{M}\mathbf{e}_0) = \mathbf{M}^2\mathbf{e}_0,$$

$$\mathbf{e}_3 = \mathbf{M}\mathbf{e}_2 = \mathbf{M}(\mathbf{M}^2\mathbf{e}_0) = \mathbf{M}^3\mathbf{e}_0.$$

The error at each iteration equals $\mathbf{M}$ raised to the iteration number times the initial error:

$$\mathbf{e}_k = \mathbf{M}^k\mathbf{e}_0. \tag{20}$$

If $\mathbf{M}^k$ approaches zero as k increases, $\mathbf{e}_k$ approaches zero as k increases. If $\mathbf{M}^k$ diverges as k increases, then except for special choices of $\mathbf{e}_0$ (like $\mathbf{e}_0 = \mathbf{0}$), $\mathbf{e}_k$ will diverge. Since the norm of a product is less than or equal to the product

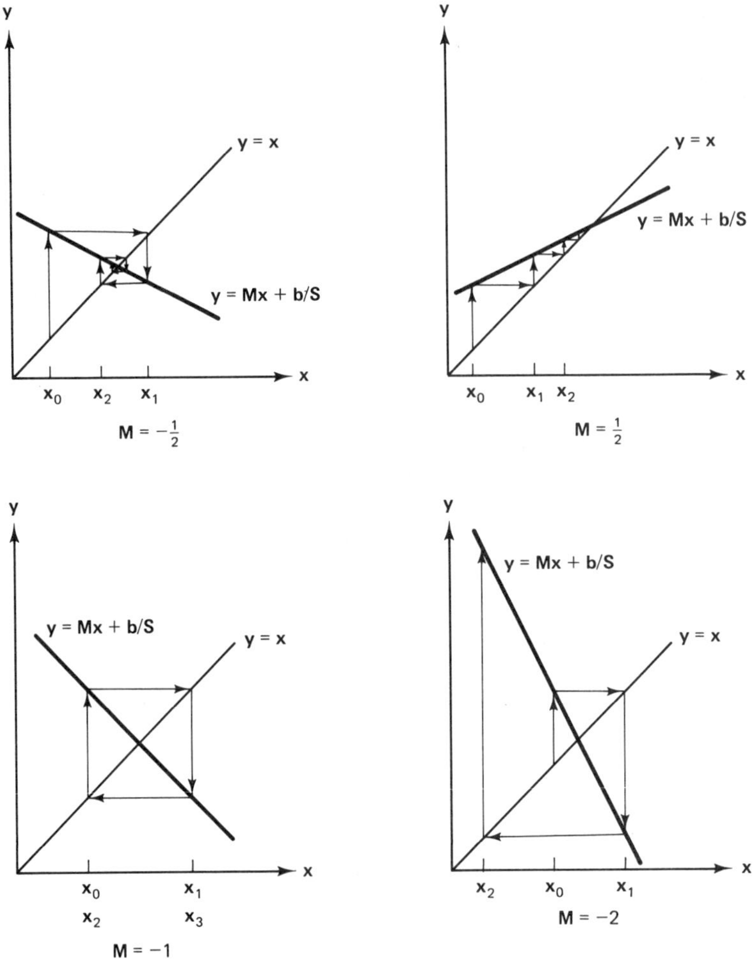

Figure 7-1 Iteration cobwebs.

of the norms, the norm of $\mathbf{e}_k$ satisfies the inequalities

$$\|\mathbf{e}_k\| \leqq \|\mathbf{M}^k\| \|\mathbf{e}_0\| \leqq \|\mathbf{M}\|^k \|\mathbf{e}_0\|. \tag{21}$$

If $\|\mathbf{M}\| < 1$, then $\|\mathbf{M}\|^k$ goes to zero as k increases and by (21), the norm of the error goes to zero as k increases.

Let us apply (21) to Jacobi's iteration for system (4). By (5) and (6), the corresponding $\mathbf{M}$ is

$$\mathbf{M} = \mathbf{S}^{-1}\mathbf{T} = -\begin{bmatrix} 2 & 0 & 0 \\ 0 & 3 & 0 \\ 0 & 0 & 2 \end{bmatrix}^{-1} \begin{bmatrix} 0 & 1 & 0 \\ 1 & 0 & 1 \\ 0 & 1 & 0 \end{bmatrix}.$$

To invert a diagonal matrix, we invert the diagonal elements, giving us

$$\mathbf{M} = -\begin{bmatrix} \frac{1}{2} & 0 & 0 \\ 0 & \frac{1}{3} & 0 \\ 0 & 0 & \frac{1}{2} \end{bmatrix} \begin{bmatrix} 0 & 1 & 0 \\ 1 & 0 & 1 \\ 0 & 1 & 0 \end{bmatrix} = -\begin{bmatrix} 0 & \frac{1}{2} & 0 \\ \frac{1}{3} & 0 & \frac{1}{3} \\ 0 & \frac{1}{2} & 0 \end{bmatrix}.$$

Using the max-norm, we have $\|\mathbf{M}\|_\infty = \frac{2}{3}$ and by (21),

$$\|\mathbf{e}_k\|_\infty \leqq (\tfrac{2}{3})^k \|\mathbf{e}_0\|_\infty. \tag{22}$$

In other words, the max-norm of the error at step k is bounded by $(\frac{2}{3})^k$ times the initial error. Note that this bound slightly overestimates the actual error. If the Jacobi iterations of Table 7-2 were continued for many iterations, we would find (on the average) that the error is multiplied by the factor $1/\sqrt{3} \approx .58$ in each iteration rather than by the factor $\frac{2}{3} \approx .67$ appearing in (22).

For a second example, consider the iteration

$$\begin{bmatrix} x_1^{\text{new}} \\ x_2^{\text{new}} \end{bmatrix} = \begin{bmatrix} \frac{1}{2} & 1 \\ 0 & \frac{1}{4} \end{bmatrix} \begin{bmatrix} x_1^{\text{old}} \\ x_2^{\text{old}} \end{bmatrix} + \begin{bmatrix} -3 \\ 3 \end{bmatrix}. \tag{23}$$

Since the coefficient matrix $\mathbf{S}$ for $\mathbf{x}^{\text{new}}$ is the identity $\mathbf{I}$, we have

$$\mathbf{M} = \mathbf{I}^{-1}\mathbf{T} = \mathbf{T} = \begin{bmatrix} \frac{1}{2} & 1 \\ 0 & \frac{1}{4} \end{bmatrix}.$$

Even though $\|\mathbf{M}\|_\infty = \frac{3}{2}$, which is greater than 1, we see in Table 7-9 that the iterations still converge to the solution $x_1 = 2$ and $x_2 = 4$ of the equation $\mathbf{Sx} = \mathbf{Tx} + \mathbf{b}$:

$$\begin{bmatrix} x_1 \\ x_2 \end{bmatrix} = \begin{bmatrix} \frac{1}{2} & 1 \\ 0 & \frac{1}{4} \end{bmatrix} \begin{bmatrix} x_1 \\ x_2 \end{bmatrix} + \begin{bmatrix} -3 \\ 3 \end{bmatrix}.$$

A precise estimate for the error in any iteration is obtained from the diagonalization of $\mathbf{M}$. Recall from Chapter 6 that if $\mathbf{M}$ is not defective, there exists a matrix $\mathbf{X}$ whose columns are the eigenvectors of $\mathbf{M}$ and there exists a diagonal matrix $\mathbf{\Lambda}$ whose diagonal elements are the eigenvalues $\lambda_1, \lambda_2, \cdots, \lambda_n$ of $\mathbf{M}$ such that

$$\mathbf{M} = \mathbf{X\Lambda X}^{-1}.$$

We now show that

$$\|\mathbf{x}_k - \mathbf{x}\|_p \leqq c_p(\mathbf{X})\rho(\mathbf{M})^k \|\mathbf{x}_0 - \mathbf{x}\|_p, \tag{24}$$

TABLE 7-9 ITERATION (23) STARTING FROM $x_1 = 0$ AND $x_2 = 0$

Iteration	x_1	x_2
0	0.00	0.00
1	−3.00	3.00
2	−1.50	3.75
3	0.00	3.94
4	0.94	3.98
5	1.45	4.00
6	1.72	4.00
7	1.86	4.00
8	1.93	4.00
9	1.96	4.00
10	1.98	4.00
11	1.99	4.00
12	2.00	4.00

where $c_p(\mathbf{X}) = \|\mathbf{X}\|_p\|\mathbf{X}^{-1}\|_p$ is the p-norm condition number of $\mathbf{X}$ and

$$\rho(\mathbf{M}) = \text{maximum}\{|\lambda_1|, |\lambda_2|, \cdots, |\lambda_n|\}$$

is the largest absolute eigenvalue of $\mathbf{M}$. The parameter $\rho(\mathbf{M})$ is called the **spectral radius** of $\mathbf{M}$.

To establish (24), we utilize equation (20), which relates the error at step k to the initial error: The error at step k is equal to $\mathbf{M}^k$ times the initial error. Our first step is to express the powers of $\mathbf{M}$ in terms of its diagonalization. Observe that

$$\mathbf{M}^2 = \mathbf{MM} = (\mathbf{X\Lambda X}^{-1})(\mathbf{X\Lambda X}^{-1}) = \mathbf{X\Lambda}(\mathbf{X}^{-1}\mathbf{X})\mathbf{\Lambda X}^{-1} = \mathbf{X\Lambda}^2\mathbf{X}^{-1},$$
$$\mathbf{M}^3 = \mathbf{MM}^2 = (\mathbf{X\Lambda X}^{-1})(\mathbf{X\Lambda}^2\mathbf{X}^{-1}) = \mathbf{X\Lambda}(\mathbf{X}^{-1}\mathbf{X})\mathbf{\Lambda}^2\mathbf{X}^{-1}) = \mathbf{X\Lambda}^3\mathbf{X}^{-1}.$$

Therefore, the kth power of $\mathbf{M}$ and the matrices $\mathbf{\Lambda}$ and $\mathbf{X}$ satisfy the relation

$$\mathbf{M}^k = \mathbf{X\Lambda}^k\mathbf{X}^{-1}. \tag{25}$$

Taking norms, we have

$$\|\mathbf{M}^k\| = \|\mathbf{X\Lambda}^k\mathbf{X}^{-1}\| \leqq \|\mathbf{X}\|\,\|\mathbf{\Lambda}\|^k\|\mathbf{X}^{-1}\|. \tag{26}$$

Since the p-norm of a diagonal matrix is the largest diagonal element in absolute value (see Exercise 3-3.3 and Problem 3-16), $\|\mathbf{\Lambda}\| = \rho(\mathbf{M})$ and relations (21) and (26) combine to give (24). Our observations concerning the iteration (16) can be summarized as follows:

> *If* $\|\mathbf{M}\| = \|\mathbf{S}^{-1}\mathbf{T}\| < 1$, *then the iteration (16) converges to the solution of (18) and the error* $\mathbf{e}_k = \mathbf{x}_k - \mathbf{x}$ *satisfies the bound (21), which may overestimate the actual error. A sharper error bound is formulated in terms of the spectral radius* $\rho(\mathbf{M})$. *If the spectral radius of* $\mathbf{M}$ *is greater than 1, then* $\mathbf{\Lambda}^k$ *diverges as k increases. By (25)* $\mathbf{M}^k$ *diverges when* $\mathbf{\Lambda}^k$ *diverges and except for special starting guesses, the error* $\mathbf{e}_k = \mathbf{M}^k\mathbf{e}_0$ *also diverges as k increases. Conversely, when the spectral radius of* $\mathbf{M}$ *is less than 1, both* $\mathbf{\Lambda}^k$ *and* $\mathbf{M}^k$ *tend to zero as k increases and the iteration error* $\mathbf{e}_k = \mathbf{M}^k\mathbf{e}_0$ *tends to zero. Moreover, by (24) the error at step k is bounded by a constant times* $\rho(\mathbf{M})^k$.

Based on these observations, one can analyze the convergence of iterative methods for various classes of matrices. Three important results are stated below:

1. If the matrix $\mathbf{A}$ is row diagonally dominant, both Jacobi and Gauss-Seidel converge (see Exercise 7-3.4).
2. If ω is greater than 2 or less than 0, SOR always diverges.
3. If $\mathbf{A}$ is symmetric and positive definite and ω is between 0 and 2, SOR always converges.

With regard to SOR, the best choice for the relaxation parameter ω is now evident: The best ω will make the spectral radius of $\mathbf{M} = \mathbf{S}^{-1}\mathbf{T}$ as small as possible. For a special (but important) class of matrices, the consistently ordered matrices, Young has determined the optimal relaxation parameter. First let us define a consistently ordered matrix. A $p \times p$ block matrix is **consistently ordered** if the set of integers between 1 and p can be decomposed into distinct subsets

$S_1, S_2, \cdots, S_d$ with the following property: If $\mathbf{A}_{ij}$ is an off-diagonal element that is not completely zero and S_k is the set containing i, then either $j > i$ and j is a member of S_{k+1} or $j < i$ and j is a member of S_{k-1}. Taking $S_1 = \{1\}$, $S_2 = \{2\}, \cdots, S_p = \{p\}$, we see that the block tridiagonal matrix (14) is consistently ordered: The only nonzero off-diagonal elements in row i are $\mathbf{A}_{i,i+1}$ and $\mathbf{A}_{i,i-1}$. If $\mathbf{A}_{ij}$ is a nonzero off-diagonal element, then either $j = i + 1$ or $j = i - 1$. In the first case, $j > i$ and j is a member of S_{i+1}. In the second case, $j < i$ and j is a member of S_{i-1}.

By Young's theory, if the coefficient matrix of a linear system is symmetric, positive definite, and consistently ordered, the optimal ω for the corresponding block SOR iteration is given by

$$\omega_{\text{opt}} = \frac{2}{1 + \sqrt{1 - \mu}}, \tag{27}$$

where μ is the spectral radius of the matrix $\mathbf{M} = \mathbf{S}^{-1}\mathbf{T}$ associated with Gauss-Seidel iterations. Young also shows that μ is the square of the spectral radius of the $\mathbf{M}$ corresponding to Jacobi iterations while the spectral radius of the SOR $\mathbf{M}$ associated with ω_{opt} is $\omega_{\text{opt}} - 1$.

To apply Young's formula for the optimal ω, we must estimate the spectral radius for Gauss-Seidel's $\mathbf{M}$. The spectral radius of a matrix can be estimated with the power method. The power iterations converge to the eigenvector corresponding to the dominant eigenvalue and the magnitude of this eigenvalue is the spectral radius. For the iteration $\mathbf{S}\mathbf{x}_{k+1} = \mathbf{T}\mathbf{x}_k + \mathbf{b}$, the power iterations corresponding to the matrix $\mathbf{M} = \mathbf{S}^{-1}\mathbf{T}$ are available for free! Subtracting the relation $\mathbf{S}\mathbf{x}_k = \mathbf{T}\mathbf{x}_{k-1} + \mathbf{b}$ from $\mathbf{S}\mathbf{x}_{k+1} = \mathbf{T}\mathbf{x}_k + \mathbf{b}$, we obtain $\mathbf{S}\mathbf{d}_{k+1} = \mathbf{T}\mathbf{d}_k$, where $\mathbf{d}_k = \mathbf{x}_k - \mathbf{x}_{k-1}$ is the iteration difference. Multiplying by $\mathbf{S}^{-1}$ yields $\mathbf{d}_{k+1} = \mathbf{M}\mathbf{d}_k$. Since $\mathbf{d}_k$ is multiplied by $\mathbf{M}$ to obtain $\mathbf{d}_{k+1}$, the iteration differences are power iterations for $\mathbf{M}$. Referring to Chapter 6, the Rayleigh quotient $\mathbf{d}_k^T\mathbf{d}_{k+1}/\mathbf{d}_k^T\mathbf{d}_k$ converges to the dominant eigenvalue when there is a unique eigenvalue with largest magnitude. Typically, the Gauss-Seidel $\mathbf{M}$ corresponding to a symmetric, positive definite, consistently ordered matrix has a unique positive dominant eigenvalue so the Rayleigh quotient corresponding to the iteration difference can be used to estimate the spectral radius.† On the other hand, in situations where there are two dominant eigenvalues, we saw in Chapter 6 that they are approximated by the roots of the quadratic

$$\lambda^2 + c_1\lambda + c_2 = 0, \tag{28}$$

with c_1 and c_2 the solution to

$$\begin{bmatrix} \mathbf{d}_{k+1} \cdot \mathbf{d}_{k+1} & \mathbf{d}_{k+1} \cdot \mathbf{d}_k \\ \mathbf{d}_{k+1} \cdot \mathbf{d}_k & \mathbf{d}_k \cdot \mathbf{d}_k \end{bmatrix} \begin{bmatrix} c_1 \\ c_2 \end{bmatrix} = - \begin{bmatrix} \mathbf{d}_{k+2} \cdot \mathbf{d}_{k+1} \\ \mathbf{d}_{k+2} \cdot \mathbf{d}_k \end{bmatrix}.$$

To summarize, the spectral radius of $\mathbf{M}$ can be estimated while the iteration $\mathbf{S}\mathbf{x}_{k+1} = \mathbf{T}\mathbf{x}_k + \mathbf{b}$ is executed. When $\mathbf{M} = \mathbf{S}^{-1}\mathbf{T}$ has a unique dominant eigenvalue,

† When using this method to estimate the spectral radius of $\mathbf{M}$, be sure to evaluate the iteration difference $\mathbf{d}_k = \mathbf{x}_k - \mathbf{x}_{k-1}$ with sufficient accuracy. As the iterations converge, $\mathbf{x}_k$ and $\mathbf{x}_{k-1}$ approach each other and $\mathbf{d}_k$ is the difference of nearly equal numbers. By the theory of Chapter 1, the relative error in the computed $\mathbf{d}_k$ approaches infinity as $\mathbf{x}_k$ approaches $\mathbf{x}_{k-1}$. The estimated spectral radius is only near the true spectral radius when k is large and $\mathbf{x}_k$ is not too close to $\mathbf{x}_{k-1}$.

it is approximated by the Rayleigh quotient corresponding to the iteration differences. When $\mathbf{M}$ has two dominant eigenvalues, they are approximated by the roots of (28). For the system (4) and the Gauss-Seidel iterations in Table 7-5, the Rayleigh quotient $\mathbf{d}_k^T\mathbf{d}_{k+1}/\mathbf{d}_k^T\mathbf{d}_k$ approaches $\frac{1}{3}$ and the optimal relaxation parameter is

$$\omega_{\text{opt}} = \frac{2}{1 + \sqrt{1 - \frac{1}{3}}} \approx 1.10102.$$

Although the power iteration $\mathbf{d}_{k+1} = \mathbf{M}\mathbf{d}_k$ often converges slowly when $\mathbf{M}$ is large, convergence can be accelerated using Aitken's extrapolation. Other techniques for estimating ω_{opt} are developed in the book [H3] by Hageman and Young.

Although three particular splittings $\mathbf{S} - \mathbf{T}$ of the coefficient matrix $\mathbf{A}$ were emphasized in Section 7-2, almost any splitting can be employed as long as we remember two things: First, $\mathbf{S}$ must be chosen so that it is easy to solve the system $\mathbf{S}\mathbf{x}^{\text{new}} = \mathbf{T}\mathbf{x}^{\text{old}} + \mathbf{b}$ for the new $\mathbf{x}$. Second, $\mathbf{T}$ must be chosen so that the spectral radius of $\mathbf{S}^{-1}\mathbf{T}$ is less than 1. For example, when $\mathbf{A}$ is the sum $\mathbf{B} + \mathbf{C}$, where $\mathbf{B}$ is a tridiagonal matrix and $\mathbf{C}$ is a matrix whose elements are small, it is natural to choose $\mathbf{S} = \mathbf{B}$ and $\mathbf{T} = -\mathbf{C}$. Since $\mathbf{S}$ is tridiagonal, it is easy to solve the system $\mathbf{S}\mathbf{x}^{\text{new}} = \mathbf{T}\mathbf{x}^{\text{old}} + \mathbf{b}$ for the new $\mathbf{x}$. And if the elements of $\mathbf{T}$ are small enough, $\|\mathbf{S}^{-1}\mathbf{T}\|$ is less than 1.

Exercises

7-3.1. Consider the system of equations

$$\begin{bmatrix} 10 & -25 \\ -1 & 10 \end{bmatrix}\begin{bmatrix} x_1 \\ x_2 \end{bmatrix} = \begin{bmatrix} 5 \\ 7 \end{bmatrix}.$$

The matrix $\mathbf{M} = \mathbf{S}^{-1}\mathbf{T}$ corresponding to the Jacobi scheme is

$$\mathbf{M} = \begin{bmatrix} 0 & 2.5 \\ .1 & 0 \end{bmatrix}.$$

(a) Since $\|\mathbf{M}\|_1 = 2.5$, does the inequality $\|\mathbf{e}_k\|_1 \leqq \|\mathbf{M}\|_1^k\|\mathbf{e}_0\|_1$ tell us anything about the convergence of Jacobi iterations?

(b) Show that $\|\mathbf{e}_k\|_1 \leqq 5 \times .5^k\|\mathbf{e}_0\|_1$. Will the iterations converge or diverge? [*Hint:* Utilize the first inequality in (21) and observe that $\|\mathbf{M}^2\|_1 < 1$ and $\mathbf{M}^k = (\mathbf{M}^2)^{k/2}$ if k is even.]

To summarize, the iteration (16) converges if $\|(\mathbf{S}^{-1}\mathbf{T})^p\| < 1$ for some $p \geqq 1$.

7-3.2. Consider the system $\mathbf{S}\mathbf{x} = \mathbf{T}\mathbf{x} + \mathbf{b}$ and suppose that $\|\mathbf{S}^{-1}\mathbf{T}\| < 1$. Show that

$$\|\mathbf{x}\| \leqq \frac{\|\mathbf{S}^{-1}\mathbf{b}\|}{1 - \|\mathbf{S}^{-1}\mathbf{T}\|}.$$

7-3.3. Suppose that a 1000×1000 matrix $\mathbf{A}$ can be expressed as the sum $\mathbf{B} + \mathbf{C}$, where $\mathbf{B}$ is a symmetric positive definite tridiagonal matrix and $\mathbf{C}$ is sparse. If the system $\mathbf{A}\mathbf{x} = \mathbf{b}$ is solved using subroutines FACT and SOLVE, then from Section 7-1, we know that the computing time can be more than 3 minutes on an IBM 370 model 3090 computer. Assuming that $\|\mathbf{B}^{-1}\mathbf{b}\| = 1$, $\|\mathbf{B}^{-1}\mathbf{C}\| = .5$, and $\mathbf{C}$ has at most 5000 nonzero entries which are stored using one of the sparse matrix schemes discussed in Section 2-4.5, estimate the time to compute an approximation $\mathbf{x}_k$ for which $\|\mathbf{x}_k - \mathbf{x}\| \leqq .001$, where the $\mathbf{x}_k$ are generated by the iteration

$$\mathbf{B}\mathbf{x}^{\text{new}} = \mathbf{C}\mathbf{x}^{\text{old}} - \mathbf{b}$$

with the starting guess $\mathbf{x}_0 = \mathbf{0}$. (*Hint:* Use Exercise 7-3.2 to estimate $\|\mathbf{e}_0\|$.)

7-3.4. A matrix is **row diagonally dominant** if in each row, the diagonal element is larger in absolute value than the absolute sum of off-diagonal elements. In other words, a $n \times n$ matrix is row diagonally dominant if

$$|a_{ii}| > \sum_{\substack{j=1 \\ j \neq i}}^{n} |a_{ij}| \qquad \text{for } i = 1 \text{ to } n.$$

(a) Show that Jacobi's iteration always converges when the coefficient matrix is row diagonally dominant. (*Hint:* Demonstrate that $\|\mathbf{M}\|_\infty < r$, where r is the maximum ratio between the absolute sum of off-diagonal elements in each row and the corresponding diagonal element.)

(b) Show that Gauss-Seidel iterations always converge when the coefficient matrix is row diagonally dominant. (*Hint:* Given a vector $\mathbf{x}$, let $\mathbf{y}$ satisfy $\mathbf{Sy} = \mathbf{Tx}$, where $\mathbf{A} = \mathbf{S} - \mathbf{T}$ is the Gauss-Seidel splitting of $\mathbf{A}$. Show that $|y_i| \leqq r\|\mathbf{x}\|_\infty$ for $i = 1, 2, \cdots, n$. It follows that $\|\mathbf{y}\|_\infty = \|\mathbf{S}^{-1}\mathbf{Tx}\|_\infty = \|\mathbf{Mx}\|_\infty \leqq r\|\mathbf{x}\|_\infty$ and $\|\mathbf{M}\| \leqq r$.)

7-4. ITERATIVE REFINEMENT

Iterative refinement is a technique for improving the accuracy in the computed solution to a linear system. This technique uses a low-precision factorization and a high-precision residual to compute a high-precision solution. Letting $\mathbf{x}^{\text{old}}$ be the current approximation to $\mathbf{x}$, one iteration of iterative refinement has three steps:

(a) Compute the residual $\mathbf{r} = \mathbf{b} - \mathbf{Ax}^{\text{old}}$ in high precision. (b) Solve $\mathbf{Ae} = \mathbf{r}$ in low precision. (c) Set $\mathbf{x}^{\text{new}} = \mathbf{x}^{\text{old}} + \mathbf{e}$ in high precision.	(29)

The iterations can be stopped if the 1-norm of the current $\mathbf{e}$ is greater than or equal to the 1-norm of the previous $\mathbf{e}$. Since the system $\mathbf{Ae} = \mathbf{r}$ is solved in each iteration, the matrix $\mathbf{A}$ is factored once at the start. The arrays storing $\mathbf{e}$, $\mathbf{r}$, and the factorization of $\mathbf{A}$ can be single precision, but the array storing $\mathbf{x}$ must be double precision and the computation of $\mathbf{b} - \mathbf{Ax}^{\text{old}}$ must use high-precision arithmetic. Letting $\tilde{\mathbf{L}}\tilde{\mathbf{U}}$ denote the computed low-precision factorization of $\mathbf{A}$, we see that (29) is essentially the splitting scheme (2), where $\mathbf{S} = \tilde{\mathbf{L}}\tilde{\mathbf{U}}$ and $\mathbf{T} = \tilde{\mathbf{L}}\tilde{\mathbf{U}} - \mathbf{A}$.

Let us use the theory of Section 7-3 to determine when iterative refinement converges. If $\tilde{\mathbf{A}}$ denotes the product $\tilde{\mathbf{L}}\tilde{\mathbf{U}}$, the $\mathbf{M}$ matrix corresponding to iterative refinement is given by

$$\mathbf{M} = \mathbf{S}^{-1}\mathbf{T} = \tilde{\mathbf{A}}^{-1}(\tilde{\mathbf{A}} - \mathbf{A}).$$

Taking norms, we obtain

$$\|\mathbf{M}\| = \|\tilde{\mathbf{A}}^{-1}(\tilde{\mathbf{A}} - \mathbf{A})\| \leqq \|\tilde{\mathbf{A}}^{-1}\|\,\|\tilde{\mathbf{A}} - \mathbf{A}\|.$$

Noting that

$$\|\tilde{\mathbf{A}}^{-1}\|\,\|\tilde{\mathbf{A}} - \mathbf{A}\| = \|\tilde{\mathbf{A}}\|\,\|\tilde{\mathbf{A}}^{-1}\|\frac{\|\tilde{\mathbf{A}} - \mathbf{A}\|}{\|\tilde{\mathbf{A}}\|} = c(\tilde{\mathbf{A}})\frac{\|\tilde{\mathbf{A}} - \mathbf{A}\|}{\|\tilde{\mathbf{A}}\|},$$

where $c(\tilde{\mathbf{A}})$ is the condition number of $\tilde{\mathbf{A}}$, we conclude that $\|\mathbf{M}\| < 1$ and iterative refinement converges if the product between the condition number of $\tilde{\mathbf{A}}$ and the relative error $\|\tilde{\mathbf{A}} - \mathbf{A}\|/\|\tilde{\mathbf{A}}\|$ is less than 1. Consequently, an ill-conditioned system must be factored more precisely than a well-conditioned system to ensure that the convergence condition

$$c(\tilde{\mathbf{A}})\frac{\|\tilde{\mathbf{A}} - \mathbf{A}\|}{\|\tilde{\mathbf{A}}\|} < 1$$

is satisfied.

The ultimate accuracy in the solution computed by iterative refinement depends on the accuracy of the residual. If the residual is computed in double precision, the accuracy in the iteration limit is comparable to the accuracy obtained by solving $\mathbf{Ax} = \mathbf{b}$ with double-precision versions of FACT and SOLVE. If the residual is computed in quadruple precision, the accuracy in the iteration limit is comparable to the accuracy obtained by solving $\mathbf{Ax} = \mathbf{b}$ with quadruple precision versions of FACT and SOLVE. The computation of $\mathbf{r} = \mathbf{b} - \mathbf{Ax}^{\text{old}}$ introduces two errors: (a) the error in the stored values of a_{ij} and b_i, and (b) the rounding errors that result when $\mathbf{A}$ is multiplied by $\mathbf{x}^{\text{old}}$ and subtracted from $\mathbf{b}$. If the coefficient matrix $\mathbf{A}$ and the right side $\mathbf{b}$ are stored in single-precision arrays but double-precision arithmetic is used to compute $\mathbf{r} = \mathbf{b} - \mathbf{Ax}^{\text{old}}$, then iterative refinement essentially computes a very accurate solution to the equation $\text{fl}(\mathbf{A})\mathbf{x} = \text{fl}(\mathbf{b})$, where $\text{fl}(\mathbf{A})$ and $\text{fl}(\mathbf{b})$ are the single-precision floating-point representations of $\mathbf{A}$ and $\mathbf{b}$, respectively. To obtain double-precision accuracy in the computed solution, both double-precision coefficients and double-precision arithmetic must be employed. We will illustrate the errors (a) and (b) using the **Hilbert matrix**, a notoriously ill-conditioned matrix.

The coefficient in row i and column j for the Hilbert matrix is

$$a_{ij} = \frac{1}{i + j - 1}.$$

Let us apply iterative refinement to the system $\mathbf{Ax} = \mathbf{b}$, where $\mathbf{A}$ is the 5×5 Hilbert matrix and $\mathbf{b}$ is the vector with every component equal to 1:

$$\begin{aligned}
x_1 + \tfrac{1}{2}x_2 + \tfrac{1}{3}x_3 + \tfrac{1}{4}x_4 + \tfrac{1}{5}x_5 &= 1,\\
\tfrac{1}{2}x_1 + \tfrac{1}{3}x_2 + \tfrac{1}{4}x_3 + \tfrac{1}{5}x_4 + \tfrac{1}{6}x_5 &= 1,\\
\tfrac{1}{3}x_1 + \tfrac{1}{4}x_2 + \tfrac{1}{5}x_3 + \tfrac{1}{6}x_4 + \tfrac{1}{7}x_5 &= 1,\\
\tfrac{1}{4}x_1 + \tfrac{1}{5}x_2 + \tfrac{1}{6}x_3 + \tfrac{1}{7}x_4 + \tfrac{1}{8}x_5 &= 1,\\
\tfrac{1}{5}x_1 + \tfrac{1}{6}x_2 + \tfrac{1}{7}x_3 + \tfrac{1}{8}x_4 + \tfrac{1}{9}x_5 &= 1.
\end{aligned} \tag{30}$$

In our first program, we store the coefficients and the right side in single-precision arrays A and B, respectively, and we compute the residual using double-precision arithmetic:

```
        DO 10 I = 1,N
10         R(I) = B(I)
        DO 20 J = 1,N
           DO 20 I = 1,N
20            R(I) = R(I) - A(I,J)*X(J)
```

If R and X are double-precision arrays, the loop surrounding the statement

```
20            R(I) = R(I) - A(I,J)*X(J)
```

computes the residual in double precision using the single-precision coefficients A(I,J). In a second program, the single-precision array element A(I,J) is replaced by the actual coefficient 1/(I + J − 1) and the residual is computed by the following code:

```
      DO 10 I = 1,N
10       R(I) = 1.
      DO 20 J = 1,N
         DO 20 I = 1,N
20          R(I) = R(I) - X(J)/(I + J - 1)
```

If X and R are double-precision arrays, this code computes the residual with double-precision accuracy.

Table 7-10 compares the computed solutions. As expected, the solution computed by program 2 is more accurate than the solution computed by program 1. In going from program 1 to program 2, the error associated with coefficient storage is reduced by computing the high-precision residual using the actual coefficient $a_{ij} = 1/(i + j - 1)$ instead of the single-precision array element A(I,J). An accurate solution to $\mathbf{Ax} = \mathbf{b}$ can also be achieved with subroutines FACT and SOLVE if the single-precision declaration "REAL" is changed to either "DOUBLE PRECISION" or "QUADRUPLE PRECISION." However, changing arrays to double precision or quadruple precision multiplies storage by two or by four and increases the running time. For a computer where high-precision arithmetic is implemented with software, the increase in time can be significant, and for a computer with limited memory, the increase in storage can be significant. In these cases, it is better to refine the solution iteratively rather than increase the program precision.

Exercises

7-4.1. Suppose that after factoring **A** and solving the linear system $\mathbf{Ax} = \mathbf{b}$, we are asked to solve a nearby system $\mathbf{Cx} = \mathbf{d}$. State a condition that guarantees the convergence of the iteration $\mathbf{Sx}^{\text{new}} = \mathbf{Tx}^{\text{old}} + \mathbf{d}$, where $\mathbf{S} = \mathbf{A}$ and $\mathbf{T} = \mathbf{A} - \mathbf{C}$.

7-4.2. For some linear systems, the error associated with coefficient storage can be eliminated if the coefficients are rational numbers and the equations are multiplied by the common denominator. For example, if equation (30) is multiplied by 2520, then every coefficient of the new system is an integer and the largest coefficient is 2520. Since the four-digit integer 2520 can be stored exactly in single precision on almost every computer, the error associated with coefficient storage has been eliminated. Use iterative refinement to compute a high-precision solution to the

TABLE 7-10 SOLUTION COMPUTED BY ITERATIVE REFINEMENT

k	x_k	Program 1	Program 2
1	5	4.95	5.000000000004
2	−120	−119.05	−120.000000000080
3	630	625.99	630.000000000347
4	−1120	−1114.01	−1120.000000000525
5	630	627.10	630.000000000257

5×5 Hilbert system $\mathbf{Ax} = \mathbf{b}$, where every component of $\mathbf{b}$ is 2520 and $a_{ij} = 2520/(i + j - 1)$. Compare your solution to the correct solution in Table 7-10.

7-5. ACCELERATION TECHNIQUES

In the SOR method, a parameter ω is adjusted to make the spectral radius of the iteration matrix $\mathbf{M} = \mathbf{S}^{-1}\mathbf{T}$ as small as possible. We now consider acceleration techniques where the iteration matrix is fixed but the structure of the iteration is altered to increase the convergence speed. An iterative method to solve $\mathbf{Ax} = \mathbf{b}$ has the form

$$\mathbf{Sx}_{k+1} = \mathbf{Tx}_k + \mathbf{b}, \tag{31}$$

where $\mathbf{A} = \mathbf{S} - \mathbf{T}$ is a splitting for the coefficient matrix. A three-term acceleration scheme based on iteration (31) has the form

$$\mathbf{Sx}_{k+1} = \alpha_k[\beta_k(\mathbf{Tx}_k + \mathbf{b}) + (1 - \beta_k)\mathbf{Sx}_k] + (1 - \alpha_k)\mathbf{Sx}_{k-1}, \tag{32}$$

where α_k and β_k are parameters that characterize the acceleration scheme, $\mathbf{x}_{-1} = \mathbf{0}$, and $\mathbf{x}_0$ is the starting guess. Observe that iteration (32) reduces to iteration (31) if $\alpha_k = 1$ and $\beta_k = 1$ for every k. On the other hand, by an appropriate choice for the α_k and the β_k, iteration (32) often converges quicker than iteration (31).

With the substitution $\mathbf{A} = \mathbf{S} - \mathbf{T}$, the equation $\mathbf{Ax} = \mathbf{b}$ can be rearranged to obtain $\mathbf{Sx} = \mathbf{Tx} + \mathbf{b}$. Therefore, if $\mathbf{x}_k$ is equal to $\mathbf{x}$, (31) implies that $\mathbf{Sx}_{k+1} = \mathbf{Tx} + \mathbf{b} = \mathbf{Sx}$. This shows that if the kth iteration is equal to the solution $\mathbf{x}$, then $\mathbf{x}_{k+1} = \mathbf{x}$ and the iterations repeat—each subsequent iteration is equal to $\mathbf{x}$. The accelerated iteration (32) has a similar property. If $\mathbf{x}_k = \mathbf{x}_{k-1} = \mathbf{x}$, then (32) implies that

$$\mathbf{Sx}_{k+1} = \alpha_k[\beta_k(\mathbf{Tx} + \mathbf{b}) + (1 - \beta_k)\mathbf{Sx}] + (1 - \alpha_k)\mathbf{Sx}. \tag{33}$$

When $\mathbf{Tx} + \mathbf{b}$ is replaced with $\mathbf{Sx}$, (33) reduces to $\mathbf{Sx}_{k+1} = \mathbf{Sx}$ or $\mathbf{x}_{k+1} = \mathbf{x}$. In other words, if two successive iterations for the accelerated scheme are equal to $\mathbf{x}$, each subsequent iteration is equal to $\mathbf{x}$.

For **Chebyshev acceleration**, the parameters α_k and β_k are

$$\alpha_0 = 1, \quad \alpha_1 = \frac{2}{2 - \nu},$$

$$\alpha_{k+1} = \frac{4}{4 - \nu\alpha_k} \quad \text{for } k > 0,$$

$$\beta_k = \frac{2}{2 - l - s} \quad \text{for } k \geqq 0,$$

where l denotes the most positive eigenvalue of $\mathbf{M} = \mathbf{S}^{-1}\mathbf{T}$, s denotes the most negative eigenvalue of $\mathbf{M}$, and

$$\nu = \left(\frac{l - s}{2 - l - s}\right)^2.$$

The Chebyshev acceleration parameters are derived in the following way (see [H3]): By (32) the error $\mathbf{e}_k = \mathbf{x}_k - \mathbf{x}$ has the form $p(\mathbf{M})\mathbf{e}_0$ where p is a polynomial of degree at most k. In Chebyshev acceleration, the α_k and the β_k are chosen so that for the corresponding polynomial p, the maximum of $|p(x)|$ over x between l and u is as small as possible.

To employ Chebyshev acceleration, the spectral radius of **M** must be less than 1, the eigenvalues of **M** must be real, and estimates for the extreme eigenvalues l and s of **M** must be known. To demonstrate that the accelerated iteration (32) can be substantially faster than the original iteration (33), let us consider the tridiagonal system $\mathbf{Ax} = \mathbf{b}$ of n equations, where

$$\mathbf{A} = \begin{bmatrix} 2 & -1 & & & \\ -1 & 2 & \cdot & & \\ & \cdot & \cdot & \cdot & \\ & & \cdot & \cdot & \cdot \\ & & & \cdot & \cdot & -1 \\ & & & & -1 & 2 \end{bmatrix} \quad \text{and} \quad \mathbf{b} = = \begin{bmatrix} 2 \\ 2 \\ \cdot \\ \cdot \\ \cdot \\ 2 \end{bmatrix}$$

Jacobi's **M** corresponding to this **A** is

$$\mathbf{M} = \frac{1}{2}\begin{bmatrix} 0 & 1 & & & \\ 1 & 0 & \cdot & & \\ & \cdot & \cdot & \cdot & \\ & & \cdot & \cdot & \cdot \\ & & & \cdot & \cdot & 1 \\ & & & & 1 & 0 \end{bmatrix}. \tag{34}$$

By Problems 6-1 and 6-2 we know that $l = -s \approx 1 - (\pi/(n+1))^2/2$. When n is 20 and the starting guess is $\mathbf{x}_0 = \mathbf{0}$, the one norm difference in successive iterations is less than .001 after 89 iterations of the accelerated iteration while 870 iterations of the standard iteration (31) are needed to obtain the same iteration difference. The accelerated iteration is much faster than the standard iteration. In general, estimates for the extreme eigenvalues of **M** can be obtained using either the power method, special tricks, like those used above for the matrix (34), or more sophisticated procedures such as those found in [H3].

The convergence speed of the Chebyshev scheme is related to the parameter ν. It can be shown that if **M** is not defective, the error at step k satisfies the inequality

$$\|\mathbf{x}_k - \mathbf{x}\|_p \leqq 2c_p(\mathbf{X})\|\mathbf{x}_0 - \mathbf{x}\|_p \frac{r^{k/2}}{1 + r^k}, \tag{35}$$

where $c_p(\mathbf{X})$ is the p-norm condition number for the eigenvector matrix **X** in the diagonalization of **M** and

$$r = \frac{1 - \sqrt{1 - \nu}}{1 + \sqrt{1 - \nu}}.$$

In contrast, by (24) the error in the standard iteration (31) satisfies the inequality

$$\|\mathbf{x}_k - \mathbf{x}\|_p \leqq c_p(\mathbf{X})\|\mathbf{x}_0 - \mathbf{x}\|_p \rho^k, \tag{36}$$

where ρ is either $|s|$ or $|l|$, whichever is larger.

Let us compare the error estimate (35) to the error estimate (36) when l is near 1 and $l = -s = 1 - \Delta$, where Δ is a small positive quantity. By the definition of ρ, ν and r, we have $\rho = 1 - \Delta$, $\nu = (1 - \Delta)^2$ and

$$r = \frac{1 - \sqrt{2\Delta - \Delta^2}}{1 + \sqrt{2\Delta - \Delta^2}}.$$

Since Δ^2 is much smaller than Δ, $r \approx 1 - 2\sqrt{2\Delta}$. Using the error estimates

(35) and (36), we can determine the number of iterations required by either (31) or (32) to achieve a given error. Omitting the arithmetic, it can be shown that the Chebyshev accelerated iteration (32) is roughly $\sqrt{2/\Delta}$ times faster than the original iteration (31). In particular, for the matrix (34), we have

$$\Delta \approx \frac{1}{2}\left(\frac{\pi}{n+1}\right)^2,$$

and (32) is roughly $2(n+1)/\pi$ times faster than (31).

The **conjugate gradient method** is an acceleration technique that does not require estimates for the eigenvalues. The conjugate gradient method corresponds to taking $\mathbf{M} = \mathbf{I} - \mathbf{A}$ and making the following choices for the acceleration parameters: $\alpha_0 = 1$,

$$\beta_k = \frac{\|\mathbf{r}_k\|^2}{\mathbf{r}_k^T\mathbf{A}\mathbf{r}_k} \qquad \text{for } k \geqq 0$$

and

$$\alpha_k = \left(1 - \frac{\beta_k}{\beta_{k-1}}\frac{\|\mathbf{r}_k\|^2}{\|\mathbf{r}_{k-1}\|^2}\frac{1}{\alpha_{k-1}}\right)^{-1} \qquad \text{for } k \geqq 1,$$

where $\mathbf{r}_k = \mathbf{b} - \mathbf{A}\mathbf{x}_k$ is the residual at step k and $\|\cdot\|$ denotes the Euclidean norm. Although the conjugate gradient method can be considered an acceleration technique like Chebyshev acceleration, the theoretical basis for the conjugate gradient method is quite different from the theoretical basis for Chebyshev acceleration. The conjugate gradient method actually has its roots in optimization theory and it is guaranteed to converge only when $\mathbf{A}$ is symmetric and positive definite. In the field of optimization, the conjugate gradient method is often expressed in the following way:

$$\begin{array}{lll}
\mathbf{p}_k & \leftarrow \mathbf{A}\mathbf{d}_k, & \\
\mathbf{x}_{k+1} & \leftarrow \mathbf{x}_k + \alpha_k\mathbf{d}_k, & \text{where } \alpha_k = \mathbf{r}_k^T\mathbf{r}_k/\mathbf{d}_k^T\mathbf{p}_k, \\
\mathbf{r}_{k+1} & \leftarrow \mathbf{r}_k - \alpha_k\mathbf{p}_k, & \\
\mathbf{d}_{k+1} & \leftarrow \mathbf{r}_{k+1} + \beta_k\mathbf{d}_k, & \text{where } \beta_k = \mathbf{r}_{k+1}^T\mathbf{r}_{k+1}/\mathbf{r}_k^T\mathbf{r}_k.
\end{array} \tag{37}$$

If $\mathbf{x}_0$ is the starting guess for $\mathbf{x}$, the starting values for $\mathbf{d}$ and $\mathbf{r}$ are $\mathbf{d}_0 = \mathbf{r}_0 = \mathbf{b} - \mathbf{A}\mathbf{x}_0$. Since each iteration of the conjugate gradient method computes the product between the matrix $\mathbf{A}$ and the vector $\mathbf{d}_k$ and since the time τ to compute this product is usually much greater than the time to compute either $\mathbf{x}_{k+1}$, $\mathbf{r}_{k+1}$, or $\mathbf{d}_{k+1}$, the time to perform k iterations of the conjugate gradient method is often proportional to $k\tau$.

There are several different versions of the conjugate gradient method corresponding to different but equivalent ways of writing α_k and β_k. Another common choice is

$$\alpha_k = \frac{\mathbf{d}_k^T\mathbf{r}_k}{\mathbf{p}_k^T\mathbf{d}_k} \qquad \text{and} \qquad \beta_k = \frac{-\mathbf{p}_k^T\mathbf{r}_{k+1}}{\mathbf{p}_k^T\mathbf{d}_k}. \tag{38}$$

In [R1] Reid performs some numerical experiments to compare the formulas (38) to those in (37). His experiments indicate that the number of iterations needed to achieve a given error tolerance differs by at most one for these two different sets of formulas. Note though that (38) requires the computation of one more inner product than the formulas in (37) since the value for $\mathbf{r}_{k+1}^T\mathbf{r}_{k+1}$ in (37) can be saved for the next iteration.

As stated above, the conjugate gradient method can be viewed as an optimization algorithm. If $\mathbf{A}$ is symmetric and positive definite, you will show in Exercise 7-5.1 that the quadratic

$$\mathbf{x}^T\mathbf{A}\mathbf{x} - 2\mathbf{b}^T\mathbf{x} \tag{39}$$

attains its minimum at $\mathbf{x} = \mathbf{A}^{-1}\mathbf{b}$. Therefore, minimizing the quadratic (39) is equivalent to solving the linear system $\mathbf{A}\mathbf{x} = \mathbf{b}$. It can be shown that the $\mathbf{x}_{k+1}$ generated by the conjugate gradient method minimizes (39) over $\mathbf{x}$'s of the form

$$\mathbf{x}_0 + c_0\mathbf{r}_0 + c_1\mathbf{r}_1 + \cdots + c_k\mathbf{r}_k,$$

where the c_i denote arbitrary constants. And if $\mathbf{A}$ is $n \times n$, then at step n, if not sooner, $\mathbf{x}_n$ is equal to $\mathbf{A}^{-1}\mathbf{b}$. Unlike Chebyshev acceleration where the iterations merely approach the solution to $\mathbf{A}\mathbf{x} = \mathbf{b}$ as k increases, the conjugate gradient method computes (in theory) the exact solution in a finite number of iterations. Of course, numerical errors pollute the computations and $\mathbf{x}_n$ just approximates the true solution $\mathbf{x} = \mathbf{A}^{-1}\mathbf{b}$. For this reason, it is often observed that a better approximation to the true solution is obtained if slightly more than n iterations are performed. (Although in practical applications, n can be very large and far fewer than n iterations are performed.)

When analyzing the convergence of the conjugate gradient method, it is convenient to measure the error using a special norm defined in terms of the $\mathbf{A}$ matrix. The $\mathbf{A}$-norm of a vector is defined by

$$\|\mathbf{x}\|_{\mathbf{A}} = \sqrt{\mathbf{x}^T\mathbf{A}\mathbf{x}}.$$

The fundamental rule governing the error $\mathbf{x}_k - \mathbf{x}$ in each iteration of the conjugate gradient method can be stated as follows: If $\mathbf{A}$ is symmetric and positive definite and $\lambda_1, \lambda_2, \cdots, \lambda_n$ denote the eigenvalues of $\mathbf{A}$, we have

$$\|\mathbf{x}_{k+1} - \mathbf{x}\|_{\mathbf{A}} \leqq \underset{1 \leqq i \leqq n}{\text{maximum}} |1 - \lambda_i p_k(\lambda_i)| \, \|\mathbf{x}_0 - \mathbf{x}\|_{\mathbf{A}} \tag{40}$$

for any polynomial p_k of degree at most k.

Recall that for the eigenproblem, multiple eigenvalues can lead to numerical complications. When solving a linear system using the conjugate gradient method, multiple eigenvalues are good! Suppose that $\mathbf{A}$ has just m distinct eigenvalues. Constructing a polynomial $p_{m-1}(x)$ such that $1 + xp_{m-1}(x)$ vanishes at each distinct eigenvalue, it follows from (40) that $\mathbf{x}_m = \mathbf{x}$. Hence, if $\mathbf{A}$ has just m distinct eigenvalues, the conjugate gradient method converges in m steps rather than n steps.

Making special choices for the polynomial p_k in (40) leads to various estimates for the error in the conjugate gradient method. Assuming that the eigenvalues are ordered such that

$$\lambda_1 \geqq \lambda_2 \geqq \cdots \geqq \lambda_n$$

and letting p_k be the polynomial for which the maximum absolute value of $1 + xp_k(x)$ on the interval $[\lambda_n, \lambda_1]$ is as small as possible leads to the following estimate

$$\|\mathbf{x}_k - \mathbf{x}\|_{\mathbf{A}} \leqq 2\left(\frac{1 - \sqrt{\lambda_n/\lambda_1}}{1 + \sqrt{\lambda_n/\lambda_1}}\right)^k \|\mathbf{x}_0 - \mathbf{x}\|_{\mathbf{A}}. \tag{41}$$

This estimate tends to be overly pessimistic if $\mathbf{A}$ has one eigenvalue much larger than the other eigenvalues. Another estimate for the error is obtained by choosing p_{k-1} so that $1 + xp_{k-1}(x)$ vanishes at the $k - 1$ largest eigenvalues of $\mathbf{A}$ and at $x = (\lambda_n + \lambda_k)/2$. For this choice, it can be shown that

$$\|\mathbf{x}_k - \mathbf{x}\|_{\mathbf{A}} \leqq \frac{\lambda_k - \lambda_n}{\lambda_k + \lambda_n}\|\mathbf{x}_0 - \mathbf{x}\|_{\mathbf{A}}. \tag{42}$$

Finally, choosing p_k so that $1 + xp_k(x)$ vanishes at the $j - 1$ largest eigenvalues of $\mathbf{A}$ while $1 + xp_k(x)$ is small on the interval $[\lambda_n, \lambda_j]$, it can be shown that

$$\|\mathbf{x}_k - \mathbf{x}\|_{\mathbf{A}} \leqq 2\left(\frac{\lambda_j - \lambda_n}{\lambda_j + \lambda_n}\right)\left(\frac{1 - \sqrt{\lambda_n/\lambda_j}}{1 + \sqrt{\lambda_n/\lambda_j}}\right)^{k-j}\|\mathbf{x}_0 - \mathbf{x}\|_{\mathbf{A}}. \tag{43}$$

To illustrate the convergence of the conjugate gradient method, let us consider a coefficient matrix $\mathbf{A}$ of the form $\mathbf{I} + \mathbf{H}$, where $\mathbf{H}$ is the 20×20 Hilbert matrix. Recall that the coefficients of the Hilbert matrix are $h_{ij} = 1/(i + j - 1)$. The Hilbert matrix has one eigenvalue near 1 and many eigenvalues near zero. Adding $\mathbf{I}$ to $\mathbf{H}$ adds one to each eigenvalue. Thus $\mathbf{A}$ has one eigenvalue near 2 and many eigenvalues near 1. Consequently, the ratio λ_j/λ_{20} is near 1 for j greater than 1. Referring to (43) and taking $j = 2$, we expect that the error in $\mathbf{x}_k$ approaches zero quickly as k increases. To investigate this property, we solve the equation $(\mathbf{I} + \mathbf{H})\mathbf{x} = \mathbf{0}$, where each component of the starting guess $\mathbf{x}_0$ is 1. Of course, the solution is $\mathbf{x} = \mathbf{0}$, so the error in $\mathbf{x}_k$ is equal to $\mathbf{x}_k - \mathbf{0} = \mathbf{x}_k$. Table 7-11 gives the 1-norm of the error in each iteration. Observe that just a few iterations of the conjugate gradient method yield a good approximation to the true solution.

The convergence speed of the conjugate gradient method can be increased using preconditioning. Preconditioning means that we change variables to obtain a new equation whose coefficient matrix has a better eigenvalue distribution than that of the original coefficient matrix. Starting with the equation $\mathbf{Ax} = \mathbf{b}$, we substitute $\mathbf{x} = \mathbf{Ly}$ to obtain $\mathbf{ALy} = \mathbf{b}$. Premultiplying by $\mathbf{L}^T$ yields the equivalent equation $\mathbf{L}^T\mathbf{ALy} = \mathbf{L}^T\mathbf{b}$. Applying the conjugate gradient method to this equivalent equation and then changing from the variable $\mathbf{y}$ back to the original variable $\mathbf{x}$, the preconditioned conjugate gradient iteration is expressed in the form (see

TABLE 7-11
CONJUGATE GRADIENT METHOD APPLIED TO $(\mathbf{I}+\mathbf{H})\mathbf{X} = \mathbf{0}$

k	$\|\mathbf{x}_k\|_1$
1	$.52 \times 10^{1}$
2	$.71 \times 10^{0}$
3	$.18 \times 10^{-1}$
4	$.52 \times 10^{-4}$
5	$.14 \times 10^{-7}$
6	$.29 \times 10^{-12}$

Exercise 7-5.2):

$$
\begin{array}{lll}
\mathbf{p}_k & \leftarrow & \mathbf{A}\mathbf{d}_k, \\
\mathbf{x}_{k+1} & \leftarrow & \mathbf{x}_k + \alpha_k \mathbf{d}_k, \qquad \text{where } \alpha_k = \mathbf{q}_k^T \mathbf{r}_k / \mathbf{d}_k^T \mathbf{p}_k, \\
\mathbf{r}_{k+1} & \leftarrow & \mathbf{r}_k - \alpha_k \mathbf{p}_k, \\
\mathbf{q}_{k+1} & \leftarrow & \mathbf{P}^{-1} \mathbf{r}_{k+1}, \\
\mathbf{d}_{k+1} & \leftarrow & \mathbf{q}_{k+1} + \beta_k \mathbf{d}_k, \qquad \text{where } \beta_k = \mathbf{q}_{k+1}^T \mathbf{r}_{k+1} / \mathbf{q}_k^T \mathbf{r}_k.
\end{array}
\tag{44}
$$

Here $\mathbf{P} = (\mathbf{L}\mathbf{L}^T)^{-1}$ is called the **preconditioning matrix**. If $\mathbf{x}_0$ is the starting guess for $\mathbf{x}$, the starting residual is $\mathbf{r}_0 = \mathbf{b} - \mathbf{A}\mathbf{x}_0$, the starting direction is $\mathbf{d}_0 = \mathbf{P}^{-1}\mathbf{r}_0$, and $\mathbf{q}_0 = \mathbf{d}_0$. Note that the preconditioned conjugate gradient method is identical to the standard conjugate gradient algorithm when $\mathbf{P} = \mathbf{I}$. The preconditioned scheme also satisfies the error estimate (40); however, the relevant eigenvalues are those of $\mathbf{A}\mathbf{P}^{-1}$ rather than the eigenvalues of $\mathbf{A}$. When $\mathbf{P}$ is $\mathbf{A}$, $\mathbf{A}\mathbf{P}^{-1} = \mathbf{I}$. Since the identity matrix has just one distinct eigenvalue, the conjugate gradient method converges in one iteration when $\mathbf{P} = \mathbf{A}$. Of course, there is a catch: When using the preconditioner $\mathbf{P} = \mathbf{A}$, we must solve the equation $\mathbf{A}\mathbf{q}_{k+1} = \mathbf{r}_{k+1}$ for the unknown $\mathbf{q}_{k+1}$. Solving this equation is just as difficult as solving the original equation $\mathbf{A}\mathbf{x} = \mathbf{b}$. In general, an appropriate preconditioner for $\mathbf{A}\mathbf{x} = \mathbf{b}$ is any "simple" matrix that approximates the "essential structure" of $\mathbf{A}$. If $\mathbf{A}$ has the following structure:

$$
\mathbf{A} = \begin{bmatrix}
3 & -1 & & & & -1 \\
-1 & 3 & \cdot & & & \\
& \cdot & \cdot & \cdot & & \\
& & \cdot & \cdot & \cdot & \\
& & & \cdot & \cdot & -1 \\
-1 & & & & -1 & 3
\end{bmatrix},
\tag{45}
$$

then an appropriate preconditioner is the tridiagonal matrix $\mathbf{T}$ with 3's on the diagonal and with -1's off the diagonal. To test this preconditioner, let us solve a system of 20 equations $\mathbf{A}\mathbf{x} = \mathbf{0}$, where the first component of the starting guess is 1 and the other components are zero. The one norm of the iterations appears in Table 7-12. The second column corresponds to the preconditioner

TABLE 7-12 PRECONDITIONED CONJUGATE GRADIENT METHOD FOR (45)

Iteration k	$\mathbf{P} = \mathbf{T}^{-1}$ $\lVert\mathbf{x}_k\rVert_1$	$\mathbf{P} = \mathbf{I}$ $\lVert\mathbf{x}_k\rVert_1$
0	$.10 \times 10^{1}$	$.10 \times 10^{1}$
1	$.69 \times 10^{0}$	$.76 \times 10^{0}$
2	$.16 \times 10^{0}$	$.44 \times 10^{0}$
3	$.35 \times 10^{-32}$	$.20 \times 10^{0}$
4		$.80 \times 10^{-1}$
5		$.31 \times 10^{-1}$
6		$.12 \times 10^{-1}$

$\mathbf{P} = \mathbf{T}$ while the third column corresponds to the original conjugate gradient iteration (without preconditioning).

For a second example, suppose that $\mathbf{A}$ is the sum $\mathbf{B} + s\mathbf{CC}^T$, where $\mathbf{B}$ is $n \times n$, $\mathbf{C}$ is $n \times m$ with m smaller than n, and s is a large scalar. When s is large, $\mathbf{B}$ is negligible in size compared to $s\mathbf{CC}^T$, which suggests that an appropriate preconditioner is $s\mathbf{CC}^T$. Although $s\mathbf{CC}^T$ does not have an inverse when m is smaller than n, the matrix $\mathbf{I} + s\mathbf{CC}^T$ does have an inverse and the preconditioner $\mathbf{P} = \mathbf{I} + s\mathbf{CC}^T$ is very effective. Applying the Woodbury formula, the inverse of this preconditioner can be expressed

$$\mathbf{P}^{-1} = \mathbf{I} - \mathbf{C}(s^{-1}\mathbf{I} + \mathbf{C}^T\mathbf{C})^{-1}\mathbf{C}^T.$$

Since $\mathbf{C}^T\mathbf{C}$ is $m \times m$ and m is smaller than n, it is easier to invert the $m \times m$ matrix $s^{-1}\mathbf{I} + \mathbf{C}^T\mathbf{C}$ than to invert the $n \times n$ matrix $\mathbf{I} + s\mathbf{CC}^T$.

The preconditioned conjugate gradient algorithm is also effective in the following context: Both the coefficient matrix and the right side depend on a parameter, say t, and for many different choices of t, we have to solve the system

$$\mathbf{A}(t)\mathbf{x} = \mathbf{b}(t)$$

for the unknown $\mathbf{x}$. If these systems are solved by elimination, $\mathbf{A}(t)$ is factored for each t. But if each coefficient matrix $\mathbf{A}(t)$ is near some fixed symmetric positive definite matrix $\mathbf{B}$, we can apply the preconditioned conjugate gradient method with $\mathbf{P} = \mathbf{B}$. The preconditioned conjugate gradient method just requires one factorization, the factorization of $\mathbf{B}$, and if $\mathbf{A}(t)\mathbf{B}^{-1} \approx \mathbf{I}$, the preconditioned conjugate gradient method converges rapidly.

As we have emphasized, the conjugate gradient method is only guaranteed to converge when the coefficient matrix is symmetric and positive definite. One strategy for making symmetric a nonsymmetric system $\mathbf{Ax} = \mathbf{b}$ is to multiply by $\mathbf{A}^T$, giving us the equivalent system

$$\mathbf{A}^T\mathbf{Ax} = \mathbf{A}^T\mathbf{b}. \tag{46}$$

The coefficient matrix $\mathbf{A}^T\mathbf{A}$ is symmetric since it is equal to its transpose:

$$(\mathbf{A}^T\mathbf{A})^T = \mathbf{A}^T\mathbf{A}.$$

Furthermore, if $\mathbf{A}$ is invertible, $\mathbf{A}^T\mathbf{A}$ is positive definite. Each iteration of the conjugate gradient method (37) applied to the symmetric system (46) involves computing the product $\mathbf{A}^T\mathbf{Ad}_k$. In contrast, each iteration of the original conjugate gradient method just requires the product $\mathbf{Ad}_k$. Therefore, each iteration of the conjugate gradient method applied to $\mathbf{A}^T\mathbf{Ax} = \mathbf{A}^T\mathbf{b}$ takes about twice as much time as each iteration of the conjugate gradient method applied to $\mathbf{Ax} = \mathbf{b}$. Another bad feature of the symmetrized equation is that the distribution of eigenvalues for $\mathbf{A}^T\mathbf{A}$ is often worse than the distribution of eigenvalues for $\mathbf{A}$. In other words, if $\mathbf{A}$ is symmetric, the conjugate gradient method applied to $\mathbf{A}^T\mathbf{Ax} = \mathbf{A}^T\mathbf{b}$ is slower than the conjugate gradient method applied to $\mathbf{Ax} = \mathbf{b}$.

Subroutine PRECG implements the preconditioned conjugate gradient algorithm (44). The subroutine arguments are

X:	solution
DIF:	input for subroutine WHATIS

SIZE:	input for subroutine WHATIS
NDIGIT:	desired number of correct digits
LIMIT:	maximum number of iterations
B:	right side
N:	number of unknowns
MULT:	name of subroutine to multiply **A** by a vector
PRE:	name of subroutine to solve **Px** = **y**
W:	work array with at least 2N elements

When subroutine PRECG is invoked, the first N elements of X must contain the starting guess for the solution to the linear system **Ax** = **b**. To implement the standard conjugate gradient method without preconditioning, set **P** = **I**. In the following program, we solve the equation **Ax** = **1** where **A** appears in (45), **1** is the vector with every component equal to 1, **P** = **T** (the tridiagonal part of **A**), and the starting guess is zero.

```
      EXTERNAL PROD,SOLV
      REAL B(20),D(20),U(20),X(20),W(40)
      COMMON D,U,M,N
      N = 20
      M = N - 1
      DO 10 I =1,N
         B(I) = 1.
         X(I) = 0.
         D(I) = 3.
10       U(I) = -1.
      CALL TFACT(U,D,U,N)
      CALL PRECG(X,DIF,SIZE,4,22,B,N,PROD,SOLV,W)
      CALL WHATIS(DIF,SIZE)
      END

      SUBROUTINE PROD(Y,X)
      REAL D(20),U(20),X(1),Y(1)
      COMMON D,U,M,N
      Y(1) = 3*X(1) - X(2) - X(N)
      Y(N) = 3*X(N) - X(1) - X(M)
      DO 10 I = 2,M
10       Y(I) = 3*X(I) - X(I-1) - X(I+1)
      RETURN
      END

      SUBROUTINE SOLV(Y,X)
      REAL D(20),U(20),X(1),Y(1)
      COMMON D,U,M,N
      CALL TSOLVE(Y,U,D,U,X)
      RETURN
      END
```

Exercises

7-5.1. Making the change of variables $\mathbf{x} = \mathbf{y} + \mathbf{A}^{-1}\mathbf{b}$, show that the quadratic (39), where A is a symmetric positive definite matrix, attains its minimum at $\mathbf{y} = \mathbf{0}$ or equivalently, at $\mathbf{x} = \mathbf{A}^{-1}\mathbf{b}$.

7-5.2. Verify that the conjugate gradient method applied to $\mathbf{L}^T\mathbf{A}\mathbf{L}\mathbf{y} = \mathbf{L}^T\mathbf{b}$ can be expressed in the form (44) with $\mathbf{x}_k = \mathbf{L}\mathbf{y}_k$.

7-6. THE MULTIGRID METHOD

The multigrid method exploits the connection between a physical problem and its matrix analog to accelerate the convergence of an iteration. The fundamental principles that justify the multigrid method will be illustrated using the boundary-value problem

$$-\frac{d^2x(t)}{dt^2} = b(t), \qquad 0 \leqq t \leqq 1, \tag{47}$$

with the boundary condition $x(0) = x(1) = 0$. In equation (47), $b(t)$ is a given function and $x(t)$ is an unknown function to be computed. Using the finite-difference approximation developed in Section 1-5, equation (47) is transformed into a matrix equation. The interval [0, 1] is partitioned into N subintervals, each subinterval of width $\Delta t = 1/N$. Letting $t_i = i\Delta t$ denote the grid points and letting x_i denote $x(t_i)$, the second derivative of x evaluated at t_i has the approximation

$$\frac{d^2x(t_i)}{dt^2} \approx \frac{x_{i+1} - 2x_i + x_{i-1}}{(\Delta t)^2}.$$

Making this substitution into equation (47) and letting b_i denote $b(t_i)$, we obtain the following system $\mathbf{Ax} = \mathbf{b}$ of $N - 1$ equations:

$$\underbrace{N^2\begin{bmatrix} 2 & -1 & & & & \\ -1 & 2 & \cdot & & & \\ & \cdot & \cdot & \cdot & & \\ & & \cdot & \cdot & \cdot & \\ & & & \cdot & \cdot & -1 \\ & & & & -1 & 2 \end{bmatrix}}_{\mathbf{A}} \underbrace{\begin{bmatrix} x_1 \\ x_2 \\ \cdot \\ \cdot \\ \cdot \\ x_{N-1} \end{bmatrix}}_{\mathbf{x}} = \underbrace{\begin{bmatrix} b_1 \\ b_2 \\ \cdot \\ \cdot \\ \cdot \\ b_{N-1} \end{bmatrix}}_{\mathbf{b}}. \tag{48}$$

Abstractly, the solution to equation (47) can be expressed in terms of the eigenpairs for the boundary-value problem

$$-\frac{d^2y(t)}{dt^2} = \lambda y(t), \qquad y(0) = y(1) = 0. \tag{49}$$

This boundary-value problem has an infinite number of eigenvalues λ_1, λ_2, λ_3, $\cdots$ and corresponding eigenfunctions $y_1(t)$, $y_2(t)$, $y_3(t)$, $\cdots$. In particular, the jth eigenpair is

$$\lambda_j = (j\pi)^2 \qquad \text{and} \qquad y_j(t) = \frac{1}{\sqrt{2}} \sin j\pi t.$$

In terms of these eigenpairs, the solution to (47) is

$$x(t) = \sum_{j=1}^{\infty} \frac{1}{\lambda_j} c_j y_j(t), \tag{50}$$

where

$$c_j = \int_0^1 y_j(t) b(t)\, dt.$$

Typically, the c_j approach zero as j increases. Since λ_j is proportional to j^2, $1/\lambda_j$ also approaches zero as j increases and the coefficients of $y_j(t)$ in the expansion (50) tend quickly to zero. Hence, a good approximation to $x(t)$ is obtained by summing the first few terms in the series (50). The expansion (50) is called a **Fourier expansion** of x. The terms in (50) corresponding to small j will be called low-frequency components of x while the terms corresponding to large j will be called high-frequency components. Since both c_j and $1/\lambda_j$ approach zero as j increases, the coefficient c_j/λ_j corresponding to low-frequency components of the solution is much larger than the coefficient corresponding to high-frequency components of the solution.

Let us apply Jacobi's scheme to (48). The Jacobi iteration can be expressed $\mathbf{x}_{k+1} = \mathbf{M}\mathbf{x}_k + \mathbf{g}$, where

$$\mathbf{M} = \frac{1}{2}\begin{bmatrix} 0 & 1 & & & & \\ 1 & 0 & \cdot & & & \\ & \cdot & \cdot & \cdot & & \\ & & \cdot & \cdot & \cdot & \\ & & & \cdot & \cdot & 1 \\ & & & & 1 & 0 \end{bmatrix} \quad \text{and} \quad \mathbf{g} = \frac{1}{2N^2}\mathbf{b}.$$

If $\mathbf{x}$ denotes the solution to (48), then by (20) the error $\mathbf{e}_k = \mathbf{x}_k - \mathbf{x}$ is given by $\mathbf{e}_k = \mathbf{M}^k\mathbf{e}_0$. The error is now expressed in terms of the eigenpairs of $\mathbf{M}$. If $\mu_1 < \mu_2 < \mu_3 < \cdots$ denote the eigenvalues of $\mathbf{M}$, there exist corresponding eigenvectors $\mathbf{y}_1, \mathbf{y}_2, \mathbf{y}_3, \cdots$ such that

$$\mathbf{e}_0 = \mathbf{y}_1 + \mathbf{y}_2 + \mathbf{y}_3 + \cdots.$$

It can be shown that these eigenvectors are nearly multiples of the continuous eigenfunctions $y_j(t)$ evaluated at the grid points when N is large. Since $\mathbf{M}\mathbf{y}_j = \mu_j\mathbf{y}_j$, it follows that $\mathbf{M}^k\mathbf{y}_j = \mu_j^k\mathbf{y}_j$ and

$$\mathbf{e}_k = \mu_1^k\mathbf{y}_1 + \mu_2^k\mathbf{y}_2 + \mu_3^k\mathbf{y}_3 + \cdots.$$

Referring to Problems 6-1 and 6-2, the small eigenvalues of $\mathbf{M}$ have the approximation

$$\mu_j \approx \frac{\lambda_j}{2N^2} - 1 = \frac{(j\pi)^2}{2N^2} - 1.$$

When j is much smaller than N, μ_j decreases in magnitude as j increases or as N decreases. Therefore, as k increases, μ_2^k approaches zero faster than μ_1^k, μ_3^k approaches zero faster than μ_2^k, and so on. In summary the Jacobi scheme, as well as many other iterative schemes, has the property that the error in high-frequency components of the solution decays to zero much faster than the error in low-frequency components of the solution. Moreover, the error in low-frequency components decays to zero much faster when N is small and the mesh is coarse. In the multigrid method, low-frequency components of the solution are computed using a coarse mesh and high-frequency components are computed using a fine mesh.

In our first statement of the multigrid method, we work with two different meshes. The mesh corresponding to small N is called the coarse mesh and the mesh corresponding to large N is called the fine mesh. We start with the fine mesh system $\mathbf{Ax} = \mathbf{b}$ and we perform several iterations of an iterative method such as Jacobi's scheme. These fine mesh iterations are often called the smoothing phase since the error in the high-frequency (wiggly) components is being removed. After several iterations using the fine mesh, we obtain an approximation $\mathbf{y}$ to $\mathbf{x}$. The residual corresponding to $\mathbf{y}$ is $\mathbf{r} = \mathbf{b} - \mathbf{Ay}$. If $\mathbf{z}$ denotes the solution to $\mathbf{Az} = \mathbf{r}$, then $\mathbf{x} = \mathbf{y} + \mathbf{z}$. Since the error in the high-frequency components of $\mathbf{y}$ is relatively small, the vector $\mathbf{z} = \mathbf{x} - \mathbf{y}$ reflects the error in the low-frequency components of $\mathbf{y}$. To compute $\mathbf{z}$, we employ a coarse mesh since the error in low-frequency components decays quickly when the mesh is coarse. To obtain an equation for $\mathbf{z}$ on the coarse mesh, we first use linear interpolation (see Figure 7-2) to construct a continuous function $r(t)$ which is equal to r_i at $t = t_i$. The continuous function r is equal to the ith component of $\mathbf{r}$ at the grid point t_i, and between each pair of adjacent grid points, r is linear. Now consider the following boundary-value problem associated with the linear system $\mathbf{Az} = \mathbf{r}$:

$$-\frac{d^2z(t)}{dt^2} = r(t), \qquad z(0) = z(1) = 0. \tag{51}$$

If this boundary-value problem is solved using the coarse mesh discretization, then the iterations converge rapidly since N is small. Moreover, evaluating the coarse mesh solution at the grid points for the fine mesh yields an approximation to $\mathbf{z}$. Adding the approximate $\mathbf{z}$ to $\mathbf{y}$ gives us a better approximation to $\mathbf{x}$. This entire process can be repeated. That is, using the fine mesh and the current estimate of $\mathbf{x}$ for the starting guess, we perform several Jacobi iterations to obtain a better approximation $\mathbf{y}$ to $\mathbf{x}$. Forming the residual $\mathbf{r} = \mathbf{b} - \mathbf{Ay}$, we solve (51) using a coarse mesh discretization. Finally, the computed $\mathbf{z}$ is added to $\mathbf{y}$ giving us a better estimate of $\mathbf{x}$.

In practice the multigrid method is implemented using a sequence of nested meshes. For example, when solving equation (47), we may use meshes corre-

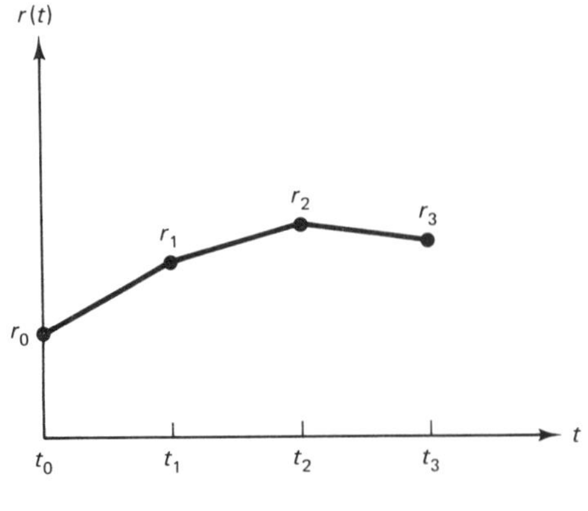

Figure 7-2 Linear interpolation between grid points.

sponding to $N = 2$, $N = 4$, $N = 8$, $\cdots$, $N = 2^F$. That is, we work with meshes corresponding to $N = 2^k$ intervals, where k is any integer between 1 and F. In stating the multigrid method for multiple meshes, the linear system corresponding to the kth mesh will be denoted $\mathbf{A}_k\mathbf{x}_k = \mathbf{b}_k$, so that our goal is to solve $\mathbf{A}_F\mathbf{x}_F = \mathbf{b}_F$.

1. Initialize $k = F$ and $\mathbf{r}_F = \mathbf{b}$.
2. Apply several iterations of an iterative method to $\mathbf{A}_k\mathbf{x}_k = \mathbf{r}_k$.
3. If the convergence in step 2 is slow, let $\mathbf{r}_{k-1}$ denote $\mathbf{r}_k - \mathbf{A}_k\mathbf{x}_k$ evaluated on mesh $k - 1$, replace the value of k with $k - 1$, initialize $\mathbf{x}_k = \mathbf{0}$, and branch to step 2.
4. If the convergence in step 2 is fast, continue the iterations until the error is below some specified tolerance. If $k = F$, stop. Otherwise, replace the value of k with $k + 1$, set $\mathbf{x}_k^{\text{new}} = \mathbf{x}_k^{\text{old}} + \mathbf{x}_{k-1}$, and branch to step 2.

(52)

As a concrete illustration of the multigrid method, we present a program (stored under the name MGRID in NAPACK) to solve (47) in the special case $b(t) = 1$ for $0 \leqq t \leqq 1$. This program employs meshes consisting of $N = 2^k$ intervals, where k is between 1 and 8. The array X stores $\mathbf{x}_8$ down to $\mathbf{x}_1$ and the array R stores $\mathbf{b}$ followed by the residuals $\mathbf{r}_7$ down to $\mathbf{r}_1$. Since there are $N + 1$ unknowns $x_0, x_1, \cdots, x_N$ corresponding to N mesh intervals, $\mathbf{x}_1$ has $2^1 + 1 = 3$ elements, $\mathbf{x}_2$ has $2^2 + 1 = 5$ elements, and $\mathbf{x}_8$ has $2^8 + 1 = 257$ elements. The dimension of the X array is

$$\sum_{k=1}^{8} 2^k + 1 = 3 + 5 + \cdots + 257 = 518.$$

In general, if the multigrid method is applied to (47) using F meshes, the dimension of array X is

$$\sum_{k=1}^{F} 2^k + 1 = 2^{F+1} + F - 2.$$

In the program that follows, four Gauss-Seidel iterations are used for the smoothing phase. To test for convergence, the iteration difference is monitored: The iterations have "converged" when the 1-norm difference between two successive iterations is less than 10^{-4}. The convergence is viewed as "fast" if the 1-norm difference between two successive iterations decays by at least the factor .7 each iteration.

```
      SOLVE -X"(T) = 1, X(0) = X(1) = 0
         BY THE MULTIGRID METHOD

   REAL R(518),X(518),D0,D1,S,T,U
   INTEGER I,IT,J,K,L,LL,M,N,NL
   K = 8
   N = 2**K
   IT = 4
   T = .0001
   U = .7
```

```
      *** INPUT RIGHT SIDE ***

      DO 10 I = 1,N
10       R(I) = 1.

      *** INITIALIZE ***

      S = (1./N)**2
      DO 20 I = 1,N
20       R(I) = S*R(I)
      M = N
      L = 1
      LL = N
      NL = N + N + K - 2
      DO 30 I = 1,NL
30       X(I) = 0.
      D1 = 0.
40    J = 0

      *** GAUSS-SEIDEL ITERATION ***

50    D0 = D1
      D1 = 0.
      J = J + 1
      I = L
60    I = I + 1
         S = .5*(X(I-1)+X(I+1)+R(I))
         D1 = D1 + ABS(S-X(I))
         X(I) = S
      IF ( I .LT. LL ) GOTO 60
      WRITE(6,70) D1
70    FORMAT(' DIF:',F20.10)
      IF ( J .LT. IT ) GOTO 50
      IF ( D1 .LT. T ) GOTO 100
      IF ( D1/D0 .LT. U ) GOTO 50
      IF ( N .EQ. 2 ) GOTO 50

      *** COARSER MESH (SLOW CONVERGENCE) ***

      I = LL + 2
80    L = L + 2
      I = I + 1
      IF ( L .GT. LL ) GOTO 90
         X(I) = 0.
         R(I) = 4*(R(L)+X(L-1)-2*X(L)+X(L+1))
      GOTO 80
90    N = N/2
      WRITE(6,*) 'NUMBER OF MESH INTERVALS:',N
      LL = LL + N + 1
      L = L + 1
      GOTO 40
```

```
      *** FINER MESH (FAST CONVERGENCE) ***

100   IF ( N .EQ. M ) GOTO 120
      I = L - 3
      J = LL
110      X(I) = X(I) + X(J)
         X(I+1) = X(I+1) + .5*(X(J)+X(J+1))
         I = I - 2
         J = J - 1
      IF ( J .GT. L ) GOTO 110
      N = N + N
      WRITE(6,*) 'NUMBER OF MESH INTERVALS:',N
      LL = L - 2
      L = I
      X(I+1) = X(I+1) + .5*X(J+1)
      GOTO 40

      *** PRINT ANSWER ***

120   M = N + 1
      DO 130 I = 1,M
         J = I - 1
         S = J/FLOAT(N)
130      WRITE(6,140) J,S,X(I)
140   FORMAT(I5,F10.5,F15.6)
      END
```

The boundary-value problem (47) and the program above illustrate the ideas involved in the multigrid method. Although Gaussian elimination may be the fastest way to solve a tridiagonal system, the multigrid method may be faster than Gaussian elimination when the coefficient matrix is derived from the discretization of a boundary-value problem in two or three space dimensions. The multigrid method has been the focus of much research in recent years—as of May, 1986, McCormick's multigrid bibliography contains more than 600 citations. Some basic multigrid references include [B9 and B12] by Brandt, [H1] by Hackbush, and the proceedings [H2] of the Köln-Porz multigrid conference. Other references appear at the end of the chapter.

REVIEW PROBLEMS

7-1. Let $\mathbf{A}$ be an arbitrary 3×3 matrix and let $\mathbf{b}$ be an arbitrary vector with three components:

$$\mathbf{A} = \begin{bmatrix} a_{11} & a_{12} & a_{13} \\ a_{21} & a_{22} & a_{23} \\ a_{31} & a_{32} & a_{33} \end{bmatrix} \qquad \text{and} \qquad \mathbf{b} = \begin{bmatrix} b_1 \\ b_2 \\ b_3 \end{bmatrix}.$$

For the function $\mathbf{f}(\mathbf{x}) = \mathbf{Ax} - \mathbf{b}$, show that the Jacobi, Gauss-Seidel, and SOR iterations presented in Section 7-2 are identical to the Jacobi, Gauss-Seidel, and SOR iterations appearing in Section 4-4.

7-2. Using Jacobi iterations and the starting guess $\mathbf{x} = \mathbf{0}$, solve the following system of equations:

$$\begin{bmatrix} 6 & 1 & 2 & 0 & 1 \\ 2 & 8 & 1 & 2 & 2 \\ 1 & -2 & 8 & 1 & 0 \\ 0 & 0 & -1 & 9 & 2 \\ 1 & 1 & 0 & -1 & 7 \end{bmatrix} \begin{bmatrix} x_1 \\ x_2 \\ x_3 \\ x_4 \\ x_5 \end{bmatrix} = \begin{bmatrix} 10 \\ 15 \\ 8 \\ 10 \\ 8 \end{bmatrix}.$$

Count the number of iterations until $\|\mathbf{x}_k - \mathbf{x}\|_1$ is less than .001. (*Hint:* The exact solution is $x_i = 1$ for every i.) Now evaluate the 1-norm of Jacobi's $\mathbf{M}$ and use (21) to estimate the number of iterations until $\|\mathbf{x}_k - \mathbf{x}\|_1$ is less than .001.

7-3. Consider the equation $\mathbf{Ax} = \mathbf{b}$, where $\mathbf{b}$ is the vector with every component equal to 1 and where $\mathbf{A}$ is the 100×100 symmetric matrix with every diagonal element equal to 3, with every superdiagonal and subdiagonal element equal to 1, and with every other element equal to .001. Solve this equation using iteration (3) where every element of the splitting matrix $\mathbf{T}$ is equal to $-.001$. (*Note:* For any vector $\mathbf{y}$, each component of $\mathbf{Ty}$ is equal to $-.001$ times the sum of $\mathbf{y}$'s components. Since the product between $\mathbf{T}$ and a vector is easily computed, you do not need to store the elements of $\mathbf{T}$.)

7-4. Consider the system $\mathbf{Ax} = \mathbf{b}$, where $\mathbf{A}$ is a $n \times n$ symmetric tridiagonal matrix which is stored using an array D for the diagonal elements and an array U for the superdiagonal elements. Modify the program in Table 7-3 to implement Gauss-Seidel iterations for a tridiagonal system. Test your program using the system of 100 equations given by

$$\begin{aligned} a_{i+1,i} = a_{i,i+1} &= -1 \quad \text{for } i = 1, \cdots, 99, \\ a_{ii} &= 2 \quad \text{for } i = 1, \cdots, 100, \\ b_1 = b_{100} &= 1, \\ b_i &= 0 \quad \text{for } i = 2, \cdots, 99, \end{aligned} \tag{53}$$

where the starting guess is $\mathbf{x}_0 = \mathbf{0}$. Stop the iterations when the iteration difference satisfies the inequality $\|\mathbf{x}^{\text{new}} - \mathbf{x}^{\text{old}}\|_1/\|\mathbf{x}^{\text{new}}\|_1 \leqq .001$. What is the 1-norm relative error in your computed solution? (*Hint:* The correct solution is $x_i = 1$ for every i.)

7-5. You saw in Problem 7-4 that the iteration difference $\|\mathbf{x}^{\text{new}} - \mathbf{x}^{\text{old}}\|$ is often a poor measure of the iteration error. This discrepancy between the iteration difference and the iteration error is related to the system's poor conditioning. For an ill-conditioned system, a relatively small iteration difference does not imply a relatively small error. In this exercise, we develop a stopping criterion for iterative methods. The error $\mathbf{e}_k = \mathbf{x}_k - \mathbf{x}$ at step k can be expressed

$$\mathbf{e}_k = \mathbf{x}_k - \mathbf{x} = \mathbf{x}_k - \mathbf{x}_{k+1} + \mathbf{x}_{k+1} - \mathbf{x} = -\mathbf{d}_{k+1} + \mathbf{e}_{k+1}, \tag{54}$$

where $\mathbf{d}_k = \mathbf{x}_k - \mathbf{x}_{k-1}$ is the iteration difference. From the analysis in Section 7-3, it follows that the norm of the error decays on the average by the factor $\rho(\mathbf{M})$ each iteration. Taking the norm of (54), applying the triangle inequality, and replacing $\|\mathbf{e}_{k+1}\|$ by $\rho(\mathbf{M})\|\mathbf{e}_k\|$, we conclude that

$$\|\mathbf{e}_k\| \lesssim \frac{\|\mathbf{d}_{k+1}\|}{1 - \rho(\mathbf{M})}.$$

Therefore, if

$$\|\mathbf{d}_{k+1}\| < (1 - \rho(\mathbf{M}))\|\mathbf{x}_k\|10^{-t}, \tag{55}$$

then the relative error $\|\mathbf{e}_k\|/\|\mathbf{x}_k\|$ is approximately less than 10^{-t}. An algorithm for estimating $\rho(\mathbf{M})$ is developed in Section 7-3. Incorporate the convergence criterion

(55) with $t = 2$ in the program developed for Problem 7-4 and verify that the relative error in the computed solution is on the order of .01.

7-6. If $\|\mathbf{M}\| < 1$, the stopping criterion (55) can be replaced by

$$\|\mathbf{d}_{k+1}\| < (1 - \|\mathbf{M}\|)\|\mathbf{x}_k\|10^{-t}$$

since $\rho(\mathbf{M}) \leqq \|\mathbf{M}\|$. Taking $t = 3$ and using the ∞-norm, apply this stopping criterion to the Jacobi iterations in Problem 7-2.

7-7. Using SOR, solve the tridiagonal system $\mathbf{Ax} = \mathbf{b}$ given by (53). Try the following five values for ω: 1, 1.5, 1.8, 1.9, 1.94. Count the number of iterations until the 1-norm relative error is less than or equal to .01. We observed in Section 7-3 that SOR diverges if $\omega > 2$. Since ω_{opt} is often just less than 2, the optimal SOR scheme skirts that fine line separating brilliant success from total disaster.

7-8. The $N \times N$ tridiagonal matrix with 2's on the diagonal and -1's off the diagonal has the property that $1 - \mu = CN^p$, where C is a slowly varying function of N and μ is the spectral radius of the corresponding Gauss-Seidel $\mathbf{M}$. Using the Rayleigh quotient approximation to μ described in Sec. 7-3, estimate μ when N is 20 and when N is 40. Then estimate C and p using the technique developed in Chapter 1, equations (30) and (31). Utilizing the asymptotic relation

$$\mu \approx 1 - CN^p,$$

estimate ω_{opt} when N is 100 and when N is 1000.

7-9. By (41) each iteration of the conjugate gradient method decreases the error by about the factor

$$\frac{1 - \sqrt{\lambda_n/\lambda_1}}{1 + \sqrt{\lambda_n/\lambda_1}}, \tag{56}$$

and by (35), each iteration of the Chebyshev accelerated scheme decreases the error by about the factor

$$\left(\frac{1 - \sqrt{1 - \nu}}{1 + \sqrt{1 - \nu}}\right)^{1/2}. \tag{57}$$

Consider the $n \times n$ tridiagonal matrix $\mathbf{A}$ with 2's on the diagonal and with -1's off the diagonal and the corresponding Jacobi $\mathbf{M}$. Estimate the ratio between (56) and (57) when n is large. (Remember, the eigenvalues of $\mathbf{A}$ are relevant to the convergence of the conjugate gradient method while the eigenvalues of $\mathbf{M}$ are relevant to the convergence of the Chebyshev accelerated scheme.)

7-10. Suppose that zero is an eigenvalue of $\mathbf{A}$ and consider the splitting $\mathbf{A} = \mathbf{S} - \mathbf{T}$, where $\mathbf{S}$ is invertible. Show that $\rho(\mathbf{S}^{-1}\mathbf{T}) \geqq 1$.

7-11. Given a square matrix $\mathbf{A}$, construct a sequence of square matrices using the recurrence

$$\mathbf{X}_{k+1} = \mathbf{X}_k + \mathbf{X}_k(\mathbf{I} - \mathbf{AX}_k).$$

Show that $\rho(\mathbf{I} - \mathbf{AX}_0) < 1$ if and only if $\mathbf{A}$ is invertible and the $\mathbf{X}_k$ approach $\mathbf{A}^{-1}$. [*Hint:* Show that $\mathbf{I} - \mathbf{AX}_k = (\mathbf{I} - \mathbf{AX}_0)^{2^k}$. Also, from the analysis in Section 3-5, we know that the matrix $\mathbf{I} - \mathbf{B}$ is invertible whenever $\rho(\mathbf{B}) < 1$.]

REFERENCES

Some excellent references for iterative methods applied to linear systems include the book [H3] by Hageman and Young, the book [V2] by Varga, and the book [Y2] by Young. Some conference proceedings which contain many helpful articles include [B7], [B21], [D24], [R2], and [R9]. As their names suggest, Jacobi and Gauss-Seidel iterations can be traced back to Gauss, Jacobi, and Jacobi's student

Seidel. The successive overrelaxation iteration was simultaneously introduced by Frankel [F7] and Young [Y1]. The result that SOR diverges for ω greater than 2 or for ω less than 0 is due to Kahan [K1]. The result that SOR converges if the coefficient matrix is symmetric and positive definite and ω is between 0 and 2 is due to Ostrowski [O6]. The conjugate gradient method was developed by Hestenes and Stiefel [H13]. A comprehensive treatment of conjugate direction methods is given by Hestenes in the book [H12]. The estimate (40) appears in [S13], the estimate (41) appears in [D2], and the estimate (42) appears in [L4]. The preconditioner $\mathbf{P} = \mathbf{I} + s\mathbf{C}\mathbf{C}^T$ is discussed in [H6]. A technique for preconditioning a Toeplitz system using a circulant matrix appears in [S16]. The efficient implementation of SSOR and incomplete factorization preconditioners is studied in [B3] and [E1]. Some generalizations of the conjugate gradient method which apply to nonsymmetric systems are developed in [E2], [E5], [S2], and [S3]. Besides those references cited in Section 7-6, other references for the multigrid method include [B4], [B10], [B13], [H11], [M1], [M3], [M5], and [N1]. With regard to the iterative scheme incorporated in the multigrid method, both the conjugate gradient and the Gauss-Seidel schemes are often effective. There is also recent evidence (see [K9]) that the SOR scheme can be very effective when the relaxation parameter ω is slightly larger than 1. Despite the conclusion of Problem 7-10, iterative methods can still be applied to a system with a singular coefficient matrix—see the paper [R8] by Rose. Problem 7-11 essentially appears in [D1].

8

Numerical Software

8-1. EISPACK, LINPACK, MINPACK, AND NAPACK

In this chapter we survey software and subroutine packages that are relevant to applied numerical linear algebra. For an excellent study of the sources and development of mathematical software, see the collection of papers [C7] edited by Cowell. We begin with LINPACK, a collection of Fortran subroutines that analyze and solve linear equations. The software for LINPACK can be obtained from either

National Energy Software Center (NESC)
Argonne National Laboratory
9700 South Cass Avenue
Argonne, IL 60439
Phone: 312-972-7250

or

IMSL Distribution Services
2500 ParkWest Tower One
2500 CityWest Boulevard
Houston, Texas 77042-3020
Phone: 713-782-6060 (In U.S. and outside Texas: 1-800-222-IMSL)
Telex: 791923 IMSL INC HOU

If you reside in a country included in the European Organization for Economic Cooperation and Development, the software can be obtained without charge from

NEA Data Bank
B.P. No. 9 (Bat. 45)
F-91191 Gif-sur-Yvette
France

LINPACK, which is thoroughly documented in [D8], contains subroutines to factor a matrix, to solve a factored system, to estimate the condition number, to compute the determinant, to update a Cholesky decomposition after various perturbations in the coefficient matrix, and to compute the singular value decomposition. Each LINPACK subroutine has a name composed of five letters that describes the subroutine's purpose. The first letter, which indicates the level of precision and the type of data, has the following meaning:

S: Real single-precision computation
D: Real double-precision computation
C: Complex single-precision computation
Z: Complex double-precision computation

The next two letters indicate the form of the matrix or its decomposition:

GE: General matrix
GB: General band matrix
PO: Symmetric positive definite matrix
PP: Symmetric positive definite matrix stored in compressed mode
PB: Symmetric positive definite band matrix
SI: Symmetric indefinite matrix
SP: Symmetric indefinite matrix stored in compressed mode
HI: Hermitian indefinite matrix
HP: Hermitian indefinite matrix stored in compressed mode
GT: General tridiagonal matrix
PT: Positive definite tridiagonal matrix
TR: Triangular matrix
CH: Cholesky decomposition
QR: Orthogonal triangular decomposition
SV: Singular value decomposition

The final two letters indicate the nature of the computation:

FA: Compute a **LU** factorization
CO: Compute a **LU** factorization and estimate condition number
DC: Compute a decomposition
SL: Use the results of FA, CO, or DC to solve a linear system
DI: Compute the determinant, inverse, or inertia
UD: Update a Cholesky decomposition

DD: Downdate a Cholesky decomposition

EX: Update a Cholesky decomposition after an interchange

For example, the subroutine SGEFA computes the single-precision factorization of a general matrix while the subroutine SGESL solves a general factored system. The subroutine DGBFA computes the double-precision factorization of a band matrix while the subroutine DGBSL solves a factored band system using double-precision arithmetic.

LINPACK makes use of a more general package called the basic linear algebra subprograms or the BLAS. Although the BLAS are included in LINPACK, these subroutines appear separately in Algorithm 539 of the Collected Algorithms of the ACM and, like all the algorithms published in the Collected Algorithms of the ACM since about 1975, Algorithm 539 can be obtained from the IMSL Distribution Service at the address given previously. The operations performed by the BLAS on vectors include the following: Find the index of the largest vector component, compute the norm of a vector, copy one vector to another vector, interchange two vectors, compute a dot product, add a vector to the product between a scalar and another vector, and apply a Givens transformation. Currently, there is a proposal [D6] for extending the BLAS to include matrix-vector operations which can be implemented efficiently using vector and parallel computers. The proposed Extended BLAS (or Level 2 BLAS) will include subroutines to compute the following: the sum of a vector and a scalar-matrix-vector product, the sum of matrix and either a rank one correction or a rank two correction, and the solution to a triangular system of equations.

One of the most widely used packages to compute the eigenvalues and the eigenvectors of a matrix is called EISPACK. The software for EISPACK can be obtained from either NESC, IMSL, or the NEA Data Bank (the addresses appear at the beginning of the chapter). Documentation for EISPACK is given in [S12] and [G1]. There have been three editions of EISPACK. Edition 2 expanded the capabilities of Edition 1 by including subroutines to process real symmetric band matrices, to solve the generalized eigenproblem, and to compute the singular value decomposition. Edition 3 fine-tunes Edition 2 by eliminating the machine dependent constants, by reducing the probability of overflow or underflow, and by modifying some subroutines to decrease their execution time. In solving an eigenproblem, a sequence of subroutines is invoked to perform the following tasks:

1. When the matrix is nonsymmetric, it can be scaled (or balanced) so that rounding errors are reduced in subsequent calculations. A nonsymmetric tridiagonal matrix with the property $a_{i,i+1}\, a_{i+1,i} \geqq 0$ for every i can be symmetrized using a balancing transformation.
2. A nonsymmetric matrix can be reduced to upper Hessenberg form and a symmetric matrix can be reduced to tridiagonal form using a similarity transformation. Generalized eigenproblems may be reduced to standard eigenproblems or to quasi-triangular form.
3. Eigenvalues and possibly the eigenvectors of an upper Hessenberg matrix or a symmetric tridiagonal matrix are computed using either a **QR** type method or a Sturm sequence method.

4. Eigenvectors associated with the original matrix (before the step 2 reduction) are computed using results from step 3.
5. If the matrix was balanced in step 1, the inverse balancing transformation must be applied to the eigenvectors computed in step 4.

Since the EISPACK subroutine names are not as systematic as the LINPACK subroutine names, we give a brief description of each subroutine contained in EISPACK. These subroutines are partitioned into the five categories described above. The EISPACK subroutines related to the singular value decomposition are listed after the fifth category. Subroutines documented in [G1] are flagged with a superscript 1, subroutines documented in [S12] are flagged with a superscript 2, and subroutines documented in both [G1] and [S12] are flagged with a superscript 3.

TABLE 8-1 BALANCE MATRIX

Subroutine	Matrix structure	Action
BALANC[1]	General	Apply balancing transformation
CBAL[1]	Complex general	Apply balancing transformation
FIGI[1]	Tridiagonal	Use a balancing transformation to symmetrize a nonsymmetric tridiagonal matrix for which $a_{i,i+1}\, a_{i+1,i} \geqq 0$ for every i
FIGI2[1]	Tridiagonal	Similar to FIGI except that the balancing transformation is also output

TABLE 8-2 REDUCE MATRIX

Subroutine	Matrix structure	Action
BANDR[2]	Symmetric band	Reduce to symmetric tridiagonal form; optionally output the similarity transformation
COMHES[1]	Complex general	Reduce to upper Hessenberg form using elimination
CORTH[1]	Complex general	Reduce to upper Hessenberg form using Householder matrices
ELMHES[1]	General	Reduce to upper Hessenberg form using elimination
ELTRAN[1]	General	Use the output of ELMHES to construct the similarity transformation that generates the upper Hessenberg form
HTRID3[1]	Complex Hermitian	Reduce to symmetric tridiagonal matrix using Householder matrices; input matrix stored in compressed mode
HTRIDI[1]	Complex Hermitian	Reduce to symmetric tridiagonal matrix using Householder matrices
ORTHES[1]	General	Reduce to upper Hessenberg form using Householder matrices
ORTRAN[1]	General	Use the output of ORTHES to construct the similarity transformation that generates the upper Hessenberg form

Subroutine	Matrix structure	Action
QZHES[2]	General	Reduce the generalized eigenproblem to standard form, where one matrix is upper Hessenberg and the other matrix is upper triangular
QZIT[2]	Upper Hessenberg	Given the generalized eigenproblem $\mathbf{Ax} = \lambda\mathbf{Bx}$, where **A** is upper Hessenberg and **B** is upper triangular, reduce **A** to quasi-upper triangular form using the **QZ** algorithm
REDUC[2]	Symmetric	Reduce the symmetric generalized eigenproblem $\mathbf{Ax} = \lambda\mathbf{Bx}$, where **B** is positive definite to the standard symmetric eigenproblem using the Cholesky factorization of **B**
REDUC2[2]	Symmetric	Reduce the eigenvalue problem $(\mathbf{AB})\mathbf{x} = \lambda\mathbf{x}$, where both **A** and **B** are symmetric and either **A** or **B** is positive definite to the standard symmetric eigenproblem using the Cholesky factorization
TRED1[3]	Symmetric	Reduce to symmetric tridiagonal form using Householder matrices
TRED2[3]	Symmetric	Reduce to symmetric tridiagonal form using Householder matrices; the similarity transformation that yields the tridiagonal form is also constructed
TRED3[3]	Symmetric	Reduce to symmetric tridiagonal form using Householder matrices; input matrix stored in compressed mode

TABLE 8-3 COMPUTE EIGENVALUES

Subroutine	Matrix structure	Action
BISECT[3]	Symmetric tridiagonal	Determine eigenvalues that lie on specified interval using Sturm sequences
BQR[2]	Symmetric band	Determine some eigenvalues using the **QR** method
COMLR[1]	Complex upper Hessenberg	Compute all eigenvalues using modified **LR** algorithm
COMLR2[1]	Complex upper Hessenberg	Compute all eigenvalues and eigenvectors using modified **LR** method
COMQR[1]	Complex upper Hessenberg	Compute all eigenvalues using **QR** algorithm
COMQR2[1]	Complex upper Hessenberg	Compute all eigenvalues and eigenvectors using **QR** algorithm
HQR[1]	Upper Hessenberg	Compute all eigenvalues using the implicit **QR** method
HQR2[1]	Upper Hessenberg	Compute all eigenvalues and eigenvectors using the implicit **QR** method
IMTQL1[1]	Symmetric tridiagonal	Compute eigenvalues using the implicit **QL** method
IMTQL2[1]	Symmetric tridiagonal	Compute the eigenvalues and eigenvectors using the implicit **QL** method; if the eigenpairs of a symmetric matrix are desired, input the similarity transformation computed by TRED2

Subroutine	Matrix structure	Action
IMTQLV[1]	Symmetric tridiagonal	Compute eigenvalues using implicit **QL** method while preserving the input matrix
QZVAL[2]	Quasi-upper triangular	Compute the eigenvalues for the generalized eigenproblem $\mathbf{Ax} = \lambda\mathbf{Bx}$, where $\mathbf{A}$ is quasi-upper triangular and $\mathbf{B}$ is upper triangular
RATQR[1]	Symmetric tridiagonal	Determine extreme eigenvalues using the **QR** method with Newton corrections
TQL1[3]	Symmetric tridiagonal	Compute all eigenvalues using the **QL** algorithm
TQL2[3]	Symmetric tridiagonal	Compute all eigenvalues and eigenvectors using the **QL** method; if the eigenpairs of a symmetric matrix are desired, input the similarity transformation computed by TRED2
TQLRAT[3]	Symmetric tridiagonal	Determine all eigenvalues using Reinsch's version of the **QL** algorithm
TRIDIB[3]	Symmetric tridiagonal	Compute those eigenvalues between specified indices using the Sturm sequence property
TSTURM[3]	Symmetric tridiagonal	Compute those eigenvalues in a specified interval using the Sturm sequence property; the corresponding eigenvectors are computed using the inverse power iteration

TABLE 8-4 COMPUTE EIGENVECTORS

Subroutine	Matrix structure	Action
BANDV[2]	Symmetric band	Given approximate eigenvalues, use inverse power iteration to obtain corresponding eigenvectors
CINVIT[1]	Complex upper Hessenberg	Given approximate eigenvalues, use inverse power iteration to obtain corresponding eigenvectors
COMBAK[1]	Complex general	Given eigenvectors of upper Hessenberg matrix computed by COMHES, compute corresponding eigenvectors of original matrix
CORTB[1]	Complex general	Given eigenvectors of upper Hessenberg matrix computed by CORTH, compute corresponding eigenvectors of original matrix
ELMBAK[1]	General	Given eigenvectors of the upper Hessenberg matrix output by ELMHES, compute corresponding eigenvectors of original matrix
HTRIB3[1]	Complex Hermitian	Given eigenvectors of the real symmetric tridiagonal matrix output by HTRID3, compute the corresponding eigenvectors of the original matrix
HTRIBK[1]	Complex Hermitian	Given eigenvectors of the real symmetric tridiagonal matrix output by HTRIDI, compute the corresponding eigenvectors of the original matrix
INVIT[1]	Upper Hessenberg	Compute eigenvector corresponding to given eigenvalue using inverse power iteration
ORTBAK[1]	General	Given eigenvectors of the upper Hessenberg matrix output by ORTHES, compute the corresponding eigenvectors of the original matrix

Subroutine	Matrix structure	Action
QZVEC[2]	Quasi-upper triangular	Given the eigenvalues for the generalized eigenproblem $\mathbf{Ax} = \lambda\mathbf{Bx}$, where $\mathbf{A}$ is quasi-upper triangular and $\mathbf{B}$ is upper triangular, compute the corresponding eigenvectors
REBAK[2]	Symmetric	Given the eigenvectors of the symmetric matrix output by REDUC or REDUC2, compute the eigenvectors corresponding to the original generalized eigenproblem
REBAKB[2]	Symmetric	Given the eigenvectors of the symmetric matrix output by REDUC2, compute the eigenvectors corresponding to the original eigenproblem $(\mathbf{AB})\mathbf{x} = \lambda\mathbf{x}$
TINVIT[3]	Symmetric tridiagonal	Compute the eigenvectors corresponding to given eigenvalues using the inverse power iteration
TRBAK1[3]	Symmetric	Given the eigenvectors of the symmetric tridiagonal matrix output by TRED1, compute the corresponding eigenvectors of the original symmetric matrix
TRBAK3[3]	Symmetric	Given the eigenvectors of the symmetric tridiagonal matrix output by TRED3, compute the corresponding eigenvectors of the original symmetric matrix

TABLE 8-5 INVERT BALANCING TRANSFORMATION

Subroutine	Matrix structure	Action
BAKVEC[1]	Nonsymmetric tridiagonal	Invert the balancing transformation made by FIGI
BALBAK[1]	General	Invert the balancing transformation made by BALANC
CBABK2[1]	Complex general	Invert the balancing transformation made by CBAL

TABLE 8-6 COMPUTE SINGULAR VALUE DECOMPOSITION

Subroutine	Matrix structure	Action
MINFIT[2]	General	For the linear system $\mathbf{Ax} = \mathbf{b}$, compute the singular value decomposition $\mathbf{A} = \mathbf{QSP}^T$ and the vector $\mathbf{Q}^T\mathbf{b}$
SVD[2]	General	Compute the singular value decomposition

For example, all the eigenvalues of a general real matrix can be computed using the following sequence of subroutines: BALANC, ORTHES, and HQR. All the eigenpairs of a general real matrix can be computed using the sequence BALANC, ORTHES, HQR2, ORTBAK, and BALBAK. Since the balancing steps are optional, the abbreviated sequence ORTHES, HQR2, and ORTBAK also works although the rounding errors may be more severe than when the matrix is balanced. The eigenvalues of a symmetric matrix stored in the compressed storage mode can be computed using TRED3 followed by TQLRAT. To assist in using EISPACK, 13 "drivers" are provided with the package. These drivers call commonly used sequences of subroutines.

Moler has developed an interactive computer program for matrix computations that provides easy access to the LINPACK and EISPACK subroutines. His program, called MATLAB [M8], can be obtained from

The Math Works, Inc.
20 North Main Street, Suite 250
Sherborn, Massachusetts 01770
Phone: 617-653-1415

For nonlinear systems of equations, a helpful software package is MINPACK. The MINPACK project is a research effort whose goal is to develop a systematized collection of quality optimization software. The first step in this research effort is the subroutine package MINPACK-1 (see [M10]) which solves nonlinear systems of equations

$$f_i(x_1, x_2, \cdots, x_n) = 0 \qquad \text{for } i = 1, 2, \cdots, n$$

and nonlinear least squares problems of the form

$$\text{minimize} \sum_{i=1}^{m} f_i(x_1, x_2, \cdots, x_n)^2.$$

The software and documentation for MINPACK-1 are available from the NESC and from IMSL (the addresses appear at the beginning of this chapter). When using the MINPACK software, the user has the option of providing the Jacobian matrix or having the software approximate the Jacobian matrix using finite differences. When the Jacobian is provided, more accurate numerical results are often obtained. If the user provides a subroutine to compute the Jacobian matrix, he or she can use the MINPACK subroutine CHKDER to test for errors in the Jacobian code. Subroutine CHKDER estimates the Jacobian using finite differences and compares the estimated Jacobian to the user's Jacobian. For systems of equations, MINPACK implements a hybrid method due to Powell. For the least squares problem, a version of the Levenberg-Marquardt algorithm is employed.

Finally, we discuss NAPACK, the subroutine package developed by the author in conjunction with this book. NAPACK contains subroutines to process matrices and equations. Each subroutine name consists of a stem that describes the action of the subroutine preceded by a prefix. If there is no prefix in front of the stem, the subroutine usually applies to a general real matrix. The prefixes, listed below, describe the data and the matrix type.

B: Band matrix
C: Complex matrix
E: Upper Hessenberg matrix
H: Symmetric band matrix
I: Symmetric matrix (symmetric pivoting)

K: General matrix (complete pivoting)
O: Circulant matrix
P: Tridiagonal matrix (partial pivoting)
S: Symmetric matrix
T: Tridiagonal matrix

The stems that allow one or more prefixes are the following:

Stem	Prefixes	Action
BAL	C	Balance the matrix
CON	B,E,H,I,K,P,S,T	Estimate condition number
DET	B,E,H,I,K,P,S,T	Compute the determinant
DIAG	C,E,H,S,T	Compute the diagonalization
FACT	B,E,H,I,K,P,S,T	Compute the **LU** factorization
HESS	C,H,S	Reduce to upper Hessenberg form (insert A prefix to also balance)
MULT	B,E,H,O,S,T	Multiply matrix by vector
PACK	C,R	Rearrange elements of an array so that elements of a square matrix are stored sequentially (use R prefix if matrix is rectangular)
POWER	C,M	Compute dominant eigenpairs by the power method (use M prefix to compute several eigenpairs)
SIM	C,H,S	Compute the similarity transform used in the reduction to either Hessenberg or tridiagonal form
SOLVE	B,E,H,I,K,O,P,S,T	Solve a factored system of equations
TRANS	B,E,K,P,T	Solve the transpose of a factored system
VALS	C,E,H,O,S,T	Compute eigenvalues
VECT	C,E,H,S,T	Compute eigenvector corresponding to given eigenvalue
VERT	B,C,E,H,I,K,O,P,S,T	Invert a matrix

The subroutines in NAPACK that do not have prefixes are the following:

Subroutine	Action
ADDCHG	Add one vector to another and evaluate 1-norm of the increment and 1-norm of the sum
BASIS	Compute an orthonormal basis for a collection of vectors
BIDAG	Reduce a matrix to bidiagonal form
BIDAG2	Reduce a matrix to bidiagonal form and evaluate all or part of the orthogonal matrices used in the reduction process
CG	Compute an unconstrained minimum for a multivariate function using the (preconditioned) conjugate gradient method
CZERO	Compute zeros of a (complex) polynomial
FFC	Conjugate fast Fourier transform
FFT	Fast Fourier transform
LANCZ	Perform an iteration of Lanczos method to reduce a matrix to tridiagonal form

Subroutine	Action
NEWTON	Apply one step of Newton's method to the characteristic polynomial for a tridiagonal matrix
NORM1	Estimate 1-norm of a matrix
NULL	Compute an orthonormal basis for the space perpendicular to a given collection of vectors
OVER	Compute the least squares solution to an overdetermined linear system
PRECG	Solve a linear system using preconditioned conjugate gradients
PSEUDO	Compute regularized pseudoinverse
QR	**QR** factor a matrix
QUASI	Use a quasi-Newton method to solve a nonlinear system
ROOT	Solve a scalar equation
RSOLVE	Compute the regularized solution to a linear system
SDIAG2	Same as SDIAG except the input matrix is not stored in compressed format
SING	Compute the singular value decomposition of a general matrix
SINGB	Compute singular value decomposition of a bidiagonal matrix
SLICE	Compute the eigenvalues contained on a given interval for a tridiagonal matrix whose cross-diagonal products are nonnegative
STOPIT	Test for convergence
TVAL	Compute the kth smallest or the kth largest eigenvalue of a tridiagonal matrix whose cross-diagonal products are nonnegative
UNDER	Compute the least squares (minimum norm) solution to an underdetermined linear system
UPDATE	Equate one vector to another and evaluate 1-norm of the difference and 1-norm of the updated vector
WHATIS	Print iteration number, iteration difference, and stopping criterion

Additional documentation for these subroutines appears in Appendix 1, in the subroutine comment statements, and throughout the text. The software for NAPACK can be ordered from

Professor William W. Hager
Department of Mathematics
Pennsylvania State University
University Park, PA 16802
Phone: 814-865-3873
Current cost: $60

Individual programs in NAPACK can be obtained by sending an electronic mail message to Psulib at Pennsylvania State University. Currently, this library can be accessed through the Arpanet, the Bitnet, and Uucp. On the Bitnet, the address is "psulib@psuvax1.bitnet." An alternative address is "psulib@psuvax1.psu.edu." To obtain a subroutine from NAPACK, say subroutine SOLVE, the contents of the electronic mail message is the following:

send solve from napack

Psulib will return to you by electronic mail a copy of subroutine solve. Psulib utilizes the software distribution system "Netlib" developed by Dongarra and

Grosse [D10]. More information concerning the structure of Netlib requests appears in Section 8-3.

8-2. SPARSE MATRIX PACKAGES

This section discusses packages for processing a sparse matrix. Although this discussion focuses on techniques to solve a sparse system of equations, we point out that a helpful package for sparse eigenproblems and for sparse singular value problems is LANCZOS, a collection of routines developed by Cullum and Willoughby and documented in [C9]. There are two general strategies for solving a sparse linear system: (a) apply an interative method, or (b) apply Gaussian elimination after permuting equations and unknowns to reduce the fill-in. ITPACK is a collection of iterative methods which is tailored to systems whose coefficient matrix is sparse, symmetric, and positive definite. ITPACK 2C (Algorithm 586 in the Collected Algorithms of the ACM) can be obtained from the IMSL Distribution Service at the address given in Section 8-1. ITPACK combines basic iterative procedures such as Jacobi's method, successive overrelaxation, and symmetric successive overrelaxation with acceleration procedures such as Chebyshev and conjugate gradient acceleration. The programs automatically determine the appropriate acceleration parameters. References related to the ITPACK algorithms include [H3], [K5], [K6], [Y2], and [Y3].

Although ITPACK can be applied to a linear system whose coefficient matrix is mildly nonsymmetric, it is most successful in solving a system whose coefficient matrix is symmetric and positive definite. For nonsymmetric linear systems, the software package PCGPAK can be utilized. This package can be ordered from

Scientific Computing Associates, Inc.
246 Church Street, Suite 408
New Haven, CT 06510
Phone: 203-777-7442
Cost: $1000

PCGPAK implements three related schemes: ORTHOMIN, the restarted generalized conjugate residual method, and the generalized minimal residual method. These schemes, described in [E2], [E5], [S2], and [S3], can also be combined with a preconditioner to accelerate the convergence. Three preconditioners included in PCGPAK are the incomplete **LU** factorization, the modified incomplete **LU** factorization, and the symmetric successive overrelaxation factorization.

SPARSPAK is a collection of subroutines which implement direct methods to solve a sparse linear system or a sparse constrained linear least squares problem of the form

$$\underset{\mathbf{x}\varepsilon\Omega}{\text{minimize}} \|\mathbf{D}(\mathbf{A}\mathbf{x} - \mathbf{a})\|_2,$$

where

$$\Omega = \{\mathbf{x}:\mathbf{x} \text{ minimizes } \|\mathbf{E}(\mathbf{B}\mathbf{x} - \mathbf{b})\|_2\}.$$

Since partial pivoting is not performed in SPARSPAK, the package is most successful when the coefficient matrix is symmetric and positive definite. To solve a linear system using SPARSPAK, the user supplies the structure of the coefficient matrix and the package reorders the equations and the unknowns to reduce the fill-in during Gaussian elimination. The user can specify one of five different reordering algorithms:

1. Reverse Cuthill-McKee algorithm
2. One-way dissection algorithm
3. Refined quotient tree algorithm
4. Nested dissection algorithm
5. Minimum degree algorithm

After the equations are reordered, the user supplies the nonzero coefficients and the package factors the coefficient matrix. Then the user supplies the right side and the package computes the solution to the linear system. SPARSPAK is documented in the book [G5] as well as in the paper [G6] and the reports [G7], [G8], [G9], and [G10]. The software for SPARSPAK can be obtained from

SPARSPAK
Waterloo Research Institute
University of Waterloo
Waterloo, Ontario
Canada, N2L 3G1

Another package for directly solving a sparse linear system without partial pivoting is the Yale Sparse Matrix Package (YSMP). YSMP contains subroutines for processing both symmetric and nonsymmetric coefficient matrices. Although a subroutine to implement the minimum degree algorithm is included in the package, any ordering algorithm can be utilized. YSMP is documented in [E3] and [E4]. Further information concerning the package can be obtained from either Eisenstat or Schultz at the following address:

Professor Stanley C. Eisenstat
Professor Martin H. Schultz
Department of Computer Science
Yale University
P.O. Box 2158
New Haven, CT 06520
Phone: 203-436-8160

YSMP can be ordered from

YSMP Librarian
Department of Computer Science
Yale University
P.O. Box 2158
New Haven, CT 06520

Several subroutines within the Harwell Subroutine Library can be used to process a sparse linear system. Some examples are the subroutines MA27, MA28, MA30, MA31, MA32, MA37, and ME28. These codes solve a wide range of sparse matrix problems. MA31 uses conjugate gradients preconditioned by a partial factorization and the others employ a direct method based on Gaussian elimination. An excellent general reference for these direct methods is the book [D20] by Duff, Erisman, and Reid. The code MA27 (see [D22]) performs some pivoting and is also appropriate for a problem whose coefficient matrix is not positive definite. Codes MA28 (see [D14]) and MA30 solve unsymmetric systems using a technique that both preserves sparsity and is numerically stable. MA32, documented in [D15], [D18], and [D19], uses a frontal technique to solve an unsymmetric linear system and will, if requested, hold the factorization on auxiliary storage. Using auxiliary storage, systems with more than 100,000 equations can be solved. A unique feature of MA32 is that it allows input either by equations or by elements. A very efficient version of MA32 for the CRAY supercomputers, is also available. MA37 (see [D23]), which like MA27 employs a multifrontal technique, is particularly efficient if the coefficient matrix is structurally symmetric. ME28 (see [D16]) solves a complex sparse system of linear equations. The Harwell library also contains several routines for manipulating a sparse matrix. For example, MC09 computes a matrix-vector product, MC10, MC12, and MC19 scale a sparse matrix. MC20 sorts the nonzero elements of a sparse matrix, MC22 permutes a sparse matrix, MC13 (see [D21]) and MC23 permute a sparse matrix to block triangular form, and MC24 calculates an *a posteriori* bound on stability. Copies of these subroutines and their documentation can be obtained from

> Mr. S. Marlow
> Building 8.9
> Harwell Laboratory
> Didcot, Oxon, OX11 ORA, England
> Phone: (0235) 24141, extension 3430; international: 44-235-24141

Information concerning the algorithms used in the Harwell codes mentioned above can be obtained from

> Dr. I. S. Duff
> Building 8.9
> Harwell Laboratory
> Didcot, Oxon, OX11 ORA, England
> Phone: 44-235-24141, extension 2670 or 2051

Other useful packages for manipulating a sparse matrix include ACM Algorithms 408 and 601, which perform operations such as the addition, multiplication, and transposition of sparse matrices, and ACM Algorithms 508, 509, and 582, which reduce the bandwidth and profile of a matrix using row and column permutations. Also, ACM Algorithm 533 (the subroutine package NSPIV) can be used to solve a nonsymmetric system of equations using partial pivoting. The software for algorithms published in the *ACM Transactions on Mathematical Software* can be obtained from IMSL Distribution Services.

In some applications involving sparse matrices, the coefficient matrix is Toeplitz or circulant. A matrix is Toeplitz or circulant if along each diagonal, every coefficient is the same. A 4×4 Toeplitz matrix has the form

$$\mathbf{A} = \begin{bmatrix} a_0 & a_1 & a_2 & a_3 \\ a_{-1} & a_0 & a_1 & a_2 \\ a_{-2} & a_{-1} & a_0 & a_1 \\ a_{-3} & a_{-2} & a_{-1} & a_0 \end{bmatrix}.$$

Thus a 4 $\times$ 4 Toeplitz matrix is described in terms of the seven scalars a_{-3}, a_{-2}, a_{-1}, a_0, a_1, a_2, and a_3, which form the seven diagonals of **A**. In many important applications, most of the a_i are zero so that the matrix is sparse. A 4 $\times$ 4 circulant matrix has the form

$$\mathbf{A} = \begin{bmatrix} a_1 & a_2 & a_3 & a_4 \\ a_4 & a_1 & a_2 & a_3 \\ a_3 & a_4 & a_1 & a_2 \\ a_2 & a_3 & a_4 & a_1 \end{bmatrix}.$$

A 4 $\times$ 4 circulant matrix is described in terms of the four scalars a_1, a_2, a_3, and a_4. Notice that each row of a circulant matrix is obtained from the preceding row by shifting the coefficients to the right and wrapping the last coefficient around to the start of the next row. One of the beauties of a circulant matrix is that it is essentially diagonalized by a Fourier transform. That is, a $n \times n$ circulant matrix **A** constructed from the scalars $a_1, a_2, \cdots, a_n$ has the diagonalization $\mathbf{A} = \mathbf{Q}\mathbf{\Lambda}\mathbf{Q}^*$, where **Q** and **Λ** are defined by

$$q_{ij} = \frac{1}{\sqrt{n}} W^{(i-1)(j-1)}, \qquad W = e^{2\pi \mathbf{i}/n}, \qquad \lambda_{ii} = \sqrt{n} \sum_{j=1}^{n} q_{ij} a_j.$$

Thus the solution to a circulant system $\mathbf{Ax} = \mathbf{b}$ can be expressed $\mathbf{x} = \mathbf{Q}\mathbf{\Lambda}^{-1}\mathbf{Q}^*\mathbf{b}$. Since computing a product such as $\mathbf{Q}^*\mathbf{b}$ or computing a diagonal element λ_{ii} is essentially the same as computing a Fourier transform, the fast Fourier transform can be used to quickly solve a circulant system when n is a power of 2 or when n has many factors. If n is exactly a power of 2, the computing time is proportional to $n \log_2 n$. The package TOEPLITZ, documented in [A7] and distributed by the IMSL Distribution Services, contains subroutines to solve a linear system whose coefficient matrix is Toeplitz or circulant. Also, NAPACK contains routines to process a circulant matrix using the fast Fourier transform.

Many sparse matrix equations arise from the discretization of partial differential equations. Some packages tailored to partial differential equations are summarized below. A catalog for sparse matrix software was prepared in conjunction with the Sparse Matrix Symposium held in Fairfield Glade, Tennessee, October 24–27, 1982 (also see the article [D17] by Duff). The Sparse Matrix Symposium catalog can be obtained from either:

Robert C. Ward
Union Carbide Corporation Nuclear Division
P.O. Box Y, Oak Ridge, Tennessee, 37830

or

Michael T. Heath
Oak Ridge National Laboratory
P.O. Box Y, Building 9207, Oak Ridge, Tennessee, 37830.

Some of the information presented below is extracted from this catalog as well as from the article [B8] by Boisvert and Sweet.

CMMPAK: Solve the Helmholtz equation with Dirichlet or Neumann boundary conditions on general bounded two-dimensional domains. The capacitance matrix method [P11] is used to transform the domain to a rectangle where fast direct methods are applied. There is no charge for the software which can be obtained from W. Proskurowski, Department of Mathematics, University of Southern California, Los Angeles, CA 90007.

D03EAF: Solve Laplace's equation on an arbitrary two-dimensional domain bounded internally or externally by one or more closed contours on which either the value of the solution or its normal derivative is prescribed. The software is available as a part of the NAG library, which is marketed by Numerical Algorithms Group Inc., 1250 Grace Court, Downers Grove, IL 60515 (phone: 312-971-2337) and Numerical Algorithms Group Ltd., NAG Central Office, 7 Banbury Road, Oxford OX2 6NN, England (international phone: 44-865-511245).

ELLPACK: Solve linear second-order elliptic partial differential equations in two and three space dimensions. General domains can be handled in two dimensions and rectangular domains can be handled in three dimensions. The package utilizes part of the Yale Sparse Matrix Package, ITPACK, and ACM algorithm 533. Documentation is given in [R7]. The software can be obtained from John R. Rice, Mathematical Science 428, Purdue University, West Lafayette, IN 47907 (phone: 317-494-6007).

FFT9: Solve $au_{xx} + bu_{yy} + cu = f$, where a, b, and c are constants, the domain is a rectangle, and the boundary conditions are Dirichlet. This is ACM Algorithm 543 [H19], which can be obtained from IMSL Distribution Services.

FISHPAK: Solve linear separable elliptic equations in two and three space dimensions. Both the cyclic reduction algorithm (see [B24], [S19], and [S21]) and Swarztrauber's FFTPAK (fast Fourier transform) package are utilized. Documentation for the package is given in ACM Algorithm 541 [S20]. The software, which costs about $300, can be obtained from NCAR Program Library, National Center for Atmospheric Research, P.O. Box 3000, Boulder, CO 80307. For further information, contact Paul N. Swarztrauber or Roland A. Sweet.

GRIDPACK: Solve problems (such as discretized partial differential equations) defined on grids using the multigrid method. Documentation is provided with the package and a general description of the methods is given in [B11]. There is no charge for the software, which can be obtained from the Department of Applied Mathematics, The Weizmann Institute of Science, Rehovot, Israel, 76100. For further information, contact Dr. Dan Ophir at the Weizmann Institute (international phone: 972-54-83545).

HELM3D: Solve the Helmholtz equation on a general bounded three-dimensional domain with Dirichlet boundary conditions. The capacitance matrix technique is used to reduce the problem to a cube where fast direct methods are utilized. This is ACM Algorithm 572 [O1], which can be obtained from the IMSL Distribution Services.

ICCG: Solve a linear variable coefficient self-adjoint problem on a rectangular domain with mixed or periodic boundary conditions. The discretized linear system is solved by the conjugate gradient method using an incomplete Cholesky factorization for the preconditioner. The package is documented in [V1] and the software can be obtained from ACCU-Reeks nr. 29, Academic Computer Centre, Budapestlaan 6, de Uithof–Utrecht, The Netherlands.

MADPACK: Solve systems of linear equations using multigrid or aggregation–disaggregation methods. A wide class of partial differential equations can be solved, regardless of the discretization method. Some documentation is given in [D12] and [D13]. For additional information, contact Craig Douglas, IBM T. J. Watson Research Center, P.O. Box 218, Yorktown Heights, NY 10598 (phone: 914-945-1475).

MG00: Solve linear variable coefficient equations of the form

$$au_{xx} + bu_{yy} + cu = f \quad \text{or} \quad \nabla \cdot (p \nabla u) + cu = f$$

on rectangular domains with Dirichlet, Neumann, or mixed boundary conditions. The package is documented in [F1] and the software can be obtained without charge from GMD, Postfach 1240, D-5205 St. Augustin 1, West Germany.

PDE2D: Solve second-order quasi-linear homogeneous partial differential equations in two space variables using either the conjugate gradient method or successive overrelaxation. The package is documented in [R4] and the software can be obtained without charge from Robert Renka, Department of Computer Sciences, North Texas State University, P.O. Box 13886, Denton, TX 76203.

PLTMG and PLTMGC: Solve nonlinear elliptic systems with mixed boundary conditions on bounded two-dimensional domains. The discrete equations are solved using the multigrid method and a coarse user-supplied mesh is refined adaptively. PLTMGC uses the continuation method to compute the solution as a function of a scalar parameter, while PLTMG solves a single equation (with no parameter dependence). The methods employed by PLTMG are described in [B1] and [B5], while PLTMGC is described in [B2]. The software can be obtained without charge from R. E. Bank, Department of Mathematics, University of California at San Diego, La Jolla, CA 92093.

SLDGL: Solve nonlinear elliptic and parabolic systems of partial differential equations on rectangular domains in two and three space dimensions. The package, described in [S6], can be obtained from W. Schonauer, Rechenzentrum der Universität Karlsruhe, D-7500 Karlsruhe, West Germany.

TWODEPEP: Solve nonlinear systems of elliptic equations with nonlinear Dirichlet or Neumann boundary conditions on general bounded two-dimensional domains. Parabolic problems and eigenvalue problems can also be handled. The package uses a "block" elimination technique to solve linear equations and Newton's method for nonlinear problems. The Cuthill-McKee algorithm is used to number nodes and a special bandwidth reduction algorithm is used to decrease the bandwidth of the Jacobian. Storage outside main memory is organized using the frontal method. An overview of the algorithms used in TWODEPEP appears in [S9] and the software can be obtained from IMSL. The cost of the software is about \$1500 per year (\$900 for universities).

Besides these rather general partial differential equation packages, there are also various packages of applications-oriented partial differential equations software. Some examples are the structural analysis systems NASTRAN [M4],

AKSA [S7], and SAP [W7]. For a comprehensive discussion of finite element structural mechanics programs, see the book [F9] by Fredriksson and Mackerle as well as the article [N2] by Noor.

8-3. SOFTWARE LIBRARIES

In this section we discuss some of the organizations that provide mathematical software. If you are just interested in obtaining a specific program or a related set of programs from a software package, then "Netlib" [D10] is one possible source. As of January 1, 1987, Netlib contains the following packages:†

Software	Description
ALLIANT	Programs collected from Alliant users
APOLLO	Programs collected from Apollo users
BENCHMARK	Various benchmark programs
BIHAR	Solve biharmonic equation in rectangular and polar coordinates
BMP	Brent's multiple precision package
CONFORMAL	Schwarz-Christoffel codes by Trefethen, Bjorstad, and Grosse
CORE	Machine constants, BLAS
DOMINO	C-language programs for communication and scheduling of multiple tasks
EISPACK	Compute the eigenpairs of a matrix
ELEFUNT	Programs to test the elementary functions provided with Fortran compilers
ERRATA	Corrections to numerical books
FISHPACK	Solve separable elliptic partial differential equations; codes by Swarztrauber and Sweet
FITPACK	A package for splines under tension
FFTPACK	Fast Fourier transform
FMM	Routines from the book *Computer Methods for Mathematical Computations* by Forsythe, Malcolm, and Moler
FNLIB	Fullerton's special function library
GO	Routines that have been widely used but are not available through the standard libraries (Golden Oldies)
HARWELL	Currently contains just the sparse matrix routine MA28 from the Harwell library
HOMPACK	Solve systems of nonlinear equations by homotopy methods
ITPACK	Solve linear systems using iterative methods; codes by Young and Kincaid
LANCZOS	Lanczos programs by Cullum and Willoughby for computing eigenpairs of real symmetric matrices, for computing eigenpairs of Hermitian matrices, and for computing singular values and vectors of real, rectangular matrices
LASO	Scott's Lanczos program for computing a few eigenvalues of a large (sparse) symmetric matrix
LINPACK	Analyze and solve linear systems; programs by Dongarra, Bunch, Moler, and Stewart
LP	Data for a set of linear programming test problems

† The subroutine package NAPACK is scheduled to appear in Netlib during the Fall, 1987.

Software	Description
MACHINES	Information on high-performance computers that are available or are soon to be available
MICROSCOPE	Alfeld's tool for checking smoothness of functions
MINPACK	Solve nonlinear systems of equations and nonlinear least squares problems; codes by Moré, Garbow, and Hillstrom
MISC	Contains various pieces of software collected over time (includes Netlib program)
ODEPACK	Ordinary differential equations solver by Hindmarsh
PARANOIA	Kahan's program to explore the floating-point system on your computer
PCHIP	Piecewise Hermite cubic interpolation; codes by Fritsch and Carlson
PORT	Public subset of the PORT library
PPPACK	Subroutines from the book *A Practical Guide to Splines* by Carl de Boor
QUADPACK	Numerical quadrature for one-dimensional integrals; codes by Piessens, de Donker, and Kahaner
SIAM	Typesetting macros corresponding to the SIAM journal format
SLATEC	Machine constants and error handling package from the Slatec library
SPECFUN	Fortran programs for special functions
TOEPLITZ	Solve systems of equations with coefficient matrix of Toeplitz or circulant form
TOMS	Collected algorithms of the ACM
Y12M	A program to solve a large sparse linear system

To obtain a program from one of these packages or to obtain more information concerning the contents of a package, send an electronic message to Netlib at the Argonne National Laboratory using either the Arpanet, Csnet, Telenet, or PhoneNet or send a message to Netlib at Bell Laboratories using the UNIX network. For example, to send electronic mail to Netlib at Argonne National Laboratory using the Arpanet, the appropriate command on many systems is

mail netlib@anl-mcs.arpa

On the UNIX network, the corresponding address is "research!netlib." To obtain a list of the software packages currently stored in Netlib, the contents of your message should be the following:

send index

Netlib will return to you by electronic mail a copy of the current Netlib index. To obtain a detailed list of each routine contained in a particular package, you can request the index for that package. For example, to obtain a list of the routines in LINPACK, send the following message to Netlib:

send index for linpack

The keyword facility can be used to find all the routines relevant to a particular subject. For example, sending the message "find eigenvalue" to Netlib will produce a list of routines in Netlib for solving eigenvalue problems. To get a particular subroutine from some library, send a message of the form: "send x from y." For example, the following message will produce the LINPACK routine SGEFA:

send sgefa from linpack

Netlib will not only send you the subroutine SGEFA, but also any subroutine that SGEFA invokes. To obtain both SGEFA and SGESL, send the following message to Netlib:

send sgefa sgesl from linpack

To get a particular subroutine such as SGEFA without the invoked subroutines, the following message can be used:

send only sgefa from linpack

A single piece of mail may include several requests, however, put each request on a separate line. When a subroutine comes in more than one precision, the double-precision version is sent whenever the precision is not specified. To obtain a single-precision version, add the prefix S to the library. For example, if you request the subroutine HQR from EISPACK, you will receive the double-precision version. To obtain the single-precision version, request subroutine HQR from library SEISPACK.

Another large library for mathematical software is maintained by the International Mathematical and Statistical Libraries, better known as IMSL. IMSL markets and distributes entire packages such as B-SPLINE, EISPACK, EDA, ELEFUNT, GRAPHPAK, LINPACK, LLSQ, MINPACK, QUADPACK, ROSEPACK, TOEPLITZ, TWODEPEP, and PROTRAN. In addition, copies of ACM algorithms can be obtained from IMSL. Finally, IMSL maintains an extensive library which can be obtained under a license agreement. Some of the areas encompassed by this library include statistical analysis, numerical approximations, linear algebra, and differential equations. The IMSL library and documentation can be obtained from

IMSL
2500 ParkWest Tower One
2500 CityWest Boulevard
Houston, TX 77042-3020
Phone: 713-782-6060 (In U.S. and outside Texas: 1-800-222-IMSL)
Telex: 791923 IMSL INC HOU

A large software library is also maintained by the Numerical Algorithms Group, abbreviated NAG. This library contains subroutines covering the following areas:

Simultaneous linear equations
Eigenvalues and eigenvectors of matrices
Nonlinear optimization
Operations research
Nonlinear equations
Quadrature
Numerical differentiation
Ordinary differential equations
Partial differential equations
Integral equations

Curve and surface fitting
Fast Fourier transforms
Special functions
Basic statistics
Statistical distribution functions
Correlation and regression analysis
Analysis of variance
Random number generators
Sorting
Error trapping
Mathematical constants and machine parameters

The NAG library can be obtained from

Numerical Algorithms Group Ltd.
NAG Central Office
7 Banbury Road
Oxford, OX2 6NN, England
Phone: Oxford(0865) 511245, international: 44-865 511245
Telex: 83147 Ref NAG

or

Numerical Algorithms Group Inc.
1250 Grace Court
Downers Grove, IL 60516
Phone: 312-971-2337

Another large library with high-quality numerical software is the Harwell Subroutine Library. Harwell will provide copies of the subroutines and thorough documentation. The cost of Harwell subroutines, particularly for universities and for research institutions is very low, but unlike IMSL and NAG, no support is included in the license agreement. Information regarding the Harwell Subroutine Library can be obtained from

Mr. S. Marlow
Building 8.9
Harwell Laboratory
Didcot, Oxon, OX11 0RA, England
Phone: (0235) 24141, extension 3430; international: 44-235-24141

Finally, we wish to note that the computing departments of Sandia National Laboratory at Albuquerque, Los Alamos National Laboratory, Air Force Weapons Laboratory at Albuquerque, Sandia National Laboratory at Livermore, Lawrence Livermore National Laboratory, and the National Bureau of Standards have organized a large collection of software called SLATEC. Also, Bell Laboratories has developed a mathematical library called PORT which is described in [F4] and [F5].

To conclude this section, we provide a list of ACM algorithms which are relevant to applied numerical linear algebra. As mentioned earlier, the algorithms published in the *ACM Transactions on Mathematical Software* can be obtained from the IMSL Distribution Services as well as from Netlib.

MATRIX-VECTOR MANIPULATIONS

67	Store a symmetric matrix using compressed storage mode
230	Apply a permutation to specified rows or columns of a matrix
302	Compute the transpose of a rectangular matrix in place
380	Compute the transpose of a rectangular matrix in place
467	Compute the transpose of a rectangular matrix in place
513	Compute the transpose of a rectangular matrix in place
539	Basic linear algebra subprograms

SYSTEMS OF LINEAR EQUATIONS

16	Crout factorization with pivoting
17	Solve a tridiagonal system of equations without pivoting
24	Solve a tridiagonal system of equations without pivoting
43	Crout factorization with pivoting
107	Solve a system of equations using Gaussian elimination without pivoting
126	Solve a system of equations using Gaussian elimination without pivoting
127	A general orthogonalization algorithm
135	Solve a system of equations using Crout's method, row equilibrium, row interchanges, and iterative improvement
195	Solve a system when the coefficient matrix has band structure
220	Solve a linear system using the Gauss-Seidel iteration
238	Solve a linear system using the conjugate gradient method
287	Factor a matrix with integer arithmetic
288	Find the smallest positive integer k such that an integer solution to $\mathbf{Ax} = k\mathbf{b}$ exists
290	Compute the exact solution to $\mathbf{Ax} = \mathbf{b}$ when $\mathbf{A}$ and $\mathbf{b}$ are integer
298	Compute the square root of a positive definite matrix
319	Given the Cholesky factorization of $\mathbf{A}$, compute the factorization of the modified matrix $\mathbf{A} + \mathbf{XBX}^T$
328	Compute the solution to an overdetermined linear system which minimizes the l_∞ norm of the residual
338	Evaluate the fast Fourier transform
348	Scale a matrix using integer programming
406	Compute the exact solution to $\mathbf{Ax} = \mathbf{b}$ when $\mathbf{A}$ and $\mathbf{b}$ are integer
423	Solve a linear system using Gaussian elimination and partial pivoting
432	Solve $\mathbf{AX} + \mathbf{XB} = \mathbf{C}$
470	Solve a linear system whose coefficient matrix is almost tridiagonal
478	Determine the solution to an overdetermined linear system which minimizes the l_1 norm of the residual
495	Compute the solution to an overdetermined linear system which minimizes the l_∞ norm of the residual
512	Solve a symmetric positive definite quindiagonal system of equations
522	Compute the exact solution to an integer system of linear equations using congruence techniques
544	Solve a weighted least squares problem using modified Gram-Schmidt with iterative refinement
546	Solve a linear system where the coefficient matrix is almost block diagonal
551	Determine the solution to an overdetermined linear system which minimizes the l_1 norm of the residual

SYSTEMS OF LINEAR EQUATIONS (cont.)

552 Determine the solution to a linear system which minimizes the l_1 norm of the residual subject to a system of equality and inequality constraints
563 Determine the solution to a linear system which minimizes the l_1 norm of the residual subject to a system of equality and inequality constraints
576 Solve a linear system using Gaussian elimination and a new pivoting strategy
578 Solve a linear system using Gaussian elimination and partial pivoting in a paged virtual store
580 Update the **QR** factorization when rows or columns are added or deleted or when a rank one change is added
583 LSQR: Solve sparse linear equations and least squares problems
587 Determine the solution to a linear system which minimizes the Euclidean norm of the residual subject to a system of equality and inequality constraints
603 COLROW and ARCECO: Fortran package for solving almost block diagonal linear systems by modified alternate row and column elimination
633 An algorithm for linear dependency analysis for multivariate data
634 Routines for fitting multinomials in a least squares sense
635 An algorithm for the solution of systems of complex linear equations in the l_∞ norm with constraints on the unknowns
641 Exact solution of general integer systems of linear equations

SPARSE MATRICES

408 Perform operations on a matrix stored in sparse mode
508 Reduce the bandwidth and profile of a sparse matrix using row and column permutations
509 Reduce the bandwidth and profile of a sparse matrix using row and column permutations
529 Symmetrically permute the rows and columns of a matrix to obtain block lower triangular form
533 NSPIV: Solve a sparse system of linear equations with partial pivoting
575 Permute rows of a matrix to obtain nonzero diagonal elements
582 Reduce the bandwidth and profile of a sparse matrix using row and column permutations
586 ITPACK 2C: A Fortran package for solving large sparse linear systems by adaptive accelerated iterative methods
601 Multiplications and additions involving sparse matrices
636 Fortran subroutines for estimating sparse Hessian matrices

MATRIX INVERSION

42 Invert a general matrix without pivots
50 Invert a finite segment of the Hilbert matrix
51 Adjust the inverse of a matrix when an element is perturbed
58 Invert a matrix using Gauss-Jordan elimination
66 Invert a symmetric matrix
120 Invert a matrix using Gaussian elimination
140 Invert a matrix in place
150 Invert a symmetric matrix
166 Compute a single row of the inverse of a matrix using a Monte Carlo technique
197 Compute $\mathbf{A}^{-1}\mathbf{B}$ when $\mathbf{A}$ is positive definite
231 Invert a matrix using Gauss-Jordan elimination and complete pivoting
325 Adjust the inverse of a symmetric matrix when two symmetric elements are changed
640 Efficient calculation of frequency response matrices from state space models

DETERMINANTS

41	Evaluate a determinant by factoring the matrix
159	Evaluate a determinant by the combinatorial definition
170	Evaluate a determinant of a matrix whose elements are polynomials
224	Evaluate a determinant by factoring the matrix (corrects algorithm 41)
269	Evaluate a determinant using a row equilibrium and factorization with pivots

TEST PROBLEMS

52	Generate a matrix with known inverse and known eigenvalues
274	Generate test matrices from the Hilbert matrix
564	Generate a discrete l_1 minimization problem with known solution
566	Generate test problems for systems of equations, least squares problems, and unconstrained minimization problems

NONLINEAR SYSTEMS

2	Compute a root of a single equation using a secant-like method
4	Compute a root of a single equation using the bisection method
15	Compute a root of a single equation using a secant-like method
25	Compute real roots of a single equation using Muller's method
26	Compute a root of a single equation using a secant-like method
139	Compute integer solutions to $ax + by = c$, where a, b, and c are given integers
196	Compute real and complex roots of a single equation using Muller's method
314	Solve a system of equations using an inverse interpolation method
315	Solve a least squares problem using a damped Newton iteration
316	Solve a system of equations using Brown's modified Newton method
365	Determine complex zeros of a transcendental equation for which the corresponding function is analytic
378	Solve a system of equations using a Newton-like method
443	Solve $xe^x = c$
502	Compute the solution to a nonlinear system as a function of a parameter using Newton's method and an Adams integration scheme
554	Solve a nonlinear system using Brent's method
555	Solve a nonlinear system using a continuation method
573	Solve the nonlinear least squares problem using a variation of Newton's method
598	Solve $\mathbf{AX}^2 + \mathbf{BX} + \mathbf{C} = \mathbf{0}$
617	DAFNE: Solve a nonlinear system using a least squares approach and an integration scheme for differential equations
618	Estimate sparse Jacobians
631	Find a bracketed zero by Larkin's method of rational interpolation

ZEROS OF POLYNOMIALS

3	Compute polynomial zeros using Bairstow's method
30	Compute polynomial zeros using the Bairstow-Newton iteration
59	Compute polynomial zeros using a resultant procedure
75	Find the rational zeros of a polynomial with integer coefficients
78	Find the rational zeros of a polynomial with integer coefficients
105	Compute the zeros of a polynomial whose roots are real and simple using the Newton-Maehly method

ZEROS OF POLYNOMIALS (cont.)

174 Determine bounds on the zeros of a polynomial
256 Compute polynomial zeros using the modified Graeffe method
283 Compute all the roots of a polynomial at once
326 Compute the zeros of quadratic, cubic, and quartic polynomials
340 Compute polynomial zeros using a root squaring and resultant procedure
419 Compute the zeros of a polynomial with complex coefficients using the Jenkins-Traub algorithm
429 Determine regions containing the zeros of a polynomial
493 Compute the zeros of a polynomial with real coefficients using the Jenkins-Traub algorithm

EIGENVALUES AND EIGENVECTORS

85 Compute the eigenpairs of a symmetric matrix using a modified Jacobi method
104 Reduce a symmetric band matrix to tridiagonal form
122 Reduce a symmetric matrix to tridiagonal form using Givens rotations
183 Reduce a symmetric band matrix to tridiagonal form
253 Compute the eigenvalues of a real symmetric matrix by the **QR** method
254 Compute the eigenpairs of a real symmetric matrix by the **QR** method
270 Compute eigenvector given the eigenvalue
297 Solve the generalized eigenproblem $\mathbf{Ax} = \lambda\mathbf{Bx}$, where both **A** and **B** are symmetric and either $\pm\mathbf{A}$ or $\pm\mathbf{B}$ is positive definite
343 Compute eigenvalues of a general real matrix using the **QR** method; compute eigenvectors using inverse iteration
358 Compute the singular value decomposition of a complex matrix
384 Compute eigenpairs of a real symmetric matrix by reducing it to tridiagonal form and applying the **QR** method
405 Solve the generalized eigenproblem when the matrices are rectangular
464 Compute the eigenpairs of a real symmetric tridiagonal matrix using Reinsch's method
496 Solve the generalized eigenproblem with complex coefficient matrices using the **LZ** algorithm
506 Reduce a real upper Hessenberg matrix to quasi-triangular form with eigenvalues ordered in magnitude
517 Compute the condition numbers of matrix eigenvalues without computing eigenvectors
530 Compute the eigenpairs of a real skew-symmetric matrix or a real tridiagonal matrix with constant diagonal
535 Solve the generalized eigenproblem with complex coefficient matrices using the **QZ** algorithm
538 The power method for generalized symmetric matrices
560 Compute the Jordan normal form of a complex matrix
570 The power method for real sparse matrices
581 An improved algorithm for computing the singular value decomposition
589 SICEDR: Iterative refinement for eigenpairs
590 DSUBSP and EXCHQZ: Given a reduced generalized eigenproblem, obtain an equivalent problem where the eigenvalues achieve a specified ordering

ORDINARY DIFFERENTIAL EQUATIONS

9 Integrate a first-order system of differential equations using the Runge-Kutta method
194 Find a simple zero for the first component of the solution to a first-order system of differential equations
218 Integrate a first-order system of differential equations using the Kutta-Merson method

ORDINARY DIFFERENTIAL EQUATIONS (cont.)

407 Integrate a system of ordinary differential equations using a multistep predictor-corrector method
461 Compute a cubic spline approximation to the solution of a second-order functional differential boundary-value problem
497 Integrate a first-order system of functional differential equations
504 GERK: Integrate a first-order system of differential equations using the Runge-Kutta-Fehlberg (4,5) method
534 STINT: Integrate a stiff system of differential equations
569 COLSYS: Solve a multipoint boundary-value problem for a mixed-order system of differential equations using a collocation method
596 Find the solution curve corresponding to a system of $n - 1$ equations in n unknowns using a prediction-correction type method

PARTIAL DIFFERENTIAL EQUATIONS

392 Solve the initial value problem for a quasilinear system of two equations in two independent variables and two unknowns
460 Compute optimum parameters for alternating direction implicit procedures
494 PDEONE: An interface program which uses centered difference approximations to convert one-dimensional systems of partial differential equations into a system of ordinary differential equations
527 Generalized marching method to solve linear systems resulting from the discretization of separable or constant coefficient elliptic partial differential equations on a rectangular domain
540 PDECOL: Solve coupled systems of nonlinear partial differential equations in one space and one time dimension; finite element collocation techniques used for the spatial dimension
541 Solve separable elliptic partial differential equations
543 FFT9: Fast solution of Helmholtz-type partial differential equations
553 M3RK: An explicit time integrator for semidiscrete parabolic equations
565 PDETWO/PSETM/GEARB: Solve systems of time-dependent nonlinear partial differential equations defined over a two-dimensional rectangular region
572 Solve the Helmholtz equation with Dirichlet boundary conditions on general bounded three-dimensional regions
593 Solve the Helmholtz equation in nonrectangular planar regions using the capacitance matrix method and fast solvers
621 Solve systems of time-dependent nonlinear partial differential equations defined over a two-dimensional rectangular region with low storage requirements
629 Solve Laplace's equation in three dimensions using an integral equation approach
637 GENCOL: Collocation on general domains with bicubic Hermite polynomials
638 INTCOL and HERMCOL: Collocation on rectangular domains with bicubic Hermite polynomials

Appendix 1

Subroutine Names and Arguments in NAPACK

Below we list the subroutine names and arguments for NAPACK. These subroutines are organized first by the relevant chapter and then in alphabetical order by the stem of the name. The section of a chapter which is relevant to a particular subroutine appears as a superscript following the subroutine name. A brief description of each subroutine is given in Section 8-1 while a detailed description of each subroutine argument appears in the comment statements that accompany the subroutine. The minimum dimension for each argument that is an array appears in parentheses after the argument. When an input parameter for one subroutine is an output parameter from another subroutine, then the minimum dimensions are only indicated in the subroutine which outputs the array. For example, the A argument of FACT is used as input for subroutines DET, SOLVE, CON, and TRANS. Below, we only indicate the minimum dimension for A in subroutine FACT.

CHAPTER 1 INTRODUCTION

ADDCHG(DIF,SIZE,NEW[N],CHG[N],N)
STOPIT(DIF,SIZE,NDIGIT,LIMIT)[2]
UPDATE(DIF,SIZE,NEW[N],OLD[N],N)[6]
WHATIS(DIF,SIZE)[2]

CHAPTER 2 ELIMINATION SCHEMES

DET(I,A)[9]
BDET(I,A)
EDET(I,A)
HDET(I,A)
IDET(I,A)

CHAPTER 2 ELIMINATION SCHEMES (cont.)

- KDET(I,A)
- PDET(I,A)
- SDET(I,A)
- TDET(I,A)

FACT(A[3 + N*(N + 1)],LA,N)[3]
- BFACT(A[5 + N*(2*L + U + 2)],LA,N,L,U)[4.2]
- EFACT(A[1 + N*(N + 5)/2],N)[12]
- HFACT(A[4 + N*(H + 1)],LA,N,H)[4.4]
- IFACT(A[5 + N*(N + 7)/2],N)[4.3]
- KFACT(A[2 + N*(N + 2)],LA,N)[3]
- PFACT(A[3 + 4*N],LA,N)[4.1]
- SFACT(A[N*(N + 1)/2],N,W[N])[4.3]
- TFACT(L[N],D[3 + N],U[N],N)[4.1]

MULT(Y[N],X[N],A[M,N],LA,M,N)
- BMULT(Y[N],X[N],A[1 + L + U,N],LA,N,L,U)
- CEMULT(Y[N],X[N],A[N*(N + 3)/2 − 1],N) − Y,X,A (complex)
- EMULT(Y[N],X[N],A[N*(N + 3)/2 − 1],N)
- HMULT(Y[N],X[N],A[H + 1,N],LA,N,H)
- OMULT(Y[N],X[N],A[N],N,W) − Y,X,A,W (complex)
- SMULT(Y[N],X[N],A[N*(N + 1)/2],N)
- TMULT(Y[N],X[N],L[N],D[N],U[N],N)

PACK(A[N,N],LA,N)[3]
- CPACK(A[N,N],LA,N)[3] − A (complex)
- RPACK(A[M,N],LA,M,N)[3]

SOLVE(X[N],A,B[N])[3]
- BSOLVE(X[N],A,B[N])[4.2]
- ESOLVE(X[N],A,B[N])[12]
- HSOLVE(X[N],A,B[N])[4.4]
- ISOLVE(X[N],A,B[N])[4.3]
- KSOLVE(X[N],A,B[N])[3]
- OSOLVE(X[N],A,N,B[N],W[N]) − X,A,B,W (complex)
- PSOLVE(X[N],A,B[N])[4.1]
- SSOLVE(X[N],A,B[N])[4.3]
- TSOLVE(X[N],L,D,U,B[N])[4.1]

TRANS(X[N],A,B[N])
- BTRANS(X[N],A,B[N])
- ETRANS(X[N],A,B[N])
- KTRANS(X[N],A,B[N])
- PTRANS(X[N],A,B[N])
- TTRANS(X[N],L,D,U,B[N])

VERT(V[N,N],LV,N,W[N − 1])[7]
- BVERT(V[N,N],LV,A)
- CVERT(V[N,N],LV,N,W[N − 1]) − V (complex)
- EVERT(V[N,N],LV,A)
- HVERT(V[.5*N*(N + 1)],A)
- IVERT(V[3 + .5*N*(N + 9)],W[2*N − 2])
- KVERT(V[N,N],LV,N,W[2*N])
- OVERT(A[N],N,W[N]) − A,W (complex)
- PVERT(V[N,N],LV,A)
- SVERT(V[.5*N*(N + 1)],N,W[N])
- TVERT(V[N,N],L,D,U)

CHAPTER 3 CONDITIONING

CON(A,B[N])[4]
BCON(A,B[N])[4]
ECON(A,B[N])[4]
HCON(A,B[N])[4]
ICON(A,B[N])[4]
KCON(A,B[N])[4]
PCON(A,B[N])[4]
SCON(A,B[N])[4]
TCON(L,D,U,B[N])[4]

CHAPTER 4 NONLINEAR SYSTEMS

CZERO(Z[N],P[N],N,W[roughly 4N to 6N])[8] – Z,P,W (complex)—the storage needed for W depends on the difficulty associated with computing the zeros
QUASI(X[N],H[N,N],LH,N,DIF,SIZE,NDIGIT,LIMIT,SUB,W[3*N])[5]
ROOT(Y,Z,T,F)[1]

CHAPTER 5 LEAST SQUARES

BASIS(B[M,N],LB,N,A,C)[8]
NULL(B[M,N],LB,N,A,C)[8]
OVER(X[M],A,B[M])[7]
UNDER(X[N],A,B[M])[7]
QR(A[3 + N + M*N + MIN(M,N)],LA,M,N)[5]

CHAPTER 6 EIGENPROBLEMS

BAL(A[N,N],LA,N,D[N],W[2*N])
CBAL(A[N,N],LA,N,D[N],W[2*N]) – A (complex)
BIDAG(D[N],B[M],A[M,N],LA,M,N)[11]
BIDAG2(D[N],B[M],Q[M,?],LQ,IQ,P[N,?],LP,IP,A[M,N],LA,M,N)[11]—dimensions of Q and P depend on the values of IQ and IP
DIAG(E[N],V[N,N],LV,A[N*(N + 7) – 2],LA,N)[6] – E,V (complex)
CDIAG(E[N],V[N,N],LV,A[1 + N*(N + 2)],LA,N)[6] – E,V,A (complex)
CEDIAG(E[N],V[N,N],LV,A[N*(N + 3)/2 – 1],N,W[4*N])[6] – E,V,A (complex)
EDIAG(E[N],V[N,N],LV,A[(N + 1)*(N + 2) – 4],N,W[4*N])[6] – E,V (complex)
HDIAG(E[N],V[N,N],LV,A,LA,N,H,W[(H + 1)*(N – .5*H + 6) – 12])[6]
SDIAG(E[N],V[N,N],LV,A[N*(N + 3)/2],N,W[2*N])[6]
SDIAG2(E[N],V[N,N],LV,N,W[2*N])[6]
TDIAG(E[N],V[N,N],LV,L[N],D[N],U[N],N)[6]
HESS(A[2 + N*(N + 1)],LA,N,W[N])[7]
AHESS(A[1 + N*(N + 2)],LA,N,W[N])[7]
CHESS(A[2 + N*(N + 1)],LA,N,W[N])[7] – A,W (complex)
CAHESS(A[1 + N*(N + 2)],LA,N,W[N])[7] – A,W (complex)
SHESS(D[N],U[N],A[N*(N + 1)/2],N)[8]
HHESS(D[N],U[N],A[H + 1,N],LA,N,H,W[(H + 1)*(N – .5*H + 6) – 12])
LANCZ(Q[N],D[N],U[N],P[N],J,N,MULT,W[2*N])[10]
NEWTON(T,S,K,D[N],U[N],N)[10]
POWER(EX,X[N],EY,Y[N],N,DIF,SIZE,NDIGIT,LIMIT,MULT)[4] – EX,X,EY,Y (complex)
CPOWER(E,X[N],N,DIF,SIZE,NDIGIT,LIMIT,MULT,W[N])[4] – E,X,W (complex)
MPOWER(EX,X[N],EY,Y[N],N,NE,DIF,SIZE,NDIGIT,LIMIT,MULT,W[NE*(N + 2) – 2])[4] – EX,X,EY,Y (complex)

CHAPTER 6 EIGENPROBLEMS (cont.)

PSEUDO(A[MQ,MP],LA,Q[MQ,M],LQ,MQ,D[M],P[MP,M],LP,MP,R)[11] where M = MIN(MQ,MP)

SIM(P[N,N],LP,A)[7]

CSIM(P[N,N],LP,A)[7] – P (complex)

HSIM(P[N,N],LP,D[N],U[N],A[H + 1, N],LA,N,H,W[(H + 1)*(N – .5*H + 6) – 12])[7]

SSIM(P[N,N],LP,A)[7]

SING(Q[M,?],LQ,IQ,S[N],P[N,?],LP,IP,A[M,N],LA,M,N,W[MAX(M,3*MIN(M,N) – 1)])[11]— dimensions of Q and P depend on the values of IQ and IP

SINGB(D[N],N,U[N],IU,Q[M,?],LQ,MQ,IQ,P[N,?],LP,MP,IP,E[N],F[N])

SLICE(E[K],K,Y,Z,L[N],D[N],U[N],N,W[N + 9K])[9]

RSOLVE(X[MP],B[M],Q[MQ,M],LQ,MQ,D[M],P[MP,M],LP,MP,R)[11] where M = MIN(MP,MQ)

TVAL(E,K,L[N],D[N],U[N],N,W[N])[9]

VALS(E[N],A[N*(N + 7) – 2],LA,N,V[1 + N*(N + 2)])[6] – E (complex)

CVALS(E[N],A[1 + N*(N + 2)],LA,N,V[1 + N*(N + 2)])[6] – E,A,V (complex)

CEVALS(E[N],A[N*(N + 3)/2 – 1],N,W[4*N])[6] – E,A (complex)

EVALS(E[N],A[(N + 1)*(N + 2) – 4],N,W[4*N])[6] – E (complex)

HVALS(E[N],D[N],U[N],A[H + 1,N],LA,N,H,W[(H + 1)*(N – .5*H + 6) – 12])[6]

OVALS(A[N],N,W[N]) – A,W (complex)

SVALS(E[N],D[N],U[N],A[1 + N*(N + 3)/2],N)[8]

TVALS(E[N],L[N],D[N],U[N],N,W[N])[8]

VECT(E,X[N],V,W[M*(N*(N + 9) – 6)])[6] – E,X (complex), M = 1 (or 1/2) for complex (or real) eigenvalue

CVECT(E,X[N],V,W[N*(N + 9)/2 – 3])[6] – E,X,W (complex)

EVECT(E,X[N],A[N*(N + 7)/2 – 2],N)[6] – X (complex)

CEVECT(E,X,A[N*(N + 7)/2 – 2],N)[6] – E,X,A (complex)

HVECT(E,X[N],A[H + 1,N],LA,N,H,W[N*(3H + 2)])[8]

SVECT(E,X[N],D[N],U[N],A,W[4*N])[8]

TVECT(E,X[N],L[N],D[N],U[N],N,W[4*N])[9]

CHAPTER 7 ITERATIVE METHODS

PRECG(X[N],DIF,SIZE,NDIGIT,LIMIT,B[N],N,MULT,PRE,W[2*N])[5]

Appendix 2

Solutions to Exercises

Chapter 1

1-1.1. $x_2 = 1$ and $x_3 \approx 2.56$.

1-1.2. The iterations converge to the root $x \approx -1.13$.

1-2.1. Should consider relative error not absolute error.

1-3.1. On most computers and calculators, the computed X is not zero since dividing 1 by 41 yields a number which cannot be stored using a finite number of binary or decimal digits. That is, float(1/41) does not equal 1/41. Hence, float(float(1/41)*41) may not equal 1 and the computed difference 1. − (1./41.)*41. is typically a small number, say ε. The final step in the evaluation of X multiplies ε by 10^{18}. Often, the computed value for ε is about b^{-n}. If the factor 10^{18} is changed to b^n, then the value of X is on the order of 1. Since 1./64. can be expressed using a finite number of binary or decimal digits, float(float(1/64)*64) = 1 and the computed value for X is zero when the "41" is changed to "64." (41 seems to be the smallest positive integer with the property that the computed value of (1./41.)*41. is not equal to 1 on computers like the VAX 780 that round a 24-bit mantissa.)

1-3.2. A = .015. If the computer works in base b, where b is 2 or a power of 2, the final A is b^{-n}.

1-3.3.

```
        X = 1.
   10   S = X
        X = X/2.
        IF ( X .GT. 0. ) GOTO 10
        WRITE(6,20) S
   20   FORMAT (E20.10)
```

Note that you may have to turn off the underflow check on your compiler before running the code. If your program terminates due to underflow, try putting the following statement at the start of your program:

CALL ERRSET(208,256, −1, −1,0).

1-4.1. Since c is small relative to b, the expression $\sqrt{b^2 + c}$ minus b is the difference between nearly equal quantities. Similarly, the expression $c/(\sqrt{b^2 + c} - b)$ minus $2b$ is the difference between nearly equal quantities. The second expression for a does not involve subtracting nearly equal numbers.

1-5.1. $N = 2$: $y_1 = \frac{1}{8}e^{-y_1}$. $N = 3$: $2y_1 - y_2 = \frac{1}{9}e^{-y_1}$ and $-y_1 + 2y_2 = \frac{1}{9}e^{-y_2}$.

1-5.2. $y_{i-1} - 2y_i + y_{i+1} = (\Delta t)^2 e^{-t_i}$, $i = 1, 2, \cdots, N - 1$.

1-5.3. $y_{i-1} - (2 + \Delta t + 2(\Delta t)^2)y_i + (1 + \Delta t)y_{i+1} = (\Delta t)^2 e^{-t_i}$, $i = 1, 2, \cdots, N - 1$.

1-5.5. Assuming that the error associated with the evaluation of the numerator in (23) is $5|x(t)|E$, the best Δt is $|60x(t)E/x^{(4)}(t)|^{1/4}$.

1-6.1. $y_i^{\text{new}} = \frac{1}{2}[y_{i-1}^{\text{new}} + y_{i+1}^{\text{old}} + (\Delta t)^2 e^{-y_i^{\text{old}}}]$, where $i = 1$ to $N - 1$ and $y_0 = y_N = 0$ or $y_i^{\text{new}} = \frac{1}{2}[y_{i-1}^{\text{old}} + y_{i+1}^{\text{new}} + (\Delta t)^2 e^{-y_i^{\text{old}}}]$, where $i = N - 1$ down to 1 and $y_0 = y_N = 0$.

1-6.2. In the loop where OLD is initialized to zero, initialize an array B where B(I) = S*EXP(−D*I). For the equation in Exercise 1-5.2, change the expressions for NEW(1) and NEW(I) to the following: NEW(1) = .5*(OLD(2) − B(1)) and NEW(I) = .5*(OLD(I − 1) + OLD(I + 1) − B(I)). For the equation in Exercise 1-5.3, initialize C0 = 1./(2. + D + S + S) and C1 = 1. + D and change the expression for NEW(1) and NEW(I) to the following: NEW(1) = C0*(C1*OLD(2) − B(I)) and NEW(I) = C0*(OLD(I−1) + C1*OLD(I+1) − B(I)).

1-8.3. Limit is 2.

Review Problems:

1-1. First method works. $x_6 = 1.5213 \cdots$ is correct to five places.

1-2. Relative error corresponding to the starting condition $I_{20} = .2$ is about 10^{-14}. Relative error corresponding to the starting condition $I_{20} = 0$ is much less than 10^{-14}.

1-3. B is base, N is number of digits in the mantissa, and R is 1 if the machine rounds.

1-6. $x_+ = \dfrac{-2c}{b + \sqrt{b^2 - 4c}}$ and $x_- = \dfrac{2c}{-b + \sqrt{b^2 - 4c}}$.

Note that (new) $x_+ = c$ over (old) x_-. For your computer program, note that the roots are complex and equation (40) of Chapter 1 can be used when $b^2 < 4c$. If $b^2 \geqq 4c$ and $b > 0$, then $x_- = (-b - \sqrt{b^2 - 4c})/2$ and $x_+ = c/x_-$. If $b^2 \geqq 4c$ and $b < 0$, then $x_+ = (-b + \sqrt{b^2 - 4c})/2$ and $x_- = c/x_+$. If $b = c = 0$, then $x_+ = x_- = 0$. Another way to improve your code is to rewrite the formulas for $x_\pm$ whenever $|b|$ is large enough or small enough that the computation of b^2 generates an overflow or an underflow. For example, if the potential for underflow or overflow exists and $|b| \geqq 2\sqrt{|c|}$, equation (40) of Chapter 1 can be rewritten

$$x_\pm = \frac{-b \pm |b|\sqrt{1 - \sigma r^2}}{2},$$

where $r = 2\sqrt{|c|}/b$ and σ is $+1$ is $c > 0$ while σ is -1 if $c < 0$. If the potential for underflow or overflow exists and $|b| < 2\sqrt{|c|}$, then equation (40) of Chapter 1 can be written

$$x_\pm = \sqrt{|c|}(-r \pm \sqrt{r^2 - \sigma}),$$

where $r = b/(2\sqrt{|c|})$ and σ is $+1$ if $c > 0$ while σ is -1 if $c < 0$.

1-7. The relative error in the computed y is greater than 1 whenever the computed y is negative. Using equation (14) of Chapter 1 with $z = 1$ and with $m = 6$, the diameter of the region where the relative error in the computed y is at least 1 is about .4.

1-8. $-(1 + \Delta t + 2(\Delta t)^2)y_1 + (1 + \Delta t)y_2 = 3(\Delta t)^2 e^{t_1}$ and $y_{i-1} - (2 + \Delta t + 2(\Delta t)^2)y_i + (1 + \Delta t)y_{i+1} = 3(\Delta t)^2 e^{t_i}$ for $i = 2$ to $N - 1$, where

$y_N = 0$. Computed solution after rounding is $y_1 = -0.94$, $y_2 = -0.93$, $y_3 = -0.90$, $y_4 = -0.86$, $y_5 = -0.79$, $y_6 = -0.70$, $y_7 = -0.58$, $y_8 = -0.43$, $y_9 = -0.24$.

1-9. $[1.5 + 9(\Delta t)^2]y_1 - y_2 - .5y_{N-1} = 10(\Delta t)^2 \sin t_1$,
$[1.5 + 9(\Delta t)^2]y_{N-1} - y_{N-2} - .5y_1 = 10(\Delta t)^2 \sin t_{N-1}$,
$[2 + 9(\Delta t)^2]y_i - y_{i-1} - y_{i+1} = 10(\Delta t)^2 \sin t_i$ for $i = 2$ to $N - 2$.
Computed solution after rounding is $y_1 = 0.59$, $y_2 = 0.95$, $y_3 = 0.95$, $y_4 = 0.59$, $y_5 = 0.00$, $y_6 = -0.59$, $y_7 = -0.95$, $y_8 = -0.95$, $y_9 = -0.59$.

1-10.

```
      SUBROUTINE AIKEN(Z,N)
      DOUBLE PRECISION Z(1),A,B,C
      M = N + N - 1
      K = 1
10    K = K + 2
      M = M - 2
      L = K + M
      C = Z(K-2)
      B = Z(K-1)
      A = Z(K)
20    Z(K+M) = A - (A-B)**2/(C-B-B+A)
      K = K + 1
      IF ( K .GE. L ) GOTO 30
      C = B
      B = A
      A = Z(K)
      GOTO 20
30    IF ( M .GT. 1 ) GOTO 10
      RETURN
      END
```

1-12. About 73,000,000 trapezoids.

Chapter 2

2-1.1. (a) $\begin{bmatrix} 0 & 2 & -1 \\ 1 & -1 & 1 \\ 1 & 1 & 0 \end{bmatrix}$ **(b)** 14 **(c)** $\begin{bmatrix} 1 & 2 & 3 \\ 2 & 4 & 6 \\ 3 & 6 & 9 \end{bmatrix}$ **(d)** $\begin{bmatrix} 1 \\ 0 \\ 0 \\ 1 \end{bmatrix}$

2-1.2. (a) $$\begin{bmatrix} 1 & 0 \\ 3 & 1 \end{bmatrix}\begin{bmatrix} 2 & 4 \\ 0 & -3 \end{bmatrix} = \begin{bmatrix} 1 & 0 \\ 3 & 1 \end{bmatrix}\begin{bmatrix} 2 & 0 \\ 0 & -3 \end{bmatrix}\begin{bmatrix} 1 & 2 \\ 0 & 1 \end{bmatrix}$$

(b) $$\begin{bmatrix} 1 & 0 & 0 \\ 4 & 1 & 0 \\ 2 & 3 & 1 \end{bmatrix}\begin{bmatrix} 4 & 12 & 16 \\ 0 & -3 & -6 \\ 0 & 0 & -2 \end{bmatrix} = \begin{bmatrix} 1 & 0 & 0 \\ 4 & 1 & 0 \\ 2 & 3 & 1 \end{bmatrix}\begin{bmatrix} 4 & 0 & 0 \\ 0 & -3 & 0 \\ 0 & 0 & -2 \end{bmatrix}\begin{bmatrix} 1 & 3 & 4 \\ 0 & 1 & 2 \\ 0 & 0 & 1 \end{bmatrix}$$

(c) $$\begin{bmatrix} 1 & 0 & 0 & 0 \\ 3 & 1 & 0 & 0 \\ 2 & -1 & 1 & 0 \\ 1 & 2 & 4 & 1 \end{bmatrix}\begin{bmatrix} 2 & 4 & 6 & 8 \\ 0 & -2 & 4 & -4 \\ 0 & 0 & -1 & 1 \\ 0 & 0 & 0 & 3 \end{bmatrix}$$

$$= \begin{bmatrix} 1 & 0 & 0 & 0 \\ 3 & 1 & 0 & 0 \\ 2 & -1 & 1 & 0 \\ 1 & 2 & 4 & 1 \end{bmatrix}\begin{bmatrix} 2 & 0 & 0 & 0 \\ 0 & -2 & 0 & 0 \\ 0 & 0 & -1 & 0 \\ 0 & 0 & 0 & 3 \end{bmatrix}\begin{bmatrix} 1 & 2 & 3 & 4 \\ 0 & 1 & -2 & 2 \\ 0 & 0 & 1 & -1 \\ 0 & 0 & 0 & 1 \end{bmatrix}$$

(d) $\begin{bmatrix} 1 & 0 & 0 & 0 \\ 2 & 1 & 0 & 0 \\ 0 & -2 & 1 & 0 \\ 0 & 0 & 3 & 1 \end{bmatrix} \begin{bmatrix} 2 & 2 & 2 & 4 \\ 0 & -1 & -1 & -7 \\ 0 & 0 & 1 & -9 \\ 0 & 0 & 0 & 29 \end{bmatrix}$

$$= \begin{bmatrix} 1 & 0 & 0 & 0 \\ 2 & 1 & 0 & 0 \\ 0 & -2 & 1 & 0 \\ 0 & 0 & 3 & 1 \end{bmatrix} \begin{bmatrix} 2 & 0 & 0 & 0 \\ 0 & -1 & 0 & 0 \\ 0 & 0 & 1 & 0 \\ 0 & 0 & 0 & 29 \end{bmatrix} \begin{bmatrix} 1 & 1 & 1 & 2 \\ 0 & 1 & 1 & 7 \\ 0 & 0 & 1 & -9 \\ 0 & 0 & 0 & 1 \end{bmatrix}$$

2-1.3. mn

2-1.4. lmn

2-1.5. (**AB**)**C**

2-1.6. $km(l + n)$ multiplications and additions to compute (**AB**)**C** and $ln(k + m)$ multiplications and additions to compute **A**(**BC**). In the special case cited, **A**(**BC**) requires fewer multiplications and additions.

2-2.1. **(a)** $x_1 = 3$, $x_2 = -1$. **(b)** $x_1 = 3$, $x_2 = -2$, $x_3 = 1$. **(c)** $x_1 = 2$, $x_2 = -1$, $x_3 = 2$, $x_4 = -1$. **(d)** $x_1 = -1$, $x_2 = 1$, $x_3 = -1$, $x_4 = 1$. **(e)** $x_1 = -1$, $x_2 = 1$, $x_3 = -1$, $x_4 = 1$. **(f)** $x_1 = 1$, $x_2 = -1$, $x_3 = 1$, $x_4 = -1$, $x_5 = 1$.

2-2.2. Row-oriented program:

```
i = 1 to n
    j = 1 to i − 1
        b_i ← b_i − l_ij b_j
    next j
  b_i ← b_i/l_ii
next i
i = n − 1 down to 1
    j = i + 1 to n
        b_i ← b_i − u_ij b_j
    next j
next i
```

Column-oriented program:

```
j = 1 to n
    b_j ← b_j/l_jj
    i = j + 1 to n
        b_i ← b_i − l_ij b_j
    next i
next j
j = n down to 2
    i = 1 to j − 1
        b_i ← b_i − u_ij b_j
    next i
next j
```

2-2.3. Solve $\mathbf{Lz} = \mathbf{b}$ by forward elimination, solve $\mathbf{Dy} = \mathbf{z}$ (just divide each component of $\mathbf{z}$ by the corresponding diagonal element of $\mathbf{D}$), and solve $\mathbf{Ux} = \mathbf{y}$ by back elimination. A column-oriented program which accomplishes this is the following:

```
j = 1 to n − 1
    i = j + 1 to n
        b_i ← b_i − a_ij b_j
  next i
next j
i = 1 to n
    b_i ← b_i/a_ii
next i
j = n down to 2
    i = 1 to j − 1
        b_i ← b_i − a_ij b_j
    next i
next j
```

2-2.4. **(a)** Components 1 through $j - 1$. **(b)** Components 1 through k. **(c)** Column j of the product $\mathbf{L}_1\mathbf{L}_2$ between two lower triangular matrices is the product between the lower triangular matrix $\mathbf{L}_1$ and a vector whose first $j - 1$ components are zero. Therefore, the first $j - 1$ elements in column j of the product matrix are zero. Since this holds for any j, the product $\mathbf{L}_1\mathbf{L}_2$ is lower triangular.

2-2.5. **(a)** Components 1 through $j - 1$. **(b)** If $\mathbf{Lx} = \mathbf{e}_j$, then $\mathbf{x} = \mathbf{L}^{-1}\mathbf{e}_j$. Since every component of $\mathbf{e}_j$ is zero except for component j which is 1, $\mathbf{x}$ is column j of $\mathbf{L}^{-1}$. By part (a), components 1 through $j - 1$ of column j are zero. Since this holds for any j, $\mathbf{L}^{-1}$ is lower triangular.

2-3.1. The first column of A inside the subroutine is (1.,2.), the second column is (3.,4.), the third column is (5.,6.), the fourth column is (7.,8.), and the fifth column just contains the single element "9" stored in A(1,5). Inside the subroutine, A(1,3) is the first element in column 3, namely "5".

2-3.2.

```
      SUBROUTINE PACK(A,LA,N)
      REAL A(1)
      INTEGER H,I,J,K,L,LA,N,O
      H = LA - N
      IF ( H .EQ. 0 ) RETURN
      IF ( H .GT. 0 ) GOTO 10
      WRITE(6,*) 'ERROR: LA ARGUMENT IN PACK'
      WRITE(6,*)'MUST BE .GE. N ARGUMENT'
      STOP
10    I = 0
      K = 1
      L = N
      O = N*N
20    IF ( L .EQ. O ) RETURN
      I = I + H
      K = K + N
      L = L + N
      DO 30 J = K,L
30       A(J) = A(I+J)
      GOTO 20
      END
```

2-3.3.

```
      REAL A(100),B(10)
10    READ (5,*) N
      IF ( N .LE. 0 ) STOP
      M = N*N
      READ (5,*) (A(I), I=1,M)
      CALL FACT(A,N,N)
      READ (5,*) (B(I), I=1,N)
      CALL SOLVE(B,A,B)
      DO 30 I = 1,N
         WRITE(6,20) I,B(I)
20       FORMAT (I7,E20.8)
30    CONTINUE
      GOTO 10
      END
//DATA.INPUT DD *
2
2. 6. 4. 9.
2. 9.
3
4. 16. 8. 12. 45. 15. 16. 58. 12.
4. 16. 6.
```

And so on. The final value for N is 0.

2-3.4. Column elimination with column pivoting is equivalent to multiplying the right side of **A** by a sequence of pivot and elimination matrices to obtain a lower triangular matrix: $\mathbf{AP}_1\mathbf{U}_1\mathbf{P}_2\mathbf{U}_2 \cdots \mathbf{P}_{n-1}\mathbf{U}_{n-1} = \mathbf{L}$. Hence,

$$\mathbf{A} = \mathbf{LU}_{n-1}^{-1}\mathbf{P}_{n-1}^{-1} \cdots \mathbf{U}_1^{-1}\mathbf{P}_1^{-1}.$$

The equation $\mathbf{Ax} = \mathbf{b}$ is equivalent to $\mathbf{Ly} = \mathbf{b}$, where

$$\mathbf{y} = \mathbf{U}_{n-1}^{-1}\mathbf{P}_{n-1}^{-1} \cdots \mathbf{U}_1^{-1}\mathbf{P}_1^{-1}\mathbf{x}.$$

Solving for **x** gives $\mathbf{x} = \mathbf{P}_1\mathbf{U}_1\mathbf{P}_2\mathbf{U}_2 \cdots \mathbf{P}_{n-1}\mathbf{U}_{n-1}\mathbf{y}$. To compute **x**, first solve the lower triangular system $\mathbf{Ly} = \mathbf{b}$ for **y**, then apply each elimination step and each pivot (in the reverse order) to **y** obtaining **x**. More precisely,

$$\mathbf{x} = \mathbf{P}_1\mathbf{U}_1\mathbf{P}_2\mathbf{U}_2 \cdots \mathbf{P}_{n-1}\mathbf{U}_{n-1}\mathbf{y}.$$

2-3.5. Forward elimination gives us the upper triangular system

$$\begin{aligned} 2x_1 + x_2 - 2x_3 &= 4, \\ -x_2 + 3x_3 &= -1, \\ 0 &= 0. \end{aligned}$$

The last equation "0 = 0" is trivially satisfied. Letting x_3 have any value, the first two equations can be solved for x_1 and x_2. For example, if $x_3 = 0$, the second equation implies that $x_2 = 1$ and by the first equation,

$$x_1 = (4 - x_2 + 2x_3)/2 = 1.5.$$

When the "8" is changed to a "7," the last equation in the eliminated system is $0 = -1$. Since 0 does not equal -1, the equations are inconsistent and there is no solution.

2-4.1. $x_1 = 1$, $x_2 = 0$, $x_3 = -1$, $x_4 = 2$.

2-4.2.
$$\begin{bmatrix} 2 & 1 & 0 & 0 \\ 4 & -2 & 1 & 0 \\ 0 & 4 & 0 & 1 \\ 0 & 0 & 4 & 7 \end{bmatrix} = \begin{bmatrix} 1 & 0 & 0 & 0 \\ 2 & 1 & 0 & 0 \\ 0 & -1 & 1 & 0 \\ 0 & 0 & 4 & 1 \end{bmatrix}\begin{bmatrix} 2 & 1 & 0 & 0 \\ 0 & -4 & 1 & 0 \\ 0 & 0 & 1 & 1 \\ 0 & 0 & 0 & 3 \end{bmatrix}$$

Note that $l_{i+1,i} = a_{i+1,i}/u_{ii}$.
$d_{i+1} \leftarrow d_{i+1} - lu/d_i$ for $i = 1$ to $n - 1$
$b_i \leftarrow b_i - lb_{i-1}/d_i$ for $i = 2$ to n
$b_n \leftarrow b_n/d_n$
$b_i \leftarrow (b_i - ub_{i+1})/d_i$ for $i = n - 1$ down to 1

2-4.3. $x_1 = 1$, $x_2 = 0$, $x_3 = -1$, $x_4 = 2$.

2-4.4. For a (l, u)-band matrix, $\mathrm{I} = i - j + u + 1$ and $\mathrm{J} = \min(i, j)$. If $i \leqq j$, then $i = \mathrm{J}$ and $j = \mathrm{J} - \mathrm{I} + u + 1$. If $i > j$, then $i = \mathrm{I} + \mathrm{J} - u - 1$ and $j = \mathrm{J}$.

2-4.5. **L** is $(l + u, 0)$ and **U** is $(0, u)$.

2-4.6. **P** is $(l_1 + l_2, u_1 + u_2)$.

2-4.7.
```
      SUBROUTINE PEFACT(L1,L2,D,U1,U2,N)
      REAL L1(1),L2(1),D(1),U1(1),U2(1)
      IF ( N .EQ. 1 ) RETURN
      J = 1
      K = 2
10    IF ( K .GE. N ) GOTO 20
         I = J
         J = K
         K = K + 1
         T = L1(I)/D(I)
         L1(I) = T
```

```
      D(J) = D(J) - T*U1(I)
      U1(J) = U1(J) - T*U2(I)
      T = L2(I)/D(I)
      L2(I) = T
      L1(J) = L1(J) - T*U1(I)
      D(K) = D(K) - T*U2(I)
      GOTO 10
20    T = L1(J)/D(J)
      L1(J) = T
      D(K) = D(K) - T*U1(J)
      RETURN
      END

      SUBROUTINE PESOLV(X,L1,L2,D,U1,U2,N,B)
      REAL B(1),L1(1),L2(1),D(1),U1(1),U2(1),X(1)
      IF ( N .GT. 1 ) GOTO 10
      X(1) = B(1)/D(1)
      RETURN
10    X(1) = B(1)
      X(2) = B(2) - L1(1)*X(1)
      J = 1
      K = 2
20    IF ( K .GE. N ) GOTO 30
         I = J
         J = K
         K = K + 1
         X(K) = B(K) - L2(I)*X(I) - L1(J)*X(J)
         GOTO 20
30    X(N) = X(N)/D(N)
      J = N
      K = N - 1
      X(K) = (X(K)-U1(K)*X(N))/D(K)
40    IF ( K .LE. 1 ) RETURN
         I = J
         J = K
         K = K - 1
         X(K) = (X(K)-U2(K)*X(I)-U1(K)*X(J))/D(K)
         GOTO 40
      END
```

2-4.8. $x_1 = 3$, $x_2 = -24$, and $x_3 = 30$.

2-4.9.

```
      SUBROUTINE SMODE(A,LA,N)
      REAL A(1)
      J = 0
      K = 1
      L = N
      M = N
10    DO 20 I = K,L
20       A(I) = A(I+J)
      K = L + 1
      M = M - 1
      L = L + M
      J = J + LA - M
```

```
      IF ( M .GT. 0 ) GOTO 10
      RETURN
      END
```

2-4.10. Assume that $j \geqq i$ (otherwise by symmetry, i and j can be interchanged). $I = (2n - i + 2)(i - 1)/2 + j - i + 1$. i is the integer part of $1.5 + n - \sqrt{n^2 + n + 2.25 - 2I}$ and $j = I + i - 1 + (2n - i + 2)(1 - i)/2$.

2-4.13. Take $\mathbf{x} = \mathbf{e}_i$.

2-4.16. **(a)** Suppose an off-diagonal element a_{ij} is the largest element in magnitude. Examine $\mathbf{x}^T\mathbf{A}\mathbf{x}$, where $\mathbf{x} = \mathbf{e}_i \pm \mathbf{e}_j$.

2-4.17. Each diagonal element of $\mathbf{C}$ is the square root of the corresponding diagonal element of $\mathbf{D}$.

2-4.18. $x_1 = 2$, $x_2 = -2$, $x_3 = 0$, $x_4 = 1$, $x_5 = 3$.

2-4.19. For a h-band matrix, $I = |i - j| + 1$ and $J = \min(i, j)$. Suppose that $i \geqq j$ (otherwise by symmetry, i and j can be interchanged). $i = I + J - 1$ and $j = J$.

2-4.20. $b_n \leftarrow b_n - l_i b_i$ for $i = 1$ to $n - 1$
$b_n \leftarrow b_n / d_n$
$b_i \leftarrow (b_i - u_i b_n)/d_i$ for $i = n - 1$ down to 1

2-6.1. **(a)** $L = (3n + 1)n/2$ and the asymptotic parameter is $1.5n^2$.
(b) $L = n(n + 1)(n + 2)/6$ and the asymptotic parameter is $n^3/6$.
(c) $L = n(n + 1)(2n + 1)/6$ and the asymptotic parameter is $n^3/3$.

2-6.2. B(I) = (DT*DT)*SIN((I − 1)*3.14159/(2.*FLOAT(N))), C = 1./(2. + DT − DT*DT), D = 1 + DT.

2-6.3. $n^3/6$.

2-7.1. After rounding, the solution is $x_1 = .0414$, $x_2 = .0732$, $x_3 = .0957$, $x_4 = .1092$, $x_5 = .1136$, $x_6 = .1092$, $x_7 = .0957$, $x_8 = .0732$, $x_9 = .0414$.

2-7.2. Defining $\mathbf{P} = \mathbf{A}^{-1}\mathbf{B}$ and letting $\mathbf{p}_i$ and $\mathbf{b}_i$ denote column i of $\mathbf{P}$ and $\mathbf{B}$, respectively, we have $\mathbf{p}_i = \mathbf{A}^{-1}\mathbf{b}_i$ or $\mathbf{A}\mathbf{p}_i = \mathbf{b}_i$. Hence, $\mathbf{p}_i$ is the solution $\mathbf{x}$ to $\mathbf{A}\mathbf{x} = \mathbf{b}_i$.

```
      CALL TFACT(L,D,U,N)
      DO 10 J = 1,M
10       CALL TSOLVE(P(1,J),L,D,U,B(1,J))
```

2-7.3. Again, approach 2 is faster since a general matrix can be factored three times faster than it can be inverted.

2-8.3. The (i, j) element of $\mathbf{a}_k\mathbf{b}_k^T$ is the product between element i of $\mathbf{a}_k$ and element j of $\mathbf{b}_k^T$. Thus the (i, j) element of $\mathbf{a}_k\mathbf{b}_k^T$ is $a_{ik}b_{kj}$. Summing over k yields

$$\left(\sum_{k=1}^{m} \mathbf{a}_k\mathbf{b}_k^T\right)_{ij} = \sum_{k=1}^{m} a_{ik}b_{kj}.$$

Observe that this sum is equivalent to row i of $\mathbf{A}$ times column j of $\mathbf{B}$. Since row i of $\mathbf{A}$ times column j of $\mathbf{B}$ is also the (i, j) element of $\mathbf{AB}$, we are done.

2-8.5. Note that $\mathbf{x} = \mathbf{A}^{-1}\mathbf{b} = \mathbf{B}^{-1}\mathbf{b} + \mathbf{B}^{-1}\mathbf{U}(\mathbf{I} - \mathbf{V}^T\mathbf{B}^{-1}\mathbf{U})^{-1}\mathbf{V}^T\mathbf{B}^{-1}\mathbf{b}$. 1. **LU** factor $\mathbf{B}$ and solve $\mathbf{Bw} = \mathbf{b}$ for $\mathbf{w}$. 2. Compute the product $\mathbf{y} = \mathbf{V}^T\mathbf{w}$. 3. Compute $\mathbf{T} = \mathbf{B}^{-1}\mathbf{U}$ using the algorithm of Exercise 2-7.2. 4. Compute $\mathbf{I} - \mathbf{V}^T\mathbf{T}$ and call the resulting matrix $\mathbf{S}$. 5. **LU** factor $\mathbf{S}$ and solve $\mathbf{Sz} = \mathbf{y}$ for $\mathbf{z}$. 6. Compute the product $\mathbf{w} = \mathbf{Tz}$. 7. Set $\mathbf{x} = \mathbf{v} + \mathbf{w}$.

2-9.1. Determinant is zero.

2-9.2. $a_{11}a_{22}a_{33} + a_{12}a_{23}a_{31} + a_{13}a_{21}a_{32} - a_{13}a_{22}a_{31} - a_{11}a_{23}a_{32} - a_{12}a_{21}a_{33}$.

2-9.3. For the first class of matrices, the determinants are $.34 \times 10^{-22}$, $.55 \times 10^{-46}$, $.75 \times 10^{-94}$, $.70 \times 10^{-190}$, which approach zero. For the second class of matrices, the determinants are $.50 \times 10^{26}$, $.12 \times 10^{51}$, $.34 \times 10^{99}$, $.15 \times 10^{196}$, which approach infinity.

2-10.1. NEW(2:N − 1) = OLD(1:N − 2) + OLD(3:N)
NEW(2:N − 1) = .5*NEW(2:N − 1)

2-10.2. Assume that x_{ij} is stored in array element $i + jn$. Initialize the array IV such that each element is zero except for the elements stored in the following locations which are one: (i, j) for $j = 1$ or $j = n$ and $i = 1$ to n, (i, j) for $i = 1$ or $i = n$ and $j = 1$ to n.

```
M = N*N
L = M - N - 1
X(N+2:L) = OLD(N+1:L-1) + OLD(N+3:L+1)
Y(N+2:L) = OLD(2:L-N) + OLD(N+N+2:L+N)
Z(N+2:L) = X(N+2:L) + Y(N+2:L)
Z(N+2:L) = .25*Z(N+2:L)
CALL MASK(Z,OLD,IV,NEW,M)
```

2-10.3.

```
P(1:N) = D(1:N)*X(1:N)
Y(2:N) = L(1:N-1)*X(1:N-1)
P(2:N) = P(2:N) + Y(2:N)
Y(1:N-1) = U(1:N-1)*X(2:N)
P(1:N-1) = P(1:N-1) + Y(1:N-1)
```

2-11.1. Factor:

$k = 1$ to $n/2 - 1$
  $l_k \leftarrow l_k/d_k$
  $d_{k+1} \leftarrow d_{k+1} - l_k u_k$
next k
$k = n - 1$ down to $n/2 + 1$
  $u_k \leftarrow u_k/d_{k+1}$
  $d_k \leftarrow d_k - l_k u_k$
next k

Forward Elimination:

$b_{k+1} \leftarrow b_{k+1} - l_k b_k$ for $k = 1$ to $n/2 - 1$
$b_k \leftarrow b_k - u_k b_{k+1}$ for $k = n - 1$ down to $n/2 + 1$

Back Substitution:

$m \leftarrow n/2$
$s \leftarrow d_m d_{m+1} - l_m u_m$
$t \leftarrow (b_m d_{m+1} - b_{m+1} u_m)/s$
$b_{m+1} \leftarrow (b_{m+1} d_m - b_m l_m)/s$
$b_m \leftarrow t$
$b_k \leftarrow (b_k - u_k b_{k+1})/d_k$ for $k = m - 1$ down to 1
$b_k \leftarrow (b_k - l_{k-1} b_{k-1})/d_k$ for $k = m + 2$ to n

2-11.2. $x_1 = 3$, $x_2 = 5$, $x_3 = 6$, $x_4 = 6$, $x_5 = 5$, $x_6 = 3$.

Review Problem

2-10. Asymptotic parameter for the loop ending at statement 20 is about $.4n^3$. Asymptotic parameter for the loop ending at statement 30 is about $1.25n^3$.

Chapter 3

3-2.1. **(a)** Since $|x_i| \geqq 0$ for every i, $\|\mathbf{x}\|_\infty$ = maximum $\{|x_1|, |x_2|, \cdots, |x_n|\} \geqq 0$, and if $\|\mathbf{x}\|_\infty = 0$, then $\mathbf{x} = \mathbf{0}$.
(b) $\|\alpha\mathbf{x}\|_\infty$ = maximum $\{|\alpha x_1|, |\alpha x_2|, \cdots, |\alpha x_n|\}$ = $|\alpha|$ maximum $\{|x_1|, |x_2|, \cdots, |x_n|\} = |\alpha|\,\|\mathbf{x}\|_\infty$.
(c) Let m denote an index with the property that $|x_m + y_m|$ = maximum $\{|x_1 + y_1|, |x_2 + y_2|, \cdots, |x_n + y_n|\} = \|\mathbf{x} + \mathbf{y}\|_\infty$. Since $|x_m + y_m| \leqq |x_m| + |y_m|$, where $|x_m| \leqq \|\mathbf{x}\|_\infty$ and $|y_m| \leqq \|\mathbf{y}\|_\infty$, it follows that $\|\mathbf{x} + \mathbf{y}\|_\infty \leqq \|\mathbf{x}\|_\infty + \|\mathbf{y}\|_\infty$.

3-2.2. Since $\|\alpha\mathbf{x}\| = |\alpha|\,\|\mathbf{x}\|$, it follows that if the inequality $\|\mathbf{x}\|_\infty \leqq \|\mathbf{x}\|_p \leqq \|\mathbf{x}\|_1$ holds when $\|\mathbf{x}\|_p = 1$, the inequality holds for every $\mathbf{x}$. In the special case $\|\mathbf{x}\|_p = 1$, we have

$$\frac{d}{dp}\|\mathbf{x}\|_p = \frac{1}{p}\left(\sum_{i=1}^{n} |x_i|^p \log |x_i|\right),$$

which is nonpositive since $|x_i| \leqq 1$ for every i.

3-3.3. **(a)** $\|\mathbf{A}\|_2 = 2$. **(e)** The p-norm of a diagonal matrix is the absolute largest diagonal element.

3-3.4. $\|\mathbf{A}\|_2 = 1$.

3-3.5. Utilizing property (d), where $\mathbf{x}$ is an eigenvector, we have $|\lambda|\,\|\mathbf{x}\| = \|\lambda\mathbf{x}\| = \|\mathbf{Ax}\| \leqq \|\mathbf{A}\|\,\|\mathbf{x}\|$. Canceling the factor $\|\mathbf{x}\|$ from each side of this inequality yields the relation $|\lambda| \leqq \|\mathbf{A}\|$.

3-3.6. For the 1-norm, $\mathbf{y} = \mathbf{e}_m$ if the largest absolute column sum occurs in column m. For the ∞-norm, let m denote a row index corresponding to the largest absolute row sum. Then $y_j = 1$ if $a_{mj} \geqq 0$ and $y_j = -1$ otherwise.

3-3.7. Minimizing $\|\mathbf{Ay}\|$ over the unit sphere is equivalent to minimizing the ratio $\|\mathbf{Az}\|/\|\mathbf{z}\|$ over arbitrary nonzero $\mathbf{z}$. Putting $\mathbf{z} = \mathbf{A}^{-1}\mathbf{x}$ yields:

$$\text{minimum}\left\{\frac{\|\mathbf{Az}\|}{\|\mathbf{z}\|} : \mathbf{z} \neq \mathbf{0}\right\} = \text{minimum}\left\{\frac{\|\mathbf{x}\|}{\|\mathbf{A}^{-1}\mathbf{x}\|} : \mathbf{x} \neq \mathbf{0}\right\}$$

$$= 1/\text{maximum}\left\{\frac{\|\mathbf{A}^{-1}\mathbf{x}\|}{\|\mathbf{x}\|} : \mathbf{x} \neq \mathbf{0}\right\} = 1/\|\mathbf{A}^{-1}\|$$

3-3.8. Since the matrix $\mathbf{A}^T\mathbf{A}$ is symmetric, (10) and (11) combine to give $\|\mathbf{A}\|_2^2 = \|\mathbf{A}^T\mathbf{A}\|_2$.

3-3.9. Let $\mathbf{x}$ denote an eigenvector which achieves the maximum in (11) and let λ denote the corresponding eigenvalue. By (11), we have $\|\mathbf{S}\|_2 = |\mathbf{x}^T\mathbf{S}\mathbf{x}| = |\lambda\mathbf{x}^T\mathbf{x}| = |\lambda|\,|\mathbf{x}^T\mathbf{x}| = |\lambda|$. Since $|\lambda| \leqq \|\mathbf{S}\|$ by Exercise 3-3.5, it follows that $\|\mathbf{S}\|_2 \leqq \|\mathbf{S}\|$.

3-4.1. Condition number is 1.

3-4.2. $c(\mathbf{AB}) = \|\mathbf{AB}\|\,\|(\mathbf{AB})^{-1}\|$. Since $(\mathbf{AB})^{-1} = \mathbf{B}^{-1}\mathbf{A}^{-1}$ and since the norm of a product is less than or equal to the product of norms, $\|\mathbf{AB}\| \leqq \|\mathbf{A}\|\,\|\mathbf{B}\|$ and $\|(\mathbf{AB})^{-1}\| \leqq \|\mathbf{A}^{-1}\|\,\|\mathbf{B}^{-1}\|$. Combining these inequalities yields $c(\mathbf{AB}) \leqq c(\mathbf{A})c(\mathbf{B})$.

3-4.3. If $\mathbf{x}$ is an eigenvector corresponding to the eigenvalue λ, then $\mathbf{Ax} = \lambda\mathbf{x}$. Multiplying by $\mathbf{A}^{-1}$ and dividing by λ gives $\mathbf{A}^{-1}\mathbf{x} = (1/\lambda)\mathbf{x}$. Thus the eigenvalues of $\mathbf{A}^{-1}$ are the reciprocals of the eigenvalues of $\mathbf{A}$. By Exercise 3-3.5, the norm of a matrix is bounded from below by the norm of each eigenvalue. In particular, $\|\mathbf{A}\| \geqq |\lambda_b|$ and $\|\mathbf{A}^{-1}\| \geqq 1/|\lambda_s|$. Combining these inequalities, $c(\mathbf{A}) \geqq |\lambda_b|/|\lambda_s|$.

3-4.4. $c_1 \approx 7 \times 10^6$, $\|\mathbf{b}\|_1 = 10201$, and $\|\mathbf{r}\|_1 = \approx .0375$, $c_1\|\mathbf{r}\|/\|\mathbf{b}\| \approx 26$. True relative error in the computed solution is about .012. Ratio between the estimated relative error and the true relative error is about $26/.012 \approx 2000$.

3-4.5. Minimum ratio is $1/c(\mathbf{A})$. The minimizing $\mathbf{x}$ is $\mathbf{x} = \mathbf{A}^{-1}\mathbf{e}_j$, where column j achieves the maximum absolute column sum for $\mathbf{A}^{-1}$, and the minimizing $\boldsymbol{\delta}\mathbf{b}$ is $\boldsymbol{\delta}\mathbf{b} = \mathbf{Ae}_k$, where column k achieves the maximum absolute column sum for $\mathbf{A}$. Thus $\mathbf{b} = \mathbf{Ax} = \mathbf{e}_j$ and $\delta\mathbf{b}$ is a multiple of column k from $\mathbf{A}$.

3-5.1. $k = 2$.

3-5.2. Since $\|\mathbf{A}\| < 1$, $\mathbf{S} = (\mathbf{I} - \mathbf{A})^{-1}$ is given by equation (26) of Chapter 3. Taking the norm of (26), applying the triangle inequality, and recalling that the norm of a product is bounded by the product of the norms, we have

$$\|\mathbf{S}\| \leqq 1 + \|\mathbf{A}\| + \|\mathbf{A}\|^2 + \cdots = 1/(1 - \|\mathbf{A}\|).$$

3-5.3. Since $\|\mathbf{A}\| = \|\mathbf{B}^{-1}\delta\mathbf{B}\| \leqq \|\mathbf{B}^{-1}\|\|\delta\mathbf{B}\|$, it follows that $\|\mathbf{A}\| < 1$ for $\delta\mathbf{B}$ sufficiently small. If $\|\mathbf{A}\| < 1$, then $\mathbf{I} - \mathbf{A}$ is invertible with inverse given by equation (26) of Chapter 3. If each term in the product $\mathbf{B}(\mathbf{I} - \mathbf{A})$ is invertible, then the product is invertible and $(\mathbf{B} + \delta\mathbf{B})^{-1} = [\mathbf{B}(\mathbf{I} - \mathbf{A})]^{-1} = (\mathbf{I} - \mathbf{A})^{-1}\mathbf{B}^{-1}$. By the analysis of Section 3-5, $(\mathbf{I} - \mathbf{A})^{-1} = \mathbf{I} + \mathbf{E}$, where $\mathbf{E} = \mathbf{A} + \mathbf{A}^2 + \cdots$ and $\|\mathbf{E}\| \leqq \|\mathbf{A}\|/(1 - \|\mathbf{A}\|)$. Substituting $(\mathbf{I} - \mathbf{A})^{-1} = \mathbf{I} + \mathbf{E}$ yields $(\mathbf{B} + \delta\mathbf{B})^{-1} = (\mathbf{I} - \mathbf{A})^{-1}\mathbf{B}^{-1} = (\mathbf{I} + \mathbf{E})\mathbf{B}^{-1} = \mathbf{B}^{-1} + \mathbf{E}\mathbf{B}^{-1}$. Since both $\mathbf{A}$ and $\mathbf{E}$ approach zero as $\delta\mathbf{B}$ tends to zero, we conclude that $(\mathbf{B} + \delta\mathbf{B})^{-1}$ tends to $\mathbf{B}^{-1}$ as $\delta\mathbf{B}$ tends to zero.

3-5.4. Since $\|\mathbf{A}\| < 1$, $(\mathbf{I} - \mathbf{A})^{-1}$ is given by the series (26) of Chapter 3. Since the elements of $\mathbf{A}$ are nonnegative, each term in this series (as well as the sum) is nonnegative.

Chapter 4

4-1.1. **(a)** $x = 0$ and $x \approx .950$. **(b)** $x = 0$ and $x \approx \pm 1.90$. **(c)** $x \approx 1.13$.

4-1.2. **(a)** x_2 is near $\pm\infty$. **(b)** x_3 is near $\pm\infty$. **(c)** x_{k+1} is near $\pm\infty$. Whenever $f(x_k)$ is relatively close to zero in the secant method, the iterations should stop.

4-1.3. $z = 1 \pm 2\mathbf{i}$ and $z = -3 \pm \mathbf{i}$, where $\mathbf{i} = \sqrt{-1}$.

4-2.1. **(a)** Monotone convergence. **(b)** Oscillating convergence. **(c)** Monotone convergence.

4-2.2. $r = 1/2$ for example.

4-2.3. **(a)** Third order. **(b)** First order. **(c)** Second order.

4-2.4. Since the derivative of the inverse function g^{-1} is $1/g'$, it follows that the derivative of $g^{-1}(x)$ at $x = \alpha$ is $1/g'(\alpha)$. Hence the derivative of g^{-1} is less than 1 in magnitude whenever the derivative of g is greater than 1 in magnitude.

4-2.5. With the starting guess $x_0 = 2$, we have $x_1 \approx 2.3$. Defining $f(x) = 3 - \log x - x$, observe that $f(2) \approx .3$ and $f(2.3) \approx -.13$. Since the sign of $f(x)$ at $x = 2$ and at $x = 2.3$ are opposite, there exists a zero of f on the interval [2, 2.3]. For the function $g(x) = 3 - \log x$, we have $g'(x) = -1/x$. Hence, for x between 2 and 2.3, $|g'(x)| \leqq \frac{1}{2}$. From the discussion in the text, the iterations $x_{k+1} = g(x_k)$ converge to a root of $x = g(x)$ on the interval [2, 2.3] and the error e_k at iteration k satisfies the inequality $e_k \leqq .5^k e_0$. Since the root lies on the interval [2, 2.3], it follows that $\alpha \geqq 2$, $e_0 \leqq 2.3 - 2 = .3$, and the relative error is at most 10^{-8} when $.5^k \times .3/2 \leqq 10^{-8}$. Thus $k \geqq 23.8$ and the relative error is at most 10^{-8} when $k = 24$.

4-2.8. $F(x) = f(x)^3$.

4-3.1. **(a)** $\begin{bmatrix} 1 + x_2 & x_1 - \sin x_2 \\ 1 & 2x_2 \end{bmatrix}$ **(b)** $\begin{bmatrix} 2(x_1 + x_2) & 2x_1 & \cos x_3 \\ 1 & 1 & 1 \\ -\sin x_1 & 0 & e^{x_3} \end{bmatrix}$.

Neither of the matrices in (a) and (b) has an inverse at $\mathbf{x} = \mathbf{0}$. Since a Newton iteration involves the inverse of the Jacobian, Newton's method cannot be started from $\mathbf{x} = \mathbf{0}$.

4-3.2. $x_1 = x_2 \approx .567$.

4-3.3. After rounding, the solution is $x_1 = .0414$, $x_2 = .0732$, $x_3 = .0957$, $x_4 = .1092$, $x_5 = .1136$, $x_6 = .1092$, $x_7 = .0957$, $x_8 = .0732$, $x_9 = .0414$.

4-3.4. $x_1 \approx .49998$, $x_2 \approx .02000$, $x_3 \approx -.49950$ and $x_1 \approx .49998$, $x_2 \approx -.02000$, $x_3 \approx -.50050$. When $x_2 = 0$, the last two columns of the Jacobian are multiples of each other and the Jacobian does not have an inverse. Since a Newton iteration involves the inverse of the Jacobian, Newton's method cannot be applied when the x_2 component of the starting guess vanishes.

4-4.1. $\mathbf{x}_{k+1} = \mathbf{x}_k - \mathbf{A}^{-1}(\mathbf{A}\mathbf{x}_k - \mathbf{b}(\mathbf{x}_k))$. Hence, $\mathbf{P}(\mathbf{x}) = \mathbf{A}^{-1}$ and $\mathbf{f}(\mathbf{x}) = \mathbf{A}\mathbf{x} - \mathbf{b}(\mathbf{x})$.

4-4.2. Since $p_i = i(n + 1 - i)$, $p_i \geqq 0$ when i is between 1 and n. From the definition (38), we see that the ith component of $\mathbf{Ap}$ is $2p_i - p_{i-1} - p_{i+1}$. Substituting $p_i = (n + 1)i - i^2$, we obtain $(\mathbf{Ap})_i = 2$, which is greater than 0 for every i.

4-4-3. We verify properties (a)–(c). The existence of $\mathbf{y}_1$ satisfying (a) follows from Exercise 4-4.2. By Exercise 4-4.1, $\mathbf{P} = \mathbf{A}^{-1}$ and by Exercise 3-5.5, $\mathbf{P} \geqq \mathbf{0}$. Taking $\mathbf{Q} = \mathbf{A}$, the conditions $\mathbf{PQ} \leqq \mathbf{I}$ and $\mathbf{QP} \leqq \mathbf{I}$ are satisfied trivially since $\mathbf{PQ} = \mathbf{QP} = \mathbf{I}$. Hence, condition (b) holds. Finally, condition (c) is verified at the end of this section.

4-4.4. This is a special case of Exercise 4-4.5.

4-4.5. $\mathbf{x}^{\text{new}} = \mathbf{x}^{\text{old}} - \mathbf{D}(\mathbf{x}^{\text{old}})^{-1}[\mathbf{A}\mathbf{x}^{\text{old}} - \mathbf{b}(\mathbf{x}^{\text{old}})]$, where $\mathbf{A}$ and $\mathbf{b}$ are defined beneath equation (78) of Chapter 2 and $\mathbf{D}$ is the diagonal matrix whose ith diagonal element is $2 + e^{-x_i^{\text{old}}}$. In component form,

$$x_i^{\text{new}} = x_i^{\text{old}} - \frac{1}{2 + (\Delta t)^2 e^{-x_i^{\text{old}}}}[-x_{i-1}^{\text{old}} + 2x_i^{\text{old}} - x_{i+1}^{\text{old}} - (\Delta t)^2 e^{-x_i^{\text{old}}}].$$

The existence of $\mathbf{y}_1$ satisfying condition (a) is established in Exercise 4-4.2. Taking $\mathbf{Q} = \mathbf{A}$, condition (c) is established at the end of this section. Since the diagonal of $\mathbf{D}(\mathbf{x})$ is positive for every $\mathbf{x}$, $\mathbf{P}(\mathbf{x}) = \mathbf{D}(\mathbf{x})^{-1} \geqq \mathbf{0}$. Observe that $\mathbf{P}(\mathbf{x})\mathbf{Q} = \mathbf{D}(\mathbf{x})^{-1}\mathbf{A} = \mathbf{D}(\mathbf{x})^{-1}(\mathbf{D}(\mathbf{x}) + \mathbf{A} - \mathbf{D}(\mathbf{x})) = \mathbf{I} + \mathbf{D}(\mathbf{x})^{-1}(\mathbf{A} - \mathbf{D}(\mathbf{x}))$. Since $\mathbf{A} - \mathbf{D}(\mathbf{x}) \leqq \mathbf{0}$, it follows that $\mathbf{D}(\mathbf{x})^{-1}(\mathbf{A} - \mathbf{D}(\mathbf{x})) \leqq \mathbf{0}$ and $\mathbf{P}(\mathbf{x})\mathbf{Q} = \mathbf{I} + \mathbf{D}(\mathbf{x})^{-1}(\mathbf{A} - \mathbf{D}(\mathbf{x})) \leqq \mathbf{I}$. By a similar argument, $\mathbf{QP}(\mathbf{x}) \leqq \mathbf{I}$ and condition (b) holds.

4-4.6. $\mathbf{x}^{\text{new}} = \mathbf{x}^{\text{old}} - \mathbf{L}(\mathbf{x}^{\text{old}})^{-1}[\mathbf{A}\mathbf{x}^{\text{old}} - \mathbf{b}(\mathbf{x}^{\text{old}})]$, where $\mathbf{A}$ and $\mathbf{b}$ are defined beneath equation (78) of Chapter 2 and $\mathbf{L}(\mathbf{x})$ is the lower triangular matrix with -1's on the subdiagonal, with ith diagonal element equal to $2 + e^{-x_i^{\text{old}}}$, and with zeros elsewhere. In component form,

$$x_i^{\text{new}} = x_i^{\text{old}} - \frac{1}{2 + (\Delta t)^2 e^{-x_i^{\text{old}}}}[-x_{i-1}^{\text{new}} + 2x_i^{\text{old}} - x_{i+1}^{\text{old}} - (\Delta t)^2 e^{-x_i^{\text{old}}}].$$

Once again, the existence of $\mathbf{y}_1$ satisfying condition (a) is established in Exercise 4-4.2. With $\mathbf{Q} = \mathbf{A}$, condition (c) is established at the end of this section. By Exercise 3-5.5, $\mathbf{P}(\mathbf{x}) = \mathbf{L}(\mathbf{x})^{-1} \geqq \mathbf{0}$. Observe that

$$\mathbf{P}(\mathbf{x})\mathbf{Q} = \mathbf{L}(\mathbf{x})^{-1}\mathbf{A} = \mathbf{L}(\mathbf{x})^{-1}(\mathbf{L}(\mathbf{x}) + \mathbf{A} - \mathbf{L}(\mathbf{x})) = \mathbf{I} + \mathbf{L}(\mathbf{x})^{-1}(\mathbf{A} - \mathbf{L}(\mathbf{x})).$$

Since $\mathbf{A} - \mathbf{L}(\mathbf{x}) \leqq \mathbf{0}$, it follows that $\mathbf{L}(\mathbf{x})^{-1}(\mathbf{A} - \mathbf{L}(\mathbf{x})) \leqq \mathbf{0}$ and

$$\mathbf{P}(\mathbf{x})\mathbf{Q} = \mathbf{I} + \mathbf{L}(\mathbf{x})^{-1}(\mathbf{A} - \mathbf{L}(\mathbf{x})) \leqq \mathbf{I}.$$

By a similar argument, $\mathbf{QP}(\mathbf{x}) \leqq \mathbf{I}$ and condition (b) holds.

4-4.7. $$\mathbf{g}(\mathbf{x}) = \begin{bmatrix} \frac{2}{3}e^{-x_1} + \frac{1}{3}e^{-x_2} \\ \frac{1}{3}e^{-x_1} + \frac{2}{3}e^{-x_2} \end{bmatrix}.$$

Since $\|\nabla\mathbf{g}(\mathbf{x})\|_1 = \text{maximum}\,\{e^{-x_1}, e^{-x_2}\}$, $\|\nabla\mathbf{g}(\mathbf{x})\|_1 < 1$ whenever x_1 and x_2 are positive.

4-6.1. $2x = (1 + x^2)\text{arc tan } x$, $x \approx 1.391745200$.

4-6.2. LU factor $\mathbf{J}(\mathbf{x}_k)$, solve the factored system $(\mathbf{LU})\mathbf{y} = \mathbf{f}(\mathbf{x}_k)$ for $\mathbf{y}$.

4-6.3. $x_1 \approx 24.1016$, $x_2 \approx .567143$, and $x_3 \approx .567143$.

4-7.1. (1) $-.17 \times 10^{-14}$, (2) $.86 \times 10^{-8}$, (3) $-.11 \times 10^{-3}$, (4) $.11$, (5) $-.26 \times 10^{2}$, (6) $.20 \times 10^{4}$, (7) $-.76 \times 10^{5}$, (8) $.16 \times 10^{7}$, (9) $-.20 \times 10^{8}$, (10) $.16 \times 10^{9}$, (11) $-.89 \times 10^{9}$, (12) $.35 \times 10^{10}$, (13) $-.98 \times 10^{10}$, (14) $.20 \times 10^{11}$, (15) $-.30 \times 10^{11}$, (16) $.32 \times 10^{11}$, (17) $-.24 \times 10^{11}$, (18) $.12 \times 10^{11}$, (19) $-.34 \times 10^{10}$, (20) $.45 \times 10^{9}$. Zero 16 has the worst condition number.

4-7.2. (1) $.20 \times 10^{2}$, (2) $-.19 \times 10^{3}$, (3) $.11 \times 10^{4}$, (4) $-.48 \times 10^{4}$, (5) $.16 \times 10^{5}$, (6) $-.39 \times 10^{5}$, (7) $.78 \times 10^{5}$, (8) $-.13 \times 10^{6}$, (9) $.17 \times 10^{6}$, (10) $-.18 \times 10^{6}$, (11) $.17 \times 10^{6}$, (12) $-.13 \times 10^{6}$, (13) $.78 \times 10^{5}$, (14) $-.39 \times 10^{5}$, (15) $.16 \times 10^{5}$, (16) $-.48 \times 10^{4}$, (17) $.11 \times 10^{4}$, (18) $-.19 \times 10^{3}$, (19) $.20 \times 10^{2}$, (20) $-.10 \times 10^{1}$. Zero 10 has the worst condition number.

4-7.3. Estimated root is 3.05 (correct root is roughly 3.047).

Review Problems

4-1. After rounding, the solution is $x_1 = .0414$, $x_2 = .0732$, $x_3 = .0957$, $x_4 = .1092$, $x_5 = .1136$, $x_6 = .1092$, $x_7 = .0957$, $x_8 = .0732$, $x_9 = .0414$.

4-3. After rounding, the solution is $x_1 = .000337$, $x_2 = .001065$, $x_3 = .001834$, $x_4 = .002395$, $x_5 = .002599$, $x_6 = .002395$, $x_7 = .001834$, $x_8 = .001065$, $x_9 = .000337$.

4-5. **(a)** Second order. **(b)** First order.

Chapter 5

5-1.1. $ad = bc$.

5-1.2. **(a)** $ax_2 = bx_1$. **(b)** $x_1 = ac/(a^2 + b^2)$ and $x_2 = bc/(a^2 + b^2)$.

5-1.3. When both a and b are zero and c is nonzero.

5-2.1. $a = 1.9$ and $b = .5$.

5-2.2. For the data in Table 5-2, $a = 62/30 \approx 2.07$ and in general, $a = (\Sigma\, x_k y_k)/(\Sigma\, x_k^2)$.

5-2.3. $x_1 = 9$ and $x_2 = 5$.

5-2.4.

```
      L = 0
      DO 30 J = 1,N
         T = 0.
         DO 10 I = 1,M
10          T = T + A(I,J)*B(I)
         R(J) = T
         DO 30 I = J,N
            T = 0.
            DO 20 K = 1,M
20             T = T + A(K,I)*A(K,J)
            L = L + 1
30          S(L) = T
      CALL SFACT(S,N)
      CALL SSOLVE(X,S,R)
```

5-3.1. $x_1 = 13$, $x_2 = 4$, $x_3 = -7$.

5-3.2. $x_1 = 6/5$ and $x_2 = 3/5$ (make the change of variables $y_1 = 2x_1$ and $y_2 = 3x_2$).

5-3.3. $x_1 = 4$, $x_2 = 2$, $y = -2$.

5-3.4.

```
      L = 0
      DO 20 J = 1,M
         DO 20 I = J,M
            T = 0.
            DO 10 K = 1,N
```

```
10            T = T + A(K,I)*A(K,J)
            L = L + 1
20          S(L) = T
        CALL SFACT(S,M)
        CALL SSOLVE(Y,S,B)
        DO 30 I = 1,N
30         X(I) = 0.
        DO 40 J = 1,M
           T = Y(J)
           DO 40 I = 1,N
40            X(I) = X(I) + T*A(I,J)
```

5-4.1. Reflection across the plane $x_2 = x_1/2$.

5-4.2. $\mathbf{A} = \mathbf{0}$.

5-4.3. The identity $\mathbf{y}^T\mathbf{Q}^T\mathbf{Q}\mathbf{y} = \mathbf{y}^T\mathbf{y}$ implies that $\mathbf{y}^T(\mathbf{I} - \mathbf{Q}^T\mathbf{Q})\mathbf{y} = 0$ for every $\mathbf{y}$. By Exercise 5-4.2, $\mathbf{I} - \mathbf{Q}^T\mathbf{Q} = \mathbf{0}$ or $\mathbf{Q}^T\mathbf{Q} = \mathbf{I}$.

5-4.4. $\mathbf{x}^T\mathbf{x} = \mathbf{y}^T\mathbf{Q}^T\mathbf{Q}\mathbf{y} = \mathbf{y}^T\mathbf{y}$ since $\mathbf{Q}^T\mathbf{Q} = \mathbf{I}$. Since $\mathbf{x}^T\mathbf{x} = \mathbf{y}^T\mathbf{y}$, $\mathbf{x} = \mathbf{Q}\mathbf{y}$ is an orthogonal transformation.

5-4.5. $(\mathbf{Q}_1\mathbf{Q}_2)^T(\mathbf{Q}_1\mathbf{Q}_2) = \mathbf{Q}_2^T(\mathbf{Q}_1^T\mathbf{Q}_1)\mathbf{Q}_2 = \mathbf{Q}_2^T\mathbf{Q}_2 = \mathbf{I}$.

5-4.6. Since $\mathbf{Q}$ is an orthogonal matrix, $\mathbf{Q}^T\mathbf{Q} = \mathbf{I}$ and $\mathbf{Q}^T = \mathbf{Q}^{-1}$. Thus $(\mathbf{Q}^{-1})^T\mathbf{Q}^{-1} = (\mathbf{Q}^T)^T\mathbf{Q}^T = \mathbf{Q}\mathbf{Q}^T = \mathbf{I}$, which shows that $\mathbf{Q}^{-1}$ is an orthogonal matrix.

5-5.1. $\mathbf{H} = \frac{1}{13}\begin{bmatrix} -5 & -12 \\ -12 & 5 \end{bmatrix}$.

5-5.2. $\begin{bmatrix} 5 & -13 \\ 12 & 26 \end{bmatrix} = \frac{1}{13}\begin{bmatrix} -5 & -12 \\ -12 & 5 \end{bmatrix}\begin{bmatrix} -13 & -19 \\ 0 & 22 \end{bmatrix}$.

5-5.3. $a = 2\mathbf{w}^T\mathbf{x}$.

5-5.4. $j = 1$ to n

$a \leftarrow 0$

$a \leftarrow a + w_i a_{ij}$ for $i = 1$ to n

$a \leftarrow 2a$

$a_{ij} \leftarrow a_{ij} - aw_i$ for $i = 1$ to n

next j

5-5.5. $(\mathbf{Hx})_k = -\text{sign}(x_k)s$.

5-5.6. First column of $\mathbf{H}$ is $-\mathbf{x}/(\text{sign}(x_1)\|\mathbf{x}\|_2)$. If h_k denotes column k of $\mathbf{H}$, then $\mathbf{x}^T\mathbf{h}_k = 0$ for $k > 1$ and $\mathbf{x}^T\mathbf{h}_1 = -\|\mathbf{x}\|_2\text{sign}(x_1)$.

5-5.7. Since $\mathbf{w}^T\mathbf{e}_j = 0$ for $j < k$, we have $\mathbf{H}_k\mathbf{e}_j = (\mathbf{I} - 2\mathbf{w}\mathbf{w}^T)\mathbf{e}_j = \mathbf{e}_j - 2\mathbf{w}(\mathbf{w}^T\mathbf{e}_j) = \mathbf{e}_j$.

5-5.8. **(a)** $\frac{1}{13}\begin{bmatrix} 13 & 0 & 0 \\ 0 & -5 & -12 \\ 0 & -12 & 5 \end{bmatrix}$ **(b)** $\frac{1}{13}\begin{bmatrix} 5 & -12 & 0 \\ -12 & -5 & 0 \\ 0 & 0 & 13 \end{bmatrix}$

$\mathbf{w}^T = [2s(s + |x_k|)]^{-1/2}[x_1, x_2, \cdots, x_{k-1}, x_k + \text{sign}(x_k)s, 0, 0, \cdots, 0]$, where $s = \sqrt{x_1^2 + x_2^2 + \cdots + x_k^2}$.

5-5.10. $a = x_1/\|\mathbf{x}\|_2$ and $b = x_2/\|\mathbf{x}\|_2$

5-5.11. If x_2/x_1 is near zero, then $s \approx |x_1|$ and $x_1 - \text{sign}(x_1)s \approx x_1 - \text{sign}(x_1)|x_1| = x_1 - x_1 = 0$. Therefore, when x_2/x_1 is near zero, the difference $x_1 - \text{sign}(x_1)s$ is computed relatively inaccurately since nearly equal numbers are subtracted.

5-6.1. $\frac{1}{13}\begin{bmatrix} 5 & 12 \\ -12 & 5 \end{bmatrix}$.

5-6.2. $\begin{bmatrix} 5 & -13 \\ 12 & 26 \end{bmatrix} = \frac{1}{13}\begin{bmatrix} 5 & -12 \\ 12 & 5 \end{bmatrix}\begin{bmatrix} 13 & 19 \\ 0 & 22 \end{bmatrix}.$

5-6.3. $s = x_1/\sqrt{x_1^2 + x_2^2}$ and $c = x_2/\sqrt{x_1^2 + x_2^2}$.

5-6.4. For a vector with 3 components, rotations of the form $\mathbf{G}_{13}$ or $\mathbf{G}_{23}$ can be used to annihilate the third components of the vector. For a vector with n components, a rotation of the form $\mathbf{G}_{in}$, where i is between 1 and $n - 1$, can be used to annihilate the last component of the vector.

5-6.5. The text uses the sequence of rotations $\mathbf{G}_{21}$, $\mathbf{G}_{31}$, $\cdots$, $\mathbf{G}_{n1}$, $\mathbf{G}_{32}$, $\mathbf{G}_{42}$, $\cdots$, $\mathbf{G}_{n2}$, $\cdots$ to annihilate a_{21}, a_{31}, $\cdots$, a_{n1}, a_{32}, a_{42}, $\cdots$, a_{n2}, $\cdots$. Thus the coefficients beneath the diagonal are annihilated column by column starting at the subdiagonal and working down the column. As an alternative to this order, we can annihilate the coefficients row by row, starting in column 1 and working along the row, stopping at the diagonal. The following sequence of rotations can be employed: $\mathbf{G}_{21}$, $\mathbf{G}_{31}$, $\mathbf{G}_{32}$, $\mathbf{G}_{41}$, $\mathbf{G}_{42}$, $\mathbf{G}_{43}$, $\mathbf{G}_{51}$, $\mathbf{G}_{52}$, $\cdots$.

5-6.6. $\mathbf{A}^T\mathbf{A} = (\mathbf{QR})^T\mathbf{QR} = \mathbf{R}^T\mathbf{Q}^T\mathbf{QR} = \mathbf{R}^T\mathbf{R}$.

5-6.7. $(\mathbf{Q}_1\mathbf{Q}_2)^{-1} = \mathbf{Q}_2^{-1}\mathbf{Q}_1^{-1} = \mathbf{Q}_2^T\mathbf{Q}_1^T$ since $\mathbf{Q}^{-1} = \mathbf{Q}^T$ for an orthogonal matrix (see Exercise 5-4.3).

5-7.1. **(a)** $x_1 = 24$ and $x_2 = -2$. **(b)** $x_1 = 20$, $x_2 = 10$, and $x_3 = 0$.

5-8.1. Suppose that some vector, say $\mathbf{q}_k$, is a linear combination of the other vectors. Then there exists scalars x_i, $i \neq k$, such that $\mathbf{q}_k = \sum_{i \neq k} x_i\mathbf{q}_i$. Multiplying this equality by $\mathbf{q}_j^T$, where $j \neq k$, we have $\mathbf{q}_j^T\mathbf{q}_k = x_j\mathbf{q}_j^T\mathbf{q}_j$. Since $\mathbf{q}_j^T\mathbf{q}_k = 0$ and $\mathbf{q}_j^T\mathbf{q}_j = 1$, it follows that $x_j = 0$. Since each x_j is zero, $\mathbf{q}_k$ is zero, which is impossible since $\mathbf{q}_k$ is a unit vector. Hence, $\mathbf{q}_k$ is not a linear combination of the other vectors.

5-8.2. Let $\mathbf{e}_i$ denote the vector with every component zero except for component i which is 1. The computed orthonormal basis is the single vector $\mathbf{e}_1$ while the correct orthonormal basis is $\mathbf{e}_1$ through $\mathbf{e}_n$.

5-8.3. Let $\mathbf{p}_{j,k}$ denote the vector $\mathbf{p}_j$ generated by (82) just after the index i has the value k. Show that $\mathbf{p}_{j,k} = \mathbf{a}_j - \sum_{i=1}^{k} (\mathbf{q}_i^T\mathbf{a}_j)\mathbf{q}_i$. Since $\mathbf{q}_i$ is orthogonal to $\mathbf{q}_1$ through $\mathbf{q}_{i-1}$, the coefficient $\mathbf{q}_i^T\mathbf{p}_{j,i-1}$ of $\mathbf{q}_i$ in (82) reduces to $\mathbf{q}_i^T\mathbf{a}_j$, which is also the coefficient of $\mathbf{q}_i$ in (80).

5-8.4. Multiplying (80) by $\mathbf{q}_k$, where k is any integer between 1 and $j - 1$ and recalling that $\mathbf{q}_k$ is orthogonal to $\mathbf{q}_i$ for $i \neq k$, we have $\mathbf{q}_k^T\mathbf{p}_j = \mathbf{q}_k^T\mathbf{a}_j - (\mathbf{q}_k^T\mathbf{a}_j)(\mathbf{q}_k^T\mathbf{q}_k) = 0$.

5-8.5. The asymptotic running time is $\frac{1}{2}mn^2(C_1 + C_2)$, where C_1 and C_2 are the cycle times corresponding to the following two loops (which evaluate r_{ij} and update $\mathbf{q}_j$):

```
      DO 10 K = 1,M
10       T = T + Q(K,I)*A(K,J)

      DO 10 K = 1,M
10       Q(K,J) = Q(K,J) - T*Q(K,I)
```

5-8.6. The asymptotic running time is $\frac{1}{2}(mk^2 - \frac{1}{3}k^3)(C_1 + C_2)$, where C_1 and C_2 are the cycle times corresponding to the following two loops:

```
      DO 10 I = 1,M
10       T = T + A(I,L)*Q(I,J)

      DO 10 I = 1,M
10       Q(I,J) = Q(I,J) - T*A(I,L)
```

5-8.7. An orthonormal basis for the space spanned by the vectors is $\mathbf{q}_1^T \approx (.0, -.7809, -.6247)$ and $\mathbf{q}_2^T \approx (.7684, .3998, -.4998)$. A basis for the space perpendicular to the given vectors is $\mathbf{q}_1^T = (.64, -.48, .60)$.

Chapter 6

6-1.6. Let $(\lambda, \mathbf{x})$ and $(\mu, \mathbf{y})$ denote eigenpairs corresponding to distinct eigenvalues. Since $\mathbf{Qx} = \lambda\mathbf{x}$ and $\mathbf{Qy} = \mu\mathbf{y}$, we have $(\mathbf{Qy})^*\mathbf{Qx} = (\mu\mathbf{y})^*\lambda\mathbf{x}$ or $\mathbf{y}^*\mathbf{Q}^*\mathbf{Qx} = \mu^*\lambda(\mathbf{y}^*\mathbf{x})$. Since $\mathbf{Q}^*\mathbf{Q} = \mathbf{I}$, it follows that $\mathbf{y}^*\mathbf{x} = \mu^*\lambda(\mathbf{y}^*\mathbf{x})$. Multiplying by μ and recalling that $\mu\mu^* = 1$ (see Exercise 6-1.5) yields $\mu\mathbf{y}^*\mathbf{x} = \lambda\mathbf{y}^*\mathbf{x}$ or $(\mu - \lambda)\mathbf{y}^*\mathbf{x} = 0$. Since $\mu \neq \lambda$, we conclude that $\mathbf{y}^*\mathbf{x} = 0$.

6-2.1. **(a)** $\begin{bmatrix} 2 & 1 \\ 1 & -2 \end{bmatrix}\begin{bmatrix} 5 & 0 \\ 0 & -5 \end{bmatrix}\left(\frac{1}{5}\begin{bmatrix} 2 & 1 \\ 1 & -2 \end{bmatrix}\right)$

(b) $\begin{bmatrix} 1 & 1 \\ -\mathrm{i} & \mathrm{i} \end{bmatrix}\begin{bmatrix} 2 + 3\mathrm{i} & 0 \\ 0 & 2 - 3\mathrm{i} \end{bmatrix}\left(\frac{1}{2\mathrm{i}}\begin{bmatrix} \mathrm{i} & -1 \\ \mathrm{i} & 1 \end{bmatrix}\right).$

6-2.2. The eigenvalues of $\mathbf{H}$ are $\lambda = \pm 1$. The eigenvalues of $\mathbf{G}$ are $\lambda = c \pm s\mathrm{i}$. In terms of the rotation angle θ, the eigenvalues of $\mathbf{G}$ are $\lambda = \pm e^{\mathrm{i}\theta}$. The eigenvectors of $\mathbf{G}$ are $[1, \pm\mathrm{i}]$.

6-2.3. Premultiplying the identity $\mathbf{A} = \mathbf{X\Lambda X}^{-1}$ by $\mathbf{X}^{-1}$ yields $\mathbf{X}^{-1}\mathbf{A} = \mathbf{\Lambda X}^{-1}$. If $\mathbf{y}_i^T$ denotes row i of $\mathbf{X}^{-1}$, then row i of $\mathbf{X}^{-1}\mathbf{A}$ is $\mathbf{y}_i^T\mathbf{A}$ while row i of $\mathbf{\Lambda X}^{-1}$ is $\lambda_i\mathbf{y}_i^T$. Since $\mathbf{y}_i^T\mathbf{A} = \lambda_i\mathbf{y}_i^T$, $\mathbf{y}_i$ is a left eigenvector of $\mathbf{A}$ corresponding to the eigenvalue λ_i.

6-2.4. The eigenvectors of $\mathbf{I} + \mathbf{A}$ are the same as the eigenvectors of $\mathbf{A}$ while the eigenvalues of $\mathbf{I} + \mathbf{A}$ are $1 + \lambda_1, 1 + \lambda_2, \cdots, 1 + \lambda_n$, where λ_1 through λ_n denote the eigenvalues of $\mathbf{A}$. The eigenvectors of $\mathbf{I} + 2\mathbf{A}$ are the same as the eigenvectors of $\mathbf{A}$ while the eigenvalues are $1 + 2\lambda_1, 1 + 2\lambda_2, \cdots, 1 + 2\lambda_n$.

6-2.5. One eigenpair is $(1, \mathbf{w})$. All the remaining eigenvalues are zero while the corresponding eigenvectors are any basis for the space of vectors perpendicular to $\mathbf{w}$.

6-2.6. One eigenpair is $(-1, \mathbf{w})$. All the remaining eigenvalues are 1 and the corresponding eigenvectors are any basis for the space of vectors perpendicular to $\mathbf{w}$.

6-2.7. If $\mathbf{w}_i$ denotes column i of $\mathbf{W}$, then $(1, \mathbf{w}_i)$ is an eigenpair of $\mathbf{WW}^T$. The remaining eigenvalues are zero and the corresponding eigenvectors are any basis for the null space of $\mathbf{W}^T$.

6-2.8. If $\mathbf{a}_i$ denotes column i of $\mathbf{A}$, then $(0, \mathbf{a}_i)$ is an eigenpair for $\mathbf{P}$. The remaining eigenvalues are 1 and the corresponding eigenvectors are any basis for the null space of $\mathbf{A}^T$.

6-2.9. Letting $\mathbf{A}$ denote the matrix $\mathbf{WW}^T$, observe that $\mathbf{AW} = \mathbf{W}(\mathbf{W}^T\mathbf{W}) = \mathbf{W}(\mathbf{X\Lambda X}^{-1})$. Postmultiplying this identity by $\mathbf{X}$ yields $\mathbf{A}(\mathbf{WX}) = (\mathbf{WX})\mathbf{\Lambda}$. If $\mathbf{x}_i$ denotes column i of $\mathbf{WX}$ and λ_i denotes the ith diagonal element of $\mathbf{\Lambda}$, then $\mathbf{Ax}_i = \lambda_i\mathbf{x}_i$. Hence, the columns of $\mathbf{WX}$ are eigenvectors of $\mathbf{A}$ while the corresponding eigenvalues are the diagonal elements of $\mathbf{\Lambda}$. The remaining eigenvalues are zero and the corresponding eigenvectors are any basis for the null space of $\mathbf{W}^T$.

6-2.10. By Exercise 6.1.4, $\mathbf{x}_i^T\mathbf{x}_j = 0$ for $i \neq j$. Since the columns of $\mathbf{X}$ are normalized, $\mathbf{x}_i^T\mathbf{x}_i = 1$. For the matrix $\mathbf{X}^T\mathbf{X}$, the element in row i and column j is $\mathbf{x}_i^T\mathbf{x}_j$. Since $\mathbf{x}_i^T\mathbf{x}_j = 0$ for $i \neq j$ and $\mathbf{x}_i^T\mathbf{x}_j = 1$ for $i = j$, it follows that $\mathbf{X}^T\mathbf{X} = \mathbf{I}$. Thus $\mathbf{X}$ is an orthogonal matrix.

6-2.11. $u(t) = -4e^{5t} + e^{-5t}$ and $v(t) = -2(e^{5t} + e^{-5t})$. $u = 0$ and $v = 0$ is an unstable equilibrium point.

6-3.1. **(a)** Stable. **(b)** Unstable.

6-4.1. If λ_1 and λ_n denote the largest and the smallest eigenvalues of $\mathbf{A}$, then any shift

σ less than $(\lambda_1 + \lambda_n)/2$ can be used to compute the most positive eigenvalue. To compute the most negative eigenvalue, any shift σ greater than $(\lambda_1 + \lambda_n)/2$ can be employed. For the matrix in part (a) of Exercise 6-3.1, the eigenvalues lie between -9 and -1. To compute the most positive eigenvalue, we can use $\sigma = -9$ and to compute the most negative eigenvalue, we can use $\sigma = -1$. For the matrix in part (b) of Exercise 6-1.3, the eigenvalues lie between -6 and 3. To compute the most positive eigenvalue, we can use $\sigma = -6$, and to compute the most negative eigenvalue, we can use $\sigma = 3$.

6-4.2. Since $\mathbf{x}_k = \lambda_1^k\mathbf{e}_1 + \lambda_2^k\mathbf{e}_2 + \cdots + \lambda_n^k\mathbf{e}_n$, it follows that $\mathbf{x}_k \approx \lambda_1^k\mathbf{e}_1 + \lambda_2^k\mathbf{e}_2$. Similarly, $\mathbf{x}_{k+1} \approx \lambda_1^{k+1}\mathbf{e}_1 + \lambda_2^{k+1}\mathbf{e}_2$. Hence, $\mathbf{x}_{k+1} - \lambda_1\mathbf{x}_k \approx (\lambda_2^{k+1} - \lambda_1\lambda_2^k)\mathbf{e}_2 = 2\lambda_2^{k+1}\mathbf{e}_2$ and $\mathbf{x}_{k+1} + \lambda_1\mathbf{x}_k \approx 2\lambda_1^{k+1}\mathbf{e}_1$. This shows that $\mathbf{x}_{k+1} + \lambda_1\mathbf{x}_k$ is a multiple of $\mathbf{e}_1$ while $\mathbf{x}_{k+1} - \lambda_1\mathbf{x}_k$ is a multiple of $\mathbf{e}_2$.

6-4.3. The largest eigenvalue (roughly 390.2) is computed by the power method in 42 iterations when NDIGIT is 5 and the starting guess is the vector with every component equal to 1. Computing the smallest eigenvalue (roughly 9.789) by the shifted power method with shift 390.2 requires 72 iterations. Computing the smallest eigenvalue using the inverse power method requires 8 iterations. Thus the inverse power method is much faster.

6-5.1. The deflation procedure described in the text generates an upper triangular matrix $\mathbf{U}$, where $\mathbf{U} = \mathbf{P}^T\mathbf{A}\mathbf{P}$. For a symmetric matrix $\mathbf{A}$, $\mathbf{U}^T = (\mathbf{P}^T\mathbf{A}\mathbf{P})^T = \mathbf{P}^T\mathbf{A}^T\mathbf{P} = \mathbf{P}^T\mathbf{A}\mathbf{P} = \mathbf{U}$. Since $\mathbf{U}$ is upper triangular and $\mathbf{U}^T = \mathbf{U}$, it follows that $\mathbf{U}$ is a diagonal matrix. Solving for $\mathbf{A}$ in the identity $\mathbf{U} = \mathbf{P}^T\mathbf{A}\mathbf{P}$, we have $\mathbf{A} = \mathbf{P}\mathbf{U}\mathbf{P}^T$. Since $\mathbf{U}$ is diagonal and $\mathbf{P}$ is orthogonal, $\mathbf{P}\mathbf{U}\mathbf{P}^T$ is the diagonalization of $\mathbf{A}$ and the eigenvector matrix $\mathbf{P}$ is orthogonal.

6-6.1. (a) By the definition of $\mathbf{A}_k$ and $\mathbf{P}_k$, we have $\mathbf{P}_k^T\mathbf{A}_k\mathbf{P}_k = \mathbf{Q}_k^T\mathbf{Q}_{k-1}\mathbf{Q}_{k-1}^T\mathbf{X}_k\mathbf{Q}_{k-1}^T\mathbf{Q}_k = \mathbf{Q}_k^T\mathbf{X}_k\mathbf{Q}_{k-1}^T\mathbf{Q}_k$. By equation (36) of Chapter 6, $\mathbf{X}_{k+1} = \mathbf{A}\mathbf{Q}_k = \mathbf{A}\mathbf{Q}_{k-1}\mathbf{Q}_{k-1}^T\mathbf{Q}_k = \mathbf{X}_k\mathbf{Q}_{k-1}^T\mathbf{Q}_k$. Therefore $\mathbf{P}_k^T\mathbf{A}_k\mathbf{P}_k = \mathbf{Q}_k^T\mathbf{X}_{k+1}$. (b) By the definition of $\mathbf{P}_k$ and by equation (36) of Chapter 6, we have $\mathbf{P}_k\mathbf{R}_k = \mathbf{Q}_{k-1}^T\mathbf{Q}_k\mathbf{Q}_k^T\mathbf{X}_k = \mathbf{Q}_{k-1}^T\mathbf{X}_k = \mathbf{A}_k$. (c) By the definition of $\mathbf{P}_k$, by equation (36) of Chapter 6, and by part (a), we have $\mathbf{R}_k\mathbf{P}_k = \mathbf{Q}_k^T\mathbf{X}_k\mathbf{Q}_{k-1}^T\mathbf{Q}_k = \mathbf{Q}_k^T\mathbf{X}_{k+1} = \mathbf{A}_{k+1}$.

6-6.2. $\mathbf{Q}_1\mathbf{Q}_2\mathbf{R}_2\mathbf{R}_1 = \mathbf{Q}_1\mathbf{A}_2\mathbf{R}_1 = \mathbf{Q}_1\mathbf{R}_1\mathbf{Q}_1\mathbf{R}_1 = \mathbf{A}^2$. The identity $\mathbf{A}_k = \mathbf{Q}_k\mathbf{R}_k$ implies that $\mathbf{R}_k = \mathbf{Q}_k^T\mathbf{A}_k$ and $\mathbf{A}_{k+1} = \mathbf{R}_k\mathbf{Q}_k = \mathbf{Q}_k^T\mathbf{A}_k\mathbf{Q}_k$. In particular, $\mathbf{A}_3 = \mathbf{Q}_2^T\mathbf{A}_2\mathbf{Q}_2$, $\mathbf{A}_2 = \mathbf{Q}_1^T\mathbf{A}_1\mathbf{Q}_1$, and $\mathbf{A}_3 = \mathbf{Q}_2^T\mathbf{Q}_1^T\mathbf{A}\mathbf{Q}_1\mathbf{Q}_2$. For the shifted **QR** method,

$$\begin{aligned}\mathbf{Q}_1\mathbf{Q}_2\mathbf{R}_2\mathbf{R}_1 &= \mathbf{Q}_1(\mathbf{A}_2 - \sigma_2\mathbf{I})\mathbf{R}_1 = \mathbf{Q}_1(\mathbf{Q}_1^T\mathbf{A}_1\mathbf{Q}_1 - \sigma_2\mathbf{Q}_1^T\mathbf{Q}_1)\mathbf{R}_1\\ &= (\mathbf{A}_1 - \sigma_2\mathbf{I})\mathbf{Q}_1\mathbf{R}_1 = (\mathbf{A}_1 - \sigma_2\mathbf{I})(\mathbf{A}_1 - \sigma_1\mathbf{I}) = (\mathbf{A} - \sigma_2\mathbf{I})(\mathbf{A} - \sigma_1\mathbf{I}).\end{aligned}$$

6-7.1. In Section 5-5, we showed that premultiplying a matrix by an orthogonal matrix does not change the 2-norm of the matrix. Now consider the effect of postmultiplying a matrix $\mathbf{A}$ by an orthogonal matrix $\mathbf{Q}$ to obtain the product $\mathbf{B} = \mathbf{A}\mathbf{Q}$. From the definition of a matrix norm, we have

$$\|\mathbf{B}\| = \max_{\mathbf{x}\neq\mathbf{0}} \frac{\|\mathbf{A}\mathbf{Q}\mathbf{x}\|_2}{\|\mathbf{x}\|_2} = \max_{\mathbf{y}\neq\mathbf{0}} \frac{\|\mathbf{A}\mathbf{y}\|_2}{\|\mathbf{Q}^T\mathbf{y}\|_2} = \max_{\mathbf{y}\neq\mathbf{0}} \frac{\|\mathbf{A}\mathbf{y}\|_2}{\|\mathbf{y}\|_2} = \|\mathbf{A}\|_2,$$

where $\mathbf{y} = \mathbf{Q}\mathbf{x}$ and $\|\mathbf{Q}^T\mathbf{y}\|_2 = \|\mathbf{y}\|_2$ since $\mathbf{Q}^T$ is an orthogonal matrix. Since $\|\mathbf{A}\|_2 = \|\mathbf{B}\|_2 = \|\mathbf{A}\mathbf{Q}\|_2$, postmultiplying a matrix by an orthogonal matrix does not change its 2-norm. Each iteration of Householder's reduction to upper Hessenberg form can be expressed $\mathbf{A}_{k+1} = \mathbf{H}_k\mathbf{A}_k\mathbf{H}_k$. Since premultiplying and postmultiplying a matrix by orthogonal matrices does not change the 2-norm of the matrix, we conclude that $\|\mathbf{A}_{k+1}\|_2 = \|\mathbf{A}_k\|_2$. Since each iteration does not change the 2-norm, the 2-norm of the starting matrix is equal to the 2-norm of the final upper Hessenberg matrix.

6-7.2. Since each subdiagonal coefficient of $\mathbf{A}$ is nonzero, each subdiagonal coefficient of $\mathbf{A} - \lambda\mathbf{I}$ is also nonzero. Thus the first $n - 1$ columns of $\mathbf{A} - \lambda\mathbf{I}$ are linearly independent and the null space of $\mathbf{A} - \lambda\mathbf{I}$ contains at most one linearly independent vector. If $\mathbf{A}$ has a multiple eigenvalue, then there is more than one linearly independent eigenvector corresponding to this eigenvalue (since $\mathbf{A}$ is diagonalizable). Since the eigenvectors of $\mathbf{A}$ are the vectors in the null space of $\mathbf{A} - \lambda\mathbf{I}$ and since there is at most one linearly independent vector in this null space, it follows that the eigenvalues of $\mathbf{A}$ are all distinct.

6-7.3. For algorithm (59), the asymptotic parameters are $\frac{1}{2}n^3$, $\frac{1}{6}n^3$, $\frac{1}{3}n^3$, and $\frac{1}{3}n^3$ corresponding to the following loops: $b_i \leftarrow b_i + a_{ij}a_{jk}$ for $i = 1$ to n, $a_{ij} \leftarrow a_{ij} - b_i a_{jk}$ for $i = 1$ to k, $t \leftarrow t - a_{ij}a_{ik}$ for $i = k + 1$ to n, and $a_{ij} \leftarrow a_{ij} + ta_{ik} + b_i a_{jk}$ for $i = k + 1$ to n. For algorithm (60), the asymptotic parameters are $\frac{1}{3}n^3$ and $\frac{1}{2}n^3$ corresponding to the loops: $a_{ij} \leftarrow a_{ij} - a_{ik}a_{k+1,j}$ for $i = k + 2$ to n and $a_{i,k+1} \leftarrow a_{i,k+1} + a_{jk}a_{ij}$ for $i = 1$ to n.

6-8.1. $p_{11} = 1$, $p_{22} = \frac{1}{2}$, and $p_{33} = \frac{1}{6}$.

6-8.2. $p_{11} = 1$ and $p_{i+1,i+1} \leftarrow p_{ii}\sqrt{a_{i+1,i}/a_{i,i+1}}$ for $i = 1$ to $n - 1$. The diagonal of $\mathbf{A}$ and the diagonal of $\mathbf{P}^{-1}\mathbf{AP}$ are the same. The cross-diagonal products for $\mathbf{A}$ are the same as the cross-diagonal products for $\mathbf{P}^{-1}\mathbf{AP}$.

6-8.3. By the quadratic formula, the eigenvalues are $\frac{1}{2}(a + c) \pm \sqrt{\delta^2 + b^2}$. Since $(a + c)/2$ is the midpoint of the interval $[a,c]$, we use the plus sign when c is greater than a (or equivalently, $\delta < 0$) and we use the minus sign when c is less than a (or equivalently, $\delta > 0$). Hence, the eigenvalue closest to c is given by $\frac{1}{2}(a + c) - \text{sign}(\delta)\sqrt{\delta^2 + b^2}$. After some algebra, this formula for the eigenvalue closest to c reduces to the formula given in the exercise. The formula given in the exercise does not involve the subtraction of nearly equal numbers.

6-8.4. Asymptotic parameter is n.

6-9.1. $\mathbf{A} = \begin{bmatrix} 1 & 2 & 0 \\ 2 & 2 & 3 \\ 0 & 3 & 3 \end{bmatrix}$.

6-9.2. The eigenvalues for $\mathbf{A}$ and $\mathbf{B}$ are the same since the diagonal and the cross-diagonal products for each matrix are the same.

6-9.3. There is one sign change associated with the determinants given by recurrence (69) of Chapter 6 so there is one negative eigenvalue.

6-9.4. By Exercise 6-5.1, every symmetric matrix can be diagonalized and by Exercise 6-7.2, each eigenvalue of a symmetric tridiagonal matrix with nonzero subdiagonal coefficients is distinct.

6-11.1. As you saw in Exercise 6-7.1, multiplying either the left side or the right side of a matrix by an orthogonal matrix does not change the Euclidean norm. Hence, if $\mathbf{QSP}^T$ is the singular value decomposition of $\mathbf{A}$, then $\|\mathbf{A}\|_2 = \|\mathbf{S}\|_2$. By Exercise 3-3.8, $\|\mathbf{S}\|_2^2 = \|\mathbf{S}^T\mathbf{S}\|_2$. Since $\mathbf{S}^T\mathbf{S}$ is a diagonal matrix whose diagonal elements are the squares of the singular values and since the 2-norm of a diagonal matrix is the largest absolute diagonal element, $\|\mathbf{A}\|_2^2 = \|\mathbf{S}^T\mathbf{S}\|_2 = \sigma_1^2$, where σ_1 is the largest singular value. Combining these observations, $\|\mathbf{A}\|_2 = \sigma_1$.

6-11.2. $\|\mathbf{A}^{-1}\|_2 = 1/\sigma_p$, where σ_p is the smallest positive singular value.

6-11.3. By Exercises 6-11.1 and 6-11.2, $\|\mathbf{A}\|_2\|\mathbf{A}^{-1}\|_2 = \sigma_1/\sigma_p$, where σ_1 is the largest singular value and σ_p is the smallest singular value. If the condition number is 1, then $\sigma_1 = \sigma_p$. Thus all the singular values are equal. If $\mathbf{QSP}^T$ is the singular value decomposition of $\mathbf{A}$, then $\mathbf{S} = s\mathbf{I}$, where s is the common singular value and $\mathbf{A} = \mathbf{Q}(s\mathbf{I})\mathbf{P}^T = s\mathbf{QP}^T$. Since $\mathbf{P}^T$ is orthogonal and since the product of orthogonal matrices is orthogonal (see Exercise 5-4.5), $\mathbf{A}$ is s times the orthogonal matrix $\mathbf{QP}^T$.

6-11.4. Letting λ_i denote the eigenvalue of $\mathbf{A}^T\mathbf{A}$ corresponding to the eigenvector $\mathbf{p}_i$, we have $(\mathbf{AA}^T)\mathbf{q}_i = (\mathbf{AA}^T)\mathbf{Ap}_i = \mathbf{A}(\mathbf{A}^T\mathbf{A})\mathbf{p}_i = \lambda_i\mathbf{Ap}_i = \lambda_i\mathbf{q}_i$. Thus $\mathbf{q}_i$ is an eigenvector of $\mathbf{AA}^T$ corresponding to the eigenvalue λ_i. Now observe that $\mathbf{q}_i{}^T\mathbf{q}_j = (\mathbf{Ap}_i)^T(\mathbf{Ap}_j) = \mathbf{p}_i{}^T(\mathbf{A}^T\mathbf{A})\mathbf{p}_j = \lambda_j\mathbf{p}_i{}^T\mathbf{p}_j = 0$ when $i \neq j$ since $\mathbf{p}_i$ and $\mathbf{p}_j$ are columns from an orthogonal matrix.

6-11.5. If $\mathbf{QSP}^T$ is the singular value decomposition of $\mathbf{A}$, then the singular value decomposition of $\mathbf{A}^{-1} = \mathbf{PS}^{-1}\mathbf{Q}^T$ is simply $\mathbf{PS}^{-1}\mathbf{Q}^T$. Therefore, $(\mathbf{A}^{-1})^{-1} = \mathbf{Q}(\mathbf{S}^{-1})^{-1}\mathbf{P}^T$. Since $(\mathbf{S}^{-1})^{-1} = \mathbf{S}$, it follows that $(\mathbf{A}^{-1})^{-1} = \mathbf{A}$.

6-11.6. Let $\mathbf{QSP}^T$ denote the singular value decomposition of $\mathbf{A}$.
(a) $\mathbf{AA}^{-1}\mathbf{A} = \mathbf{QSP}^T\mathbf{PS}^{-1}\mathbf{Q}^T\mathbf{QSP}^T = \mathbf{QSP}^T = \mathbf{A}$.
(b) $\mathbf{A}^{-1}\mathbf{AA}^{-1} = \mathbf{PS}^{-1}\mathbf{Q}^T\mathbf{QSP}^T\mathbf{PS}^{-1}\mathbf{Q}^T = \mathbf{PS}^{-1}\mathbf{Q}^T = \mathbf{A}^{-1}$.
(c) Since $\mathbf{SS}^{-1}$ is symmetric, we have $(\mathbf{AA}^{-1})^T = (\mathbf{QSP}^T\mathbf{PS}^{-1}\mathbf{Q}^T)^T = (\mathbf{QSS}^{-1}\mathbf{Q}^T)^T = \mathbf{Q}(\mathbf{SS}^{-1})^T\mathbf{Q}^T = \mathbf{QSS}^{-1}\mathbf{Q}^T = \mathbf{QSP}^T\mathbf{PS}^{-1}\mathbf{Q}^T = \mathbf{AA}^{-1}$.
(d) By part (c), $(\mathbf{BB}^{-1})^T = \mathbf{BB}^{-1}$. Setting $\mathbf{B} = \mathbf{A}^{-1}$ yields (d).

6-11.7. Using an IBM 370 computer, the computed solutions, rounded to 5 significant digits, are the following: **(a)** $x_1 = -.5$, $x_2 = .0$, and $x_3 = .5$. **(b)** $x_1 = 2.8540$, $x_2 = -6.7080$, and $x_3 = 3.8540$. **(c)** $x_1 = -.5$, $x_2 = .0$, and $x_3 = .5$.

6-12.1. **(a)** If $a_{11} = b_{11} = 0$, then $\det(\mathbf{A} - \lambda\mathbf{B}) = 0 \times (a_{22} - \lambda b_{22}) - 0 \times \lambda b_{12} = 0$ for any λ. Thus any λ is an eigenvalue. Similarly, $\det(\mathbf{A} - \lambda\mathbf{B}) = 0$ for any λ when $a_{22} = b_{22} = 0$. **(b)** If $b_{11} = b_{22} = 0$, then $\det(\mathbf{A} - \lambda\mathbf{B}) = a_{11}a_{22}$, which is never zero when both a_{11} and a_{22} are nonzero. **(c)** If $b_{11} = 0$, then $\det(\mathbf{A} - \lambda\mathbf{B}) = a_{11}(a_{22} - \lambda b_{22})$, which only vanishes when $\lambda = a_{22}/b_{22}$. If $b_{22} = 0$ but both a_{22} and b_{11} are nonzero, then there is exactly one eigenvalue. In each of the cases (a)–(c), $\mathbf{B}$ does not have an inverse.

6-12.2. Premultiplying $\mathbf{AB}$ by $\mathbf{L}^T$ and postmultiplying by $\mathbf{L}^{-T}$, we have $\mathbf{L}^T(\mathbf{AB})\mathbf{L}^{-T} = \mathbf{L}^T\mathbf{ALL}^T\mathbf{L}^{-T} = \mathbf{L}^T\mathbf{AL}$. Since a similarity transformation does not change the eigenvalues, the eigenvalues of $\mathbf{AB}$ are the same as the eigenvalues of $\mathbf{L}^T\mathbf{AL}$. If $\mathbf{y}$ is an eigenvector of $\mathbf{L}^T\mathbf{AL}$, then the corresponding eigenvector of $\mathbf{AB}$ is $\mathbf{x} = \mathbf{L}^{-T}\mathbf{y}$.

6-13.1. If d_{mm} is the smallest diagonal element in absolute value, then

$$\|\mathbf{Dy}\|_p = \left(\sum_{i=1}^{n} |d_{ii}y_i|^p\right)^{1/p} \geqq \left(\sum_{i=1}^{n} |d_{mm}y_i|^p\right)^{1/p} = |d_{mm}|\left(\sum_{i=1}^{n} |y_i|^p\right)^{1/p} = |d_{mm}|\,\|\mathbf{y}\|_p.$$

Review Problems:

6-1. After rounding, the eigenvalues of $\mathbf{A}$ are $\lambda_1 = -1.98$, $\lambda_2 = -1.91$, $\lambda_3 = -1.80$, $\lambda_4 = -1.65$, $\lambda_5 = -1.47$, $\lambda_6 = -1.25$, $\lambda_7 = -1.00$, $\lambda_8 = -.73$, $\lambda_9 = -.45$, $\lambda_{10} = -.15$, $\lambda_{11} = .15$, $\lambda_{12} = .45$, $\lambda_{13} = .73$, $\lambda_{14} = 1.00$, $\lambda_{15} = 1.25$, $\lambda_{16} = 1.47$, $\lambda_{17} = 1.65$, $\lambda_{18} = 1.80$, $\lambda_{19} = 1.91$, and $\lambda_{20} = 1.98$.

6-3. After rounding, the computed eigenvalues are $\lambda_1 = 9.9$, $\lambda_2 = 39.8$, $\lambda_3 = 90.5$, $\lambda_4 = 163.2$, $\lambda_5 = 259.7$, $\lambda_6 = 382.3$, $\lambda_7 = 534.0$, $\lambda_8 = 718.2$, $\lambda_9 = 938.9$, $\lambda_{10} = 1200.0$, $\lambda_{11} = 1505.5$, $\lambda_{12} = 1857.9$, $\lambda_{13} = 2257.2$, $\lambda_{14} = 2698.4$, $\lambda_{15} = 3169.9$, $\lambda_{16} = 3645.4$, $\lambda_{17} = 4092.4$, $\lambda_{18} = 4464.0$, $\lambda_{19} = 4712.4$.

Chapter 7

7-2.1.

```
10     DO 20 = 1,N
20        NEW(I) = B(I)
       DO 60 J = 1,N
          T = OLD(J)
          K = J - 1
```

```
              IF ( K .EQ. 0 ) GOTO 40
              DO 30 I = 1,K
   30            NEW(I) = NEW(I) - A(I,J)*T
   40         K = J + 1
              IF ( K .GT. N ) GOTO 60
              DO 50 I = K,N
   50            NEW(I) = NEW(I) - A(I,J)*T
   60      CONTINUE
           DO 70 I = 1,K
   70         NEW(I) = NEW(I)/A(I,I)
           CALL UPDATE(DIF,SIZE,NEW,OLD,N)
           CALL STOPIT(DIF,SIZE,NDIGIT,LIMIT)
           IF ( DIF .GT . 0. ) GOTO 10
```

7-2.2.

```
   10      DO 30 I = 1,N
              T = B(I)
              DO 20 J = 1,N
   20            T = T - A(I,J)*OLD(J)
   30         NEW(I) = OLD(I) + T/A(I,I)
           CALL UPDATE(DIF,SIZE,NEW,OLD,N)
           CALL STOPIT(DIF,SIZE,NDIGIT,LIMIT)
           IF ( DIF .GT . 0. ) GOTO 10
```

7-2.3.

```
   10      DIF = 0.
           SIZE = 0.
           DO 60 I = 1,N
              T = 0.
              K = I - 1
              IF ( K .EQ. 0 ) GOTO 30
              DO 20 J = 1,K
   20            T = T + A(I,J)*BOTH(J)
   30         K = I + 1
              IF ( K .GT. N ) GOTO 50
              DO 40 J = K,N
   40            T = T + A(I,J)*BOTH(J)
   50         T = (B(I) - T)/A(I,I)
              DIF = DIF + ABS(T-BOTH(I))
              SIZE = SIZE + ABS(T)
              BOTH(I) = T
   60      CONTINUE
           CALL STOPIT(DIF,SIZE,NDIGIT,LIMIT)
           IF ( DIF .GT . 0. ) GOTO 10
```

7-3.1. **(a)** Since $\|\mathbf{M}\|_2 = 2.5$, the inequality $\|\mathbf{e}_k\|_1 \leqq 2.5^k\|\mathbf{e}_0\|_1$ does not tell us whether the iterations converge or diverge. **(b)** Since $\mathbf{M}^2$ is the diagonal matrix with each diagonal element equal to .25, $\|\mathbf{M}^2\|_1 = .25$. If k is even, then $\|\mathbf{M}^k\|_1 = \|(\mathbf{M}^2)^{k/2}\|_1 \leqq \|\mathbf{M}^2\|_1^{k/2} = .25^{k/2} = .5^k$. If k is odd, then $k - 1$ is even and $\|\mathbf{M}^k\|_1 = \|\mathbf{M}(\mathbf{M}^{k-1})\|_1 \leqq \|\mathbf{M}\|_1\|\mathbf{M}^{k-1}\|_1 \leqq \|\mathbf{M}\|_1\|\mathbf{M}^2\|_1^{(k-1)/2} = 2.5 \times .25^{(k-1)/2} = 5 \times .5^k$. These two inequalities along with equation (21) of Chapter 7 imply that $\|\mathbf{e}_k\|_1 \leqq 5 \times .5^k\|\mathbf{e}_0\|_1$. Since $.5^k$ tends to zero as k increases, the error approaches zero as k increases and the iterations converge.

7-3.2. Taking the norm of the relation $\mathbf{x} = \mathbf{S}^{-1}\mathbf{T}\mathbf{x} + \mathbf{S}^{-1}\mathbf{b}$ gives

$$\|\mathbf{x}\| \leqq \|(\mathbf{S}^{-1}\mathbf{T})\mathbf{x}\| + \|\mathbf{S}^{-1}\mathbf{b}\| \leqq \|\mathbf{S}^{-1}\mathbf{T}\|\,\|\mathbf{x}\| + \|\mathbf{S}^{-1}\mathbf{b}\|.$$

Rearranging this inequality, we find that $\|\mathbf{x}\| \leqq \|\mathbf{S}^{-1}\mathbf{b}\|/(1 - \|\mathbf{S}^{-1}\mathbf{T}\|)$.

7-3.3. After 11 iterations, the error is at most .001. Referring to Tables 2-13 and 2-14, the time to factor **B** is about .003 second for an IBM 370 model 3090 computer. If the time to multiply **C** by $\mathbf{x}^{\text{old}}$ is about 5000 microseconds and the time to solve the factored tridiagonal system is about 800 microseconds, then the time for 11 iterations is roughly $(.0008 + .005) \times 11 \approx .06$ second. Hence, the total time to make the error less than .001 is roughly .06 seconds.

7-4.1. $c(\mathbf{A})\|\mathbf{A} - \mathbf{C}\|/\|\mathbf{A}\| < 1$.

7-5.1. Substituting $\mathbf{x} = \mathbf{y} + \mathbf{A}^{-1}\mathbf{b}$, we have

$$\mathbf{x}^T\mathbf{A}\mathbf{x} - 2\mathbf{b}^T\mathbf{x} = (\mathbf{y} + \mathbf{A}^{-1}\mathbf{b})^T\mathbf{A}(\mathbf{y} + \mathbf{A}^{-1}\mathbf{b}) - 2\mathbf{b}^T(\mathbf{y} + \mathbf{A}^{-1}\mathbf{b}) = \mathbf{y}^T\mathbf{A}\mathbf{y} - \mathbf{b}^T\mathbf{A}^{-1}\mathbf{b}.$$

Since **A** is positive definite, $\mathbf{y}^T\mathbf{A}\mathbf{y} \geqq 0$ and $\mathbf{y}^T\mathbf{A}\mathbf{y} = 0$ only if $\mathbf{y} = \mathbf{0}$. Hence, the quadratic $\mathbf{y}^T\mathbf{A}\mathbf{y} - \mathbf{b}^T\mathbf{A}^{-1}\mathbf{b}$ attains its minimum at $\mathbf{y} = \mathbf{0}$. Since $\mathbf{x} = \mathbf{y} + \mathbf{A}^{-1}\mathbf{b}$, it follows that the quadratic $\mathbf{x}^T\mathbf{A}\mathbf{x} - 2\mathbf{b}^T\mathbf{x}$ attains its minimum at $\mathbf{x} = \mathbf{A}^{-1}\mathbf{b}$.

Bibliography

[A1] J. O. Aasen, On the reduction of a symmetric matrix to tridiagonal form, *BIT*, 11(1971), pp. 233–242. §2-4.3.

[A2] N. N. Abdelmalck, Roundoff error analysis for Gram-Schmidt method and solution of linear least squares problems, *BIT*, 11(1971), pp. 345–368. §5-9.

[A3] E. L. Allgower and K. Georg, Simplicial and continuation methods for approximating fixed points and solutions to systems of equations, *SIAM Rev.*, 22(1980), pp. 28–85. §4-6.

[A4] E. L. Allgower and K. Georg, Predictor-corrector and simplicial methods for approximating fixed points and zero points of nonlinear mappings, in *Mathematical Programming: The State of the Art*, A. Bachem, M. Grötschel, and B. Korte, eds., Springer-Verlag, New York, 1983. §4-6.

[A5] E. L. Allgower, K. Glashoff, and H.-O. Peitgen, eds., *Numerical Solution of Nonlinear Equations*, Springer Lecture Notes in Mathematics, Number 876, Springer-Verlag, New York, 1981. §4-9.

[A6] L. Armijo, Minimization of functions having Lipschitz continuous first-partial derivatives, *Pacific J. Math.*, 16(1966), pp. 1–3. §4-9.

[A7] O. B. Arushanian, M. K. Samarin, V. V. Voevodin, E. E. Tyrtyshnikov, B. S. Garbow, J. M. Boyle, W. R. Cowell, and K. W. Dritz, *The Toeplitz Package Users' Guide*, Report ANL-83-16, Argonne National Laboratory, Argonne, Ill., 1983. §8-2.

[B1] R. E. Bank, *PLTMG User's Guide*, June 1981 version, Technical Report, Mathematics Department, University of California at San Diego, 1982. §8-2.

[B2] R. E. Bank and T. F. Chan, PLTMGC: A multi-grid continuation program for parameterized nonlinear elliptic systems, *SIAM J. Sci. Stat. Comput.*, 7(1986), pp. 540–559. §8-2.

[B3] R. E. Bank and C. C. Douglas, An efficient implementation for SSOR and incomplete factorization preconditionings, to appear in *Appl. Numer. Math.* §7-7.

[B4] R. E. Bank and T. F. Dupont, An optimal order process for solving finite element equations, *Math. Comp.*, 36(1981), pp. 35–51. §7-7.

[B5] R. E. Bank and A. H. Sherman, An adaptive, multi-level method for elliptic boundary value problems, *Computing*, 26(1981), 91–106. §8-2.

[B6] V. Barwell and J. A. George, A comparison of algorithms for solving symmetric indefinite systems of linear equations, *ACM Trans. Math. Software*, 2(1976), pp. 242–251. §2-12.

[B7] A. Björck, R. J. Plemmons, and H. Schneider, eds., *Large-Scale Matrix Problems*, North-Holland, New York, 1981. §7-7.

[B8] R. F. Boisvert and R. A. Sweet, Mathematical software for elliptic boundary value problems, in *Sources and Development of Mathematical Software*, W. R. Cowell, ed., Prentice-Hall, Englewood Cliffs, N.J., 1984, pp. 200–263. §8-2.

[B9] A. Brandt, Multi-level adaptive techniques (MLAT) for fast numerical solution to boundary value problems, *Proceedings Third International Conference on Numerical Methods in Fluid Mechanics*, H. Cabannes and R. Temam, eds., Lecture Notes in Physics, Vol. 18, Springer-Verlag, New York, 1973, pp. 82–89. §7-6.

[B10] A. Brandt, Multi-level adaptive solutions to boundary-value problems, *Math. Comp.*, 31(1977), pp. 333–390. §7-7.

[B11] A. Brandt, Multilevel adaptive techniques (MLAT) for partial differential equations: ideas and software, *Mathematical Software III*, J. R. Rice, ed., Academic Press, New York, 1977, pp. 277–318. §8-2.

[B12] A. Brandt, Multigrid techniques: 1984 guide with applications to fluid dynamics, *GMD-Studien Nr. 85*, Gesellschaft für Mathematik und Datenverarbeitung, St. Augustin, West Germany, 1984. §7-6.

[B13] A. Brandt, S. F. McCormick, and J. Ruge, Algebraic multigrid (AMG) for sparse matrix equations, *Sparsity and Its Applications*, D. J. Evans, ed., Cambridge University Press, Cambridge, 1984. §7-7.

[B14] R. P. Brent, *Algorithms for Minimization without Derivatives*, Prentice-Hall, Englewood Cliffs, N.J., 1973. §4-9.

[B15] C. G. Broyden, A class of methods for solving nonlinear simultaneous equations, *Math. Comp.*, 19(1965), pp. 577–593. §4-5.

[B16] J. R. Bunch, Analysis of the diagonal pivoting method, *SIAM J. Numer. Anal.*, 8(1971), pp. 656–680. §2-12.

[B17] J. R. Bunch, Partial pivoting strategies for symmetric matrices, *SIAM J. Numer. Anal.*, 11(1974), pp. 521–528. §2-12.

[B18] J. R. Bunch and L. Kaufman, Some stable methods for calculating inertia and solving symmetric linear systems, *Math. Comp.* 31(1977), pp. 162–179. §2-12.

[B19] J. R. Bunch, L. Kaufman, and B. N. Parlett, Decomposition of a symmetric matrix, *Numer. Math.*, 27(1976), pp. 95–109. §2-12.

[B20] J. R. Bunch and B. N. Parlett, Direct methods for solving symmetric indefinite systems of linear equations, *SIAM J. Numer. Anal.*, 8(1971), pp. 639–655. §2-12.

[B21] J. R. Bunch and D. J. Rose, eds., *Sparse Matrix Computations*, Academic Press, New York, 1976. §7-7.

[B22] J. C. P. Bus and T. J. Dekker, Two efficient algorithms with guaranteed convergence for finding a zero of a function, *ACM Trans. Math. Software*, 1(1975), pp. 330–345. §4-9.

[B23] P. Businger and G. H. Golub, Linear least squares solutions by Householder transformations, *Numer. Math.*, 7(1965), pp. 269–276. §5-9.

[B24] B. L. Buzbee, G. H. Golub, and C. W. Neilson, On direct methods for solving Poisson's equations, *SIAM J. Numer. Anal.*, 7(1970), pp. 627–656. §§2-12 and 8-2.

[C1] S. N. Chow, J. Mallet-Paret, and J. A. Yorke, Finding zeroes of maps: Homotopy methods that are constructive with probability one, *Math. Comp.*, 32(1978), pp. 887–899. §4-9.

[C2] A. K. Cline, C. B. Moler, G. W. Stewart, and J. H. Wilkinson, An estimate for the condition number of matrix, *SIAM J. Numer. Anal.*, 16(1979), pp. 368–375. §3-6.

[C3] W. J. Cody and W. Waite, *Software Manual for the Elementary Functions*, Prentice-Hall, Englewood Cliffs, N.J., 1980. §1-9.

[C4] S. D. Conte and C. de Boor, *Elementary Numerical Analysis*, McGraw-Hill, New York, 1980. §1-9.

[C5] J. W. Cooley and J. W. Tukey, An algorithm for the machine calculation of complex Fourier series, *Math. Comp.*, 19(1965), 297–301. §4-8.

[C6] D. Coppersmith and S. Winograd, On the asymptotic complexity of matrix multiplications, *SIAM J. Comput.*, 11(1982), pp. 472–492. §2-12.

[C7] W. R. Cowell, ed., *Sources and Development of Mathematical Software*, Prentice-Hall, Englewood Cliffs, N.J., 1984. §8-1.

[C8] J. Cullum and R. A. Willoughby, A Lanczos procedure for the modal analysis of very large nonsymmetric matrices, *Proceedings of the 23rd Conference on Decision and Control*, Las Vegas, 1984, pp. 1758–1761. §6-14.

[C9] J. Cullum and R. A. Willoughby, *Lanczos Algorithms for Large Symmetric Eigenvalue Computations*, Vol. 1: Theory, Vol. 2: Programs, Birkhäuser, Boston, Mass., 1985. §§6-10 and 8-2.

[C10] J. Cullum and R. A. Willoughby, *A Practical Procedure for Computing Eigenvalues of Large Sparse Nonsymmetric Matrices*, Research Report, IBM T.J. Watson Research Center, 1985. §6-14.

[D1] G. Dahlquist and A. Björck, *Numerical Methods*, translated by N. Anderson, Prentice-Hall, Englewood Cliffs, N.J., 1974. §§1-1, 1-9, and 7-7.

[D2] J. W. Daniel, The conjugate gradient method for linear and nonlinear operator equations, *SIAM J. Numer. Anal.*, 4(1967), pp. 10–26. §7-7.

[D3] T. J. Dekker, Finding a zero by means of successive linear interpolation, in *Constructive Aspects of the Fundamental Theorem of Algebra*, B. Dejon and P. Henrici, eds., Wiley-Interscience, New York, 1969. §4-9.

[D4] J. E. Dennis and R. B. Schnabel, *Numerical Methods for Unconstrained Optimization*, Prentice-Hall, Englewood Cliffs, N.J., 1983. §4-9.

[D5] J. J. Dongarra, Algorithm 589, SICEDR: A Fortran subroutine for improving the accuracy of computed matrix eigenvalues, *ACM Trans. Math. Software*, 8(1982), pp. 371–375. §6-14.

[D6] J. J. Dongarra, J. Du Croz, S. Hammarling, and R. J. Hanson, A proposal for an extended set of Fortran basic linear algebra subprograms, *SIGNUM Newsletter*, 20(1985), pp. 2–18. An update notice on the extended BLAS, *SIGNUM Newsletter*, 21(1986), pp. 2–4. §8-1.

[D7] J. J. Dongarra and T. Hewitt, Implementing dense linear algebra algorithms using multitasking on the CRAY X-MP-4 (or approaching the gigaflop), *SIAM J. Sci. Stat. Comput.*, 7(1986), pp. 347–350. §2-11.

[D8] J. J. Dongarra, C. B. Moler, J. R. Bunch, and G. W. Stewart, *LINPACK Users'*

Guide, Society for Industrial and Applied Mathematics, Philadelphia, 1979. §§2-12 and 8-1.

[D9] J. J. Dongarra, C. B. Moler, and J. H. Wilkinson, Improving the accuracy of computed eigenvalues and eigenvectors, *SIAM J. Numer. Anal.*, 20(1983), pp. 23–45. §6-14.

[D10] J. J. Dongarra and E. Grosse, *Distribution of Mathematical Software via Electronic Mail*, Report ANL/MCS-TM-48, Mathematics and Computer Science Division, Argonne National Laboratory, Argonne, Ill., 1985. §8-3.

[D11] J. J. Dongarra, F. G. Gustavson, and A. Karp, Implementing linear algebra algorithms for dense matrices on a vector pipeline machine, *SIAM Rev.*, 26(1984), pp. 91–112. §2-10.

[D12] C. C. Douglas, *A Multi-Grid Optimal Order Elliptic Partial Differential Equation Solver*, Department of Computer Science, Yale University, New Haven, Conn., 1981. §8-2.

[D13] C. C. Douglas, A multilevel solver for boundary-value problems, *IEEE Transactions on Electron Devices*, 32(1985), pp. 1987–1991. §8-2.

[D14] I. S. Duff, *MA28—A Set of Fortran Subroutines for Sparse Unsymmetric Linear Equations*, Harwell Report R.8730, London, 1977. §8-2.

[D15] I. S. Duff, *MA32—A Package for Solving Sparse Unsymmetric Systems Using the Frontal Method*, Harwell Report R.10079, London, 1981. §8-2.

[D16] I. S. Duff, *ME28—A* Sparse Unsymmetric Linear Equation Solver for Complex Equations, *ACM Trans. Math. Software*, 7(1981), pp. 505–511. §8-2.

[D17] I. S. Duff, A survey of sparse matrix software, in *Sources and Development of Mathematical Software*, W. R. Cowell, ed., Prentice-Hall, Englewood Cliffs, N.J., 1984, pp. 165–199. §8-2.

[D18] I. S. Duff, Design features of a code for solving sparse symmetric linear systems out-of-core, *SIAM J. Sci. Stat. Comput.*, 5(1984), pp. 270–280. §8-2.

[D19] I. S. Duff, *Enhancements to the MA32 Package for Solving Sparse Unsymmetric Equations*, Harwell Report R.11009, London, 1983. §8-2.

[D20] I. S. Duff, A. M. Erisman, and J. K. Reid, *Direct Methods for Sparse Matrices*, Oxford University Press, Oxford, 1986. §§2-4.5 and 8-2.

[D21] I. S. Duff and J. K. Reid, Algorithm 529: Permutations to block triangular form, *ACM Trans. Math. Software*, 4(1978), pp. 189–192. §8-2.

[D22] I. S. Duff and J. K. Reid, The frontal solution of indefinite sparse symmetric linear systems, *ACM Trans. Math. Software*, 9(1983), pp. 302–325. §8-2.

[D23] I. S. Duff and J. K. Reid, The multifrontal solution of unsymmetric sets of linear equations, *SIAM J. Sci. Stat. Comput.*, 5(1984), pp. 633–641. §8-2.

[D24] I. S. Duff and G. W. Stewart, eds., *Sparse Matrix Proceedings, 1978*, Society for Industrial and Applied Mathematics, Philadelphia, 1979. §7-7.

[E1] S. C. Eisenstat, Efficient implementation of a class of conjugate gradient methods, *SIAM J. Sci. Stat. Comput.*, 2(1981), pp. 1–4. §7-7.

[E2] S. C. Eisenstat, H. C. Elman, and M. H. Schultz, Variational iterative methods for nonsymmetric systems of linear equations, *SIAM J. Numer. Anal.*, 20(1983), pp. 345–357. §§7-7 and 8-2.

[E3] S. C. Eisenstat, M. C. Gursky, M. H. Schultz, and A. H. Sherman, *Yale Sparse Matrix Package*, I. The symmetric codes, Report 112, Department of Computer Science, Yale University, New Haven, Conn., 1977. §8-2.

[E4] S. C. Eisenstat, M. C. Gursky, M. H. Schultz, and A. H. Sherman, *Yale Sparse Matrix Package*, II. The nonsymmetric codes, Report 114, Department of Computer Science, Yale University, New Haven, Conn., 1977. §8-2.

[E5] H. C. Elman, Iterative methods for large, sparse, nonsymmetric systems of linear equations, Ph.D. dissertation, Department of Computer Science, Yale University, New Haven, Conn., 1982. §§7-7 and 8-2.

[F1] H. Foerster and K. Witsch, On efficient multigrid software for elliptic problems on rectangular domains, *Math. Comput. Simulation*, 23(1981). §8-2.

[F2] G. E. Forsythe, M. A. Malcolm, and C. B. Moler, *Computer Methods for Mathematical Computations*, Prentice-Hall, Englewood Cliffs, N.J., 1977. §2-3.

[F3] G. E. Forsythe and C. B. Moler, *Computer Solution of Linear Algebraic Systems*, Prentice-Hall, Englewood Cliffs, N.J., 1967. §2-3.

[F4] P. Fox, The PORT mathematical subroutine library, in *Sources and Development of Mathematical Software*, W. R. Cowell, ed., Prentice-Hall, Englewood Cliffs, N.J., 1984, pp. 346–374. §8-3.

[F5] P. A. Fox, A. D. Hall, and N. L. Schryer, The PORT mathematical subroutine library, *ACM Trans. Math. Software*, 4(1978), pp. 104–126. §8-3.

[F6] J. G. F. Francis, The QR transformation: A unitary analogue to the LR transformation, *Comput. J.*, 4(1961), Part I: pp. 265–272 and Part II: pp. 332–345. §6-6.

[F7] S. P. Frankel, Convergence rates of iterative treatments of partial differential equations, *Math. Tables Aids Comput.*, 4(1950), pp. 65–75. §7-7.

[F8] J. N. Franklin, *Matrix Theory*, Prentice-Hall, Englewood Cliffs, N.J., 1968. §6-3.

[F9] B. Fredriksson and J. Mackerle, *Structural Mechanics Finite Element Computer Programs, A Decade with Finite Elements*, Advanced Engineering Corporation, Box 3044, Linköping, Sweden, 1980. §8-2.

[G1] B. S. Garbow, J. M. Boyle, J. J. Dongarra, and C. B. Moler, *Matrix Eigensystem Routines—EISPACK Guide Extension*, Springer-Verlag, New York, 1977. §8-1.

[G2] D. M. Gay, Some convergence properties of Broyden's method, *SIAM J. Numer. Anal.*, 16(1979), pp. 623–636. §4-5.

[G3] W. M. Gentleman, Least squares computations by Givens transformations without square roots, *J. Inst. Math. Appl.*, 12(1973), pp. 329–336. §5-9.

[G4] W. M. Gentleman and S. B. Marovich, More on algorithms that reveal properties of floating point arithmetic units, *Comm. ACM* 17(1974), pp. 276–277. §1-9.

[G5] A. George and J. W. H. Liu, *Computer Solution of Large Sparse Positive Definite Systems*, Prentice-Hall, Englewood Cliffs, N.J., 1981. §§ 2-4.5 and 8-2.

[G6] A. George and J. W. H. Liu, The design of a user interface for a sparse matrix package, *ACM Trans. Math. Software*, 5(1979), pp. 134–162. §8-2.

[G7] A. George, J. Liu, E. Ng, and E. Chu, *SPARSPAK: Waterloo Sparse Matrix Package—User's Guide for SPARSPAK-A,* Report CS-84-36, Department of Computer Science, University of Waterloo, Waterloo, Ontario, Canada, 1984. §8-2.

[G8] A. George and E. Ng, A brief description of SPARSPAK: Waterloo sparse linear equations package, *SIGNUM Newsletter*, 16(1981), pp. 17–20. §8-2.

[G9] A. George and E. Ng, *SPARSPAK: Waterloo Sparse Matrix Package—User's Guide for SPARSPAK-B,* Report CS-84-37, Department of Computer Science, University of Waterloo, Waterloo, Ontario, Canada, 1984. §8-2.

[G10] A. George and E. Ng, *The Design and Implementation of a Package for Sparse Constrained Least Squares Problems,* Report CS-85-39, Department of Computer Science, University of Waterloo, Waterloo, Ontario, Canada, 1985. §8-2.

[G11] W. Givens, Computation of plane unitary rotations transforming a general matrix to triangular form, *SIAM J. Appl. Math.*, 6(1958), pp. 26–50. §5-9.

[G12] H. H. Goldstine, *A History of Numerical Analysis from the 16th through the 19th Century*, Springer-Verlag, New York, 1977. §5-9.

[G13] G. H. Golub, Numerical methods for solving linear least squares problems, *Numer. Math.*, 7(1965), pp. 206–216. §5-9.

[G14] G. H. Golub and W. Kahan, Calculating the singular values and pseudo-inverse of a matrix, *SIAM J. Numer. Anal.* Ser. B, 2(1965), pp. 205–224. §6-11.

[G15] G. H. Golub and C. F. Van Loan, *Matrix Computations*, Johns Hopkins University Press, Baltimore, Md., 1983. §§2-12, 3-6, 5-9, 6-11, 6-12, and 6-14.

[G16] F. A. Graybill, *Matrices with Applications in Statistics*, Wadsworth, Belmont, Calif., 1983. §5-9.

[G17] F. G. Gustavson, Some basic techniques for solving sparse systems of equations, in *Sparse Matrices and their Applications*, D. J. Rose and R. A. Willoughby, eds., Plenum Press, New York, 1972, pp. 41–52. §2-4.5.

[H1] W. Hackbusch, *Multigrid Methods and Applications*, Springer Series in Computational Mathematics, Vol. 4, Springer-Verlag, New York, 1985. §7-6.

[H2] W. Hackbusch and U. Trottenberg, eds., *Multigrid Methods*, Lecture Notes in Mathematics, Vol. 960, Springer-Verlag, New York, 1982. §7-6.

[H3] L. A. Hageman and D. M. Young, *Applied Iterative Methods*, Academic Press, New York, 1981. §§7-3, 7-5, 7-7, and 8-2.

[H4] W. W. Hager, Condition estimates, *SIAM J. Sci. Stat. Comput.*, 5(1984), pp. 311–316. §3-4.

[H5] W. W. Hager, A modified fast Fourier transform and the Jenkins-Traub algorithm, *Numer. Math.*, 50(1987), pp. 253–261. §4-8.

[H6] W. W. Hager, Dual techniques for constrained optimization, to appear in *J. Optim. Theory Appl.*, 55(1987), pp. 37–71. §7-7.

[H7] W. W. Hager, Bidiagonalization and diagonalization, to appear in *Comput. Math. Appl.* §6-14.

[H8] S. Hammarling, A note on modifications to the Givens plane rotation, *J. Inst. Math. Appl.*, 13(1973), pp. 215–218. §5-9.

[H9] D. Heller, Some aspects of the cyclic reduction algorithm for block tridiagonal linear systems, *SIAM J. Numer. Anal.*, 13(1976), pp. 484–496. §2-12.

[H10] D. Heller, A survey of parallel algorithms in numerical linear algebra, *SIAM Rev.*, 20(1978), pp. 740–777. §2-12.

[H11] P. W. Hemker, R. Kettler, P. Wesseling, and P. M. deZeeuw, Multigrid methods: development of fast solvers, *Appl. Math. Comp.*, Vol. 13, Proceedings of the International Multigrid Conference April 6–8, 1983, Copper Mountain, Colorado, S. F. McCormick and U. Trottenberg, eds., North-Holland, New York, 1983, pp. 311–326. §7-7.

[H12] M. R. Hestenes, *Conjugate Direction Methods in Optimization*, Springer-Verlag, New York, 1980. §7-7.

[H13] M. R. Hestenes and E. Stiefel, Methods of conjugate gradients for solving linear systems, *J. Res. Nat. Bur. Standards*, 29(1952), pp. 409–439. §7-7.

[H14] N. J. Higham, Efficient algorithms for computing the condition number of a tridiagonal matrix, *SIAM J. Sci. Stat. Comput.*, 7(1986), pp. 150–165. §3-6.

[H15] M. W. Hirsch and S. Smale, On algorithms for solving $F(x) = 0$, *Comm. Pure Appl. Math.*, 32(1979), pp. 281–312. §4-9.

[H16] R. A. Horn and C. R. Johnson, *Matrix Analysis*, Cambridge University Press, Cambridge, 1985. §3-6.

[H17] A. S. Householder, Unitary triangularization of a nonsymmetric matrix, *J. Assoc. Comput. Mach.*, 5(1958), pp. 339–342. §5-9.

[H18] A. S. Householder, *The Theory of Matrices in Numerical Analysis*, Dover, New York, 1975 (originally published by Blaisdell, Waltham, Mass., 1964). §3-6.

[H19] E. N. Houstis and T. S. Papatheodorou, Algorithm 543: FFT9, fast solution of Helmholtz type partial differential equations, *ACM Trans. Math. Software*, 5(1979), pp. 431–441. §8-2.

[J1] C. G. J. Jacobi, Über eine neu Auflösungsart der bei der Methode der kleinsten Quadrate vorkommenden linearen Gleichungen, *Astronom. Nachr.*, 32(1845), pp. 297–306. §5-9.

[J2] M. A. Jenkins and J. F. Traub, A three-stage variable-shift iteration for polynomial zeros and its relation to generalized Rayleigh iteration, *Numer. Math.*, 14(1970), pp. 252–263. §4-9.

[J3] M. A. Jenkins and J. F. Traub, A three-stage algorithm for real polynomials using quadratic iteration, *SIAM J. Numer. Anal.*, 7(1970), pp. 545–566. §4-9.

[J4] A. Jennings, A compact storage scheme for the solution of symmetric linear simultaneous equations, *Comput. J.* 9(1966), pp. 281–285. §2-4.5.

[K1] W. Kahan, Gauss-Seidel methods of solving large systems of linear equations, Ph.D. dissertation, University of Toronto, 1958. §7-7.

[K2] S. Kaniel, Estimates for some computational techniques in linear algebra, *Math. Comp.*, 20(1966), pp. 369–378. §6-10.

[K3] L. Kantorovich and G. Akilov, *Functional Analysis in Normed Spaces*, Fizmatgiz, Moscow, 1959; transl. by D. Brown and A. Robertson, Pergamon Press, Oxford, 1964. §4-9.

[K4] H. B. Keller, Global homotopies and Newton methods, in *Recent Advances in Numerical Analysis*, C. de Boor and G. H. Golub, eds., Academic Press, New York, 1978, pp. 73–94. §4-9.

[K5] D. R. Kincaid and D. M. Young, *Adapting Iterative Algorithms Developed for Symmetric Systems to Nonsymmetric Systems*, Report 161, Center for Numerical Analysis, University of Texas at Austin, Austin, Tex., 1980. §8-2.

[K6] D. R. Kincaid, J. R. Respess, D. M. Young, and R. G. Grimes, Algorithm 586: ITPACK 2C, A Fortran package for solving large sparse linear systems by adaptive accelerated iterative methods, *ACM Trans. Math. Software*, 8(1982), pp. 302–322. §8-2.

[K7] V. N. Kublanovskaya, On some algorithms for the solution of the complete eigenvalue problem, *Zh. Vycisl. Mat. i Mat. Fiz.*, 1(1961), pp. 555–570. §6-6.

[K8] R. H. Kuhn and D. A. Padua, eds., *Tutorial on Parallel Processing*, IEEE Press, New York, 1981. §2-12.

[K9] C.-C. Kuo, Discretization and solution of elliptic PDEs: A transform domain approach, Ph.D. dissertation, Department of Electrical Engineering and Computer Science, Massachusetts Institute of Technology, Cambridge, Mass., 1987. §7-7.

[L1] C. Lanczos, An iteration method for the solution of the eigenvalue problem of linear differential and integral operators, *J. Res. Nat. Bur. Standards*, 45(1950), pp. 255–282. §6-10.

[L2] C. L. Lawson and R. J. Hanson, *Solving Least Squares Problems*, Prentice-Hall, Englewood Cliffs, N.J., 1974. §5-9.

[L3] A. M. Legendre, Nouvelles méthodes pour la determination des orbites des comètes, Paris, 1805. §5-9.

[L4] D. G. Luenberger, Convergence rates of a penalty-function scheme, *J. Optim. Theory Appl.*, 7(1971), pp. 39–51. §7-7.

[M1] J. F. Maitre and F. Musy, Multigrid methods: convergence theory in a variational framework, *SIAM J. Numer. Anal.*, 21(1984), pp. 657–671. §7-7.

[M2] M. Malcolm, Algorithms to reveal properties of floating-point arithmetic, *Comm. ACM* 15(1972), pp. 949–951. §1-9.

[M3] J. Mandel, S. F. McCormick, and J. Ruge, An algebraic theory for multigrid methods for variational problems, *SIAM J. Numer. Anal.*, to appear. §7-7.

[M4] C. W. McCormick, The NASTRAN program for structural analysis, in *Advances in Computational Methods in Structural Mechanics and Design*, J. T. Oden, R. W. Clough, and Y. Yamamoto, eds., University of Alabama Press, Huntsville, Ala., 1972, pp. 551–573. §8-2.

[M5] S. F. McCormick, ed., *Multigrid Methods*, Frontiers in Applied Mathematics, Vol. 5, Society for Industrial and Applied Mathematics, Philadelphia, 1986. §7-7.

[M6] R. Menzel and H. Schwetlick, Zur Lösung parameterabhängiger nichtlinearer Gleichungen mit singulären Jacobi-Matrizen, *Numer. Math.*, 30(1978), pp. 65–79. §4-9.

[M7] C. B. Moler, Matrix computations with Fortran and paging, *Comm. ACM*, 15(1972), pp. 268–270. §§2-2 and 2-3.

[M8] C. B. Moler, *MATLAB User's Guide*, Report CS81-1, Department of Computer Science, University of New Mexico, Albuquerque, N.M., 1980. §8-1.

[M9] C. B. Moler and G. W. Stewart, An algorithm for generalized matrix eigenvalue problems, *SIAM J. Numer. Anal.*, 10(1973), pp. 241–256. §6-12.

[M10] J. J. Moré, B. S. Garbow, and K. E. Hillstrom, *User Guide for MINPACK-1*, Report 80-74, Argonne National Laboratory, Argonne, Ill., 1980. §§1-9 and 8-1.

[N1] R. A. Nicolaides, On multiple grid and related techniques for solving discrete elliptic systems, *J. Comput. Phys.*, 19(1975), p. 418–431. §7-7.

[N2] A. K. Noor, Survey of computer programs for heat transfer analysis, *Finite Elements in Analysis and Design*, 2(1986), pp. 259–312. §8-2.

[O1] D. P. O'Leary and O. Widlund, ACM Algorithm 572: Solution of the Helmholtz equation for the Dirichlet problem on general bounded three-dimensional regions, *ACM Trans. Math. Software*, 7(1981), pp. 239–246. §8-2.

[O2] J. M. Ortega, *Numerical Analysis, A Second Course*, Academic Press, New York, 1972. §3-6.

[O3] J. M. Ortega and W. C. Rheinboldt, Monotone iterations for nonlinear equations with applications to Gauss-Seidel methods, *SIAM J. Numer. Anal.*, 4(1967), pp. 171–190. §4-4.

[O4] J. M. Ortega and W. C. Rheinboldt, *Iterative Solution of Nonlinear Equations in Several Variables*, Academic Press, N.Y., 1970. §§4-4 and 4-9.

[O5] J. M. Ortega and R. G. Voigt, Solution of partial differential equations on vector and parallel computers, *SIAM Rev.*, 27(1985), pp. 149–240. §2-11.

[O6] A. M. Ostrowski, On the linear iteration procedures for symmetric matrices, *Rend. Mat. e Appl.*, 14(1954), pp. 140–163. §7-7.

[O7] A. M. Ostrowski, *Solution of Equations and Systems of Equations*, Academic Press, New York, 1966. §4-9.

[P1] C. C. Paige, The computation of eigenvalues and eigenvectors of very large sparse matrices, Ph.D. dissertation, London University, 1971. §6-10.

[P2] V. Pan, How can we speed up matrix multiplication?, *SIAM Rev.*, 26(1984), pp. 393–415. §2-12.

[P3] B. N. Parlett, The Rayleigh quotient iteration and some generalizations for nonnormal matrices, *Math. Comp.*, 28(1974), pp. 679–693. §6-14.

[P4] B. N. Parlett, *The Symmetric Eigenvalue Problem*, Prentice-Hall, Englewood Cliffs, N.J., 1980. §§5-9, 6-8, and 6-14.

[P5] B. N. Parlett and J. K. Reid, On the solution of a system of linear equations whose coefficient matrix is symmetric but not definite, *BIT*, 10(1970), pp. 386–397. §2-12.

[P6] B. N. Parlett and D. S. Scott, The Lanczos algorithm with selective reorthogonalization, *Math. Comp.*, 33(1979), pp. 217–238. §6-10.

[P7] B. N. Parlett, D. R. Taylor, and Z. A. Liu, A look-ahead Lanczos algorithm for unsymmetric matrices, *Math. Comp.*, 44(1985), pp. 104–124. §6-14.

[P8] H.-O. Peitgen and H.-O. Walther, eds., *Functional Differential Equations and Approximation of Fixed Points*, Springer Lecture Notes in Mathematics, Number 730, Springer-Verlag, New York, 1979. §4-9.

[P9] G. Peters and J. H. Wilkinson, The least squares problem and pseudo-inverses, *Comput. J.*, 13(1970), pp. 309–316. §5-9.

[P10] G. Peters and J. H. Wilkinson, Practical problems arising in the solution of polynomial equations, *J. Inst. Math. Appl.*, 8(1971), pp. 16–35. §4-9.

[P11] W. Proskurowski, Numerical solution of Helmholtz's equation by implicit capacitance methods, *ACM Trans. Math. Software*, 5(1979), pp. 36–49. §8-2.

[R1] J. K. Reid, On the method of conjugate gradients for the solution of large sparse systems of linear equations, in *Large Sparse Sets of Linear Equations*, J. K. Reid, ed., Academic Press, New York, 1971, pp. 231–254. §7-5.

[R2] J. K. Reid, ed., *Large Sparse Sets of Linear Equations*, Academic Press, New York, 1971. §7-7.

[R3] C. H. Reinsch, A stable rational *QR* algorithm for the computation of the eigenvalues of a Hermitian, tridiagonal matrix, *Numer. Math.*, 25(1971), pp. 591–597. §6-8.

[R4] R. Renka, *A Portable Software Package for Nonlinear Partial Differential Equations*, Report ORNL/CSD-102, Oak Ridge National Laboratory, Oak Ridge, Tenn. §8-2.

[R5] W. C. Rheinboldt, Solution fields of nonlinear equations and continuation methods, *SIAM J. Numer. Anal.*, 17(1980), pp. 221–237. §4-9.

[R6] W. C. Rheinboldt, *Numerical Analysis of Parameterized Nonlinear Equations*, Wiley-Interscience, New York, 1986. §4-6.

[R7] J. R. Rice and R. F. Boisvert, *Solving Elliptic Problems Using ELLPACK*, Springer-Verlag, New York, 1985. §8-2.

[R8] D. J. Rose, Convergent regular splittings for singular *M*-matrices, *SIAM J. Alg. Disc. Meth.*, 5(1984), pp. 133–144. §7-7.

[R9] D. J. Rose and R. A. Willoughby, eds., *Sparse Matrices and Their Applications*, Plenum Press, New York, 1972. §7-7.

[R10] H. Rutishauser, Solution of eigenvalue problems with the LR transformation, *Nat. Bur. Stand. App. Math. Ser.*, 49(1958), pp. 47–81. §6-6.

[S1] Y. Saad, On the rates of convergence of the Lanczos and the block Lanczos methods, *SIAM J. Numer. Anal.*, 17(1980), pp. 687–706. §6-10.

[S2] Y. Saad and M. H. Schultz, GMRES: A generalized minimal residual algorithm for solving nonsymmetric linear systems, *SIAM J. Sci. Stat. Comput.*, 7(1986), pp. 856–869. §§7-7 and 8-2.

[S3] Y. Saad, and M. H. Schultz, *Conjugate Gradient-like Algorithms for Solving Nonsymmetric Linear Systems*, Report 283, Department of Computer Science, Yale University, New Haven, Conn., 1983. §§7-7 and 8-2.

[S4] M. Satyanarayanan, Multiprocessing: an annotated bibliography, *Comput. J.*, 13(1980), pp. 101–116. §2-12.

[S5] U. Schendel, *Introduction to Numerical Methods for Parallel Computers*, Wiley, New York, 1984. §2-12.

[S6] W. Schonauer, K. Raith, and G. Glotz, The SLDGL program package for the self-adaptive solution of nonlinear systems of elliptic and parabolic PDE's, in *Advances in Computer Methods for Partial Differential Equations*, Vol. IV, R. Vichnevetsky and R. S. Stepleman, eds., IMACS, New Brunswick, N.J., 1981, pp. 117–125. §8-2.

[S7] E. Schrem, Development and maintenance of large finite element systems, in *Structural Mechanics Computer Programs—Surveys, Assessments and Availability*, W. Pilkey, K. Saczalski, and H. Schaeffer, eds., University of Virginia Press, Charlottesville, Va., 1974, pp. 669–685. §8-2.

[S8] H. R. Schwarz, Tridiagonalization of a symmetric band matrix, *Numer. Math.*, 12(1968), pp. 231–241. §6-14.

[S9] G. Sewell, *TWODEPEP: A Small General Purpose Finite Element Program*, Report 8102, IMSL, Houston, Tex., 1981. §8-2.

[S10] R. I. Shrager, A rapid robust rootfinder, *Math. Comp.*, 44(1985), pp. 151–165. §4-9.

[S11] R. C. Singleton, On computing the fast Fourier transform, *Comm. ACM*, 10(1967), pp. 647–654. §4-8.

[S12] B. T. Smith, J. M. Boyle, J. J. Dongarra, B. S. Garbow, Y. Ikebe, V. C. Klema, and C. B. Moler, *Matrix Eigensystem Routines—EISPACK guide*, Springer-Verlag, New York, 1976. §8-1.

[S13] E. L. Stiefel, Kernel polynomials in linear algebra and their numerical applications, *J. Res. Nat. Bur. Standards, Appl. Math. Ser.*, 49(1958), pp. 1–22. §7-7.

[S14] G. W. Stewart, *Introduction to Matrix Computations*, Academic Press, New York, 1973. §§2-12, 3-6, and 6-14.

[S15] G. Strang, *Linear Algebra and Its Applications*, Academic Press, New York, 1980. §§2-9, 2-12, and 5-9.

[S16] G. Strang, A proposal for Toeplitz matrix calculations, *Stud. Appl. Math.*, 74(1986), pp. 171–176. §7-7.

[S17] G. Strang and G. Fix, *An Analysis of the Finite Element Method*, Prentice-Hall, Englewood Cliffs, N.J., 1973. §6-12.

[S18] V. Strassen, Gaussian elimination is not optimal, *Numer. Math.*, 13(1969), pp. 354–356. §2-12.

[S19] P. N. Swarztrauber, A direct method for the discrete solution of separable elliptic equations, *SIAM J. Numer. Anal.*, 11(1974), pp. 1136–1150. §§2-12 and 8-2.

[S20] P. N. Swarztrauber and R. A. Sweet, Algorithm 541: Efficient Fortran subprograms for the solution of elliptic partial differential equations, *ACM Trans. Math. Software*, 5(1979), pp. 352–364. §8-2.

[S21] R. A. Sweet, A cyclic reduction algorithm for solving block tridiagonal systems of arbitrary dimension, *SIAM J. Numer. Anal.*, 14(1977), pp. 706–720. §§2-12 and 8-2.

[T1] J. F. Traub, *Iterative Methods for the Solution of Equations*, Prentice-Hall, Englewood Cliffs, N.J., 1964. §4-9.

[V1] J. M. Van Kats and H. A. Van der Vorst, *Software for the Discretization and Solution of Second Order Self-Adjoint Elliptic Partial Differential Equations in Two Dimensions*, Report 10, ACCU, Utrecht, The Netherlands. §8-2.

[V2] R. S. Varga, *Matrix Iterative Analysis*, Prentice-Hall, Englewood Cliffs, N.J., 1962. §§3-6 and 7-7.

[W1] H.-J. Wacker, ed., *Continuation Methods*, Academic Press, New York, 1978. §4-9.

[W2] A. Wassyng, Solving $Ax = b$: a method with reduced storage requirements, *SIAM J. Numer. Anal.*, 19(1982), pp. 197–204. §2-12.

[W3] D. S. Watkins, Understanding the QR algorithm, *SIAM Rev.*, 24(1982), pp. 427–440. §6-6.

[W4] L. T. Watson, A globally convergent algorithm for computing fixed points of C^2 maps, *Appl. Math. Comput.*, 5(1979), pp. 297–311. §4-9.

[W5] J. H. Wilkinson, *Rounding Errors in Algebraic Processes*, Prentice-Hall, Englewood Cliffs, N.J., 1964. §§1-9, 4-7, and 6-2.

[W6] J. H. Wilkinson, *The Algebraic Eigenvalue Problem*, Oxford University Press, London, 1965. §§6-6 and 6-14.

[W7] E. L. Wilson, SAP—a general structural analysis program for linear systems, in *Advances in Computational Methods in Structural Mechanics and Design*, J. T. Oden, R. W. Clough, and Y. Yamamoto, eds., University of Alabama Press, Huntsville, Ala., 1972, pp. 625–640. §8-2.

[W8] A. Wouk, ed., *New Computing Environments: Parallel, Vector, and Systolic*, Society for Industrial and Applied Mathematics, Philadelphia, 1986. §2-12.

[Y1] D. M. Young, Iterative methods for solving partial difference equations of elliptic type, Ph.D. dissertation, Harvard University, 1950. §7-7.

[Y2] D. M. Young, *Iterative Solution of Large Linear Systems*, Academic Press, New York, 1971. §7-7 and 8-2.

[Y3] D. M. Young and D. R. Kincaid, *The ITPACK Package for Large Sparse Linear Systems*, Report 160, Center for Numerical Analysis, University of Texas at Austin, Austin, Tex., 1980. §8-2.

Index

A

B

E

F

M

N

O

P

Q